Innovation in Musi

Innovation in Music: Performance, Production, Technology and Business is an exciting collection comprising cutting-edge articles on a range of topics, presented under the main themes of artistry, technology, production and industry. Each chapter is written by a leader in the field and contains insights and discoveries not yet shared.

Innovation in Music covers new developments in standard practice of sound design, engineering and acoustics. It also reaches into areas of innovation, both in technology and business practice, even into cross-discipline areas. This book is the perfect companion for professionals and researchers alike with an interest in the Music industry.

Following are the co-founders of the Innovation in Music Conference (www.musicinnovation.co.uk), who also acted as editors for this volume.

Russ Hepworth-Sawyer is a member of the Audio Engineering Society and co-founder of the UK Mastering Section there. A former board member of the Music Producers Guild, Russ helped form their Mastering Group. Through MOTTOsound (www.mottosound.co.uk), Russ now works freelance in the industry as a mastering engineer, writer and consultant. Russ currently lectures part-time for York St John University, UK, and the University of Huddersfield, UK, and has taught extensively in UK Higher Education. He occasionally contributes in magazines such as *MusicTech*, *Pro Sound News Europe*, and *Sound On Sound*, and has written many titles for Focal Press and Routledge.

Jay Hodgson is on faculty at Western University, Canada, where he primarily teaches songwriting and record production. He is also a mastering engineer at MOTTOsound. His masters have twice been nominated for Juno Awards and topped Beatport's global techno and house charts. He was awarded a Governor General's academic medal in 2006, primarily in recognition of his research on audio recording; and his second book, *Understanding Records* (2010), was recently acquired by the Reading Room of the Rock and Roll Hall of Fame. He has other books with Oxford University Press, Bloomsbury, Continuum, Wilfrid Laurier University Press, Focal Press and Routledge.

Justin Paterson is Associate Professor of Music Technology in London College of Music, University of West London, UK. He has numerous research publications ranging through journal articles, conference presentations and book chapters, and is author of the *Drum Programming Handbook*. He is also an active music producer. Current research interests are 3D audio and interactive music, fields that he has investigated with prominent industrial organizations such as Warner Music Group. Together with Professor Rob Toulson of the University of Westminster, he developed the variPlay interactive music system.

Rob Toulson is Professor of Creative Industries: Commercial Music at the University of Westminster, London. He was previously Director of the Cultures of the Digital Economy Research Institute at Anglia Ruskin University, Cambridge, UK. Rob is a research leader in the field of commercial music, and he has collaborated with many international organizations in the music and audio industries. He is a successful music producer and studio engineer, as well as an experienced mobile app developer, having invented the novel iDrumTune percussion tuning application and the innovative variPlay interactive playback system, in collaboration with Professor Justin Paterson of the University of West London.

Perspectives on Music Production Series
Series Editors
Russ Hepworth-Sawyer
Jay Hodgson
Mark Marrington

Titles in the Series

Innovation in Music

Performance, Production, Technology
and Business

**Edited by Russ Hepworth-Sawyer,
Jay Hodgson, Justin Paterson and
Rob Toulson**

Routledge
Taylor & Francis Group
NEW YORK AND LONDON

First published 2019
by Routledge
52 Vanderbilt Avenue, New York, NY 10017

and by Routledge
2 Park Square, Milton Park, Abingdon, Oxon, OX14 4RN

Routledge is an imprint of the Taylor & Francis Group, an informa business

Library of Congress Cataloging-in-Publication Data
Names: Hepworth-Sawyer, Russ. | Hodgson, Jay. | Paterson, Justin. |
    Toulson, Rob.
Title: Innovation in music : performance, production, technology, and
    business / edited by Russ Hepworth-Sawyer, Jay Hodgson, Justin
    Paterson, and Rob Toulson.
Description: New York, NY : Routledge, 2019. | Series: Perspectives on
    music production series | Includes index.
Identifiers: LCCN 2018058031 (print) | LCCN 2019000227 (ebook) |
    ISBN 9781351016704 (pdf) | ISBN 9781351016681 (mobi) | ISBN
    9781351016698 (epub) | ISBN 9781138498211 (hardback) | ISBN
    9781138498198 (pbk.) | ISBN 9781351016711 (ebook)
Subjects: LCSH: Music and technology. | Music—Performance. | Sound
    recordings—Production and direction. | Music trade.
Classification: LCC ML55 (ebook) | LCC ML55 .I57 2019 (print) | DDC
    780.9/05—dc23
LC record available at https://lccn.loc.gov/2018058031

ISBN: 978-1-138-49821-1 (hbk)
ISBN: 978-1-138-49819-8 (pbk)
ISBN: 978-1-351-01671-1 (ebk)

Typeset in Times New Roman
by Apex CoVantage, LLC

Contents

Preface

The Innovation in Music network brings together practitioners, experts and academics in the rapidly evolving and connected disciplines of music performance, music production, audio technologies and music business. The Innovation in Music Conference in 2017 (InMusic17) was held at the University of Westminster, Regent St, London and included keynote interviews, performances and panel discussions with Imogen Heap, Talvin Singh, Mandy Parnell, Ken Scott and Peter Oxendale, among others. The network and conference acts as a forum for industry experts and professionals to mix with researchers and academics to report on the latest advances and exchange ideas, crossing boundaries between music disciplines and bridging academia with industry. After the conference, contributors were invited to submit articles for this book, *Innovation in Music: Performance, Production, Technology and Business*, in order to capture the innovative research and practices that were showcased at the event. This book therefore gives a broad and detailed overview of modern and cross-disciplinary innovations in the world of music.

The book is divided into four parts, representing innovations around each of music performance, production, technology and business. In Part One, innovative examples of music performance, stagecraft and composition are presented, with respect to different music genres from jazz to electronica. Part Two discusses innovation in music production, incorporating models for collaboration and creativity, and identifying toolsets and workflows for enabling innovative creative practice in the studio. Part Three presents a number of new technologies within the music and audio engineering realm, from experimentation with new interfaces and instrument designs to analysis of high-resolution audio and advanced methods for storing musical metadata. Finally, Part Four evaluates contemporary issues around music business, copyright, publishing and ownership, while also giving broad examples of the connections between enterprise and creativity, and collaboration bridging academia and industry.

We thank all chapter authors, conference speakers and delegates for their support, and intend that the chapters of this book will be a lasting record of contemporary music innovations and a resource of research information for the future.

Russ Hepworth-Sawyer, Jay Hodgson,
Justin Paterson and Rob Toulson (Editors)

Part One

Performance

1

Transforming Musical Performance

The Audience as Performer

Adrian York

INTRODUCTION

Audience collaboration in music performance is present in many contexts, from the pub sing-along to the call-and-response rituals of African and African-American cultures. There is a growing body of academic research that explores interactive audience collaboration using a variety of digital technologies. Many of the compositions and collaborative performances that have emerged have been driven by the affordances of these technologies. However, little research has been found with a focus on exploring the compositional and performance protocols that need to be developed to create successful interactive audience participation within an existing genre using pre-existing technology.

The major objective of this study is to create a dialogue between performer, audience, composer and technology (see Figure 1.1) drawing on Eco's conceptions of *open works*, which, according to Robey, require of the public "a much greater degree of collaboration and personal involvement than was ever required by the traditional art of the past" (Eco 1989. pp x–xi). Any performance of *The Singularity* brings together pre-programmed and random artificial intelligence (AI) machine-generated elements with live performances from the performer and the audience-performers, with each element having some form of interaction with the others. Inspiration for the composition's title and approach is drawn from technologist Ray Kurzweil, who popularized the term in his book *The Singularity Is Near* (Kurzweil, 2005). Kurzweil defines The Singularity as the moment when AI matches the level of human intelligence and notes that the future will be a dialogue with machines with AI collaborating with humans (Kurzweil, 2005, pp. 35–43). Using the compositional and performance structures of contemporary jazz as a model that is particularly suited to improvisational interaction, audience-performers use hand-held digital controllers to trigger different sections of the composition as well as pitched and non-pitched sound events to create the interaction among themselves, the technology and the performer.

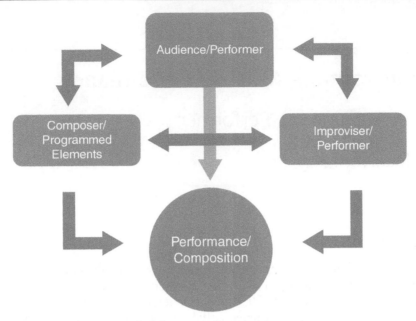

Figure 1.1 Performance Model

BACKGROUND AND RELATED WORK

Audience Participation

Performance practice and audience modes of reception and interaction vary markedly among different musical genres, cultures and sub-cultures. Pitts notes that "the traditional practices of the Western concert hall assume a relatively passive role for listeners" (Pitts, 2005, pp. 257–269), which contrasts with Williams-Jones assertion that "audience involvement and participation is vitally important in the total gospel experience"(Williams-Jones, 1975, p. 383). A typology of audience/performer relationships can therefore be divided into a dichotomy between participatory and non-participatory interactions.

Audience participation is not an uncommon element in many different fields of creative practice. *Tony and Tina's Wedding* was an immersive theatre piece from 1985 that ran for over 20 years in New York and was performed in more than 150 cities (Cassaro and Nassar, 1985).The audience played the part of the wedding guests and mingled with the characters as they ate, drank and danced. *The Rocky Horror Picture Show* film has inspired audiences to dress up as the characters, recite the script and sing along with the movie (O'Brien, 1975). There have been a variety of interactive film formats including CtrlMovie that give the audience a way of steering the narrative via real-time engagement and narrative decision-making through the medium of an app (CtrlMovie, 2017). Performance artist Marina Abramović has interacted with an audience on many occasions and in a variety of contexts. She says:

> Performance is mental and physical construction that performer make in a specific time and a space in front of audience and then energy dialogue happen. The audience and the performer make the piece together.
>
> (Abromavić, 2015)

Small proposes that "Music is not a thing at all, but an activity, something that people do" (Small, 1998, p. 2). Small describes this activity as *musicking*, which he defines as the set of relationships in the performance location between all of the stakeholders. For Small the protagonists in the performance include a cast of characters from everyone involved in the conception and production of sound to the venue management, cleaner, ice cream vendor and ticket seller. His proposal that music is an activity is at the center of this research, but it is being extended and reworked on his premise, taking the participatory element and turning it into an active musical involvement, thus trying to create a more democratic relationship between performer and audience. Nyman expands on this idea: "experimental music emphasizes an unprecedented fluidity of composer/performer/listener roles, as it breaks away from the standard sender/carrier/receiver information structure of other forms of Western music" (Nyman, 1999, pp. 23–24). Nyman is focused on the works of *experimental* composers, including John Cage, for whom the audience is expected to play an active role but only as an engaged spectator.

Nyman quotes Cage: "we must arrange our music, we must arrange our art, we must arrange everything. I believe, so that people realize that they themselves are doing it, and not that something is being done to them" (Nyman, 1999, pp. 23–24). Cage seems to be staking a claim here for the audience as participant in the creative process, but Cage does not propose any kind of active audience involvement beyond an immersive engagement with the performance or as an involuntary sound source, as in his composition *4'33"*.

In his taxonomy of research and artistic practice in the field of *Interactive Musical Participation*, Freeman creates three ranks of interactions (Freeman, 2005b, pp. 757–760). The first covers compositions in which the audience has a directly *performative* role, generating gestures that either form the whole of the soundscape or are integrated into the overall sonic and compositional architecture of the performance. The second category turns the audience into *sound transmitters* of pre-composed or curated sonic material through the medium of ubiquitous personal hand-held digital computing devices such as mobile phones. Freeman's final category sees the audience as *influencers*; this process can involve interactions as diverse as voting via hand-held digital devices and waving light sticks in the air. The data from these inputs is then analyzed and presented to the performers as some kind of visual cue that triggers a predetermined sonic gesture.

Of the three ranks identified in the taxonomy of audience interaction, it is the one that functions as a container for artifacts that gives the audience the potential for performing which is the most heterogeneous. The audience's affordances within the performance environment range from the realization of pre-composed material to the interpretation of abstract instructions and the triggering of samples. The Fluxus composer Tomas Schmit's composition *Sanitas no.35* has a performance script or score that reads as follows (Schmit, 1962):

> Empty sheets of paper are distributed to the audience. Afterwards the piece continues at least five minutes longer.

These instructions are in tune with some of the principles laid out in the *FLUXMANIFESTO ON ART AMUSEMENT* by George Maciunas, the founder and central co-ordinator of Fluxus, in 1965 (Maciunas, 1965). Maciunas says:

> He (the artist) must demonstrate self-sufficiency of the audience, He must demonstrate that anything can substitute art and anyone can do it.

Berghaus views the type of instructional compositional device proposed by Fluxus artists as an opportunity to unlock the creative potential of the audience (Berghaus and Schmit, 1994).

A more obviously active role for the audience is conceived in the 1968 recipe/performance instructions for *Thoughts on Soup*, a performance piece devised by Musica Elettronica Viva (MEV) (Rzewski and Verken, 1969). MEV was a composer's cooperative set up in 1966 in Rome by Allan Bryant, Alvin Curran, Jon Phetteplace and Frederic Rzewski for the performance of new compositions using live electronics. By 1969 the group had integrated both acoustic and found sound sources into their practice, and with *Thoughts on Soup* Rzewski calls for '*listener-spectators'* to be gradually blended in with '*player-friends'* (Rzewski and Verken, 1969, pp. 79–97). Rzewski says:

> In 1968, after having liberated the "performance", MEV set out to liberate the "audience". If the composer had become one with the player, the player had to become with the listener. We are all "musicians". We are all creators.

The recipe/performance instruction for Rzewski's *Thoughts on Soup* specifies a mix of traditional instruments and instrumentalists combining with novelty instruments (duck call, police whistle, pots and pans) for the 'listener-spectators' as well as microphones, amplifiers, mixers and speakers.

In the *Sound Pool* (1969), which Rzewski describes as "a form in which all the rules are abandoned", the audience is asked to bring along their own instruments and to perform with the MEV. In the context of *Sound Pool*, musicians are no longer elevated to the position of a star but instead work with the audience managing energies and enabling the audience to "experience the miracle" without overwhelming the audience-performers with their virtuosity. The outcome of this process is that the audience no longer exists as a discrete entity (Rzewski and Verken, 1969, pp. 79–97).

Rzewski's negation of the audience as an *alterity* creates an opportunity for the *collaborative emergences* of creativity characteristic of the processes of group improvisation (Sawyer, 2003, pp. 28–73). It is Rzewski's reshaping of the role of the audience that this research will seek to explore using digital technologies and modular compositional building blocks.

In Jean Hasse's composition *Moths* (1986), the audience become performers and are asked to whistle a variety of pitches and rhythms from a graphic score as directed by a conductor. The mass of overlapping pitches create an eerie soundscape. The score's instructions call for several minutes of rehearsal followed by three minutes of performance. Hasse reflects upon her creative process:

> Continuing a deconstructionist line of thinking, in 1986, while living in Boston, I had a chance to broaden my compositional scope away from that of a "conventional" performer, through the simple device of bringing an audience into the performance. During a concert interval years before, I had been intrigued to hear people whistling casually in the hall and wondered what it would sound like if this were formalized and even more people whistled in synchrony. My earliest sketches involved passing out whistling instructions to selected members of the audience, something akin to a Yoko Ono conceptual event. In Boston, however, when a concert appearance allowed me to develop the idea, the result became a graphic score for an entire audience to perform.
>
> (Hasse, 2017)

Hasse's 2001 piece *Pebbling* follows a similar model, with the audience clicking and rubbing together pebbles on cues from a conductor. Comparing the composition to *Moths*, Hasse comments that "it has a relatively similar graphic score, and was conducted, by gestures, to the gathered crowd. They produced a 'chattering' percussion piece, amplified by flutter echo effects arising from the cliffs—an interesting extra dimension". Hasse concludes "Ideally, audience involvement should feel somewhat natural . . . and necessary, in a variety of performance contexts " (Hasse, 2017). Hasse's analysis that audience involvement should be "natural . . . and necessary" runs counter to some of the examples of compositions within this literature review, which focus more on process and research rather than musical outcomes.

The Eos Orchestra at a fundraising banquet for the orchestra performed Terry Riley's *In C* in 2003, as detailed in Bianciardi et al. (Bianciardi et al., 2003, pp. 13–17). Pods were set up on each of the 30 banqueting tables that the audience would touch on a cue from the conductor. This haptic gesture triggered a sample of one of the 53 melodic units that make up the composition. Each trigger was synced to a main clock, with the sample coming in on the next available quaver/eighth note entry. The hardware and software system design allowed for mass audience participation, but problems were reported with the triggering instructions, as some audience members did not realize that the sample would only be triggered once the hand on the pod was removed. The conclusion that can be drawn from this research is that assumptions can too easily be made about the audience's technical understanding of a system by the system designers and implementation team who are familiar with its functionality and that clear instructions to the audience and some degree of training may be necessary.

La symphonie du millénnaire (2000) was a one-off performance at St. Joseph's Oratory in Montreal. Curated by Walter Boudreau, head of the Société de musique contemporaine du Québec, the composition involved 15 ensembles of 350 musicians performing music composed by 19 composers, all of which were built around a main theme, the Gregorian chant *Veni Creator Spiritus*. Within the audience there were 2,000 bell-ringers playing hand bells, as well as recordings of 15 church bells, two fire engine bells and the Oratory's great organ and carillon. Chénard (June 1, 2000) reports that Walter Boudreau first had the initial idea for the composition when he was 17 years old on the eastern slope of Mount Royal in Montreal. Boudreau (2000) provides detail:

> From my vantage point, the splendid panorama of Montreal gently awakening to the sounds of a thousand and one church bells unfolded before my eyes. I then imagined a kind of mega symphony that would blend the rich sounds of these bells with originally-composed music.

Chénard explains that it was only in 1997 that Boudreau started to develop the idea, prompted by the Conseil Québécois de la Musique (CQM) asking for submissions for musical ideas to celebrate the millennium. Denys Bouliane (2000), the joint artistic director of the project, emphasizes the importance of the 2,000 bell-ringers to the "participatory and 'event-full' qualities of the *Millennium Symphony*", proposing that "the resulting ritual performed on the site would thus consecrate the celebratory ethos of the work, and give it the festive mark with which we wanted to stamp it" (Bouliane, 2000). In a work of this complexity with multiple composers, pre-recorded and live elements, it is inevitable that participatory elements would have to be tightly organized and rehearsed. However, it is not a model that is suitable for this research project.

Hödl et al. developed an interactive system that allowed the audience to pan the sound of a lead guitar across the stereo image. Their conclusion was that

> balancing constraints with affordances is the key to both the audience's and musicians' acceptance of such a system and that a playful participatory design process can lead to better results in this regard. It is also shown that using smart phones opens up a large possibility space but at the same time their use has to be subtle to not distract too much from the music
>
> (Hödl et al., 2012, p. 1)

Hödl et al.'s research focused on a "rock band", and their "Findings include that musicians seem to be cautious about giving up control" (p. 1). It may have been that their reliance on 'rock' musicians rather than improvising musicians, who are more conditioned to react to unexpected changes in the musical language around them, may have contributed to their findings.

Digital Interactivity

The focus of investigation within this research is the participatory paradigm leading to the audience becoming *co-creators* through the use of what Hödl et al. describe as a *technically driven system* that affords *Interactive Musical Participation* (Hödl et al., 2012, p. 1). This process is described as being "when a spectator can take part or at least make a contribution in a live concert through a technically driven system" (p. 1).

The second rank within Freeman's taxonomy of interactive models for music performance utilizes the audience members' own hand-held digital communication devices to broadcast pre-prepared sonic gestures. The first systematic study of what has become known as distributed music was undertaken by Taylor in 2017 (Taylor, 2017). Golan Levin's 2001 composition, *Dialtones: A Telesymphony*, a composition that uses the audience's mobile phones as sound sources, is identified by Taylor as a foundational composition of this emergent genre (Levin, 2002). Levin articulates the ideas behind the composition in his artist's statement:

> The mobile phone's speakers and ringers make it a performance instrument. The buttons make it a keyboard and remote control. Its programmable rings make it a portable synthesizer. Yet, although no sacred space has remained unsullied by the interruptions of mobile phone ringtones, there is no sacred space, either, which has been specifically devoted to their free expression.
>
> (Levin, 2002)

Before the start of the performance, audience members exchanged their mobile phone number for a seat at the concert; specially composed ringtones were then downloaded onto their phones. These ringtones were triggered by musicians via a visual-musical software instrument. "Participants were lit up via a lighting system becoming an audio-visual pixel, a twinkling particle in an audio-visual substance—and the visitors, as a group, could at once be audience, orchestra and (active) score".

Involving the audience as performers can be traced back to the outputs of Musica Elettronica Viva (MEV) in the late 1960s, but one of the earliest examples of distributed music was Laurie Anderson's symphony for car horns. Entitled *An Afternoon of Automotive Transmission*, the composition was performed in 1972 by the audience at a drive-in bandstand in Vermont (Grosenick and Becker, 2001).

A much more ambitious project was Filipino composer Jose Maceda's *Ugnayan*, translated as *Interlinking* (Brown and Santos, 2010). Bringing together influences from Edgar Varèse, Pierre Schaeffer, John Cage and Karlheinz Stockhausen, Maceda created a 20-channel radio simulcast utilizing all of the radio stations in Manila in a state-sponsored cultural intervention to combine traditional Filipino instruments with a modernist musical aesthetic. The city's population was encouraged to bring their radios onto the streets to create a distributed "collaborative sound collage" (Taylor, 2017). Maceda's "Xenakis-like clouds of sounds" were realized by an ensemble performing a 100-page score of complex polyrhythms using "Kolitong (zithers), Bungbung (bamboo Horns), Ongiyung (whistle Flutes), Bangibang (yoked-shaped Wooden Bars), Balingbing (buzzers), Agung (wide-rimmed Gongs), Chinese Cymbals, Gongs and Echo Gong" (Brown and Santos, 2010). The bulk of the local population did not engage with Maceda's vision; however, he established a powerful precedent and model for distributed music in the 21st century.

CoSiMa (Collaborative Situated Media) is a project based at IRCAM in Paris that is developing a platform to turn "the smartphone in everybody's pocket into a means of collaborative production and collective expression" (CoSiMa, no date). CoSiMa has developed smartphone-based web applications such as drone, birds, monks and the rainstick, which are dependent on the motion of the device. The project is also involved in more collaborative scenarios, such as web applications *WWRY:R*, which features a selection of samples from the Queen song *We Will Rock You*, *Shaker*, which allows user-generated and recorded sounds to be uploaded and then triggered at 110 beats per minute, and *Matrix* in which a three-by-four grid of mobile phones create a matrix of loudspeakers and screens, with light and sound being triggered on and across the screens by a performer on one of the phones. A more interactive approach is utilized in a piece and application called *Weather*. This was developed by CoSiMa at the Sonar+D international conference on creativity and technology as part of the Sonar Music Festival in Barcelona in 2016. Participants use gesture to trigger sounds and visuals on the mobile phones related to four different weather states, the bird chirps associated with a sunny afternoon, wind, rain and thunder. The audience-performers' weather states create a weather profile on the server that then controls visuals appearing on a public screen and environmental sounds on the PA system. A dialogue takes place between the soundscapes generated by the audience-performers and a DJ who is playing live electronic music.

The CoSiMa platform was also used by orbe, a cross-disciplinary French research group who design accessible open environments to create novel experiences involving new media and the body. Their experiment *Collective Loops*, which comes under the framework of the *Collective Sound Checks* project, "is a collective musical experience with smartphones. Each event proposes to the participants to play together in the context of musical and playful proposals, in-group or in interaction with a performer (group, DJ, . . .)" (CoSiMa, no date).

Lee and Freeman developed a networked musical instrument application for mobile phones called *echobo* that audience members could download and perform on instantly, engaging with other audience members and generating sound that contributed to the performance. This combination of audience performance and sound transmission creates a hybrid rank within the taxonomic system of participatory performance modes. *Echobo* combines two types of instruments, one for the audience and one for the master musician who controls the harmonic structure of the piece while not generating any sounds. The master musician's chord choices are reflected in the eight note scales available to the audience on their version of the *echobo* app. "The aggregated sound results in a dense and stochastic combination of the notes in the scale and can be employed as a background harmonic texture" (Lee and Freeman, 2013).

A stand-alone acoustic musician supplies melody, in this instance a clarinet player. In the process of designing this app, Lee and Freeman proposed a set of criteria to enable a successful audience participatory experience:

1. To make participation easy
2. To collect gestures from the audience and turn them into a single musical composition
3. To drive audiences to start participation without reservation
4. To motivate people to participate and sustain the interest
5. To provide a clear relationship between their gestures and outcome in music

These principles were valuable in this research design as they provide a strong foundation for interactive composition. Audience feedback pointed to a greater sense of connection to the clarinet player rather than the other audience members/musicians and to a frustration with the rate of harmonic change as determined by the master musician as it was perceived to have limited the audience's musical expressivity.

Performer and Audience Interaction Within Jazz

Previous studies on the audience for jazz have focused both on its decline and its make-up, with little or none addressing audience interactivity. The consumption of jazz in particular is falling among millennials (18–34-year-olds), despite jazz having a strong presence in music education. According to the *Nielsen U.S. Music Year End Report* (2016), sales of jazz in 2011 represented 2.8% of all recorded music consumption in the US, falling to 1.3% in 2015 (Nielsen 2016). Live attendance at jazz concerts is also falling (NEA, 2008). However, according to Miller, many young people are now active participants in virtual music-making through game play in games such as *Guitar Band*, so engagement with the type of activities and technologies contained in this project should not feel too unfamiliar for audiences who are already familiar with gaming and smartphone technologies (Miller, 2009, pp. 395–429).

Bailey presents the performer-audience relationship as something problematic for improvising musicians, with the need for 'professionalism' leading to predictability of idiom and vocabulary (Bailey, 1993). Brand et al. provide some valuable analysis of the relationship but only in the context of a traditionally formatted jazz gig and with no mention of interactivity beyond the standardized responses of audience and musicians (Brand et al., 2012). In both studies, some hostility is expressed towards the demands of the audience and the pressures that this places musicians under. These findings suggest, albeit with a very small amount of data, that the contemporary model for jazz performance has an inbuilt tension between audience and performer.

Jazz musicians have been involved in performances using interactive audience participation. Jason Freeman created a composition called *Sketching* in 2013 for improvising musicians with audience participation via mobile phones. The audience created a graphic score collaboratively using MassMobile, a client-server smartphone participation system (Freeman, 2013). The design of the composition creates a constant *feedback loop* between the audience and performers, with the audience responding visually to the musicians and the musicians taking direction from the score. To contrast with the performance described in Hödl et al. (2012), the performers in *Sketching* were the musicians from the Georgia Tech University Jazz Ensemble, who would have been familiar with improvising and for whom being musically responsive to external stimuli was part of their artistic practice.

The interaction among musicians that is embodied in improvisatory jazz practice is extended to the audience in this research in an extension of Small's concept of *Musicking* (Small, 1998), as well as drawing on the influence of Frederic Rzewski (Rzewski and Verken, 1969). With a declining and aging jazz audience, as well as the tension between artists and audience detailed here, there is space for artistic practice that explores a greater integration of the audience into the performance.

DESIGN AND IMPLEMENTATION

Composition Construction

As with many jazz standards, as well as songs from The Great American Songbook, *The Singularity* is constructed around an AABA compositional structure, with each section being eight bars in length. The A section shifts between Bb, B and C tonalities with a passing movement through an Ab diminished chord. The B section moves between a G Phrygian and Ionian/ Lydian with the repeating final A section coming at the end of the form.

The melody is mostly based on semitone and fifth intervals and is articulated by the performer using a lead synth sound. Once the melody has been played by the performer over the AABA structure, each of the four Wiimote audience-performers takes it in turn to improvise. An improvised musical dialogue follows each of these improvisations with the performer. After all of the audience-performers have finished their improvisations, all of the performers engage in a collective improvisation. The performance ends after the performer plays the melody one final time. There are a series of programmed backings for the improvisation sections selected by the audience-performer who controls the iPhone. These did not necessarily match the AABA structure and harmonic format of the melody section.

Figure 1.2 The Singularity Score

Technical Infrastructure

The technical challenge raised by the performance of *The Singularity* (2017) was to create an infrastructure that was robust enough to:

1. withstand the stresses of live performance
2. provide powerful enough Wi-Fi and Bluetooth networks to create a stable platform for the controllers
3. provide a level of accessibility that met Lee and Freeman's (2013) five criteria
4. enable Ray Kurzweil's dialogue with machines

All sounds were generated from the Digital Audio Workstation Ableton Live (2017). The program was running on a MacBook Pro with an iPhone triggering different sections of the composition via a phone app entitled TouchOSC (2018) that sends and receives Open Sound Source control messages. Open Sound Source control is a communication protocol for electronic music instruments that is optimized for modern communication networks. TouchOSC connects via Wi-Fi to an application on the MacBook called OSCulator (2018), which transfers the control signals from TouchOSC to Ableton Live. Four Wiimote controllers also change parameters within Ableton Live, also connecting via OSCulator but using Bluetooth rather than Wi-Fi for connectivity.

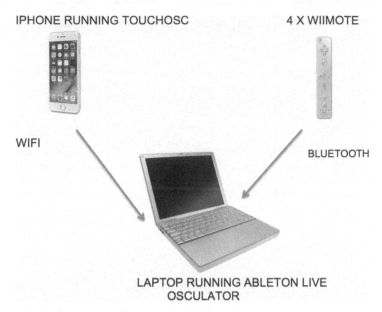

IPHONE RUNNING TOUCHOSC   4 X WIIMOTE

WIFI

BLUETOOTH

LAPTOP RUNNING ABLETON LIVE
OSCULATOR

Figure 1.3 Project Network

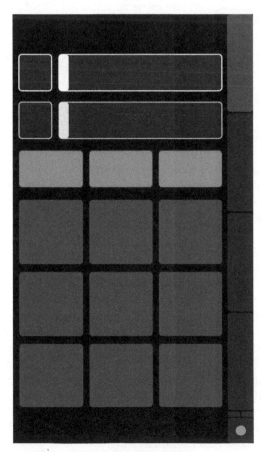

Figure 1.4 TouchOsc BeatMachine Configuration

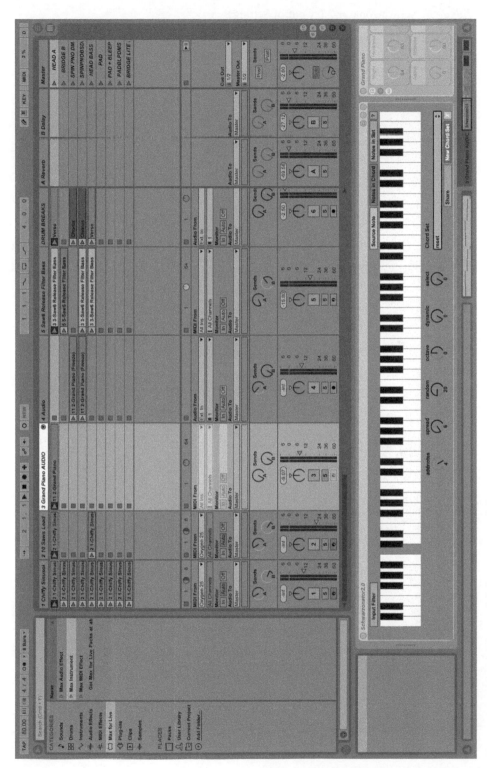

Figure 1.5 Ableton Live and Schwarzonator 2.0

Table 1.1 Schwarzonator 2.0 Control Settings

Addnotes	Spread	Random	Octave	Dynamic
4	6	26	0	0

Figure 1.6 Wiimote Controllers

Table 1.2 Wiimote Sonic Element Control

Wiimote 1	Controls pitch of 80-Elaspsych-Shy loop
Wiimote 2	Controls pitch of Electric Screamer Lead synthesizer
Wiimote 3	Controls dry/wet mix of delays on Slap 120 bpm loop
Wiimote 4	Triggers pitches Bb, C, D and F on Arp Pluck sample

There had been problems with the Wi-Fi at the venue for the pilot performance of *The Singularity*, and after consulting with the IT department it was decided that the best way to ensure a stable connection between the iPhone and the laptop was to generate a computer-to-computer network from the MacBook.

This was a very effective solution that allowed the audience-performer using TouchOSC on the iPhone to trigger Ableton clips without there being any dropout. The global quantization for these clips was set to eight bars, ensuring that each newly triggered clip entered at the end of an eight-bar passage with the previous one finishing its cycle, thereby creating smooth transitions between sections and sustaining the flow of eight-bar sections.

The Max For Life Schwarzonator 2.0 plugin was used on the piano track to generate random chord voicings built on the harmony of the composition and adding an AI element to the performance. The *addnote* function allows you to choose the density of each voicing; *spread* marks the range across the keyboard that the voicings inhabit; *random* shifts notes up and down in a random manner; *octave* shifts notes up and down; and *dynamic* adds a random element to the note's velocity.

The four Wiimote controllers were numbered and color-coded, with each of them controlling a specific sonic element.

The performer used an M-Audio Oxygen 25 keyboard to perform the melody and to create improvisations using a blend of the Chiffy Sinusoi and 10 Saws Lead synthesizer patches on Ableton Live using a USB direct connection into the Macbook Pro laptop.

Each of the elements in the technical infrastructure for the performance of *The Singularity* functioned effectively, establishing it as a good model for future research purposes.

Performance Protocols

The performer functions as the musical director/conductor/MC of the performance as well as setting up and managing the equipment and software. At the start of the performance, the performer follows the instructions listed in Figure 1.7, which involve finding five volunteer audience-performers and leading them through the performance.

Each of the audience-performers was given the following set of color-coded performance protocols matching the color of their Wiimote and technical instructions, which they were to read before the performance.

The audience-performer controlling the musical structure of the performance was given the following instructions (see Figure 1.9).

PERFORMANCE INSTRUCTIONS—P = PERFORMER. V = VOLUNTEER

1. GET 5 VOLUNTEERS
2. GIVE IPHONE TO V1 PLUS INSTRUCTIONS
3. GIVE WIIMOTES TO VS 2–5 PLUS INSTRUCTIONS
4. EXPLAIN TRACK LENGTH AND FORMAT
5. P TO CUE SOLO SECTIONS. EACH V TO GO IN TURN—SHORT
   EXPLORATION FOLLOWED BY A DIALOGUE WITH P
6. ALL TO PERFORM TOGETHER ON CUE
7. P TO FADE MASTER

Figure 1.7 Performer Instructions

THE SINGULARITY AUDIENCE/PERFORMER INSTRUCTIONS

1. MAKE SURE POWER LIGHTS ARE ON
2. AFTER MELODY SECTION START PERFORMING ON CUE FROM
   PERFORMER EACH IN TURN (AP1, AP2, AP3, AP4)
3. MAKE YOUR PEFORMANCE A SHORT EXPLORATION OF
   THE POSSIBILITIES FOLLOWED BY A DIALOGUE WITH THE
   PERFORMER
4. ON CUE FROM THE CONDUCTOR PERFORM TOGETHER UNTIL
   THE TRACK FADES

TECHNICAL INSTRUCTIONS

AP 1—PRESS BUTTON 1 ON THE WIIMOTE TO TRIGGER OR TO
STOP THE SOUND. CURL YOUR ARM HOLDING THE WIIMOTE
UP AND DOWN TO TRANSFORM THE SOUND. THIS SHOULD
INITIALLY BE DONE QUITE SLOWLY TO ALLOW FOR SONIC
EXPLORATION.

AP 2—PRESS AND HOLD BUTTON 1 ON THE WIIMOTE TO TRIG-
GER THE SOUND. CURL YOUR ARM HOLDING THE WIIMOTE
UP AND DOWN TO TRANSFORM THE SOUND. THIS SHOULD
INITIALLY BE DONE QUITE SLOWLY TO ALLOW FOR SONIC
EXPLORATION.

AP 3—PRESS BUTTON 1 ON THE WIIMOTE TO TRIGGER OR TO
STOP THE SOUND. CURL YOUR ARM HOLDING THE WIIMOTE
UP AND DOWN TO TRANSFORM THE SOUND. THIS SHOULD
INITIALLY BE DONE QUITE SLOWLY TO ALLOW FOR SONIC
EXPLORATION.

AP 4—PRESS BUTTONS 1, 2, +,—OR A ON THE WIIMOTE TO
TRIGGER THE SOUND. CURL YOUR ARM HOLDING THE WIIMOTE
UP AND DOWN TO TRANSFORM THE SOUND. THIS SHOULD
INITIALLY BE DONE QUITE SLOWLY TO ALLOW FOR SONIC
EXPLORATION.

Figure 1.8 Audience-Performer Wiimote Instructions

TOUCHOSC INSTRUCTIONS

There are 9 Purple swithches (the bottom nine squares). Each one triggers a cycle of music. When you trigger a switch it will play throught to the end of its cycle before the next one triggers. For added performance pleasure play with the yellow (top) slider when the bass synth is playing.

Figure 1.9 Audience-Performer TouchOsc Instructions

CONCLUSIONS

Evaluation

Approximately 50 audience members attended the pilot performance of *The Singularity* at the Innovation in Music Conference in September 2017. The technical infrastructure was robust despite running four Wiimotes simultaneously, and the computer-to-computer network provided a stable Wi-Fi framework allowing for Wiimote functionality from anywhere in the hall.

This performance met the five Lee and Freeman (2013) criteria in that:

1. Participation was easily accessible.
2. Gestures from the audience were turned them into a single musical composition.
3. Audience-performers had no reservations about participating.
4. Audience-performers were motivated to perform and sustained interest in their participation.
5. Audience-performers in some instances identify a clear relationship between their gestures and the musical outcomes.

Feedback on the performance was delivered verbally both from the audience-performers and from audience members both in the Q&A session that followed and in further discussions post-performance:

1. Both the audience-performers and the audience as a whole felt that there were meaningful moments of musical dialogue between the performer and the audience-performers and most obviously with the glissando Wiimote.
2. There was a sense of relief from the whole audience that the technology functioned as promised.
3. The audience-performers and the audience enjoyed the process, and there was a sense of *playfulness* and *discovery* for both groups.
4. The performance protocols worked effectively.

However, not all of the audience-performers were aware of what sounds/motifs/effects they were triggering, and there was a lack of familiarity with the layout and functionality of the Wiimote, which will inform the design of future research into the performance protocols. There is also a question as to whether there needs to be a greater emphasis on random AI-generated elements so that the performer isn't operating within a 'zone of expectation' and that with repeat performances there is always an element of surprise.

Future Work

This performance of *The Singularity* was designed as a pilot project to test out the technological infrastructure and basic premise of this research. As well as the feedback from the audience and audience-performers, the following conclusions have been drawn:

1. A modal harmonic approach should be utilized so that melodic elements triggered by the audience-performers can function across the harmonic structure of the whole piece.
2. Sounds triggered by the audience-performers should have gentle attack envelopes to avoid rhythmic incompatibility. The emerging technology of distributed synchronized playback on hand-held devices may be a way to solve rhythmic problems, but until the technology is available for use, this proposal is still speculative rather than proven.
3. Audience-performers could use distributed sound on their mobile phones, allowing for a simple scaling up of the numbers of audience-performers. OSCulator has the capacity to run more Wiimotes than were used for this performance, but this would have to be tested before scaling up in this manner.
4. The possibilities of interactive music participation with acoustic performers should be explored.
5. The audience-performers should be given a 'soundcheck' to explore the parameters and functionality of their controllers as well as the sonic possibilities.
6. Using musical textures that are less dense than those in *The Singularity* will create greater sonic clarity, allowing the audience-performers to identify their contributions with greater certainty.
7. The stability of the computer-to-computer Wi-Fi and Bluetooth networks and their range offers opportunities for distributing the Wiimotes, mobile phones and any other controllers throughout the performance space as well as the potential for using more controllers on the network.
8. Structural changes in the improvised sections of a composition should maintain the same modal harmonic framework while other elements such as rhythm, instrumentation and dynamics might change. This would create more opportunity for triggering distinct pitches by the audience-performers as in a modal context as long as the triggered pitches came from the same mode there would be no notes that clashed.

CONCLUSION

This pilot project has provided a solid basis for further research, a process that will concentrate on scaling up the amount of both performers and audience-performers and creating a musical framework that can accommodate this emerging artistic practice. The project also provides a technical infrastructure model that is both accessible and stable, making it practical for use in either artistic practice or research.

REFERENCES

Ableton (2017). *Ableton live*. Ableton. Available at: www.ableton.com/en/live/ [Accessed Nov. 2017].

Abromavić, M. (2015). *An Art made of trust, vulnerability and connection*. TED. Available at: www.ted.com/talks/marina_abramovic_an_art_made_of_trust_vulnerability_and_connection#t-220428 [Accessed 15 Apr. 2017].

Bailey, D. (1993). *Improvisation: Its nature and practice in music*. Boston: Da Capo Press.

Berghaus, G. and Schmit, T. (1994). Tomas Schmit: A fluxus farewell to perfection: An interview. *TDR (1988–)*, 38(1), pp. 79–97.

Bianciardi, D., Igoe, T. and Singer, S. (2003). Eos pods: Wireless devices for interactive musical performance. In: *Proceedings of the fifth annual conference on ubiquitous computing*. Seattle.

Bouliane, D. (2000). *Word from the artistic directors*. SMCQ. Available at: http://smcq.qc.ca/smcq/en/symphonie/mots/ [Accessed 8 June 2017].

Boudreau, W. (2000). *Word from the artistic directors*. SMCQ. Available at: http://smcq.qc.ca/smcq/en/symphonie/mots/ [Accessed 8 June 2017].

Brand, G., Sloboda, J., Saul, B. and Hathaway, M. (2012). The reciprocal relationship between jazz musicians and audiences in live performances: A pilot qualitative study. *Psychology of Music*, 40(5), pp. 634–651.

Brown, C. and Santos, R. (2010). Ugnayan CD notes. *Tzadik*.

Cassaro, N. and Nassar, M. (1985). *Tony and Tina's wedding* (Theatrical production). New York. Available at: http://tonylovestina.com/about-tony-n-tinas-wedding/ [Accessed 15 Apr. 2017].

CoSiMa. (n.d.). *CoSiMa*. Ircam. Available at: http://cosima.ircam.fr/home/ [Accessed Mar. 2017].

CtrlMovie. (n.d.). Available at: www.ctrlmovie.com [Accessed 6 June 2017].

Eco, U. (1989). *The open work* (Translated by Anna Cancogni with an Introduction by David Robey). Cambridge, MA: Harvard University Press.

Freeman, J. (2005a). *Glimmer* (Score). Available at: www.jasonfreeman.net [Accessed Apr. 30, 2018].

Freeman, J. (2005b). *Large audience participation, technology, and orchestral performance*. In *Free Sound*, 757–760. Proceedings of the International Computer Music Association. San Francisco, CA: International Computer Music Association (ICMC).

Freeman, J. (2013). *Sketching*. Available at: http://distributedmusic.gatech.edu/jason/music/sketching-2013-for-improvis/ [Accessed 30 Apr. 2018].

Grosenick, U. and Becker, I. (eds.). (2001). *Women artists in the 20th and 21st century*. Cologne: Taschen.

Hasse, J. (1986). *Moths. Visible music*. Available at: www.visible-music.com/ [Accessed 27 Apr. 2018].

Hasse, J. (2017). *Moths and pebbling* (email). Sent to Adrian York, 8 June.

Hexler. (2018). *TouchOSC*. Hexler. Available at: https://hexler.net/software/touchosc [Accessed 22 Feb. 2018].

Hödl, O., Kayali, F. and Fitzpatrick, G. (2012). *Designing interactive audience participation using smart phones in a musical performance*. Available at: https://www.researchgate.net/publication/262178344_Designing_Interactive_Audience_Participation_Using_Smart_Phones_in_a_Musical_Performance [Accessed 16 Feb. 2019].

Kurzweil, R. (2005). *The singularity is near*. New York: Penguin.

Lee, S. and Freeman, J. (2013). *Echobo: A mobile music instrument designed for audience to play*. NIME. Available at: www.nime.org/2013/program/papers/day3/poster3/291/291_Paper.pdf [Accessed 9 June 2017].

Levin, G. (2002). *Dialtones*. Audio available at: www.flong.com/storage/experience/telesymphony/, Artist's statement at www.flong.com/storage/experience/telesymphony/index.html#background [Accessed 9 June 2017].

Maciunas, G. (1965). *Fluxus Debris*. Available at: http://id3419.securedata.net/artnotart/fluxus/index.html [Accessed 19 Apr. 2017].

Miller, K. (2009). Schizophonic performance: Guitar hero, rock band, and virtual virtuosity. *Journal of the Society for American Music*, 3, pp. 395–429. doi:10.1017/S1752196309990666.

National Endowment for the Arts (NEA). Arts participation 2008—highlights National Endowment for the Arts (NEA). (2009). *Arts participation 2008—highlights from a national survey*. Washington, DC: NEA. Available at: www.arts.gov/publications/2008-survey-public-participation-arts [Accessed 26 Apr. 2018].

Nielsen. (2016). *2015 U.S. music year-end report*. Available at: www.nielsen.com/us/en/insights/reports/2016/2015-music-us-year-end-report.html [Accessed 26 Apr. 2018].

Nyman, M. (1999). *Experimental music: Cage and beyond* (Vol. 9). Cambridge: Cambridge University Press.

O'Brien, R. (1975). *The Rocky Horror P Show* (Film). 20th Century Fox.

OSCulator. (2018). About osculator. *Osculator*. Available at: https://osculator.net/ [Accessed 11 Nov. 2017].

Pitts, S. E. (2005). What makes an audience? Investigating the roles and experiences of listeners at a chamber music festival. *Music and Letters*, 86(2), pp. 257–269.

Rzewski, F. and Verken, M. (1969). Musica elettronica viva. *The Drama Review: TDR*, 14(1), (Autumn), pp. 92–97.

Sawyer, R. K. (2003). *Group creativity: Music, theater, collaboration*. Mahwah, NJ: Erlbaum.

Schmit, T. (1962). *Sanitas no. 35*. Available at: www.artnotart.com/fluxus/tschmit-sanitasno.35.html [Accessed 8 Feb. 2018].

Small, C. (1998). *Musicking: The meanings of performing and listening*. Hanover, NH: Wesleyan University Press.

Taylor, B. (2017). *A history of the audience as a speaker array*. Nime 17. Available at: homes.create.aau.dk/dano/nime17/papers/0091/paper0091.pdf [Accessed 23 July 2017].

Williams-Jones, P. (1975). Afro-American gospel music: A crystallization of the black aesthetic. *Ethnomusicology*, 19(3), pp. 373–385. doi:10.2307/850791. Available at: www.jstor.org/stable/850791.

2

Using Electroencephalography to Explore Cognitive-Cultural Networks and Ecosystemic Performance Environments for Improvisation

Tim Sayer

INTRODUCTION

The aim of this chapter is to bring together three areas of enquiry with the purpose of exploring their potential application within the realm of musical improvisation and in so doing providing a rationale for future creative developments in the area of human-computer interface design within the context of improvised music performance systems. The areas under investigation can be broadly categorized as cognitive-cultural networks, ecosystemic design and brain-computer interfaces. These three elements form a tripartite approach to the contextualization of human agency within the realm of improvised music-making and suggest a three-tier approach to the investigation of causation, in the chain of influence which affects musical behavior in this context. Much research has been undertaken, which concentrates on particular segments of the creative process, focusing primarily either on the behavior of the performer, the relationship of the performer to the means of production or the social context in which the activity exists. The motivation for taking a more holistic approach is to provide technological interventions that facilitate the development of performance environments that support improvising musicians, striving to explicate their art in a manner that satisfies a desire to create a unique musical performance—one that minimizes mechanical forms of musical behavior and utilizes pre-programmed units of musical material. This form of creative endeavor, often referred to as non-idiomatic or free music (Bailey and National Sound Archive, 1992), is rich with anecdotal evidence to support an enquiry of this nature. As a starting point, various subjective views from this field of improvisation will be presented to define the problem space and shed light on the dilemmas and frustrations experienced by practitioners. These concerns will then be subjected to brief analysis in terms of their relationship to art in the wider context of cognition, looking at cognitive evolution with specific reference to Donald's work on cognitive-cultural networks (Donald, 2008). These ideas will then be recontextualized, drawing on themes from Di Scipio's ecosystemic design principles (Di Scipio, 2003) and also passive brain-computer interaction

(BCI), to suggest a novel approach to the design of performance contexts within this field of enquiry. The theoretical themes of the chapter will be represented as a model for the development of performance architectures and performance environments. By way of an exemplar, the recently created piece *Mondrisonic* will be described as an implementation of that model.

IMPROVISATION IS NOT UNCONTENTIOUS

This investigation is very much informed by the experience of improvising musicians, and as such anecdotal evidence has been an important source of information. Given that this is such an important starting point, I think it's worth clarifying that I do not regard the content of an anecdote to represent empirical evidence of anything other than as an indication of perception. That is to say, its factual accuracy may be called into question, but unless there is a deliberate attempt to deceive, it can be regarded as a reasonable reflection of what was perceived in a given situation. Anecdotes, personal as they are, cannot escape the crudity of language as a tool to represent a domain, such as music, that could be considered in some sense meta-lingual. Anecdotes are interesting because they can often reveal a mismatch between perceived (internal) reality and the objective (external) reality. In this context, they present an opportunity for an observer to reconcile the improviser's duality in their performance, that of producer and consumer. For the musician it is a chance to offer a personal perception of a situation, which may defy an objective, logically causal explanation—things that just happen. The following quotation from Steve Lacy offers his perception of the relationship between learning and improvising and conveys what could be interpreted as an ethical stance on what can legitimately be called improvisation.

> Why should I want to learn all those trite patterns? You know, when Bud Powell made them, fifteen years earlier, they weren't patterns. But when somebody analysed them and put them into a system it became a school and many players joined it. But by the time I came to it, I saw through it—the thrill was gone. Jazz got so that it wasn't improvised anymore.
>
> (Bailey, 1993, p. 54)

What is interesting about Lacy's observation is the assertion that the pioneers of Jazz didn't play patterns and begs the question: what constitutes a pattern? Lacy seems to be suggesting that the formulation of patterns is the mechanism by which acts, that he regards as spontaneous, can be replicated. They are perhaps the product of a mimetic process for which the primary motive is 'learning' and 'copying'. What this opinion fails to address is the possibility that the 'learned' has to exist on some level in all musical improvisation, particularly improvisation at speed (Gaser and Schlaug, 2003). Borrowing from others or from an idiom certainly raises questions of authenticity, but it seems implausible to contemplate the notion that an improviser can develop their practice in a vacuum, without influence. In fact, as John Cage famously articulated, this aspect of improvised music-making can result in some artists rejecting it altogether.

> Improvisation . . . is something that I want to avoid. Most people who improvise slip back into their likes and dislikes and their memory, and . . . they don't arrive at any revelation that they are unaware of.
>
> (Cage and Turner, 1990, p. 472)

Lee Konitz's observation places more responsibility on the performer to circumvent these tendencies with an awareness of how focused attention functions, to reduce auto-responsive musical behavior. This alludes to the issues of memory, to which Cage and Lacy refer, but pulls focus on procedural (motor skills) rather than declarative memory (facts and events).

> Playing mechanically suggests a lack of real connection to what you are doing at the moment. We learn to play through things that feel good at the time of discovery. They go into the "muscular memory" and are recalled as a matter of habit.
>
> (Hamilton and Konitz, 2007, p. 109)

Konitz acknowledges here the tension within the master-slave relationship between declarative memory and procedural when engaging in an activity that is perceived to be under conscious control, suggesting that playing becomes more habitual when the executive function of declarative memory is weakened by non-attentiveness.

If we analyze the experiences of those who seem to have developed a practice that has, at least from their own perception, partially resolved the aforementioned issues, we can see an interesting subversion of episodic and semantic memory, via a reactive response to an unpredictable sequence of events. Physical reactions, when stimulated by external stimuli, can be executed with minimal need for conscious attention (Libet, 1985). The following quotations, first from Derek Bailey and then from Evan Parker, suggest that the environment is key to unlocking the creative freedom in their practice, not their learned repertoire, at any level of their memory system.

> A lot of improvisers find improvisation worthwhile. I think, because of the possibilities. Things that can happen but perhaps rarely do. One of those things is that you are 'taken out of yourself'. Something happens which so disorientates you that for a time, which might only last for a second or two, your reactions and responses are not what they normally would be. You can do something you didn't realise you were capable of or you don't appear to be fully responsible for what you are doing.
>
> (Bailey, 1993, p. 115)

> It can make a useful change to be dropped into a slightly shocking situation that you've never been in before. It can produce a different kind of response, a different kind of reaction.
>
> (Bailey, 1993, p. 128)

These statements, from two of the most influential exponents of improvisation in the post-war UK experimental music scene, echo sentiments
expressed in biographies, documentaries, articles and interviews by performers in this genre, the world over. They are bringing to the debate the
role of external context, and in so doing adding another dimension to the
path of causation that governs the musical behavior of improvising performers. This brief excursion into the frustrations and elations expressed by
improvising musicians has shown the influence of context and environment
on the subjective experience of their continuous battle to generate original
material—to evolve the music beyond that which has been played before.

CULTURAL COGNITIVE EVOLUTION

The experiences highlighted here allude to a strata of cognitive processing
which is rich in its potential to reveal points of intervention in the chain
of causation from sensory input, through perception, to manifest musical
behavior. They suggest the existence of entry points, which map to human
responses but may not necessarily feature in attentive awareness. We could
conceptualize this as a series of layers of habitual and planned action, sometimes referred to as goal-states (Cushman and Morris, 2015). In describing
the evolutionary development of cognition, Donald defines four periods
of development that map how culture and the brain interact in decision-
making. He describes this as a "cascading model", which has resonance
with this notion of layers. He suggests a process whereby hominid cognitive development retains and builds upon each earlier adaption and is a
useful lens through which to examine improvised musical behavior, probably the earliest form of human music-making (Cox and Warner, 2017). The
first period, episodic, which existed over 4 million years ago (MYA), he
describes as pure event perception, when humans existed much like other
species in the way their behavior was stimulated directly by their environment. The second period spanning 4–0.4 MYA, which he calls mimetic, is
characterized by action modeling. During this period the ability to manifest
behavior based on imitation, ritual and shared attention is developed. The
mimetic period was the first point at which human experience, and consequently behavior, was augmented by the experiences of others, purely
through observation. It was not until the third period, the mythic, some 0.5
MYA, that shared attention between individuals led to symbolic/linguistic
forms of representation and communication. These approximate periods
were mediated by neurobiological change, while the transition to the final
period, the theoretic, was stimulated over the last 2000 years, largely by
environmental and technological influences on cognition. This period is
characterized by human augmentation, both conceptual and physical. The
rate of change over this period has been unprecedented, fueled by extensive developments in the cognitive-cultural networks that move this evolution beyond the domain of the individual into the social, supported by an
extraordinary rate of technological development.

> This is a "cascade" model inasmuch as it assumes (as Darwin did) a
> basically conservative process that retains previous gains. As hominids

moved through this sequence of cognitive adaptations they retained each previous adaptation, which continued to perform its cognitive work perfectly well. . . . The first two hominid transitions-from episodic to mimetic, and from mimetic to mythic-were mediated largely by neurobiological change, while the third transition to the theoretic mode was heavily dependent on changes in external, nonbiological, or artificial, memory technology. The fully modern mind retains all of these structures, both in the individual brain as well as in the distributed networks that govern cognitive activity in cultural networks.

(Donald, 2008, p. 199)

In addition to the generic influence of cognitive-cultural networks (CCN) on the evolution of cognition, Donald applies this theory to trace the cognitive origins of art. He suggests seven main defining factors that constitute an arts practice, which can help understand its function within the evolution of cognition. In summary, these factors relate to (1) the intent to influence the mind of an audience through the reciprocal control of attention, (2) its link to a larger distributed cognitive network, (3) its ability to construct mental models and worldviews by integrating multiple sources of experience, (4) the utilization of metacognition as a form of self-reflection, (5) being technology-driven, (6) the unfixed artist's role within the distributed cognitive network, and lastly (7) aiming for a cognitive outcome by engineering a state of mind in an audience.

So, what relevance does this long view have to an exploration of improvised music-making in the 21st century. When Steve Lacy disparagingly describes the rote learning of patterns as having an undermining influence in improvisation, he is touching on a remarkably persistent mimetic facility, one that provides the basis for human self-awareness. There is a strong argument that proposes the mimetic core of hominid behavior, which Donald suggests is the basis of the evolutionary split leading to modern humans' higher cognitive abilities, has a neural correlate in the mirror neuron (Wohlschläger and Bekkering, 2002). The discovery of this physical phenomenon some 20 years ago in the brains of monkeys (Gallese et al., 1996), and now evidenced in the human brain (Decety and Grèzes, 1999), shows the basic mechanism by which the observable experiences of others can be registered partially, on a neural level, as our own. The area in which this phenomenon is observed is the premotor cortex, and it is worth noting that as these neurons are responding to audio-visual stimuli, the effect is resultant on action-related sound as well as that which is visually observable. It has been suggested that the function of the mirror neuron is strongly associated with sensorimotor associative learning, and that mirror neurons can be changed in radical ways by sensorimotor training (Lotem et al., 2017). The relevance, to the field of improvised music-making, of this layered model of cognition and the facility of mirror neurons to respond to action-related sound, is that it suggests the possibility of a cognitive-cultural network, which disrupts the regular causal flow from stimulus to behavioral response or action. Technology has the potential to initiate that disruption in a controlled and potentially creative way, via the entry points that exit within the multitude of cognitive layers that are operational when a musician is engaged in improvisation. The argument for a sensorimotor associative learning basis for mirror neurons, as opposed to a genetic adaptation designed by evolution to undertake a specific socio-cognitive function, supports the idea that this intriguing human capacity has the potential to be harnessed for creative/artistic purposes (Cook et al., 2014). Experimentation to shed light on the "action-listening" capabilities of mirror neurons has been undertaken, involving the teaching of untrained musicians to play a simple piano piece by ear. When learned pieces were listened to by the participants without any movement, the mirror neuron system became much more active than when they were exposed to an equally familiar but unpracticed piece of music. This research supports "the hypothesis of a 'hearing—doing' system that is highly dependent on the individual's motor repertoire" (Lahav et al., 2007). This suggests that emancipation from mechanistic improvisation will inevitably require an intervention, which subverts the neural infrastructure that supports "action-listening" in humans, but this subversion might come from the performance environment, as an external stimulus rather than a fully attentive action.

> Individual decisions are made in the brain. Human brains, however, are closely interconnected with, and embedded in, the distributed networks of culture from infancy. These networks may not only define the decision-space, but also create, install, and constrain many of the cognitive processes that mediate decisions.
>
> (Donald, 2008, p. 191)

ECOSYSTEMIC DESIGN

The potential to introduce computer technology into improvised performance has enabled the possibility of building interfaces that are active, not just reactive. In this sense they can respond to but also initiate interaction between a performer and their performance environment in accordance with a predetermined parameter map, in ways that the performer may or may not be consciously aware. Many computer-based interfaces continue to evolve tightly coupled gestural mapping using a variety of peripheral devices such as data gloves, motion detectors, velocity sensors, etc. There are also, however, opportunities, afforded by computer-based technologies, to explore the relationship between performer and sound source with the construction of responsive environments in a manner Di Scipio refers to as ecosystemic. The second theme of this chapter relates to Di Scipio's notion of ecosystemic design. He asserts that, in this paradigm, the performer and computer system exist in a relationship of 'ambient coupling', where the computer system is responsive, not purely to the performer but to the performer in the context of the performing environment.

> Notwithstanding the sheer variety of devices and computer protocols currently available, most interactive music systems—including developments over the Internet—share a basic design, namely a linear

communication flow: information supplied by an agent is sent to and processed by some computer algorithms, and that determines the output. This design implicitly assumes a recursive element, namely a loop between the output sound and agent-performer: the agent determines the computer's changes of internal state, and the latter, as heard by the agent, may affect his or her next action (which in turn may affect the computer internal state in some way, etc).

(Di Scipio, 2003, p. 270)

In relating this paradigm to the concept of cognitive-cultural networks, I have extrapolated the "interrelationship mediated by ambience", to which Di Scipio refers beyond purely the room ambience, into a parameter space one might describe as "cognitive ambience". Expanding the concept of ecosystemic design, into the realm of the human performer's cerebral sub-systems, presents an opportunity to explore a very distinctive epistemology in the relationship between mind and machine. It suggests a relationship, which taps into the 'cascade' model that Donald suggests encompasses the four phases of cognitive evolution, as they exist in modern humans engaging in artistic practice. This model asserts that musicians undertaking an improvisatory performance are coincidentally engaged in a practice that utilizes mimetic, mythic and theoretic modes of cognitive operation. They are utilizing all the cognitive apparatus, which spans the transitions from pure event perception, action modeling, shared attention, symbolic communication and technological augmentation.

Di Scipio's piece *Texture-Multiple*, for six instruments and room-dependent signal processing, was originally composed in 1993 but has been revisited by the composer a number of times since then. In the various iterations of this piece, Di Scipio would "try ideas concerning the interactions between human performance, machinery and space, that would later become central to the 'ecosystemic' pieces"(Placidi, 2010). He says of this work, "for the good or the bad, here human relationships are profoundly mediated by the technology. (Which is what happens in our daily life, nowadays)" (Placidi, 2010). Indeed, current technologies, alongside advancements in cognitive neuroscience, have extended the effects of the mediation to which he refers, to influence the way we react and respond to our environment. The cross-modal correspondences between taste and pitch being a good example, where it has been shown that an individual's perception of sweet or sour can be manipulated by sound (Crisinel and Spence, 2010). Extending the ecosystemic paradigm to include the performers' attentive behavior during performance, presents an opportunity to bear influence on the primal mimetic facility, which defines our response to audio-visual stimuli and the resultant action-model based behavior. The following extract is from Christine Anderson's, 2002 review of a performance of *Texture-Multiple* by Ensemble Mosaik in Berlin:

The computer intervenes in the instrumental action through a special technique of multiple granularization with different time-scale factors. This granularization is dependent on the resonant properties of the performance space, which is tracked by a microphone placed in the middle of the room. Mr. Di Scipio calls the resulting feedback loop an "ecological system . . . in the triangle between musician, machine, and space." In his words, the composition is not so much a piece of interactive music as an attempt to "compose interaction through which music is created." The result is a highly exciting affair, not only for the audience but also for the performers.

(Anderson, 2002, p. 83)

What is illuminated by this account is how the observations cohere three perspectives of the performance, bringing together the performance environment, the materiality of the music and the inner cognitive states of the audience and the performers. However colloquially expressed, the sentiments in this review indicate an holistic account of a performance system which includes environmental stimuli, algorithmic machine-based mediation and a reflexive human cognitive system, the only element of which was not present in Di Scipio's original definition of the ecosystemic performance paradigm being the inner working of the human mind.

BRAIN-COMPUTER INTERFACE (BCI)

The final element of this systemic triptych relates to the harvesting of the performers cerebral responses to their performance environment—the augmentation to Di Scipio's ecosystemic paradigm. Electroencephalograph (EEG) headsets are now low-cost consumer products. There are many computer-based or mobile apps, which allow direct 'brain' control of some aspect of a participant's environment. This might be an aspect of a video game, a remote-controlled car, a drone or a musical instrument. Although hundreds of different applications have now been developed, the vast majority of them share a similar feature. Fundamentally, they are control systems (George and Lécuyer, no date). After a period of training, the wearer of the BCI headset focuses their attention on achieving a task, and as a consequence their brain activity is captured, interpreted and sent as a control signal to a peripheral device. The generic term for this approach is known as active BCI (Ahn et al., 2014). This paradigm has a rich seam of research potential, and many journals and conferences have drawn from it. In relation to the two previous themes (CCN and ecosystemic environments) and the contentions outlined previously around improvised music-making, it seems plausible that the interventions in the cognitive causal sequence, from stimulus to musical behavior, could be achieved by BCI, but only if conscious intent was removed from the equation. One of the alternatives to active BCI is passive BCI (Ahn et al., 2014), where the participant perceives no sense of control over any aspect of the interaction. Using this approach the system monitors and reacts automatically to changes in mental state by quantifying the level of attention or differentiating among emotional states that are exhibited by the participant. In the context of the model suggested here, passive BCI allows for a flow of data from the performer into the performance environment, which is not the result of conscious control. In the same way that Di Scipio uses audio-capture technology to mediate between the performer and the performance space, this capture technology allows a type of mediation between the performers' low-level neurological response and their attentive aware-ness, thus suggesting the potential to subvert declarative memory, which is significantly influential in driving the auto-responsive musical behavior discussed in the opening of this chapter. The power of passive BCI in this context is to inject into the performance system a reflection of a per-former's cognitive activity, which is generated by their engagement in the musical processes which are unfolding in the performance. This injection is not a stream of control data but a by-product of brain activity generated by musical engagement, which is nonetheless reflective and responsive to the performance. Introducing this element into the model is the final triangulation point, which connects the performer to their performance environment without requiring their attentive awareness and the cognitive baggage that this entails. The mimetic facility which forms the legacy of our evolutionary development is given voice through this capture channel and has the potential to be mediated by technology in such a way that it cultivates an ability to 'surprise ourselves'.

THE MODEL

By way of an exemplar, the theoretical themes of the chapter will be repre-sented as a model for the development of new performance architectures. The model draws together the notion of cognitive-cultural networks, which Donald suggests reflects an arts practice within the context of cognitive evolution, together with Di Scipio's ecosystemic design, which is adapted to include the performers' brain activity, mediated by passive BCI. The binding concept, which holds these elements together, I suggest, is 'cogni-tive ambience', a term which encapsulates the flow of cultural influence among individual performers, in a real-time performance setting, medi-ated by their natural environment. By natural environment I am alluding to non-technological modes of communication, relying on instinct and fueled by sensory modalities. This does not of course discount technological components but augments the parameter-space to include communication between the performance entities that do not utilize digital technologies. An example of how this model could be implemented can be seen in the piece *Mondrisonic*, created for an improvising instrumental musician, a 'brain performer' wearing an EEG headset and an animated graphic score projected into the performance space. The piece was performed in pub-lic at the 4th International Performance Studies Network Conference in July 2016, with the improvising instrumentalist playing a bass clarinet and the graphic score projected onto a seven-meter-high media wall.

The arrows in Figure 2.1 show the flow of influence around the perfor-mance environment. In this implementation of the model the graphic score, which is styled on the paintings of Piet Mondrian, is a generative anima-tion, which is responsive to the brain activity of the brain-performer. The score is itself sonified, and its audio output has a very direct, perceivable

Eco-Systemic Performance Environment

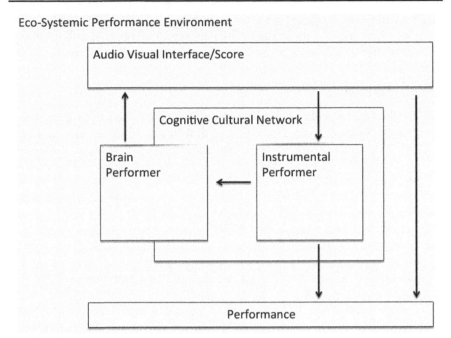

Figure 2.1 Mondrisonic Conceptual Schema

relationship to visual changes in the score. There are five tracks of audio, and each channel relates to a particular hue in the score's color pallet. The instrumental performer is therefore responding to a constantly changing mix of audio. As the piece progresses, the audio and color pallet change with each successive scene. As well as a score for the instrumentalist, the audience perceives the animation as an integral part of experiencing the piece. The particular brain activity, which is captured, is the level of attentiveness the brain performer gives to the improvisation of the instrumentalist. When this reaches a certain threshold level, a trigger is sent to the graphic score to stimulate changes in its generative output. These are perceivable by the instrumentalist, and so a curious loop is set in motion whereby the listening brain-performer is influenced by the instrumentalist, who in turn is influenced by changes in the graphic score. The causation that plays out is mediated through the 'cognitive ambience and is indicative of the type of cognitive-cultural network that Donald suggests has influenced behavior since early hominids first became self-aware. The first rendition of the piece lasted for 15 minutes and moved through five different sections, each with a different soundscape and re-mapped visual score. In each of the sections, the visual element of the score responsive to the brain performer changed. For instance, in one section, the speed of animated activity changed in response to the triggers from the brain-performer and in another the boundary between the visual elements was altered. One unforeseen characteristic of the performance was that, in embracing the principles of ecosystemic design, the brain-performer was susceptible to applying focused attention to any sonic elements in the performance space, even those not integral to the performance. This did indeed happen when ambient noise, not related to the performance, became a distraction and consequently caused a response in the audio-visual interface.

CONCLUSION

> When the individual "makes" a decision, that decision has usually been made within a wider framework of distributed cognition, and, in many instances, it is fair to ask whether the decision was really made by the distributed cognitive cultural system itself, with the individual reduced to a subsidiary role.
>
> —(Donald, 2008, p. 202)

The conceptual ideas implemented in the piece *Mondrisonic* were the first attempt to construct a simple CCN, which embraced Di Scipio's ecosystemic paradigm for the purpose of supporting improvising musicians striving to explicate their art in a manner that satisfies their desire to create music, which minimizes mechanical musical responses and maximizes originality. Of course, their perception of whether this has been achieved may fly in the face of a detailed objective analysis of their performance, but at present it is their perception that is being explored. The first experimental implementation of these ideas simultaneously violated and augmented the original ecosystemic principle, by not processing material from the acoustic environment but including brain activity generated by attentive focus on the acoustic environment. Future implementations will seek to redress this for a more holistic and faithful ecosystemic approach.

In Placidi's interview, Di Scipio makes reference to the ubiquity of technology in mediating relationships in everyday life (Placidi, 2010), a trend in which human agency is moving from one of active participation to passive involvement. Wearable devices are now available to provide feedback to the general public on general health signifiers such as sleep quality or blood pressure and suggest remedial action to avert a crisis, but it seems inevitable that in time they will detect and remedy symptoms without the need for the conscious attention of the wearer, as happens with serious medical conditions. Passive BCI has been selected for this investigation precisely to avoid the baggage of attentive awareness, primarily in this instance, conscious engagement with declarative memory. In a sense, what is proposed in this chapter is an approach that taps into two parallel epistemological traditions, the white-box approach of cognitive neuroscience and the black-box approach of experimental psychology. The first involves the monitoring, harvesting and mapping of specific neural activity onto the parameter-space of an environment designed for creative expression and the second, constructing the rules of engagement for human actors to explore during their conscious and non-conscious interactions with their environment. As Di Scipio puts it:

> The very process of "interaction" is today rarely understood and implemented for what it seems to be in living organisms (either human or

not, e.g. animal, or social), namely a by-product of lower-level inter-dependencies among system components.

(Di Scipio, 2003, p. 271)

The approach outlined here has many potential types of implementa-tion, but at its conceptual core is an exploration of how technology can facilitate other modes of human agency, other than attentive focus, in improvisatory music-making and how the concept of ambience can be extended into the realm of human thought to provide a rich domain in which to build environmental relationships. This resonates with Di Scipio's desire to "shift from creating wanted sounds via interactive means, towards creating wanted interactions having audible traces" (Di Scipio, 2003, p. 271).

REFERENCES

Ahn, M., Lee, M., Choi, J. and Jun, S. (2014). A review of brain-computer inter-face games and an opinion survey from researchers, developers and users. *Sensors*. Multidisciplinary Digital Publishing Institute, 14(8), pp. 14601–14633. doi:10.3390/s140814601.

Anderson, C. (2002). Audible interfaces festival (review). *Computer Music Jour-nal*. People's Computer Co, 26(4), pp. 83–85.

Bailey, D. (1993). *Improvisation: Its nature and practice in music*. New York: Da Capo Press.

Cage, J. and Turner, S. S. (1990). John Cage's practical utopias: John Cage in conversation with Steve Sweenery Turner. *The Musical Times*, 131(1771), pp. 469–472. doi:10.2307/1193658.

Cook, R., Bird, G., Catmur, C., Press, C. and Heyes, C. (2014). Mirror neurons: From origin to function. *Behavioral and Brain Sciences*, 37(2), pp. 177–192. doi:10.1017/S0140525X13000903.

Cox, C. 1965- and Warner, D. 1954- (2017). *Audio culture readings in modern music*. New York: Bloomsbury Academic.

Crisinel, A-S. and Spence, C. (2010). As bitter as a trombone: Synesthetic cor-respondences in nonsynesthetes between tastes/flavors and musical notes. *Attention, Perception & Psychophysics*, 72(7), pp. 1994–2002. doi:10.3758/APP.72.7.1994.

Cushman, F. and Morris, A. (2015). Habitual control of goal selection in humans. *Proceedings of the National Academy of Sciences of the United States of America*. National Academy of Sciences, 112(45), pp. 13817–13822. doi:10.1073/pnas.1506367112.

Decety, J. and Grèzes, J. (1999). Neural mechanisms subserving the perception of human actions. *Trends in Cognitive Sciences*. Elsevier Current Trends, 3(5), pp. 172–178. doi:10.1016/S1364-6613(99)01312-1.

Di Scipio, A. (2003). "Sound is the interface": From interactive to ecosys-temic signal processing. *Organised Sound*, 8(3), pp. 269–277. doi:10.1017/00000000000000000.

Donald, M. (2008). How culture and brain mechanisms interact. In: C. Engel and W. Singer, eds., *Better than conscious? Decision making, the human mind, and implications for institutions*. Cambridge, MA: MIT Press, pp. 191–207.

Gallese, V., Fadiga, L., Fogassi, L. and Rizzolatti, G. (1996). Action recognition in the premotor cortex. *Brain*. Oxford University Press, 119(2), pp. 593–609. doi:10.1093/brain/119.2.593.

Gaser, C. and Schlaug, G. (2003). Brain structures differ between musicians and non-musicians. *The Journal of Neuroscience*, 23(27), pp. 9240–9245. doi:23/27/9240 [pii].

George, L. and Lécuyer, A. (n.d.). *An overview of research on "passive" brain-computer interfaces for implicit human-computer interaction*. Available at: http://people.rennes.inria.fr/Anatole.Lecuyer/GEORGE_LECUYER_ICABB2010.pdf [Accessed 9 Sept. 2017].

Hamilton, A. and Konitz, L. (2007). *Lee Konitz: Conversations on the improviser's art*. Ann Arbor: University of Michigan Press.

Lahav, A., Saltzman, E. and Schlaug, G. (2007). Action representation of sound: Audiomotor recognition network while listening to newly acquired actions. *The Journal of Neuroscience: The Official Journal of the Society for Neuroscience*, 27(2), pp. 308–314. doi:10.1523/JNEUROSCI.4822-06.2007.

Libet, B. (1985). Unconscious cerebral initiative and the role of conscious will in voluntary action. *Behavioral and Brain Sciences*, 8(4), pp. 529–539. doi:10.1017/S0140525X00044903.

Lotem, A., Halpern, J. Y., Edelman, S. and Kolodny, O. (2017). The evolution of cognitive mechanisms in response to cultural innovations. *Proceedings of the National Academy of Sciences of the United States of America*. National Academy of Sciences, 114(30), pp. 7915–7922. doi:10.1073/pnas.1620742114.

Placidi, F. (2010). *Unidentified sound object: A conversation with Agostino Di Scipio, U.S.O. project*. Available at: http://usoproject.blogspot.co.uk/2010/06/conversation-with-agostino-di-scipio.html [Accessed 26 Jan. 2018].

Wohlschläger, A. and Bekkering, H. (2002). Is human imitation based on a mirror-neurone system? Some behavioural evidence. *Experimental Brain Research*, 143(3), pp. 335–341. doi:10.1007/s00221-001-0993-5.

3

Press Play on Tape

8-Bit Composition on the Commodore 64

Kenny McAlpine

INTRODUCTION

1982 was an auspicious year for Commodore. At the Computer Electronics Show in Las Vegas in January of that year, the company previewed its new home computer, the Commodore 64 (Mace, 1982).

The machine launched to a fanfare of Bach, courtesy of a demo program that cycled through a series of animated Christmas scenes (Bagnall, 2006, p. 249), while its sound generator, the anthropomorphically named SID chip, chirped out synthetic renditions of *In Dulce Jubilo, Jingle Bells* and *Frosty the Snowman*. The demo closed with a title sequence that was accompanied by J.S. Bach's *Two-Part Invention in A Minor*. Bach's *Invention* was picked up and used elsewhere in Commodore's marketing and provided the soundtrack for many of the C64's early television commercials (ibid., 272). It quickly became established as the machine's theme tune and concretized the idea among its potential users that the Commodore 64 was not just a games machine but also a versatile and powerful creative tool.

That was an image that Commodore had worked very hard to cultivate. The C64's predecessor, the VIC-20, had been promoted by actor William Shatner, who, in a series of television advertisements designed to highlight the machine's utility beyond the first-generation games consoles like Atari's VCS, posed a direct challenge to consumers: "Why buy just a video game?" (Shatner and Fisher, 2009, p. 185).

Why indeed? As the home computer revolution gained momentum in the 1980s, early adopters were faced with a bewildering array of options. The Acorn Atom, the Atari 400, the Oric 1 and Newbury Labs' New Brain all offered comparable levels of processing power, but nothing in the way of inter-compatibility or, crucially, software titles or market share. Sinclair's ZX Spectrum, on the other hand—the machine that really kick-started the home computer market in the UK—was hugely popular, easy-to-use and inexpensive, but it was limited in its capabilities.

Although the ZX Spectrum was a wonderful example of late-20th-century British industrial design, its creator, Sir Clive Sinclair, demanded that it was designed to a very low price point, and the design team, Rick

Dickinson and Richard Altwasser, were forced to make compromises (O'Regan, 2016, p. 136), resulting in a machine that suffered badly from color clash on its high-resolution graphics, and offered no hardware sound support: the Spectrum's single-channel 1-bit sound device, a 'beeper' speaker, was controlled directly by the main central processing unit (CPU), a Zilog Z80A processor, and so driving the speaker tied up the processor and halted any other processing tasks (McAlpine, 2016).

The Commodore 64, by contrast, seemed to have everything: (a) a powerful CPU, a variant of the 8-bit MOS Technology 6502 processor; (b) a multi-mode graphics chip, the VIC-II, which offered hardware support for sprites and scrolling; (c) 64k of RAM memory, which was a sizable step ahead of its contemporaries; and (d) its sound interface device, the SID, which gave it capabilities that were much more similar to a hardware synthesizer than they were to a home computer (McAlpine, 2018). These were the attributes on which the machine was traded, and, for those who embraced the idea of the home computer as a creative tool, they were very enticing.

"Back in the early 80s, when I was eighteen or nineteen, I was playing in bands and started touring different places", recalls Rob Hubbard, a pioneering electronic musician who is widely regarded as the originator of what has come to be known as the Commodore 64 sound.

> I was at music college in Newcastle, and I was messing about with the early synths and sequencers at the time. It was all pre-MIDI, so everything was done using control voltages. Back then, it was all very DIY. You could get a *Practical Electronics* magazine and it would have a little circuit diagram of how to build a CV converter so that you could hook up a Korg to a Moog or whatever.
>
> Back then, the music scene was as much of a hobbyist activity as home computing was, and I was really into it all. I'd a four track TEAC, so I would try to do some recorded stuff with that, and a friend of mine, he had a Revox, and a [Sequential Circuits] Pro 1 synth and [Roland] TR-808 or something. I had the TB-303. We'd hook all this stuff up, put a sync track on tape, hope that it'd stay there and try and write something. It was good, but looking back, it was clumsy.
>
> I was also reading lots of magazines, electronic magazines. And they said, "Oh well, if you're a musician you have to learn BASIC". I didn't even have a computer but then as soon as they came out I got one, and I just had to pick a 64. It was marketed as a machine with an elephant's memory and a synth at its heart. I just thought, "Well, that's the one for me!"
>
> (McAlpine, 2015a)

THE SID CHIP

The sound capabilities of the Commodore 64 were no accident. By the middle of January 1981, the VIC-20 was selling strongly, and Commodore's design engineers were keen to find a new project.

"We were fresh out of ideas for whatever chips the rest of the world might want us to do", recalls Al Charpentier, the lead engineer who was responsible for the design of many of the chips that had powered the VIC-20. "So we decided to produce state-of-the-art video and sound chips for the world's next great video game" (Wallich, 1985). Charpentier's team looked around at the current state-of-the-art and extrapolated from there to define all that the new MOS chips should be able to achieve. While Charpentier worked on the graphics, his colleague Bob Yannes began work on a new sound chip. Unlike Charpentier, however, Yannes did not look to the competition for inspiration: "I thought the sound chips on the market, including those in the Atari computers, were primitive and obviously had been designed by people who knew nothing about music", he explains (ibid.).

The design team were under real-time pressure to deliver a prototype. As they started work on the new chip set in the summer of 1981, Jack Tramiel, the bullish force at the head of Commodore, announced that he wanted a working prototype for the CES show in just six months' time. That left little time to finesse the designs.

Yannes had originally planned to have 32 different voices for his chip, using wavetable synthesis controlled by a master oscillator. "The standard way of building oscillators", he explains, "is to build one and then multiplex it until you have as many as you need. We just built an oscillator module and repeated it, because that was much faster than working out all the timing for the multiplexer" (ibid.).

Yannes, however, did manage to include components on his chip that suggest strongly that he was not really designing a video game sound chip at all, but rather a fully featured polyphonic digital synthesizer. He included an analog input pin, for example, which allowed external signals to be routed through its filter and mixed with the on-board sound channels, a feature that he borrowed from Bob Moog's Minimoog. He also included two analog-to-digital converters as input ports, which were designed to allow the SID to interface directly with potentiometers (Falconer, 1984)—useful, perhaps, as front panel controls in a hardware synth setup.

In its production-ready state, the SID's feature set was both attractive to end users and an order of magnitude more powerful than any of its competitors. While Yannes was disappointed with the end result, his colleague, Charles Winterble, noted that the chip was "already 10 times better than anything out there and 20 times better than it [needed] to be" (Bagnall, 2006, p. 237).

The SID offered three polyphonic oscillators, each with its own ADSR envelope generator and capable of producing multiple waveforms, including triangles, sawtooths, variable-width pulses, and white noise, but each of these waveforms could be augmented using the SID's sophisticated on-chip synthesis (Commodore, 1982a). Pulse width modulation, for example, a synthesis technique that was a staple of analog synthesis (McGuire and Van der Rest, 2015, p. 31), could be applied independently to each channel to create analog-like synth string effects, while three independent ring modulators allowed the instantaneous waveform on any of the three tone channels to modulate the carrier wave of the next, providing a useful way of synthesizing struck and metallic percussion (Roads, 1996, pp. 216–224).

Unusually, the SID also allowed users to hard-sync oscillators, a non-linear approach to synthesis that locks the frequency of one oscillator to that of another to create complex and harmonically rich waveforms whose timbre can be altered by varying the respective frequencies of the two signals. This type of synthesis has been in use since the early days of analog synthesis (see Russ, 2013), but is not particularly common, and, as a non-linear operation, can produce unusual and characterful results that give both the synthesis method, and, to an extent, the Commodore 64 its distinctive sound.

Rounding off its synthesis armory, the SID offered subtractive synthesis, courtesy of a single, multi-mode resonant filter, which can act on any of the voices in isolation, or in combination, and, although in principle this lent the SID enormous flexibility, its hardware implementation made filtering on the C64 something of a hit-or-miss affair.

The filter in the SID is analog, but it is controlled digitally. It uses a multi-mode VCF design, using field effect transistors (FETs) as voltage-controlled resistors to control the cutoff frequency. The resistance of those FETs, and hence the behavior of the filter cutoff, varied considerably across the production run of the chip, and so different batches of SID chips ended up sounding very different (Varga, 1996).

"The filter was the last thing that was worked on", Yannes recollects. "I ran out of time. The computer simulation said, 'This will not work very well', and it didn't. I knew it wouldn't work very well, but it was better than nothing and I didn't have time to make it better" (Wallich, 1985).

Composer Ben Daglish, who penned some of the C64's best-known game soundtracks, including the *Last Ninja* (Cale, 1987) and *Gauntlet* (Armour, 1986), recalls that the issue with the filter

> was just to do with the quality of the components; accurate frequency filtering just wasn't viable on such a small, cheap . . . chip, so Commodore made do with what they had.
>
> Basically, one could apply, say, a band-pass filter centered around X Hz, but X would vary by a fair percentage from chip to chip. I tended to use "static" filters as little as possible for exactly that reason. Generally, I'd use filter sweeps, which were pretty much guaranteed to have the same effect irrespective of the start or end frequencies.
>
> (Pouladi, 2004)

Some soundtracks, such as that of *Beach Head II* (Carver and Carver, 1985), for example, would offer users a degree of control over the filter settings to fine-tune the code for their own hardware (Collins, 2006)—not so Commodore's Japanese developers.

> The Japanese are so obsessed with technical specifications that they had written their code according to a SID spec sheet, which I had written before SID prototypes even existed. Needless to say, the specs

were not accurate. Rather than correct the obvious errors in their code, they produced games with out of tune sounds and filter settings that produced only quiet, muffled sound at the output. As far as they were concerned, it didn't matter that their code sounded all wrong, they had written their code correctly according to the spec. and that was all that mattered!

(Varga, 1996)

SOUND DRIVERS

As composers experimented with the capabilities of the SID chip and grew more familiar with its features, both planned and unplanned, soundtracks became a bigger feature of C64 games, and the role of the video game composer began to professionalize. A new breed of coder emerged, one who could fuse a thorough knowledge of music with the technical virtuosity required to optimize the expression of that music in machine code for use in video games. Those technical musicians realized fairly quickly that it made little sense to reinvent the wheel each time they wrote a new piece of music by creating bespoke code for each new tune; instead, it was easier and more efficient to create only the music data—the score—afresh each time, and reuse those bits of code that handled the timing and manipulation of the SID chip's registers. The resulting code, known as sound or music drivers, became both a source of professional pride and a form of currency for their creators.

Paul Hughes, a veteran of the games industry, began working for the UK developer Ocean in the mid-1980s. Recalls Hughes:

The '64 was my machine of choice, and after a chance meeting with a musician, Peter Clarke, in the local computer shop, we started working on soundtracks for C64 games; Peter writing the music [and] me writing the audio playback code. We did a few titles together, Scooby Doo for Elite Systems, Repton 3 for Superior Software and finally Double Take for Ocean. At this point in late 1986 I joined Ocean's in-house games development team, and Pete followed on to help and then eventually take over from [Ocean's then in-house composer] Martin Galway.

(Hughes, 2016a)

One of my small claims to fame is [that I coded the] music and sound effects drivers for Ocean. I can't take too much credit here as the real "magic" was performed by the musicians, Jonathan Dunn, Matthew Cannon and Peter Clarke. I merely wrote a bunch of code to essentially fiddle around with the SID registers whilst parsing the music data!

(Hughes, 2016b)

That fiddling, however, was a crucial part of the sound, and just as the same instrument in the hands of different musicians could sound very different, the way musicians interfaced with the hardware via their sound drivers could create a unique and very distinctive feel.

"Essentially the trick to great sounds on the SID chip was down to post modulation", explains Hughes.

> Basically once you had the basic ADSR, required waveform, and note frequency that you wanted to play, the code could then, under control of the musician, modify the frequency, the pulse widths and envelope . . . over a period of time in a variety of different ways.
>
> You could pitch bend . . . or you could set up an arpeggio table that added [a] different controlled note value to the base frequency every frame causing "wibbly" notes that sounded like [chords]. You could set up vibratos . . . or, in the case of drums, the driver would mess around with not only the frequencies but the waveforms and envelope shape every 50th of a second.
>
> (Hughes, 2016b)

While each of those approaches had, in and of themselves, a characteristic sound, the detail of their implementation imparted subtle differences. How the composer worked with the sound driver code to define the effect onset delay, or the effect depth, the effect rate, or the modulation of the control data over time, all subtly shaped the performance characteristics of the music, creating definite stylistic signatures. A Matt Gray tune, for example, sounds like a Matt Gray tune not just because of the chord sequences, the rhythmic figures and the melodic contours; it sounds like a Matt Gray tune because of Matt Gray's code.

"Yes, there's an art to it", adds Ben Daglish. "You can definitely appreciate beautiful code. You know, I often talk about the 'the art of programming'. I love to look at and hear the effects of beautiful code. That [was] a big part of [8-bit video game composition]" (McAlpine, 2015b).

Ben Daglish started his career as a music programmer while he was still at school in the early 1980s. Thanks to Sir Clive Sinclair, Britain had the highest density of microcomputers per head of population of any country in the world. The British Broadcasting Corporation (BBC) was determined that it should also have the highest rate of computer literacy in the world, and so in 1982, it launched the Computer Literacy Project, a series of themed television and radio broadcasts that were intended to explain and report on the new technology of computing (Radcliffe and Salkeld, 1983). Alongside the broadcasts, the Computer Literacy Project would launch a new home computer, the BBC Micro.

Part of the BBC's outreach work included an initiative to place BBC Micros in schools across the UK (ibid.). The organization launched an essay competition that challenged schoolchildren to think and write creatively about how computers could be used in their schools. The young Daglish entered and won, extending him the privilege of being one of the few people who were allowed routine access to 'the school computer' (McAlpine, 2015b).

Press Play on Tape

Daglish, already a very capable musician, began coding music on the machine. It was a transformative experience. The BBC's symbolic representation of the music that he had previously known only through performance made Daglish realize that musical expression could be something concrete, something that was capable of being codified and, perhaps more importantly, manipulated using logic and code.

Along with three of his school friends, Nigel Merryman, Martyn Peverly and Tony Crowther, Daglish worked in the school library, writing pieces of educational software until the pull of gaming, always more compelling than that of formal education, proved too strong. Crowther got a Commodore 64 and began to write games at home. One afternoon he mentioned to Daglish that he needed some music, ideally Chopin's *Marche Funèbre* at the end when the player died. Daglish wrote out the notes for Crowther, who took them home and typed them into the BASIC sound driver that was included at the back of the machine's user manual (Commodore, 1982b). A few weeks later, Crowther asked Daglish if he could write some more music for him, this time a rendition of Jean Michel Jarre's *Equinoxe 5*. Daglish transcribed the music, jotted it down on a piece of paper, and went round to Crowther's house to type it into the computer himself, that being quicker and more reliable than relying on his friend to re-enter the data.

That they were dealing with copyrighted material didn't even occur to the pair. Recalls Daglish:

> I had no idea that copyright existed. Quite seriously . . . I really didn't. When we wrote all the Jarre stuff and all that. . . . we had no real idea as a 14 or 15 year old kid that you couldn't just take some music that you liked, whether it was Beethoven or whether it was Jean Michel Jarre. We'd just write it down and put it in a computer game.
>
> (Burton and Bowness, 2015)

Soon, Daglish became hooked on coding music, and the pair set up a small production company, W.E.M.U.S.I.C., WE Make Use of Sound In Computers. According to Daglish, W.E.M.U.S.I.C. came into being because, "Tony and I wanted to make money writing music for games [and] needed a company. Simple as that" (Carr, 2001).

Crowther and Daglish spent the next few years developing their sound driver and working on new tracks, including covers of Led Zeppelin and much, much more Jarre. Their music was solid and well-crafted, but just after the release of their first W.E.M.U.S.I.C. demo, Daglish heard Rob Hubbard's soundtrack to *Commando* (Butler, 1985) and realized just how far he still had to travel.

He recalls:

> Back then, the "industry" . . . it was all 14 and 15 year old boys. I mean, Tony sold his first few games . . . by going into Just Micro, the local computer shop, and saying "I've made a game, do you want to sell it?" and they said, "Yeah, alright", and they took his cassette, and took it round to the back where they had a cassette duplicating

machine and they'd copy off a load, they'd photocopy a load of covers and they sell it under the counter. That's how the industry started.

So we'd been doing [our W.E.M.U.S.I.C. stuff] for about six months, a year, something like that, at which point, I heard the first bit of Rob, and Rob's a proper musician. He's like ten years older than me and he'd been playing like jazz keyboards on the cruise ships and all the rest of it, so he was like a proper, proper muso and knew his stuff, you know? And so instantly, as soon as you heard that, with the state of the player that Tony and I had been slowly working on over the last however many months, you know, suddenly we went "Ah! There you go! You can do that, can you?" It was a proper kick up the arse.

(McAlpine, 2015b)

ROB HUBBARD

Today Rob Hubbard is still recognized, nearly 30 years on from his heyday, as a pioneering electronic musician and the father of the C64 sound. Although semi-retired, he still gigs regularly, playing saxophone with a northern big band outfit, and receives regular international invitations to speak about his work.

His memories of working in the 8-bit era are mixed. On the one hand, he recognizes and celebrates the fact that he was working at a digital frontier during an explosive era of technical creativity that was borne out of possibilities that were unencumbered by rules, expectations or market forces, while on the other hand feeling a palpable sense of frustration as he recalls how that fertile creative ground was grabbed and partitioned up by the executives who moved in to commoditize video games.

He recalls vividly how, as he began his professional career as a freelance video game composer, he would often take jobs on spec over the telephone, and was given complete freedom to develop the music that he wanted. Even so, he found the process of writing music on 8-bit computers far from satisfying:

It was actually quite tedious, and quite difficult. I used to write sketches on paper, and then I would code them up in assembly language—machine code—which is all a lot of hexadecimal numbers, and then I would basically work on maybe two to four bars at a time. It was never a case of writing out, say, six minutes' worth of music on manuscript paper, and then coding the whole thing up and then trying to hope that you hadn't made any mistakes.

(McAlpine, 2017)

Hubbard was keen to bring his professional musicianship to his video game work. He brought a discipline that had been honed from years of rehearsals and gigs, and he quickly developed a reputation for turning out extremely high-quality work to what were often very punishing deadlines. He was also keen to avoid the creative trap of using the SID, with its powerful performance features, as little more than a sound generator.

You know, the thing that really got me, the thing that made me think about it was . . . the music that was around at that time was generally done by the programmers. And it was . . . so bad that I thought, well there has to be an avenue for somebody who can at least get the notes right, and also in the right time. I mean, the music [on those early games] had so many wrong notes, and it was just absolutely dreadful.

(Commodore Music, 2005)

"[Rob] was the first, and pretty much the best", continues Ben Daglish.

When he started writing for the Commodore 64, . . . the first thing he did was write a whole load of extremely cool sound routines, which did some . . . incredible things with the chip, really. Everybody else was just writing plain tones, but he was the first guy to start using wobbly chords, phasing square waves, using filters creatively, and stuff like that.

One of the things I quite often say is that just because you were working with limited sounds, then it was very difficult to get away with writing sound-effecty music, if you know what I mean; music that just relies on the way it sounds . . . to make it interesting. You had to write music that was musical. You had to write stuff that had a melody, had a good tune, was coherent as a piece of music. So that was one of the things that was good about a lot of the game music; [it] was very strong as music, it wasn't just soundtrack, if you see what I mean.

(ibid.)

Hubbard worked hard to introduce a performative aspect both to his music and to his coding. Often, he would create a short musical loop in the C64's memory, perhaps a two-bar phrase, and use a machine code monitor, a utility program that visualized the computer's instantaneous memory states, to view the status of the SID chip and manipulate its registers. He became so fluent with this symbolic approach to composition that he could play the SID chip in real time, like a Progressive Rock keyboardist sculpting a performance on a modular synth as a step sequencer loops a note sequence beneath.

Says Hubbard:

Yeah, instead of twiddling with the knobs like you would on a Moog or something like that, I was twiddling with numbers in real time, . . . changing things to the absolute maximum I could squeeze out with the little SID chip that was in the C64. . . . Then I would adjust things like the cutoff point and the note length for the release to try to sound right, all these kind of things. You'd fine tune it, just to the nth degree.

(McAlpine, 2017)

Hubbard, then, recognized that performance coding was the key to successful music-making on the C64, and he refined his sound driver over successive projects, making it more efficient with each iteration, and adding and removing features as needed. He began to think of his music procedurally, encoding, transposing and reusing different fragments in different places to save valuable memory. In fact, he became so efficient at compressing and packaging up his musical ideas that the complete five-minute score for *Master of Magic* (Darling, 1985) took up just 3K of memory. He recalls:

> That Master of Magic thing, starts off with a canon on the melody. I had an instinct that I could get this little canon going, so I write two bars, and I've got the melody. Then you take your first two bars and align the second part, and alter the third to get the fourth. You transpose them into the right registers, right, and you've then got like a guaranteed canon, that's going to sound really cool [but which is all based around this one musical idea. . .] and you can do the same thing one bar later and it'll still sound awesome.
>
> (McAlpine, 2015a)

Inevitably, Hubbard's music was influenced by the artists that he liked and listened to: Kraftwerk, Jean-Michel Jarre and Larry Fast, an American synthesist and keyboard player who had collaborated with, among others, Peter Gabriel and Rick Wakeman. With New Romanticism showing that synthesizers were a dynamic new voice in popular music, the Commodore 64 not only gave Hubbard a similar tonal palette to work with, but through his code he could control, in a very precise way, every minute detail of his work, an approach that has definite parallels with Stockhausen's notion of total control and the chance music of composers like Iannis Xenakis and John Cage (Clark, 1970).

> When the [Sequential Circuits] Prophet-5 came out I thought, "Man! If you could get your hands on one of these Prophet 5s but then control it with, like, a microprocessor and proper code, you could make it do some awesome things!" . . . That's what the C64 was. So there was this synthesizer element [with the Commodore 64 that] was very strong, but then on the compositional side, having the software control allowed you to kind of explore all kinds of other possibilities.
>
> One of the things that happened was that through the software, you could control all sorts of synth parameters. . . . You could control just about any aspect of the SID chip. But also, when you're writing code to control the actual structure of the music, it then became quite apparent that you could then take that to the next level, and try to . . . make [the music] more interactive.
>
> At the time I was starting to think about the idea of probability based controls, where you could get a complex bassline playing, and then add assignments to the notes [that would be triggered by probabilities]. And then, based upon an input from somewhere else, you would get varying degrees of intensity [or dynamics]. And then not only that, you could control certain other aspects like harmony, and what kind of

sound you would get with a lead instrument, the melody. . . . [Compositional] things like that you could then do more interactively.

(McAlpine, 2017)

Over time, however, as Hubbard explored, fairly comprehensively, the compositional landscape that the SID chip afforded, he began to find that the exhilarating challenge that had driven him to innovate so creatively during the early period of his video game career had disappeared. His creative freedom also changed as the scale of game development grew and companies became larger with well-defined management hierarchies. It became more and more difficult for him to deploy his professional judgment and write the music that he thought would work within the context of the game. Inexperienced producers, keen to make professional impact, but with little idea of how games function as creative artifacts, and little experience of working with other professional disciplines, would try and seize creative control.

I remember when I was at Electronic Arts, this producer at one point, he comes in and says to me, "Oh . . . um . . . I think the tempo should be brighter". "What d'you mean, the tempo should be brighter?" I ask. He says, "Oh, I think it's in the bassline. Could you change the bassline?"

So I wait a couple of days, call him up and say, "Ah well, I was listening to the track and you're absolutely right; the bassline needs changing. When you've got time, come by and listen to it and check it out". He comes by, has a listen and says, "Oh yeah, that's much better! I like that, a lot more!" And of course, I did nothing. I didn't change a bloody thing. But [the producers] think that they've got some kind of creative ownership and creative judgement, even though they don't have the first idea what they're talking about.

(McAlpine, 2015a)

Over time, as the market expanded and game development became more competitive, video game producers began to shoulder more of the responsibility for the market potential of games. Stuck in an uncomfortable position between the publishers who financed the games and the developers who crafted them, producers had to walk a fine line between allowing the creatives scope to produce work in which they felt professionally invested, while responding to market trends to create product that was saleable. Many responded to those pressures by ducking the responsibility for making decisions.

Hubbard stated:

Later on, when all the production teams got involved [with the music] and you got producers and executive producers and assistant producers and associate producers. . . . You got this ineffectual hierarchy. You know, "Well, I can't commit to what I think until I figure out what he thinks, and he can't figure out what to say until he can figure out what this other guy thinks".

I think that the last thing I did was some ghost-writing in 2005 or 2006. I did this one game, and it was an absolute piece of cake. The guy in charge was really good. He'd give us some direction and I knocked the stuff out for him. It was really easy to do that one. But around the same time I got another job, really similar kind of spec and everything, but it was a different producer, and this guy turned out to be an arsehole. I played the music I'd written to him and he said, "Well, I'm not sure what I want on this. Can you do something else?" So I asked him, "Well, what do you want?" and he said, "I don't know, but, well, I'll know it when I hear it!" So I ended up writing lots of different tracks for this guy and he didn't like any of them.

In the end, I ended up on the phone with this guy and I says, "Look . . . you don't have a clue what you want, and you expect me just to throw shit at the wall 'til it sticks. You can shove your job. . . . I ain't ever working for you again". I slammed the phone down and that was the last job I ever did. You can't possibly deal with somebody who's been put in a position to try and give direction when they're totally incapable of doing the job!

(McAlpine, 2015a)

THE LEGACY

By the end of the 1980s, the 8-bit era was over, and with it the characterful blip of its soundtracks. However, that culture of making music with code and sound chips is alive and well. Indeed, it is arguably stronger than ever. The chipscene, a vibrant community of musical practitioners who make music using obsolete hardware, has embraced the sound and the aesthetic of the 8-bit video game soundtrack and transformed it, fusing it with other musical styles like Reggae, Drum 'n' Bass, Rave and House, and taking it onstage, modding original hardware to provide hardware controls for the SID's filters and plugin cartridges that allow it to be controlled via MIDI (see McAlpine, 2018).

As that raw electronic sound has grown in popularity, it has been redis-covered and incorporated into the musical vocabulary of mainstream com-mercial artists, and new technologies have hit the marketplace that allow musicians to access and control the characterful voice of the SID without having to learn to code in 6502 assembly. Soft synths, like the Plogue suite of plugins (Plogue, 2009), make it easy to incorporate authentic SID sounds within a modern production environment, providing a user-friendly DAW-style interface, and the opportunity to mix and produce chip sounds as you might any other production source, while hardware devices, like Twisted Electron's Therapsid, or the Elektron SidStation, offer a tactile control interface and professional-grade inputs and outputs from genuine SID chips pulled from vintage computers.

The popular appeal of chip-style music is also beyond doubt. Hub-bard's music is performed regularly, perhaps most notably in 2005 by a full orchestra in Leipzig to a packed audience at the third Symphonic Game Music Concert (Tong, 2010). Indeed, a 2017 Kickstarter campaign, titled Project Hubbard, that was created by netlabel owner and C64 music impresario Chris Abbott, with the stated aim of creating a series of rein-terpretations of Hubbard's music alongside new work by Hubbard him-self, raised over £81,000 (Abbott, 2017). Cover bands, including Press Play on Tape and the SID80s, a chiptune supergroup, whose changing lineup has included original C64 and Commodore Amiga composers Ben Daglish, Mark Knight, Jon Hare, Matthew Cannon and Fred Gray, play—in a wonderfully self-referential and un-ironic way—performing very accomplished live covers of the original video game soundtracks that they authored in the 1980s.

It is a sound that is simultaneously current and retro, simple yet com-plex. It is a sound loaded with the positive association of childhood, that is, nevertheless, credible enough to use in mainstream music. Anamanagu-chi, for example, create upbeat, rocky music that fuses the contemporary sound of Weezer with the raw electronic sounds of chiptune (Wolinsky, 2011). More controversially, Timbaland became embroiled in a lengthy court case (Byrne, 2009) over his sampling of a SID music track by the Finnish musician Janne Suni (2002) for the Nelly Furtado song *Do It*, while the Canadian electroclash outfit Crystal Castles similarly came under fire for their use of uncleared samples taken from the 8bit Collec-tive's online archive of chip music (Dumile, 2008).

Imitation, it seems, is the sincerest form of flattery.

REFERENCES

Abbott, C. (2017). *Project Hubbard: Official Rob Hubbard kickstarter*. Available at: www.kickstarter.com/projects/c64audio/project-hubbard-official-rob-hub bard-kickstarter. [Accessed 17 Jan. 2018].

Armour, B. (1986). *Gauntlet*. Atari Games.

Bagnall, B. (2006). *On the edge: The spectacular rise and fall of commodore*. Winnipeg: Variant Press.

Burton, C. and Bowness, A. (2015). *Ben Daglish BIT Brighton 2015 interview (preview)*. Available at: www.youtube.com/watch?v=qhv6U8Wm0GY [Accessed 17 Jan. 2018].

Butler, C. (1985). *Commando*. Elite Systems.

Byrne, F. (2009). Nelly Furtado, Timbaland Sued for plagiarism. *NME*. Available at: www.nme.com/news/music/timbaland-11-1307562 [Accessed 17 Jan. 2018].

Cale, M. (1987). *The last ninja*. System 3.

Carr, N. (2001). *An interview with Ben Daglish*. Available at: www.remix64.com/interviews/interview-ben-daglish.html [Accessed 17 Jan. 2018].

Carver, B. and Carver, R. (1985). *Beach head II: The dictator strikes back*. Access Software.

Clark, R. (1970). Total control and chance in musics: A philosophical analysis. *The Journal of Aesthetics and Art Criticism*, 28(3), pp. 355–360.

Collins, K. (2006). Loops and bloops. *Soundscapes* (Vol. 8). Available at: www.icce.rug.nl/~soundscapes/VOLUME08/Loops_and_bloops.shtml [Accessed 17 Jan. 2018].

Commodore. (1982a). *Commodore 64 programmer's reference guide*. Carmel, IN: Commodore Business Machines and Howard W. Sams.

Commodore. (1982b). *Commodore 64 user's guide*. Carmel, IN: Commodore Business Machines and Howard W. Sams.

"Commodore Music". *Chiptunes*. (2005). Presented by M. Sharples Produced by D. Stowell. Flat Four Radio.

Darling, R. (1985). *Master of magic*. Mastertronic Added Dimension.

Dumile, A. (2008). *Glitch thieves: Crystal castles admit 8-bit theft*. Available at: http://drownedinsound.com/news/3491735-glitch-thieves--crystal-castles-admit-8-bit-theft [Accessed 17 Jan. 2018].

Falconer, P. (1984). *Commodore 64 sound & graphics*. Tring: Melbourne House.

Hughes, P. (2016a). About me. *Paulie's perfunctory game dev website*. Available at: www.pauliehughes.com/page4/index.html [Accessed 17 Jan. 2018].

Hughes, P. (2016b). SID music. *Paulie's perfunctory game dev website*. Available at: www.pauliehughes.com/page22/page22.html [Accessed 17 Jan. 2018].

Mace, S. (1982). Consumer electronics show wows Vegas. *InfoWorld*, 4(4), pp. 1, 5–7.

McAlpine, K. (2015a). All aboard the impulse train: An analysis of the two channel music routine in manic miner. *The Computer Games Journal*, ISSN: 2052–773X. doi:10.1007/s40869-015-0012-x.

McAlpine, K. (2015b). In-person interview with Rob Hubbard, 15 Dec. Hull, UK.

McAlpine, K. (2015c). In-person interview with Ben Daglish, 14 Dec. Matlock, UK.

McAlpine, K. (2017). In-person interview with Rob Hubbard, 9 June. Hull, UK.

McAlpine, K. (2018). *Bits and pieces: A history of chiptunes*. New York: Oxford University Press.

McGuire, S. and Van der Rest, N. (2015). *The musical art of synthesis*. Abingdon: Focal Press.

O'Regan, G. (2016). *Introduction to the history of computing: A computing history primer*. Cham: Springer.

Plogue. (2009). *Chipsounds*. Canada: Plogue.

Pouladi, A. (2004). *An interview with Ben Daglish*. Available at: www.lemon64.com/?mainurl=http%3A//www.lemon64.com/interviews/ben_daglish.php [Accessed 17 Jan. 2018].

Radcliffe, J. and Salkeld, R. (1983). *Towards computer literacy: The BBC computer literacy project 1979–1983*. London: The British Broadcasting Corporation.

Roads, C. (1996). *The computer music tutorial*. Cambridge, MA: MIT Press.

Russ, M. (2013). *Sound synthesis and sampling*. New York: Focal Press.

Shatner, W. and Fisher, D. (2009). *Up till now: The autobiography*. Bath: Windsor.

Suni, J. (2002). Acidjazzed evening. *Vandalism News*, 39. Available at: www.pouet.net/prod.php?which=37478 [Accessed 17 Jan. 2018].

Tong, S. (2010). *Sound byte: Symphonic game music concerts*. Available at: www.gamespot.com/articles/sound-byte-symphonic-game-music-concerts/1100-6275522/ [Accessed 17 Jan. 2018].

Varga, A. (1996). *Email interview with Bob Yannes*. Available at: http://sid.kubarth.com/articles/interview_bob_yannes.html [Accessed 17 Jan. 2018].

Wallich, P. (1985). Design case history: The commodore 64. *IEEE Spectrum*, Mar., pp. 48–58.

Wolinsky, D. (2011). 8-bit punks Anamanaguchi beyond the side-scrollers. *A.V. Club*. Available at: www.avclub.com/article/8-bit-punks-anamanaguchi-beyond-the-side-scrollers-58886 [Accessed 17 Jan. 2018].

4

Composing With Microsound

An Approach to Structure and Form When Composing for Acoustic Instruments With Electronics

Marc Estibeiro

INTRODUCTION

The Sea Turns Sand to Stone (2015) is a composition by the author for bass clarinet, flute, piano and electronics, in which the electronic part is generated using a software environment for granular synthesis, which manipulates sounds from the acoustic instruments. The output of the granular synthesizer is itself further manipulated by four different effects processors. A principal aim of the composition is to explore issues of structure and form in music for acoustic instruments and electronics. These are focused on the following research questions: first, how can microsound be used as part of an organizing principle in order to provide an aesthetically satisfactory sense of cohesion between the acoustic and the electroacoustic elements? Second, how can microsound be used to develop and control both small-scale and large-scale structural and formal relationships between different aspects of the composition? Before these questions can be addressed, however, it is first necessary to consider the compositional affordances of microsound more generally as well as the implications of using microsound in the context of acoustic instruments.

A DEFINITION OF MICROSOUND

In this chapter, the term "microsound" is used to refer to an approach to composition that makes creative use of granular synthesis and other FFT-based windowing techniques. Thompson defines microsound as "more than a technique, microsound is an approach to composition which places emphasis on extremely brief time scales as well as an integration of this micro-time level with the time levels of sound gestures, sections, movements and whole pieces" (2004). This is broadly the definition that will be followed in this chapter. Complex evolving sound spectra are constructed out of streams or clouds of sonic particles of very small durations, typically 100 ms or less, although grain sizes larger than 100 ms are also possible. Larger grain sizes preserve more of the acoustic spectra of the source material while shorter grain sizes tend towards wide band impulses of noise. Larger grain sizes can be combined with shorter grains to produce a wide variety of sonic possibilities and relationships to the original material. These new spectra can then be further manipulated with other electronic processes to create an elaborate network of different relationships with the source material.

THE COMPOSITIONAL AFFORDANCES OF MICROSOUND

Following on from the definition given above, microsound can be considered as more of an approach to composition than as a genre or a single technique. The emphasis is always on composing with sonic material built on a variety of time levels. Implicit in any composition which uses microsound as the primary structuring principle is an approach to time that permits the coexistence of sounds as they unfold on different time scales. Micro-events are built into gestures and textures that are then further developed into longer phrases, sections and entire compositions. In this context, definitions of gesture and texture broadly follow Smalley's definitions, where a gesture implies some form of energy-motion trajectory, spectral and morphological change, linearity and narrative, and a texture is a sound which evolves, if it evolves at all, on a more worldly or environmental scale and where internal activity is more important than forward impetus (Smalley, 1997). Gestures and textures exist along a continuum, and it is not always clear where the distinction is to be drawn between them. These micro-events can co-exist and contrast with sounds derived from other sources such as note-based events from acoustic instruments, synthesized sounds, concrète sounds from field recordings or other sound objects. These sounds can themselves be further broken down into micro-events, transformed and re-contextualized. Microsound is therefore a powerful tool for juxtaposing the recognizable with the unrecognizable or for creating a continuum from the possible to the impossible. Sounds can be divorced from their usual contexts and reframed with transformed versions of themselves.

Because of the nature of the process, environments for generating and manipulating microsound output a very large number of sonic events, and this can lead to potentially very complex control networks. The granular synthesis environment used for the *Sea Turns Sand to Stone*, for example, consists of three independent 16-voice granular synthesizers. If all three granular synthesizers function with a grain size of 10 ms, then (ignoring any limitations imposed by the signal vector size or the input/output buffer of the software environment) the resulting output would consist of 4800 grains per second. Through careful mapping of the user interface to the sound-producing engine of the software, however, the manipulation of a relatively small number of parameters can produce a huge variety of different sonic possibilities. The principal parameters which have been used for *The Sea Turns Sand to Stone* are:

- Grain size—the length of a single sonic particle
- Grain density—the number of grains per second

- Playback speed—the speed at which the software reads through the granulated soundfile to produce time stretching of time-compressing effects. Playback speed can also be reversed or set to zero in order to "freeze" the sound.
- Jitter—a variable offset of the onset time of each grain which can be manipulated to make playback of the source material progressively less linear
- Number of voices—number of simultaneous grain streams produced by a single granular synthesizer. These streams may be simultaneous or they may overlapping depending on settings for jitter.
- Number of channels—the number of independent environments for granular synthesis happening at one time. The majority of the compositions in this commentary use an environment with three granular synthesizers. This was felt to be an acceptable compromise between compositional affordances and limitations imposed by computer processing power. Each environment may use the same source material or the source material may be different to produce a variety of results.
- Relative balance of channels—controlling the relative amplitudes of the three different environments can have a significant effect on the output as contrasting spectra fade in and out
- Post-granular processing—the way in which the output of the granular synthesizer is processed, if it is processed at all, will of course have a considerable influence on the resulting sonic spectra and its relationship to the source material. It is important, therefore, that there are good aesthetic reasons for including any post-granular processing and that these are sympathetic to the overall aesthetics of the composition.

The choice of window function also has an audible effect on the output but although the software environment includes an option to change the window function, this has generally been left as a fixed value. The cosine function was the least likely to introduce unwanted artifacts into the sound, so this was chosen as the default envelope.

Most of the parameters described above can be set to static values, or they can be interpolated between different values. How these parameters have been mapped to the audio engine of the environment, and how they have been made available to the composer or the performer, will have a significant impact on both the compositional process and the performance.

Through careful manipulation of the above parameters, the environment for granular synthesis is capable of producing an enormous variety of rich and evolving sonic landscapes. Progressive application of the parameters can result in the source material appearing in the output as an electronic facsimile of the original acoustic input at one end of a continuum, and as wholly new material with little or no perceptual relationship to the original at the other end. Along this continuum, different aspects of the source material can be revealed or hidden in the output. The original material can be dramatically slowed down in order to reveal previously hidden detail, for example. Or the overall gestural shape of a sound can be preserved while dramatically changing its spectral content. Gestural sounds can quickly transform into textural sounds and back again. Sounds can appear to "dissolve" through careful manipulation of grain density.

Juxtaposing the output of the granular environment with the original acoustic material is a musically rich and aesthetically cohesive approach to composition. Microsound is, however, essentially a texture-based approach and as such typically does not conform to the aesthetics, traditions and performance practices of note-based music. Nevertheless, microsound can be used as a compositional device which extends and complements the output of the acoustic instruments.

Microsound is a particularly powerful technique for playing with source identity and context as it can act as a bridge between the real and the surreal, or between sounds which are perceived to be physically possible and sounds which are perceived to be physically impossible. When combined with other electronic techniques, it becomes a very fluid environment for re-contextualizing sonic events. Beyond its classic use as an "acoustic microscope", microsound is also a very effective means of exploring tensions and contradictions as sounds transform from the real to the imaginary, again in the sense of perceived physical possibility and impossibility. Ambiguities and contradictions arise as gestures become textures and causal relationships break down. Connections are broken and sound objects are repositioned in new contexts.

It is a contention of this chapter that microsound can be used as an effective organizing principle when combined with acoustic instruments in mixed compositions. Microsound can function as a bridge between very different traditions of electroacoustic and acoustic music. The seemingly contradictory traditions and performance practices of post-serial, pitch-based writing for acoustic instruments can be productively combined with a texture-led electroacoustic approach using microsound as a unifying factor.

DIFFERENT MODELS OF STRUCTURE AND FORM

Before we can consider the ways in which microsound can be used as an organizing principle to create a cohesion between the acoustic and the electroacoustic, it is first necessary to outline the issues raised when considering the problem of form in music, particularly in the context of mixed acoustic and electroacoustic music. It will then be possible to explore how microsound can be used to develop and control both small-scale and large-scale structural and formal relationships between different aspects of the composition.

When discussing form it is important to draw a distinction between form as a concept and the various manifestations of that concept, such as sonata form, binary form etc., which have emerged historically through the analysis of different compositional practices. One definition of form as a concept is that it is the "constructive or organising element in music" (Whittall, 2008). Another way of stating this could be that form is the way in which the smaller microstructural elements of a composition are grouped together to create an overall macrostructure. The distinction becomes problematic, however, when we consider the criteria a composer may be using, explicitly or implicitly, when ordering the material of a composition.

A composer may be choosing from perhaps three different approaches when imposing form on compositional material: a top-down schema-led model, a bottom-up material-led model and a generative or process-led model. Within those broad categories there exists the possibility of a great number of different approaches that combine ideas from each area as appropriate to a particular circumstance. The aesthetic reasons for choosing one particular approach to form over another are not always clear, however, and may be influenced by issues of genre conformity or historical precedent in ways which may not always be sympathetic to the compositional material.

Justifying compositional choices becomes even more complicated when we consider that form is not usually an isolated aspect of the composition but has an intimate relationship to the material of the composition (there may be exceptions to this: John Cage's *Imaginary Landscape No 5* is one of many examples of a composition where it could be argued that the form is imposed on the work in a highly prescriptive manner by the composer, but the content, the musical material which inhabits the form, is left to what are essentially aleatoric processes). This becomes even more problematic when we consider that there is often a somewhat circular relationship between generalizations about form and the application of formal models by composers. Formal templates and approaches to form are extracted from compositions identified as typical or exemplary in some way, and these models are then often used as examples of best practice and followed by other composers. Di Scipio (1994, 1995), Collins (2009) and Whittall (2008) provide further discussion of these ideas.

Any act of composition can also be viewed as an actualization of a *theory* of form, even if that theory is not explicitly stated, and even if the composer is not aware of the formal implications of his or her choices. In a top-down approach, composers may adopt formal templates for their compositions, which may be adaptations of existing templates or may be novel templates specific to that composer or composition. (Di Scipio, 1995) makes the point that in a top-down approach, the form of a composition pre-exists the composition, independent of the material of the composition, not only in the mind of the composer but also in the minds of the listeners as well. The composition exists as an externally conditioned idea which must be recreated in a new context. For Di Scipio, the form of the composition represents a mental *solution space* which *affords* certain actions. Di Scipio develops this idea further by stating that the composer can choose not only from the range of afforded actions but also from suggested extensions to what is explicitly afforded by the solution space. In the context of this idea, it is interesting to note that the history of the development of music technology, and by extension the history of electronic music, contains countless examples of this feedback loop where artists use technology in ways not immediately suggested by the affordances of the process or environment, and this leads to technical innovations which themselves suggest new unforeseen affordances. The history of the development of form is also driven by similar feedback loops where existing solution spaces suggest new affordances which themselves become established practices.

In a bottom-up approach to composition, composers may allow the small-scale structural elements of a composition to dictate the macro-structures. In such cases, a composer may use very strong rules or clearly defined criteria which govern the development of the material. There may be a very strong crossover with, or the process may be identical to, process- or rule-based composition. In generative or process music, the form is governed by the underlying processes embedded into an associated compositional system. In its purest form, once the rules of the system have been established, the compositional processes then dictate the nature of the material as well as the form of the overall composition, often with little or no further intervention from the composer—the material is accepted as the inevitable outcome of the process. The act of composition may also become an act of curation, where the composer selects phrase-level or larger-scale material output by the process in order to assemble the final work.

It is not always so clear, however, what criteria a composer may be using when selecting material for a composition. If the underlying criteria which determine the choice of compositional material are not explicitly recognized by the composer during the production of a composition, it follows that they are also unknown to the listener during the reception of a composition. Thus in an entirely rule-based process, the listener may or may not be aware of the underlying processes which ultimately determine the form of the work, and knowledge of such processes is rarely a prerequisite for listening. In both cases, however, both composers and listeners feel entitled to make judgements as to the degree to which a composition has been successful. Consequently, it is necessary to consider the criteria a listening community may be using to make such judgements.

Emmerson (1989) proposes a model of composition which addresses the issue of what criteria composers and listeners may be using when they evaluate a work. The model is shown in Figure 4.1. In Emmerson's model, the compositional process begins with an action, the production of sonic material, which is then tested or evaluated as being suitable or unsuitable for inclusion in the composition. If the material is unsuitable, it is either rejected or modified. The modified material then either becomes a new action or is stored in a repertoire of new actions for future use. The question, then, is what is the nature of the test used by the composer to accept or reject the material? In such cases, it may seem that composers are unconsciously imposing their own aesthetic prejudices and conditioning onto the material. For Emmerson, however, it is the existence of the action repertoire that forms the basis of the test by which the sonic material is assessed. The exact nature of the test must remain elusive (it is "unanalysable" in Emmerson's words; 1989, p. 143), but the important point is that the action repertoire is not the private property of the composer but is open to a community of interest made up of composers, performers and listeners

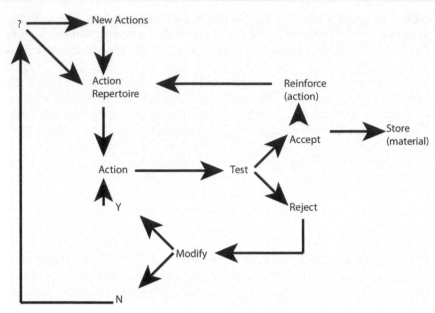

Figure 4.1 Emmerson's Model of Composition

Source: Adapted from Emmerson, 1989

whose views are trusted and valued and who collectively decide what kind of material may be included in the action repertoire.

For Di Scipio (1994), the shift from a top-down, example-based approach to form to a bottom-up, rule-based approach is a shift from an externally conditioned, analysis-based idea of form to an idea of form based on an awareness of compositional processes. Whereas in a top-down approach the form pre-exists the work, in a bottom-up approach the form emerges from an explicitly designed process and manifests itself as an epiphenomenon of some underlying structure. In electroacoustic music, a natural endpoint of a bottom-up approach is that it can be the sounds themselves that are composed through the application of rules to various synthesis processes. Sound spectra cease to function as material to fill emerging structures and instead become the structuring principle behind the composition. Spectral morphologies replace an instrumental approach to composition and thus end the dualism between form and material, between container and contents (see also Emmerson, 1986, for a further discussion of these ideas). In mixed compositions, however, the problem becomes how to unite a dualistic, content and material approach to form with a texture-centered, morphological approach.

It is of course possible, and indeed common, to combine the three broad approaches to form outlined in the previous paragraphs—top-down schema-driven, bottom-up material-driven and generative, process-driven—in order to produce a complex, multifaceted set of interrelationships among the materials of a composition which result in the final form of the piece. Mixed compositions, however, typically combine two very different traditions and approaches to form. In the following section, we will consider ways in which microsound as a process, as well as the affordances of microsound, can be applied as organizing principles when considering form in mixed compositions. To begin with, however, it is necessary to consider why mixed compositions are particularly problematic, as well as how the acoustic and the electroacoustic parts relate to each other. From this, it will be possible to explore ways in which microsound can be used as a process for bridging the two traditions.

STRUCTURE AND FORM IN CONTEXT OF MIXED ACOUSTIC AND ELECTROACOUSTIC COMPOSITIONS

A significant problem to overcome when considering issues of form specifically in relation to instrumental music with electronics is that mixed compositions combine the languages and performance practices of two often very different traditions. This is not a situation unique to mixed compositions, of course, but the issue is particularly pronounced in this case as the approaches of the different traditions often appear to contradict each other. The challenge is to find a satisfactory way to make these differences coexist.

The acoustic parts and the electroacoustic parts of a composition can relate to each other in a number of different ways. The two parts engage in a complex and shifting network of relationships in which each part may be equal, or one part may dominate the other. These relationships can also of course change during the course of the composition. Outlining these relationships can help to identify compositional strategies that can then be used to create structure and cohesion in a work. An understanding of these relationships can also be used to show how microsound can function as a compositional tool to reinforce or subvert relationships between the parts. Emmerson (1998) uses case studies to explore the ways in which the acoustic relate to the electroacoustic. Some of the ways in which the acoustic part and the electroacoustic part can relate to each other are outlined below.

The acoustic part and the electroacoustic part can be in a state of conflict or coexistence. There can be transitional or morphological relationships where events can be perceived to have their origins in one sound world before moving to the other. There can be causal relationships where events in one sound world can be perceived as causing events in the other. There can be gestural/textural relationships, which can manifest themselves through framing, layering or montage. There can be mimetic relationships where musical or extramusical relationships can emerge between the different sound worlds. There are also spatial relationships between the acoustic and the electroacoustic part. We now consider some of these relationships in more detail before showing how they can influence form and be manipulated through microsound in mixed compositions.

One of the more fundamental ways in which the parts relate to each other is through spatial relationships. These relationships can be either literal, in the sense that a sound really is coming from a certain position or has been produced by a certain sounding body, or metaphorical, where a sense of space is suggested or implied through some process or psychoacoustic phenomenon. There are different categories of spatial relationships. These can be summarized as follows:

- spatial relationships associated with movement (how is the sound perceived to be travelling in space?)
- spatial relationships associated with position (where is the sound?)
- spatial relationships associated with material (how big is the sounding body? What is it made of? How is it being excited? Etc.)
- spatial relationships associated with environment (in what sort of space is the sound world unfolding? Is it a real space, an impossible space, a changing space? Etc.).

Typically, in a live performance of a mixed composition, the acoustic part will be anchored to a fixed position on a stage and the electroacoustic part will be diffused through an array of loudspeakers. Blending the two parts can be problematic, however, because of the way in which the sounds are transmitted. The electroacoustic part typically emanates from directional speakers, whereas the acoustic part will be produced by instruments which radiate sounds in much more complex patterns (Tremblay and McLaughlin, 2009).

When developing models of form for music with mixed acoustic and electroacoustic elements, it is useful to identify a principle of internal cohesion to act as an organizing principle which will then unite the different sound worlds in a satisfactory manner. This is not necessarily straightforward, however, as the disciplines of acoustic and acousmatic music have complex and often contradictory relationships, particularly in the context of form and material. Di Scipio encapsulates these different approaches by making the distinction between composing *with* sound and composing sound (1998). In the first case, the emphasis is on the relationships (gestural, tonal, dynamic etc.) that exist between the sounds and in the second case, the emphasis is on the creation of the textures and timbres themselves and the relationships that unfold as those textures develop and interact.

In the first model, timbres are, at least to an extent, interchangeable. A phrase, for example, could be played on different instruments and still be recognizably the same. In the second model, timbre is the central focus of the composition: it is not possible to change the timbre without fundamentally changing the composition. With a great deal of crossover and a great many exceptions, acoustic instrumental music typically tends towards the first model, whereas acousmatic music tends towards the second. Therein lie some interesting tensions, but these potentially conflicting considerations need to be handled carefully. Models that combine both approaches, however, are only really satisfactory if there is a model of interaction, explicitly stated or implicit in the tradition, which unites the seemingly disparate electronic and acoustic parts.

TOWARDS A PURE SOUND/NOISE AXIS AS A MODEL FOR STRUCTURING MIXED COMPOSITIONS

One model that has been extremely successful in the structuring of instrumental music has been that of functional tonal harmony. In this model, harmonic relationships are based on hierarchies of perceived levels of stability as sounds progress through degrees of consonance and dissonance. The Finnish composer Kajia Saariaho has borrowed the ideas of consonance and dissonance from the language of tonal harmony and used them to create new models for the structuring of texture-based music (1987). Saariaho's solution to the problem of combining the different sound worlds of the acoustic and the electroacoustic parts is to develop a sound/noise axis to unite the two elements. In Saariaho's model, the concepts of consonance and dissonance are replaced with concepts of pure tone and noise. This then becomes the organizing principle behind some of her compositions. The axis allows her to create a logical timbral continuum which provides a pre-compositional framework where sounds can be placed on a theoretical hierarchical grid between pure sound, for example, a periodic sound with few or no partials—a sine wave would be the ideal, and noise—complex, aperiodic spectrally dense sounds. In Saariaho's model, timbre takes the place of harmony, with consonance being replaced by pure sound and dissonance being replaced by noise. Noisy, grainy textures take on the function of dissonance while smooth, fluid textures assume the role of consonance. The terms sensory consonance and sensory dissonance can be used to differentiate the use of the terms from their use in the context of tonal harmony.

O'Callaghan and Eigenfeldt (2010) provide a detailed examination of two of Saariaho's compositions for acoustic instruments and electronics which use this approach, namely *Verblendungen* (Saariaho (1984)), *Lichtbogen* Saariaho (1986)). Although the two pieces demonstrate different control strategies for the electronic part, *Verblendungen* uses a tape part whereas *Lichtbogen* uses live electronics featuring the processed sounds of the live instruments; they both use the same approach to sound in order to develop a structure. O'Callaghan and Eigenfeldt (2010) also propose a gesture-focused analysis of the compositions. Their analysis reaffirms Saariaho's own writings, where she discusses the use of extended instrumental techniques to create a continuum between noise and pure tone (1987).

In the acoustic parts of *Verblendungen* and *Lichtbogen*, it is the spectral quality of the instrumental gestures used in the compositions that give form to the music. Extensive use is made of extended instrumental techniques in order to shift the gestures along the sound/noise axis. For O'Callaghan and Eigenfeldt, gesture is defined as any perceptual unit or sound shape which develops over time. The use of the term to refer to a physical action that causes a sound is ignored. This is the definition that will be followed here. The variation of parameters over time can be thought of as giving "shape" to a sound and hence instigating a gesture (2010).

Saariaho's model can be easily adapted to compositions involving microsound. Indeed, granular synthesis functions as an excellent tool for shifting textures in both directions along a continuum from pure sound to noise. The composition *The Sea Turns Sand to Stone* uses Saariaho's model as the principal underlying framework upon which structure is developed. A significant difference, however, is that in the electroacoustic part, it is microsound that has been used to create the hierarchies of timbres from pure sound to noise. Saariaho's original hierarchy has also been extended to include other conceptual polarities that can exist along a continuum between consonance and

dissonance (in the context of this discussion, the terms consonance and dissonance are not used in their strict, tonal sense, but rather as terms which suggest
states of stability and instability). These concepts are shown in Figure 4.2.

Using these concepts as a guide, a schematic is created in which the
acoustic part moves from a state of sensory dissonance towards a state of
consonance, while the electronic part simultaneously moves in the opposite direction. The schematic is divided into seven sections. In each section, the ratio between consonant material and dissonant material shifts
until the last section, where the balance is effectively reversed. The overall
length of the composition is set provisionally at 7'35" and 9 seconds is
taken as a basic unit of time. The lengths of the different sections, as well
as the relationships between sensory consonance and sensory dissonance,
are shown in Table 4.1. It is worth noting that the timings function as a

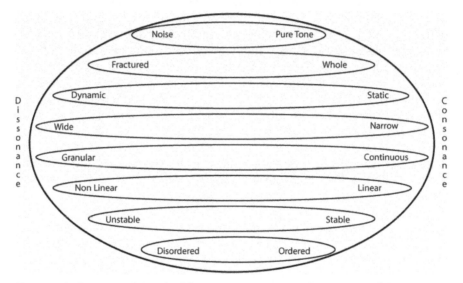

Figure 4.2 Conceptual Continuities Between Sensory Dissonance and Consonance

Source: author's own

Table 4.1 Temporal Relationships Between Different Sections in *The Sea Turns
Sand to Stone*

Length of section	Ninth of section	Ratio Dissonance: Consonance	Ratio (secs)
36"	4	8:1	32:4
45"	5	7:2	35:10
54"	6	6:3	36:18
63"	7	5:4	35:28
72"	8	4:5	32:40
81"	9	3:6	27:54
90"	10	2:7	20:70

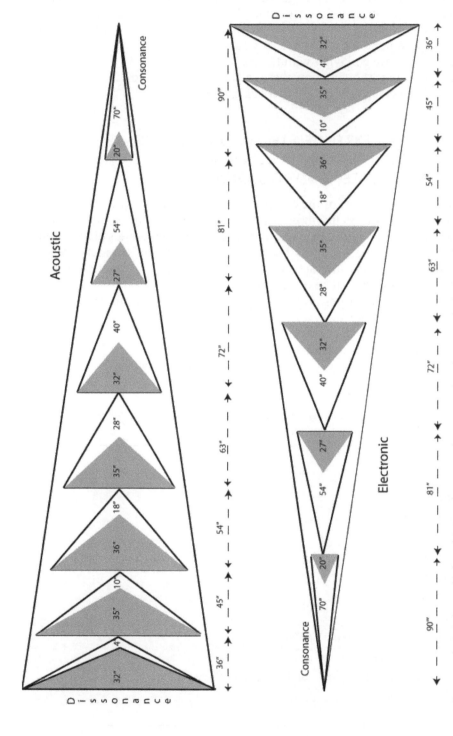

Figure 4.3 A Schematic for *The Sea Turns Sand to Stone* Showing Morphologies Between Sensory Consonance and Dissonance.

compositional aid and are not intended to be strictly adhered to during the performance. The actual length of the composition will vary from performance to performance, because the electroacoustic part is created in real time and triggered using cues in the software environment written into the score. The performers are free, therefore, to react in a more natural way than if they were playing with fixed soundfiles.

A schematic showing how these relationships apply to the overall structure of the composition is shown in Figure 4.3.

THE USE OF EXTENDED TECHNIQUES TO CREATE PERCEPTUAL CONTINUA BETWEEN SENSORY "CONSONANCE" AND "DISSONANCE"

For the acoustic part of the composition, a hierarchy of gestures has been created for each of the three instruments, starting with sounds that are perceived to be consonant and continuing through sounds that become increasingly perceived as dissonant. Following Saariaho's model, extended instrumental techniques are used extensively in the composition in order to create a suitable range of gestures.

The hierarchy of gestures used by the bass clarinet in the composition are shown in Table 4.2.

These instrumental gestures are then recorded as sound files and used as the basis for the electronic transformations. Examples of the gestures used by the bass clarinet are shown in Figures 4.4 to 4.10 below.

Table 4.2 Gestures Used by the Bass Clarinet in *The Sea Turns Sand to Stone* Ordered From Sensory Consonance to Dissonance

Low register
Senza vibrato
Ord.
Molto vibrato
Trills
Tremolo
High register
Slap tongue
Multiphonics
Tremolo between two multiphonics
Half embouchure
Morphing between air notes and half embouchure
Flutter tongue
Unpitched air notes

Figure 4.4 Bass Clarinet F2 Senza Vibrato (Cue 1)

Figure 4.5 Bass Clarinet F2 Senza Vibrato (Cue 4)

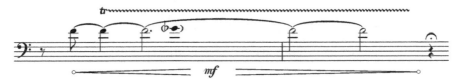

Figure 4.6 Trill (Cue 7)

Figure 4.7 Bass Clarinet Slap Tongue (Cue 10)

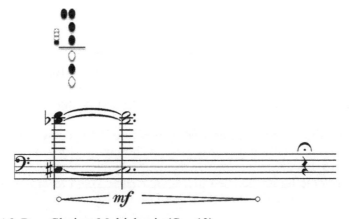

Figure 4.8 Bass Clarinet Multiphonic (Cue 13)

Figure 4.9 Bass Clarinet High Flutter Tongue (Cue 16)

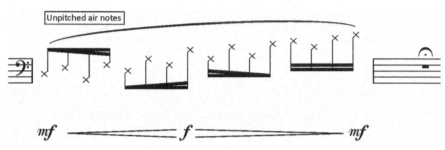

Figure 4.10 Bass Clarinet Unpitched Air Notes (Cue 19)

Table 4.3 Gestures Used by the Piano Ordered From Pure Sound to Noise

Piano chord
Piano iterative gesture
Piano pushing agitated gesture
Piano low E flat 7th harmonic
Piano harmonic then scraping gesture
Piano scraping gesture then harmonic
Piano slide bouncing off strings

Similar hierarchies of gestures are created for the piano and the flute. Table 4.3 shows the gestures used by the piano ordered from sensory consonance to dissonance.

Examples of the piano gestures used in the composition are shown in Figures 4.11 to 4.17.

Table 4.4 shows the gestures used by the flute in the composition.

Examples of the flute gestures used in the composition are shown in Figures 4.18 to 4.28.

The harmonic language used in *The Sea Turns Sand to Stone* emerges directly from the choice of material used for the gestural hierarchies.

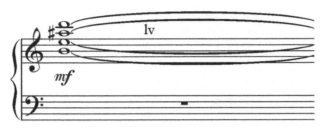

Figure 4.11 Piano Chord

Figure 4.12 Piano Iterative Gesture 1

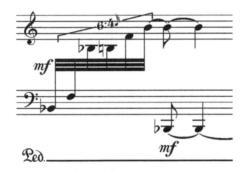

Figure 4.13 Piano Pushing Agitated Gesture

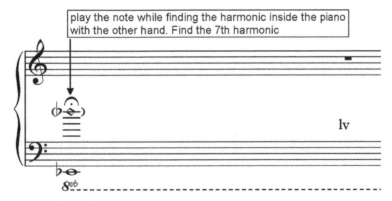

Figure 4.14 Piano Harmonic

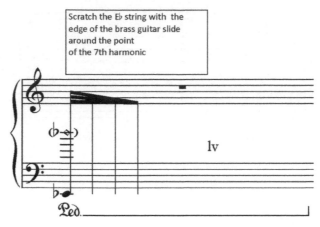

Figure 4.15 Piano Harmonic and Scraping Gesture

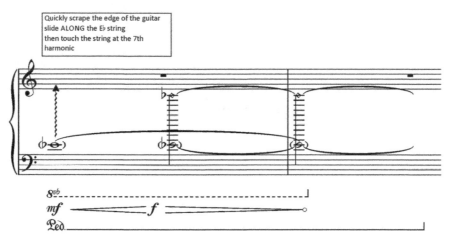

Figure 4.16 Piano Scraping Gesture Then Harmonic

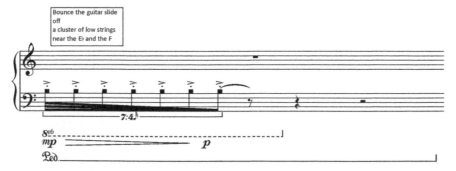

Figure 4.17 Piano Slide Bouncing Off Strings

Table 4.4 Gestures Used by the Flute Ordered From Pure Sound to Noise

Flute F4 senza vib
Flute F# 6 harmonic
Flute whistle tone F#6
Flute timbral trill
Flute tongue ram
Flute F#6 flutter tongue
Flute jet whistle

Figure 4.18 Flute F#6 Harmonic (Cue 6)

Figure 4.19 Flute Timbral Trill (Cue 12)

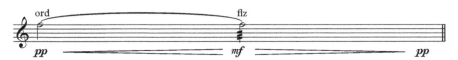

Figure 4.20 Flute Ord. to Flz (Cue 9)

Figure 4.21 Flute Short Staccato Flutter Tongue

Figure 4.22 Flute Senza Vibrato (Cue 3)

Figure 4.23 Flute Tongue Ram (Cue 15)

Figure 4.24 Flute Pizz

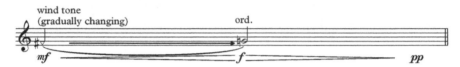

Figure 4.25 Flute Wind Tone to Ord.

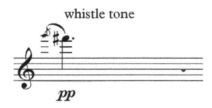

Figure 4.26 Flute Whistle Tone (Cue 9)

Figure 4.27 Flute High Flutter Tongue (Cue 18)

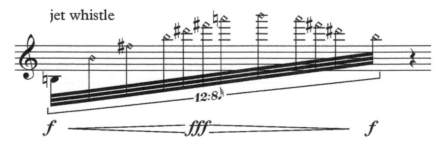

Figure 4.28 Flute Jet Whistle (Cue 21)

MAPPING AFFORDANCES FROM THE MICROSOUND ENVIRONMENT ONTO THE SENSORY CONSONANCE/ DISSONANCE AXIS

Having established a hierarchy of gestures for the instruments in the acoustic part, the next step in the compositional process is to map electronic affordances in the performance environment onto the sensory consonance/ dissonance axis. A broad overview of the environment for the electronic part is shown in Figure 4.29.

The schematic in Figure 4.29 shows three identical channels, each starting with an independent 16-voice granular synthesizer. The output of each granular synthesizer then flows through four different processors. The first process allows the elements in the output of the granular synthesizer to be reordered. The second is a delay-based pitch shifting effect which can be used to introduce comb filtering and amplitude modulation artifacts into the sound. The third is an FFT-based pitch shifter. The final effect in the chain is a spectral delay, which can be used either to give a sense of the sound inhabiting an acoustic environment or to emphasize and freeze certain frequencies in the spectrum of the sound. All of the processes after the granular synthesizer have balance controls so that the ratio of the processed to the unprocessed sounds can be adjusted.

Clearly, the performance environment for the electronics has a very large number of parameters. In the context of this composition, it would be inappropriate to expect a performer of the electronic part to be able to control the electronics in any meaningful way without significantly redesigning the interface. Indeed, the large number of user adjustable parameters could even be seen as a restriction on creativity. This is because

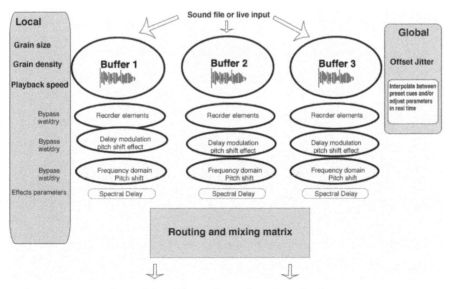

Figure 4.29 Overview of the Electronic Performance Environment

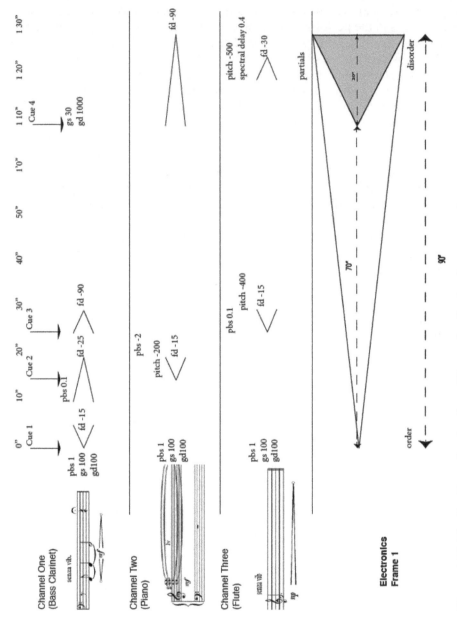

Figure 4.30 The Sea Turns Sand to Stone Electronics Section 1

constraints in interface design are as important as affordances. Creativity is often a result of what is not possible rather than what is possible. For this composition, it would be more appropriate to allow a higher level control of the parameters, where chains of events can be triggered globally by sending out multiple messages to trigger complex, carefully designed events. In this way, the benefits of having the flexibility afforded by many user-adjustable parameters can be utilized without the disadvantages of an overwhelmingly complicated environment.

Having three separate 16-voice granular synthesizers, each with its own independent effects routing, opens up many creative possibilities. For *The Sea Turns Sand to Stone*, each of the three channels is assigned to one of the three acoustic instruments. Then, using the schematic in Figure 4.3 and taking recordings of the instrumental gestures as source material, the electronic part for each of the seven sections of the composition is carefully pre-composed and mapped to cues in the Max environment. Detailed schematics of the electronic part are used to guide the compositional process. An example of one such schematic is shown in Figure 4.30.

The source material for the electronic cues has been chosen so that the instrumental gestures become increasingly dissonant as the composition progresses. The electronic processing is designed in such a way as to move the sounds increasingly further away from the recognizable instrumental gestures of the source material. At the same time, the acoustic part moves from a state of sensory dissonance and instability towards a state of sensory consonance and relative stasis. The electronic part, for example, begins with a clearly recognizable F2 played on the bass clarinet. This note is then time-stretched, beginning a gradual shift away from the source material, which continues throughout the composition.

NOTATING THE ELECTRONIC PART

The schematics for the electronic part could be thought of as compositional scores or even, to a lesser extent, analysis scores, as they contain much of the information necessary to reproduce the electronic part of the composition. As performance scores, however, they are somewhat limited, precisely because they contain too much information and it would be difficult to incorporate them with the scores for the acoustic parts written in traditional Western notation. There are, however, issues to be considered when designing scores for acoustic instruments and electronics, particularly with regard to microsound. The solution in *The Sea Turns Sand to Stone* was to use customized graphics designed to be intuitive to understand, prescriptive and representative of the sounds produced by the electronic part. These were included on the same score as the acoustic part of the composition. An example from the score is shown in Figure 4.31.

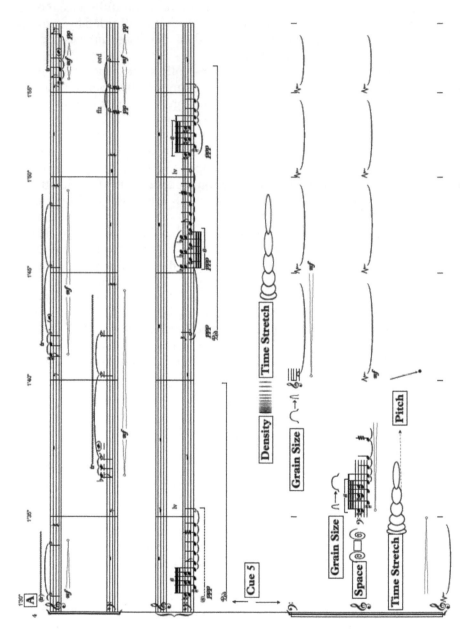

Figure 4.31 An Extract From the Score Showing Graphics Used to Notate the Electronic Part

CONCLUSION

This chapter has considered the issues raised when using microsound as an organizing principle to structure compositions for acoustic instruments with electronics. The arguments have been discussed in the context of the composition by the author of *The Sea Turns Sand to Stone* (2015). After considering the compositional affordances of microsound, the challenges of creating a coherent composition which mixes mainly note-based material from acoustic instruments with mainly texture-based electroacoustic material were discussed. The application of microsound as a technique was offered as a way of creating coherence between the different elements. Different models of form were presented, and the choices and strategies made by composers when structuring their work were discussed. The use of extended instrumental techniques in conjunction with microsound led to the creation of perceptual links between the acoustic and the electroacoustic. These ideas were then applied in the context of Saariaho's pure sound/noise axis. By extending Saariaho's model, and by using microsound as the mediating technique, a way of structuring the composition was found which was felt to be aesthetically satisfying and coherent. This approach also proved to be a powerful aid to composition. Finally, a system of graphical notation was devised for the electronic part that was intuitive to understand, prescriptive and representative of the sounds produced.

REFERENCES

Collins, N. (2009). Musical form and algorithmic composition. *Contemporary Music Review*, 28(1), pp. 103–114.

Di Scipio, A. (1994). Micro-time sonic design and timbre formation. *Contemporary Music Review*, 10(2), pp. 135–148.

Di Scipio, A. (1995). Inseparable models of materials and of musical design in electroacoustic and computer music. *Journal of New Music Research*, 24, pp. 34–50.

Di Scipio, A. (1998). Compositional models in Xenakis' electroacoustic music. *Perspectives of New Music*, 36(2), pp. 201–243.

Emmerson, S. (1986). The relation of language to materials. In: S. Emmerson, ed., *The language of electroacoustic music*. Basingstoke: Palgrave Macmillan, pp. 17–40.

Emmerson, S. (1989). Composing strategies and pedagogy. *Contemporary Music Review*, 3, pp. 133–144.

Emmerson, S. (1998). Acoustic/electroacoustic: The relationship with instruments. *Journal of New Music Research*, 27(1–2), pp. 146–164.

O'Callaghan, J. and Eigenfeldt, A. (2010). *Gesture transformation through electronics in the music of Kaija Saariaho*. Available at: ems-network.org: www.ems-network.org/IMG/pdf_EMS10_OCallaghan_Eigenfeldt.pdf [Accessed 6 Jan. 2015].

Saariaho, K. (1984). Verblendungen for orchestra and tape. Edition Wilhelm Hansen, Helsinki.

Saariaho, K. (1986). Lichtbogen for large ensemble and electronics. Edition Wilhelm Hansen, Helsinki.

Saariaho, K. (1987). Timbre and harmony: Interpolations of timbral structure. *Contemporary Music Review*, 2(1), pp. 93–133.

Smalley, D. (1997). Spectromorphology: Explaining sound shapes. *Organised Sound*, 2(2), pp. 107–126.

Thompson, P. (2004). Atoms and errors: Towards a history and aesthetics of microsound. *Organised Sound*, 9(9), pp. 207–218.

Tremblay, P. and McLaughlin, S. (2009). *Thinking inside the box: A new integrated approach to mixed music composition and performance.* University of Huddersfield Repository. Available at: http://eprints.hud.ac.uk/4081/ [Accessed 28 Apr. 2015].

Whittall, A. (2008). *Form.* Available at: Grove Music Online www.oxford musiconline.com.ezphost.dur.ac.uk/subscriber/article/grove/music/0998 1?q=form&search=quick&source=omo_gmo&pos=1&_start=1#firsthit [Accessed 25 Apr. 2015].

5

Defining and Evaluating the Performance of Electronic Music

Jenn Kirby

INTRODUCTION

The performance of electronic music is a developing field that raises many philosophical questions around what constitutes performance, what a performer requires and what an audience expects. The development of the tools and techniques associated with electronic music has opened up new possibilities for both composition and performance. While this development continually extends the opportunities for music creation, it has in some ways limited our performance possibilities. Digital instruments cannot be directly performed in the same manner as hardware instruments. The agency associated with performer and instrument comes into question. However, one might also argue that the amplification of instruments and use of effects pedals is removing agency and direct causality. Pseudo-direct causality could be considered an action or gesture that invokes a feeling of direct causality; the performer can perform a physical gesture and hear a gestural correspondence in the resulting sound. Chion defines syncresis as "the spontaneous and irresistible weld produced between a particular auditory phenomenon and visual phenomenon when they occur at the same time" (Chion, 1994). If syncresis occurs, then it may also be pseudo-direct causality, but syncresis is not always required. Physical gestures that result in an intended gestural sonic output need to be repeatable, require skill and retain the idea of craft in performance to constitute performance (Godlovitch, 2002). One can subsequently bypass the literal definition of direct causality and consider computer-mediated performance as performance, as long as it still retains these elements of performance.

I present guiding definitions in order to discuss the development of performance in electronic music and find a means for critiquing performance detached from the composition. The discussions in this chapter thereby exclude the performance of improvised works, but there may be some crossover.

The discussions are not intended as a criticism of the electronic music performance community, but rather the intention is to highlight some challenges facing the community and present ways in which they may be addressed to progress the field.

LIVE ELECTRONIC MUSIC

There are different methods of working with live electronics. They can be categorized as follows:

- Extending instruments through audio processing
- Tape/fixed media
- Electronic/digital instrument
- Sonfication through the use of controllers
- Or a combination of the above

The focus will be on the technical implementation and use of electronic and software instruments, with a focus on the use of controllers.

The degree to which an electronic/software instrument is performed by a human varies significantly. The composer and/or instrument builder must consider how the instrument is to be performed. The composer might even consider how multiple instruments could perform together in an ensemble, as can be seen in many works for laptop orchestra. Using the acoustic instrument as an analogy for an electronic instrument, the builder might consider the range of sounds that can be produced, the way in which those sounds can be controlled and manipulated, and following on from that consider the way in which they can be performed. With software instruments, the performer can produce a perfect realization of a work, one that is identical in each performance if that is what the composer wishes. For example, if the composer desires specific frequencies, a performer can read it from a score and enter it using the computer keyboard and mouse. In this scenario, however, the computer is really the performer, since the human is assisting the computer in its performance and not the other way around.

Alongside this 'input' approach, composers and performers often use controllers or hacked technology to control certain elements in an electronic music performance. At the loss of computer precision, we gain human expression. The controllers generally enable the performer to produce physical gestures that will produce data, which results in the production of a sound. For example, the mi.mu glove (Mi.Mu, 2017) tracks many elements in the movement of the performer's hand to enable him/her to control musical parameters. A performance by Imogen Heap (Heap, 2014) demonstrates pseudo-direct causality as it does not directly cause the sound, but it is developed in a way that feels like the performer has direct control of the sound output. This is often achieved through the use of haptics and tactility. The interest in hacked controllers in the DIY community has led to the development of many commercial products. The outputs from both DIY and commercial communities appears to prioritize providing the performer with expression and control over the sound. Composers and performers are creating musical gestures that will produce sounds electronically in an analogous musically meaningful way to how a performer produces sound from an acoustic instrument. For example, Stanford Laptop Orchestra's performance of *Monk-Wii See, Monk-Wii Do*

(Dahl and Berger,2008) illustrates a tactile-kinesthetic-sonic connection. The piece is performed using a wii-mote. It may seem that the differing factor between acoustic and electronic performances is technology; however, all instruments are technological (Gluck, 2007). The issue is really one of performer agency. Godlovitch defines four aspects of agency in performance: causation, intention, skill and intended audience (Godlovitch, 2002). One might consider that tactility, as in the wii-mote example, might be required for performer agency; however, "the first gesture-controlled electronic musical instrument" (Holmes, 2008), the Theremin, establishes agency and pseudo-direct causality without any physical touch. Therefore, tactility is not required but often present.

PERFORMANCE

Music is inherently gestural and arguably audio-visual. Bergeron and McIves Lopes, in their study on interpreting musical expression from sight and sound, found that audio and vision were amalgamated by the observer in their interpretation of the music. They found that "body movement conveyed roughly the same structural information as sound" (Bergeron and McIves Lopes, 2009). With instrumental music, a performance can often be visualized without being seen. This is largely due to causal listening, whereby the listener is aware of the source causing the sound and therefore is highly likely to visualize that source (Kane, 2003). The visual associations from acoustically produced sounds do not directly correspond to electronically produced sounds. In the composition of electronic music, a composer often designs an electronic instrument using synthesis methods or manipulated field recordings and therefore can be without a discernible causal sound source. This means the performance situation offers a new scope for gesture and the creation of an imagined sound source identification based on acoustic and instrumental traditions. What is meant by performance in an instrumental concert is well understood, although what can be considered performance in electronic music is less defined and precisely what this chapter aims to address. The use of controllers allows for the creation of a relationship between what the listener sees and hears—or it at least allows for the creation of a symbiotic relationship. Many concerts of electronic music focus on the acousmatic, actively disengaging with the visual; however, audio-visual performance elements of electronic music is a growing field that is well supported by the rise of laptop orchestras around the world. For example, Stanford Laptop Orchestra's performance of Ge Wang's *Twilight* (Wang, 2013) shows performers rising to their feet as we hear a swell in dynamics and quickly go to silence as they crouch back down.

Performance or Presentation

Traditionally speaking, in acoustic music the composer composes the piece and the performer performs it, but this is not always the case in electronic music. It is important to note that the origins of electronic music involved a composer composing and a machine performing, or a composer performing a machine. Human performers are not required in the production or performance of electronic music. "The electronic composer produces his score direct from his ear to the ear of the listener, sanitary, even sterile sometimes. The tide is away from the performer as it is away from the instrument" (Bowers and Kunin, 1967). There were many composer-performers working with live electronics and performing electronic music in the earlier days of electronic music, even if it is thought of as only a recent development. For example, Gordon Mumma performed *Hornpipe*, using custom-built circuitry to process the sounds from a French Horn and sounds from the performance space (Mumma, 1971).

The skills employed in performance differ from those employed in composition. Although the composition and the performance of a work can be developed in tandem, their considerations differ greatly. In electronic music, once the composition process is complete, the work is often 'made' performable. For example, certain effects and filters are used to offer some control over a sound that is already sounding. Another approach is the triggering of sounds and tweaking of volumes and EQ. In these scenarios, a piece of music is being presented and is aided by a performer. The result is that the performance closer resembles a presentation of the work, rather than a live performance. Direct causality is brought into question because it is often unclear to what extent the performer's actions are affecting the sound output. This is a question around the connection between the performer's input and sound output.

Another issue surrounds the audience's interpretation of this connection. That is, the performer input–sound output relationship cannot be understood by the audience. This is most common when the performer is using a laptop and the screen is hidden from the audience. This issue can arise in popular music and avant garde music settings. The prevalence of it leads to remarks such as "he/she is just on Facebook" or "it isn't even plugged in", etc. Performers are acutely aware of this perception, and many have responded to these remarks by adding more performative elements to their performance, and others by adding more equipment. Many performers make use of lighting and projections, and many elaborate on existing gestures for pressing buttons. However, these responses do not address the issues highlighted; these responses have answered the wrong question: "how do I look like I am performing?" instead of "how do I perform?" This is not to say that the easy option has been chosen, but rather that other means were not known or available. Scenarios where there is little skill required for the performance of a work could be better considered as "presentations" of a work, rather than a performance. That is not to say that it is not valid, but rather that there is no room for advancing one's performance skills, but only advancing one's presentation of a work. The focus in these scenarios has been placed on the composition of the work. Where human performance is not required or is minimal, defining this as presentation may be more useful. This also may deter negative comments around presentations of works and enable composers to unapologetically present their work without a need to introduce superficial performance elements. A concert may be made up of presentations of works as well as performances.

In contrast, others have developed performance techniques and tools that require skill to perform. These often involve the use of expressive performative instruments and controllers. When composers begin to respond to the question of "how do I perform my music in a musically meaningful way?", the audience begins to see and hear interesting performances. The audience also begins to see virtuosity in the performance of live electronic music. A key component in acoustic performance is gesture. Physicality in acoustic music performance is required. This is not the case in electronic music. However, composers must consider if it is required for the audience and the performer to understand it as performance.

Audio-Visual Gesture

Gesture in acoustic music performance has existed for hundreds of years and is well understood and accepted by composers, performers and audiences. One does not need to have held a violin to understand a down-bow gesture or a vibrato gesture. When the audience sees the gesture, they can predict a corresponding sound. This audio-visual relationship is so important to the audience that when they hear something that they do not understand, they often look to see how it is being produced, so that they can create that audio-visual connection. With acousmatic listening, it is often difficult not to visualize a corresponding gesture producing the sound when the source is known.

Sounds can be inherently gestural. Where this is the case, the composer may use it to influence how the sound is performed. The composer can synthesize causal listening by creating a physical gesture to match the sound output. Here the gesture of the physical movement and the gesture of the sound must harmonize if audio-visual syncresis is to be achieved. Considering the envelope of the sound, a sound with a sharp attack performed with a gentle fluid movement, is not likely to harmonize. However, if the sharp attack is met with staccato movements from the performer, the audience may be able to connect the gesture of the movement with the gesture of the sound. Many of these gestures can be borrowed from the well-understood acoustic instrument traditions. If a sound is percussive, even if it is not an acoustic sound, a striking gesture would seem to easily translate. A bowing gesture considered simply as a movement in one direction at a constant speed produces a sustained sound. The physical and sound gesture relationship can again be mapped in electronic music performance, where expressive controllers are utilized. To illustrate this point, I have produced a video featuring gestural mapping using a Gametrak controller and a microphone as a controller (Kirby, 2017). The Gametrak

> is a small base station that sits on the floor. Two retractable wires are fed through what are essentially two analog sticks, and connect to the player's hands with little gloves. By interpolating the angles of

the wires and the degree of extension, the Gametrak is able to judge movement in three dimensions.

(Block, 2016)

The video demonstrates six gestural mappings:

1. 360 Percussion
2. Voice Timpani
3. Speech Bassline
4. Guitar Tether
5. Plucked Tether
6. Bowed Tether

360 Percussion

This instrument uses the Gametrak controller, and the spatial positioning of the arms of the performer holding the controller determines the pitch produced from a downward strike. The speed of the strike also determines the amplitude of the sounding note.

Voice Timpani

The voice, specifically pitch and note duration, controls the sound output of a timpani. High notes trigger rolls, and each lower note triggers single strikes.

Speech Bassline

This gesture again makes use of the voice; for each pitch produced by the vocalist, a corresponding pitch will be produced on the software instrument that sounds similar to a plucked bass guitar.

Guitar Tether

A swinging arm gesture triggers a note on a software instrument. The word guitar refers to the gesture more than the synthesized sonic output.

Plucked Tether

By plucking the string on the tether, a note is produced. The pitch is controlled by the height of the string on the left hand.

Bowed Tether

Similar to the plucked note, except a bowing gesture is used to sustain a sound, and again pitch is determined by the height of the string. In contrast to a stringed instrument, such as a double-bass, the higher pitch can be found by holding the string up higher, and the lower pitches are further down by lowering the string.

The purpose of creating these gestural mappings is to provide demonstrations as to extended means of syncresis, pseudo-direct causality, performer agency and skill. It is not to recreate what already exists in the instrumental world, but to borrow and build upon those traditions. Conversely, this gestural mapping could be used as a tool for subversion, whereby there is intentionally a mismatch of gestures. The composer could choose to disrupt the audience's audio-visual mappings. These performance considerations need to be contemplated and designed during the compositional stages, as to do so after the fact would likely require changes to the composition.

CONSIDERATIONS AND EVALUATIONS

Suggested Performance Considerations and Evaluations

It is helpful to make a distinction between presented electronic music and performed electronic music, in order to continue to advance the performance of electronic music and develop a skilled approach considering virtuosity.

Before a performance of an electronic piece can be evaluated and critiqued independently of the composition, one might consider what a good performance in electronic music looks like. When composing a performed electronic music piece, I suggest the following contemplations:

- Is it performable?
- Is it translatable?
- Is it good?

These are highly subjective questions deserving of further consideration and qualification.

Is It Performable?

This question first needs to determine if a performer is even required. If the piece can be performed by a computer alone, then it is not performable. Assuming that a performer is required, then the composer must consider the degree of control available to the performer. What skills might the performer need to develop in order to perform the piece? If the piece can be performed instantly without error, then there is no opportunity to develop skill and virtuosity. There is also no opportunity for performer's interpretation. Can the performer perform the piece poorly? If he/she cannot, then he/she likely cannot perform it well either. If practice makes better, then it is performable.

Is It Translatable?

Can the audience interpret the gestures and make sense of them? If pseudo-direct causality occurs and there are elements of syncresis, then the performer's actions and intentions have translated and been understood by an audience. As with the violin, the audience does not have to know in any detail how it is being performed, but they can discern from the performance that the performer has command over the instrument. Therefore, a connection has been made.

Is It Good?

This is the most important question and the most difficult to answer. When we as a community of composers, performers, musicologists and audience members bypass the gadgetry and the novelty, can the performance be critiqued and evaluated? When we can measure the performance independent of the tools and the composition, then it can be truly considered performance. When we can measure different performances of the work and different performers' interpretations, it is possible to evaluate, and therefore it is possible to progress the field of electronic music performance.

Suggested Composition Considerations

The performance of the work is best considered throughout the composition process. This is particularly relevant when the composition of the work involves the use or the creation of a new software or hardware instrument. Although it is very useful to consider all the comparisons with acoustic performance, many differences need to be addressed. These differences allow for greater possibilities and greater challenges. Typically, acoustic performance is transforming the abstract instructions into concrete sound. Either the desired sound is known and the performer, with skill, knows the action to produce the sound, or occasionally the action or gesture is notated and the performer mimics the action that results in the production of the desired sound. When the gesture is notated, it generally accompanies text describing the desired sound. This approach may best fit for the notation or instruction of performed electronic music. Knowing the physical and sound gestures will enable the performer to learn to produce the sound, and so there is agency and intention.

Avoiding Determinism

When designing a work for performance or designing an instrument, it would be intuitive to only map the parameters that you wish the performer to control. However, consider the notation of scores for acoustic instruments; typically pitch, dynamic and timbre are notated. Although vibrato is occasionally notated, it is generally left to the performer's discretion. Even considering a crescendo, there are many ways that it can be performed. Much is left to the performer's discretion and interpretation. We can consider these as improvised or non-deterministic elements. When designing a new instrument, I advise introducing scope for improvised elements in the performance of the work. This allows the performer to interpret the piece beyond the notation. These improvised elements may help address the question of whether a performer is needed or it can be performed by a computer. If the piece requires the performer to interpret non-notated elements, then a performer is required.

CONCLUSION

This chapter attempts to define and propose evaluative means for the performance of electronic music. Given the diversity in electronic music, it would be naïve to suggest a framework for electronic music performance. However, I propose guiding considerations in order to ensure that the performance is musically meaningful to both performer and audience. The composer's aesthetic in terms of performance is not discussed. Instead, some focus is placed on the performer—agency and skill as well as expectation and understanding from the audience. The idea is not to introduce similar etiquette of acoustic performance into electronic performance, but to enable the development of musicianship in electronic music performance. The gestural mapping examples are fundamentally based on gesture in the performance of acoustic music. However, they present some methods of enabling syncresis and pseudo-direct causality in an electronic music context. The aim is to develop virtuosity as there is in all other aspects of music. In the same way that composers might aspire to have their acoustic work performed by a certain performer, they might also wish for an electronic music performer to realize their work, for the same reasons. By developing our understanding of what constitutes a good electronic music performance, we can develop upon it, introduce new ways of performing that builds upon and surpasses the traditions of acoustic music performance.

REFERENCES

Bergeron, V. and MvIves Lopes, D. (2009). Hearing and seeing musical expression. *Philosophy and Phenomenological Research*, 78(1), pp. 1–16.

Block, G. (2006). *Exclusive gametrak interview with developer In2Games*. Available at: http://uk.ign.com/articles/2006/04/14/exclusive-gametrak-interview-with-developer-in2games.

Bowers, F. and Kunin, D. (1967). *The electronics of music*. Aspen [Internet], 4(5). Available at: www.ubu.com/aspen/aspen4/electronics.html.

Chion, M. (1994). *Audio-vision: Sound on screen*. New York: Columbia University Press.Dahl, L. and Berger, M. (2008). *Stanford laptop orchestra—Monk-Wii* See, Monk-Wii Do. [Internet] Available at: https://www.youtube.com/watch?v=SZIAe7yRQn0

Gluck, R. (2007). *Live electronic music performance: Innovations and opportunities*. Haifa, Israel: Tav+ Music, Arts, Society.

Godlovitch, S. (2002). *Musical performance: A philosophical study* [Internet]. Taylor & Francis. Available at: https://books.google.co.uk/books?id=aqaEAgAAQBAJ.

Heap, I. (2014). *Mi.Mu. Imogen Heap's gesture-controlled digital music gloves* [Internet]. Available at: www.youtube.com/watch?v=h4V8uklXCjg.

Holmes, T. (2008). *Electronic and experimental music*. 3rd ed. New York and London: Routledge.

Kane, B. (2003). L'Objet Sonore maintenant: Pierre Schaeffer, sound objects and the phenomenological reduction. *Organised Sound*, 12(1), pp. 15–24.

Kirby, J. (2017). *Gestural mapping design demo* [Internet]. Available at: www.youtube.com/watch?v=cJn1zX8VyyI.

Mi.Mu. (2017). *mi.mu* [Internet]. Available at: http://mimugloves.com/.

Mumma, G. (1971). *Hornpipe* [Internet]. The Sonic Arts Union. Available at: http://ubu.com/sound/sau.html.

Wang, G. (2013). *Stanford laptop orchestra twilight* [Internet]. Available at: www.youtube.com/watch?v=chA-4GRCb-I.

6

Perspectives on Musical Time and Human-Machine Agency in the Development of Performance Systems for Live Electronic Music

Paul Vandemast-Bell and John Ferguson

INTRODUCTION

This chapter investigates musical time in live electronic music and discusses the authors' technological systems that are used to foster experimental process and create novel artistic work. By standardizing the way in which we perceive musical time, much commercial software misses opportunities to explore rhythm and groove. The goal of this chapter is to elucidate the role of human-machine agency while encouraging experimentation beyond perceived genre boundaries. The authors reflect upon their individual practices with the following aims:

1. Demonstrate how open-source and proprietary software can be used to move beyond an approach to musical time that relies on a single unilateral grid.
2. Create expressive groove that embraces human- and machine-generated rhythms.
3. Discuss the creation of meaningful interaction with computer-based technology.
4. Provide opportunities to experience and explore a broad continuum of rhythmic possibilities.
5. Disseminate the software tools and audio-visual documentation developed to articulate the ideas presented in this chapter (Pure data examples and Clyphx Ableton Live Script).

While commercial music production often revolves around metronomic timing, and the industry-standard quantization grid can often steer producers towards chronometric exactness, this is at odds with expressive human timing. In a radio broadcast entitled *The Art of the Loop*, electronic musician Matthew Herbert makes a critical comparison between early consumer hardware sampling technology and the modern-day Digital Audio Workstation (DAW). The former supposedly offered more in the way of creative expression and individuality due to a simple, physical interface that facilitated open-ended experimentation with sound and rhythm. In addition, he claims that manufacturers tended not to dictate how the technology was to be used. Herbert contrasts this with modern music production software, which he suggests is overtly grid-based and specifically designed to "fit people's expectations", restricting creative expression because "it closes down the circle of possibilities" (Herbert, 2014).

BACKGROUND

The authors were active as a live improvisation duo called Tron Lennon between 2004 and 2009. Tron Lennon explored turntables, samplers, drum machines, hardware effect processors, electric guitar, modified electro-mechanical systems, hacked electronic toys and appropriated objects (Tron Lennon, 2008). In uncovering hidden and unintended potentials in seemingly fixed media, Tron Lennon exposed instability, and through the exploration of indeterminate and dysfunctional systems, they sought out unpredictability as a strategy to probe, provoke and generate creative response. Their approach often focused on hardware and usually situated lively performance gesture alongside machine-generated repetition using loops and electronic pulses.

VANDEMAST-BELL

For over a decade Vandemast-Bell's practice-led research has been concerned with finding a place for the human body in electronic music. During his PhD studies on live electronic music (Bell, 2009), he began investigating how the human body could intervene in recordings by inscribing physical gesture into sound. In Tron Lennon he ventured into the world of experimental, analog Turntablism motivated by artists such as eRikm, DJ Sniff and Christian Marclay: Marclay's ideas on "embalmed" music and bringing music "back to life" (Marclay, 2007, p. 327) through gestural interaction were an important conceptual framework for his work. Later, he became interested in uses of digital repetition within improvised performance, influenced by Robert van Heumen's use of STEIM's LiSa X software, which Heumen utilized for real-time sampling and sound manipulation. In response to this, Vandemast-Bell began exploring digital DJ tools and utilized Pioneer CDJs for experimental looping. This CDJ setup still offered the potential for gestural intervention via the CDJ jog wheel (replicating the turntable platter), but the focus of his research began to shift, and he became interested in using digital repetition *against itself* to produce complex rhythms based on the phase techniques pioneered by Steve Reich (2005) and later developed by Brian Eno (2004). Presently, Vandemast-Bell's research merges his interest in four-on-the-floor techno music and digital looping experiments, embracing the continuum of rhythmic possibilities. Notions of *gesture* and *physical inscription* have given way to auditory perception and ideas from cognitive psychology. Inspired by West African drumming sensibilities, he seeks to understand and challenge conventional modes of listening. He has developed the Dynamic Looping System (DLS), which is an attempt to circumvent the restrictive, ubiquitous DAW grid and bridge the gap between experimental and club-oriented techno music. The place of the human body in his work can be found in the improvised rhythms produced during his live performance with the DLS, which can be witnessed in his duo work with audio-visual artist Michael Brown (Time.lus, 2017a).

FERGUSON

Ferguson's work attempts to chart an idiosyncratic zone within the continuum of what it is to be a live musician in the 21st century. Using a variety of technologies, both commercial and bespoke, common conceptual threads include: touching at a distance, negotiating inertias, setting processes in motion and intervening within established trajectories. Less about being in control of a situation and more about ways to find lifelike resonances with which to interact, the relationship among imagination, expectation and material is often at the foreground. Focusing on real-time interaction and the multiple connotations of *performing technologies* (who or what is performing who or what?) could seem to undermine what might be perceived as the musician's autonomy. However, he is not attempting to remove his own agency from the creative process; this is not in any sense a chance-based approach, but one that involves maximum attention and involvement. Ferguson describes his practice as post-digital because it deliberately explores the features and quirks of digital systems; it pushes beyond the digital to digital-analog hybrid systems, and it seeks renewal through continuous engagement with varied and ever-evolving technologies. As a post-digital electronic musician, he takes a variety of approaches:

1. Machine-assembled dislocation (MAD) is a hybrid computer instrument that extends an electric guitar via two Nintendo Wii controllers, a Keith McMillen SoftStep, OSCulator and Ableton software Live/Max4Live. The guitar itself is relatively unmodified, but the combination of pressure-sensitive foot-pedals and Wii controllers offer rich possibilities.
2. Feral technologies is an exploration of circuit bending, hardware hacking and self-made electronic instruments that foregrounds disassembled electronic commodities and avoids the never-ending ("it could be anything!") flexibility of computer systems.
3. Push experiments are working in an entirely improvised/live scenario and aiming to explore techniques and timbres associated with a variety of vernacular electronic music practices.

Ferguson uses Ableton software Live and hardware controller Push to keep in touch with commercial DAWs and because making music is a different kind of art to making instruments. The origin of these projects can be traced back to early experiments as part of Tron Lennon.

Ferguson's current work revolves around free and open-source software, including projects that use embedded systems via single-board computers and instruments that utilize mobile devices with hand-built Arduino-based additions. In these projects the digital and analog collide and physicality is often at the foreground.

PURSUING EXPERIMENTAL RHYTHMS

A determining factor in our ability to appreciate musical rhythm has to do with the notion of *entrainment*. Developed in cognitive psychology, it describes how we attune ourselves to external cyclical patterns through a "form of synchronization [that] resembles a phase-lock between the oscillation formed by expected sounds (in the music) and the oscillation formed by our *anticipatory attending*" (Zeiner-Henriksen, 2010, p. 126). Butler (2014) expands on this idea, stating that "as the oscillators synchronize with the stimulus rhythm, anticipation becomes stronger" (Butler, 2014, p. 196). This insight goes some way towards explaining why repetition is not only attractive to us, but also why many are drawn to four-to-the-floor bass drum patterns. As we listen to highly repetitive music, we become increasingly familiar with the rhythmic patterns, and our ability to predict what will arrive next grows in magnitude. We synchronize ourselves, physically and mentally, to the music, making it seem as if it were a part of us. By contrast, we experience *reactive attending* when unpredictable sounds and/or unanticipated patterns accost us. Some examples might include a drummer who has difficulty keeping time to a beat or a DJ who is unable to hold a mix together. In both instances our attention is disrupted because we cannot anticipate what will arrive next: the internal/external oscillations are out-of-sync. And yet, sensibilities akin to reactive attending can be witnessed in the drumming practices of some African traditions. For example, West African polyrhythmic drumming foregrounds the use of polyrhythm in which "there are always at least two rhythms going on" (Chernoff, 1981, p. 42). For listeners who are more accustomed to musics controlled by meter, the manner in which the rhythms appear to conflict in polyrhythmic drumming will often lead to a sense of confusion and unease because a primary meter does not exist. Meter produces a regular pulse or beat that is, for the most part, unwavering (especially where the music has been sequenced in a DAW), and the instruments all play together in such a way that they coincide at pre-defined moments. Specific beats are accented within a measure to produce a dominant rhythm that conforms to the underlying beat. Though syncopation can be used to produce contrasting rhythms that intentionally go against the main rhythm, or as Chernoff puts it, "a shifting of the 'normal' accents to produce an uneven or irregular rhythm" (ibid.), it does so within the limitations of meter. However, when a fundamental meter is missing, it can seem as though the musicians in polyrhythmic drumming ensembles are not listening to one another, and it becomes difficult to comprehend "how anyone can play at all" (ibid.). In alluding to the relationship between the lead drummer in the ensemble and the audience, Davis (2008) describes how "almost violently syncopated 'off-beat' lines crisscross and interfere with the other rhythms, pushing and pulling at the dancers-listener's precarious internal sense of the beat" (Davis,

2008, p. 61). Unlike those who require a unifying meter to make sense of the music, those engaged in West African polyrhythmic drumming must maintain an awareness that is "constantly open to productive chaos" (ibid.). This "productive chaos" may offer new and exciting modes of listening moving beyond the predictable (and sometimes stagnant) nature of much commercial and club-oriented techno music. It could be argued that polyrhythm essentially seeks to forge new understandings for it "impels the listener to explore a complex space of beats, to follow any of a number of fluid, warping, and shifting lines of flight" (Davis, 2008, p. 56). In light of Davis's observations, it is interesting to note that what is considered *reactive attending*, from the perspective of meter-dominated music, appears to be precisely what is valued in polyrhythmic music.

In the 1990s, experimental rhythm production exploded in electronic dance music. The emergence of Intelligent Dance Music (IDM) and Glitch saw a concerted effort to "expand [techno] music's tendrils into new areas" (Cascone, 2000, p. 15). Neill (2007) recounts, with fervor, how "beat scientists" such as Aphex Twin and Squarepusher stretched genre distinctions and advanced the intricate, micro programming techniques of Jungle and D&B through "complex quantization structures" (Neill, 2007, p. 388). For Neill, beat science brought high art practices, consistent with the experimental researches of academia, into the mainstream. These artists pushed digital technologies to their limits, heralding the beginning of something new in popular music. Moritz Von Oswald and Mark Ernestu, better known as Basic Channel, also expanded the rhythmic possibilities of underground minimal techno, deploying shifting stereo delays and modulated EQ filters in their Dub techno creations. *Radiance* (Basic Channel, 1994) exemplifies subtle, percussive rhythms that push and pull against the listener's attention, producing "endlessly inflected, fractal mosaics of flicker-riffs and shimmer-pulses" (Reynolds, 2013, p. 627). Though revered within the underground Dub techno and Microhouse scenes, Basic Channel could not lay claim to the same kind of commercial success as the likes of Aphex Twin or Squarepusher. Though the work of the 1990s beat scientists suggested a "cultural change" (Neill, 2007, p. 390), it appears not to have transpired. Meter-dominated musics of predominantly 4/4 measure continues to prevail in both commercial and club-oriented electronic musics.

EXPLORING MUSICAL TIME USING INCLUDED PURE DATA EXAMPLES

This chapter includes example patches created using Pure data (Pd) and a demonstration movie that provides an overview of each patch (Ferguson, 2017a). The aim is to offer practical and experimental starting points for exploration beyond conventions that rely on a single overarching grid or time line. It is also hoped that these patches allow users to experiment with the aforementioned notions of *anticipatory* and *reactive attending*. However, please note these patches are not an attempt to mimic the characteristics of West African polyrhythmic drumming. Pure data (vanilla) is required to open and run the patches, and this free and open-source software is available from Miller Puckette's website (Puckette, 2017).

Pd Example Zero: Come Out

0_comeout.pd is a tribute to the work of Steve Reich, especially *Come Out* from 1966, which involves a spoken phrase being captured onto analog tape via microphone. This phrase is then copied and two near-identical tape loops are created. Both versions are played back simultaneously using two tape machines. Playback of the material on these tape machines initially occurs in unison but slips out of synchronization gradually. Repetition and a regular pulse are at the foreground of this work, but gradual change is constant, leading to musical development over time. *0_comeout. pd* approximates *Come Out* by starting with the simultaneous playback of one audio sample using two separate [tabplay~] objects, delaying the play message of one object by five milliseconds at each repetition. This approach lacks the gradual change and variation that would occur across the time duration of the recorded loop when using tape (due to the subtle nuances of an analog/mechanical system). However, it does demonstrate the creative potential of a single audio sample with two playback devices and a short delay on the triggering of one device. Overall, this patch is useful when repetition and musical development over time is desirable.

Pd Example One: Slice Live Audio Into Sixteen

The phrase *musical time* used in the title of this chapter borrows from Joel Ryan's *Music and Machines IX: Resistant Material* presentation at Newcastle University in January 2009. This introduced an idea that became important to the authors' work: "musicians make time" (Ryan, 2009), one aspect of this is outlined as follows:

> [M]usicians immediately solve what have to be the most complicated of calculations. Where most people need a PC or a pen and paper to divide eight by nine, a musician can instantly tap out an intricate time relation to make a new rhythm which is both graceful and correct.
>
> (Norman et al., 1998)

Human agency is clearly at the foreground of Ryan's thinking about musical time. This was a contentious topic for Tron Lennon, because although immediacy and tactility was desirable, much of their explorations were around repetition and the use of computers to automate and intervene in human gesture. In short, the tension between human timing/agency and what Ryan calls *machine time* can be configured as a productive tension. On this point, the Pure data patch *1_slice_live_audio_into_16.pd* foregrounds the notion that musicians make time: this patch is set up to record audio for whatever length of time a button is held down and then play back immediately on release of the button. On the one hand, the immediacy of this approach to live sampling makes for a tactile and responsive system for capturing and looping audio. On the other hand, the main purpose of this patch is to slice audio into 16 equal steps, so each step can then be triggered via a linear or random sequencer, and the length of the sequence can also be constrained. Thinking broadly about the differences between tape and digital systems: *0_comeout.pd* foregrounds mechanical/analog characteristics (two tape machines drifting out of synchronization), *1_slice_live_audio_into_16.pd* looks beyond analog characteristics and celebrates one of the hallmarks of a digital system: random access to material in a live and immediate scenario. Although live/instant random access to recorded sound is almost impossible with a mechanical/analog system, those interested in the history of advanced electro-mechanical machines that go beyond the presumed linearity of tape should see the Phonogène (Roads, 2004, p. 61). Overall, this patch is useful when repetition and musical development over time is desirable, but in comparison to the previous example this variation occurs around commonly understood divisions of time.

Pd Examples Two and Three: Random and Euclidean Elements of Circles Sequencer

As with the approach to live sampling in *1_slice_live_audio_into_16.pd*, the starting point remains the length in time of the most recently captured audio sample (thus retaining the idea that musicians make time). However, rather than dividing this duration into 8 or 16, it is divided into 192 segments and used to trigger percussion via the following divisions: whole, half, quarter, eighth, sixteenth, or triplets based on quarter, eighth, sixteenths. A constrained random process chooses the specific subdivisions, and the regularity of change in time is also chosen via a second constrained random process, some elements of this were inspired by the rhythmic organization used by Brown 2016 in his software Ripples (Brown, 2015). This patch might be thought of as an alternative form of sequencing, though rather than entering individual hits, the musician controls the density and timing using only two controls. The third Pure data example *3_quick_tidy_euclid_hack.pdf* is a Euclidean rhythm generator. This relies on first choosing a maximum resolution (16 in this case) and the number of steps that will trigger an output, then calculating the greatest common divisor of these numbers to drive beats and silences; the basic result is that the output/hits are always spread as evenly as possible. In this example, each generator is represented by two circles of 8 steps and two generators are placed alongside each other to make the interaction more visual: four circles represent dual 16-step Euclidean sequencers, and the filled bangs (black dots) show where and when the output/hits fall for each iteration of the loop. One control is used to govern how many hits sound in each time period for each generator, and there is also an offset button that randomizes an offset and thus rotates the established pattern. For a comprehensive discussion, see Toussaint's paper *The Euclidean Algorithm Generates Traditional Musical Rhythms* (2005), underlying Pd implementation by Stutter (2009).

Pd Examples Four and Five: Prime Numbers and Aesthetic Durations

Pure data examples four and five each contain multiple sequencers. Like examples one, two and three, the sequencers in each patch are governed and synchronized by one time duration; each sequencer iteration is divided into a different number of steps. Example four bases these divisions on prime numbers; the following step lengths are available: two, three, five, seven, eleven, thirteen, seventeen, nineteen. Mathematical inspiration is dropped in example five, and divisions instead are chosen by aesthetic whim: steps of sixteen, fifteen, fourteen, thirteen are available. The output of each sequencer step can be routed to a snare, hi-hat, or bass drum, and the volume level of each step is controlled via graphical linear arrays. These examples were inspired by the work of Belfast musician James Joys, who uses Reason software to create similar sequences to those discussed here; for an example, see *Glyphic Bloom* (2013), particularly the tracks "Bone Dried", "Lands End" and "Little Kingdoms".

DLS (DYNAMIC LOOPING SYSTEM)

Vandemast-Bell's DLS is an idiosyncratic performance system that makes use of nativeKONTROL's ClyphX MIDI Remote Scripts for Ableton Live (The Beatwise Network, 2017). A custom Ableton Push 1 controller mapping interacts with the Clyphx script to enhance Live's looping potential, and a Novation Launch Control XL is utilized for mixing, effects and dynamic processing (Vandemast-Bell, 2017a). Dynamic looping refers to the use of real-time loops to produce both synchronous and asynchronous rhythms in an attempt to investigate the rhythmic continuum and embrace experimental practices in live, club-oriented techno music. Real-time looping is by no means a new concept, and there are many commercial DJ applications such as Native Instruments' Traktor offering considerable scope for loop exploration. However, the principal concern of applications such as Traktor is "perfect sync" (Native Instruments, 2017). Synchronization algorithms allow DJs to perform all manner of sonic trickery without the burden of timing issues, but this is at the expense of rhythmic exploration. There is little opportunity for rhythmic complexity beyond the main beat (though the tools do exist within Traktor's advanced looping panel) because deviation from the unifying meter simply has no place within the current DJ paradigm. Likewise, deliberately derailing a mix in order to explore alternative rhythms is unusual within club culture. Furthermore, inadvertently losing synchronization during segues will generally cause unrest in audiences for reasons outlined earlier concerning anticipatory/reactive attending. Experimentation is possible, but it takes place within clearly defined parameters by layering tracks, remixing stems and manipulating time-based effects and dynamic processors.

Commercial music software tends to shoehorn users into formulaic methods of working through prescribed grids. Nevertheless, the perceived limitations placed on creative expression by manufacturers through the design of their DAW interfaces are not necessarily prescriptive. Whereas some artists opt for more flexible tools (MaxMSP, Pure data) to circumvent the stifling restrictions of grid-based software, others have sought ways to adapt the software with which they are most familiar. For example, Ableton Live users can make use of MIDI Remote Scripts to communicate with Live's Application Programming Interface (API) and change behavioral aspects of the software. Scripts such as Clyphx make it possible for those without programming skills to extend the creative potential of Live. In the DLS, Clyphx facilitates the creation, and exploration, of loops in a dynamic, real-time fashion by creating flexibility within an otherwise prohibitive DAW grid. The system has been specifically designed for loop length and loop position manipulation of individual audio clips in Live's Session view. Consequently, clips can move beyond meter and grid to achieve alternative rhythms. There is no definitive rationale for determining how loops should be constructed or combined; this takes place *in the moment* and is informed by attentive listening and the inevitable risk-taking and experimentation indicative of improvised music-making. A multitude of combinations are possible on the rhythmic continuum, and the aim is to explore all possible eventualities, which might be thought of as foregrounding both *anticipatory* and *reactive attending*. A typical performance will involve the use of between four and twelve original techno tracks. These are full-length tracks of between six to eight minutes, mixed and mastered to a professional standard by Vandemast-Bell (Blake Law, 2017). Up to four loops can play back simultaneously and their individual length/position parameters changed via Clyphx X-Control Actions, a type of X-Trigger designed for use with external MIDI controllers.

The Ableton Push 1 controller was chosen to interface with Live, offering performance satisfaction in layout, tactility and visual feedback. However, Push is not essential for the system to operate, so other MIDI controllers could easily be utilized. To mix between loops and to apply time-based effects and dynamic processors, a Novation Launch Control XL is used with a custom mapping. The ClyphX documentation provides an extensive list of Actions to deploy. The syntax is entered into the relevant section of the UserSetting.txt file, located in Live's MIDI Remote Scripts directory. The custom DLS script is available to download and explore (Vandemast-Bell, 2017b). Push employs a pad/button matrix comprising eight columns and ten rows. Pads are MIDI-mapped via the inclusion of raw MIDI messages within a Clyphx Action List. Loops can be created in standard divisions of the beat (sixteen, eight, quarter, etc.) to produce synchronous rhythms, or their lengths can be increased/decreased by a factor of 1/3 to create more complex and/or asynchronous rhythms. Pitch transposition functionality has also been added to offer further scope for sound manipulation. The raw MIDI messages in the Action Lists also include velocity information for visual feedback, an important aspect of system design enabling the performer to easily distinguish between the functions of different pads. For examples of a performance with the DLS, see Vandemast-Bell (2016) and Time.lus (2017b).

CIRCLES

Ferguson's Circles is a wooden box that contains a microphone attached to a single-board computer and two microcontrollers, and bespoke software is written in Pure data and Arduino; for a discussion, see Ferguson (2016). Physical layout and initial functionality is inspired by Peter Bussigel's *n*dial, an "8-step sequencer that goes around instead of along", which has been described as "a program for playing in time and out of order" (Bussigel, 2014). As one perspective on musical time and human-machine agency, this phrase neatly summarizes what is most engaging about these interfaces. Similarities between *n*dial and Circles: (1) both allow performers to randomize and restrict the steps and step length of their sequences; (2) the circular layout of the controls on each instrument highlights the repetition that is inherent when digital audio is looped; and (3) randomization allows various iterations of the same material to be performed (complex music from simple materials). Main differences are that (1) when the central button is pressed, Circles records one sample of a duration equal to the length of time that the button is pressed, and this is then sliced into eight steps; *n*dial on the other hand detects transients and prepares one-shot samples which are sequenced independently; (2) although both instruments can work with pre-recorded or live-sampled audio, it is probably fair to suggest that *n*dial is aimed at the navigation of predetermined sound worlds, whereas Circles is better suited to live sampling—i.e. capturing source material in real time (similar to *Pd Example One: Slice Live Audio Into Sixteen*) where the length of each recording is what defines the base musical time for any interactions that follow; and (3) simplicity and restriction is at the heart of the *n*dial, the hardware interface for Circles is more multi-layered, with eight preset states offering many potentials.

RELATIONSHIP BETWEEN CIRCLES AND THE INCLUDED PD EXAMPLES

Although a patch very similar to *Pd Example One: Slice Live Audio Into Sixteen* is at the heart of Circles, *Pd Example Zero: Come Out* is also relevant. This is because Circles captures mono audio, but the output is two-channel pseudo-stereo (i.e. the sequencing for each channel is independent). This means the left output can play a random sequence while the right output plays a linear sequence; the randomization of each channel is also seeded independently. Overall, this can give the illusion of stereo movement with occasional mono output when the two sequencers align. When developing Circles, some experimentation with multiple sound sources was explored, but this was found to lead to a rather episodic performance without much consistency, so the ability to generate and trigger simple bass drum and/or hi-hat pulses independently of the audio sampling was developed, as outlined in relation to *Pd Example Two: Random Sequencer* earlier; for an in-concert example where just these processes are used, see Ferguson (2015).

Circles also uses two Euclidean generators: the first can be switched to trigger the same synthesized hi-hat mentioned previously and/or segments of the most recent audio sample; the second can either trigger the same synthesized bass drum mentioned earlier and/or a synthesized bass line. This means that the audio sample can be triggered either via a linear/random sequence or via the Euclidean Generator. And both percussion elements can be triggered via constrained random and/or Euclidean methods simultaneously. Working in a situation where the same audio materials are being triggered by very different methods makes for an engaging process, and the addition of simple mix balancing and reverb, distortion and equalization effects makes for a rich experience.

For an example of a performance with Circles where both Euclidean and constrained random is in use, see Ferguson (2017b). Or, for something a little different that perhaps also makes the use of an Euclidean generator clearer (due to clarity of visual representation), see *Drum Thing* (2017c), which celebrates the automation of percussion objects using computer-controlled solenoids; this is a development of the Circles patch and is a small-scale sketch of what will become a larger project.

CONCLUSION

It has been argued that conventions of grid-based musical time have limited the scope of rhythmic eventualities in electronic music production. The emergence of beat science in the 1990s signaled the arrival of rhythmic complexity in electronic dance music and hinted at an exciting future for musical time, but this was not to be; formulaic, 4/4 timing remains omnipresent. In response, the DLS (Dynamic Looping System) and Circles were designed to explore alternative modes of rhythm and groove. The DLS demonstrates how proprietary software (Ableton Live) and MIDI Remote Scripts (nativeKONTROL Clyphx) can be utilized to move beyond standard DAW grids, extending looping and rhythmic potential. Circles highlights the flexibility of free and open-source software such as Pure data and explores rhythmic relations that celebrate traditional notions of groove, but also the creative possibility of fracturing this groove. The arguments put forward in this chapter are underscored by the documents of artistic research that each practitioner puts forward, as well as the Pure data patches and Ableton Live script that are included.

REFERENCES

Basic Channel. (1994). *Radiance* [Vinyl]. Berlin: Basic Channel.
Bell, P. (2009). *Interrogating the live: A DJ perspective*. PhD Thesis. University of Newcastle upon Tyne. Available at: https://theses.ncl.ac.uk/dspace/handle/10443/875. [Accessed 25 Jan. 2017].
Blake Law. *SoundCloud* [online]. Available at: https://soundcloud.com/blake-law [Accessed 19 Oct. 2017].

Brown, A. R. *Ripples* [online]. Available at: http://explodingart.com/wp/
blog/2015/11/25/ripples/ [Accessed 18 Oct. 2017].

Bussigel, P. (2014). *Onism.* [online]. Available at: http://bussigel.com/pb/projects/
onism/ [Accessed 24 Oct. 2017].

Butler, M. J. (2014). *Playing with something that runs*. Oxford: Oxford University Press.

Cascone, K. (2000). The aesthetics of failure: "Post-Digital" tendencies in contemporary computer music. *Computer Music Journal*, 24(4), pp. 12–18.

Chernoff, J. M. (1981). *African rhythm and African sensibility: Aesthetics and
social action in African musical idioms*. Chicago: Chicago University Press.

Davis, E. (2008). Roots and wires remix: Polyrhythmic tricks and the black electronic. In: P. D. Miller, ed., *Sound unbound: Sampling digital music and culture*. Cambridge, MA: MIT Press, pp. 53–72.

Eno, B. (2004). *Ambient 1: Music for airports* [CD]. Virgin.

Ferguson, J. R. (2015). *Circles* [video online]. Available at: https://vimeo.
com/150366611 [Accessed 20 Oct. 2017].

Ferguson, J. R. (2016). *Circles* [online]. Available at: http://users.sussex.ac.
uk/~thm21/ICLI_proceedings/2016/Practical/Performances/10_Ferguson_
Circles_Performance.pdf [Accessed 12 Oct. 2017].

Ferguson, J. R. (2017a). *Circles @ Tilde~ new music and sound art festival, Melbourne* [video online]. Available at: https://vimeo.com/200625002 [Accessed
20 Oct. 2017].

Ferguson, J. R. (2017b). *"Drum thing"—as presented at World Science Festival
2017 in Brisbane* [video online]. Available at: https://vimeo.com/210756928
[Accessed 28 Oct. 2017].

Ferguson, J. R. (2017c). *Perspectives on musical time* [online]. Available at:
https://github.com/Feral-Technologies/Perspectives-on-Musical-Time
[Accessed 10 Oct. 2017].

Herbert, M. (2014). *The art of the loop*. BBC Radio 4.

James J. (2013). *Glyphic bloom* [online]. Available at: https://jamesjoys.band-
camp.com/album/glyphic-bloom-lp [Accessed 18 Oct. 2017].

Marclay, C. (2007). DJ culture. In: C. Cox and D. Warner, eds., *Audio culture:
Readings in modern music*. New York: Continuum, pp. 327–330.

Native Instruments. *Traktor Pro 2* [online]. Available at: www.native-instru-
ments.com/en/products/traktor/dj-software/traktor-pro-2/ [Accessed 28 Sept.
2017].

Neill, B. (2007). Breakthrough beats: Rhythm and the aesthetics of contemporary
electronic music. In: C. Cox and D. Warner, eds., *Audio culture: Readings in
modern music*. New York: Continuum, 2007, pp. 386–391.

Norman, S. J., Waisvisz, M. and Ryan, J. (1998). *Touchstone* [online]. Available
at: http://steim.org/archive/steim/texts.php?id=2 [Accessed 12 Oct. 2017].

Puckette, M. *Software by Miller Puckette* [online]. Available at: http://msp.ucsd.
edu/software.html [Accessed Oct. 2017].

Reich, S. (2005). *Come out* [CD]. Nonesuch.

Reynolds, S. (2013). *Energy flash: A journey through rave music and dance culture*. London: Faber and Faber Ltd.

Roads, C. (2004). *Microsound*. Cambridge, MA: MIT Press.

Ryan, J. (2009). *Joel Ryan—music and machines IX* [video online]. Available at: https://vimeo.com/3392802 [Accessed 12 Oct. 2017].

Stutter. (2009). *Euclid.pd* [online]. Available at: www.pdpatchrepo.info/hurleur/euclid.pd [Accessed 12 Oct. 2017].

The Beatwise Network. (2017). *ClyphX* [online]. Available at: http://beatwise.proboards.com/board/5 [Accessed 9 Oct. 2017].

Time.lus. (2017a). *About* [online]. Available at: https://timelus.wordpress.com/. [Accessed 12 Aug. 2017].

Time.lus. (2017b). *Live performance* [online]. Available at: https://timelus.wordpress.com/live-performance/ [Accessed 12 Aug. 2017].

Toussaint, G. (2005). *The Euclidean algorithm generates traditional musical rhythms* [online]. Available at: http://cgm.cs.mcgill.ca/~godfried/publications/banff.pdf [Accessed 18 Oct. 2017].

Tron Lennon. (2008). *Tron Lennon* [online]. Available at: https://vimeo.com/user595185 [Accessed 10 Aug. 2017].

Vandemast-Bell, P. (2016). *NIME 2016 "Deformed EDM" Paul Vandemast-Bell* [online]. Available at: https://vimeo.com/176701166 [Accessed 10 Jan. 2018].

Vandemast-Bell, P. (2017a). *DLS ClyphX script installation notes* [online]. Available at: https://timelus.files.wordpress.com/2018/01/dls-clyphx-script-installation-notes.docx [Accessed 12 Jan. 2018].

Vandemast-Bell, P. (2017b). *Paul Vandemast-Bell* [online]. Available at: https://timelus.wordpress.com/paulvandemast-bell/ [Accessed 12 Jan. 2018].

Zeiner-Henriksen, H. T. (2010). Moved by the groove: Bass drum sounds and body movements in electronic dance music. In: A. Danielsen, ed., *Musical rhythm in the age of digital reproduction*. New York: Routledge, pp. 121–140.

7

Visual Agency and Liveness in the Performance of Electronic Music

Tim Canfer

INTRODUCTION

Ever since its invention, electronic music has struggled to fit into the cultural standard model of performance. An extreme example is one of the first types of electronic music: musique concrète, tape music or acousmatic music, in its traditional form an exclusively studio-created work. As Emmerson states, "the idea of 'recording a performance' in this genre is musically meaningless as different interpretations are seen as venue specific" (2007, p. 31). This begs the questions, what are the criteria that determine if an event counts as a musical performance? And perhaps more importantly, what are the criteria that determine if an event *does not* count as a musical performance? This topic is discussed fully in Section "Visual Agency".

Another example is *Switched on Bach*, by Wendy Carlos released in 1968, one of the most popular and influential albums of electronic music, which was so meticulously produced that its unique sound was impossible to recreate live (Pinch and Trocco, 2002). This serves as an early, but typical example of the very common issues in attempting to translate music created in a recording studio to a live environment, while also attempting to achieve authentic performance.

A rough distinction is made here, as much socially as musicologically, between popular electronic music, for example, electronic dance music (EDM), typified by work by the producer Deadmau5, and electronic art music, which tends to be generalized as electroacoustic music, typified by work from the composer Karlheinz Stockhausen. While this is a very rough distinction with much overlap, popular electronic music has its roots in dancing and tends to have a meter, or groove, whereas electroacoustic music has its roots in academia, tends to be experienced in installations, and its form is more of a continuous soundscape. Because of this distinction, there tends to be quite different scenarios for performance, such as nightclubs and venues for popular electronic music and concert halls and art installations for electroacoustic music. These environments have a significant impact on the different nature, form and purpose of a performance.

One of the most notable developments in the evolution of electronic instruments has been the technologies that have converged into the extremely flexible performance tool that is the modern laptop computer. Eric Lyon offers a summary of the performance problem in the context of laptop performance:

> A shopworn criticism of laptop performance of computer music asserts that the performer's actions are almost entirely unseen, leaving the audience unable to determine if the performer is actively creating the music or rather playing back a digital audio file while checking e-mail to keep busy.
>
> (1992, p. 631)

This is a good summary of the main point of this paper, which is that a lack of visual agency and liveness creates a barrier to effective musical expression and communication between performer(s) and audience in the performance of electronic music. Visual agency is defined here as the visual demonstration of specific actions that produce a particular result. While the agent responsible for these actions need not necessarily be a human, a degree of intentionality is required to achieve effective musical expression and communication. The original example in this case is of course a performing musician demonstrating that they are actively producing music. The term agency is used because it allows the inclusion or exclusion of intentionality so that machine influence can be engaged with. It is taken for granted that machines have no consciousness, so they are (as yet) incapable of intent, and as an aside, it is hard to imagine that we would be able to relate to whatever machine expression may manifest itself as.

The term agency is also used to encompass wider concepts of causality, starting with Godlovitch's definitions of primary causation and indirect causation (1998). It is argued here that a definition of musical performance based only on direct causality (or primary causality as per Godlovitch; this is explored further in Section "Visual Agency") excludes a vast range of events that *do* count as authentic musical performance. To allow for a wider understanding of modern live performance from both audience and performer perspective, this chapter introduces the concept of virtual causation and discusses the merit of machine influence on performance.

Liveness is discussed to contextualize live performance with regards to dominant media forms, namely, broadcast recordings. This serves initially to (re)affirm the place of live performance for contemporary electronic music and to propose a model of liveness from a performance perspective framing musical and visual elements in categories of live, non-live and pseudo-live. The problem is identified as a lack of visual agency and liveness in the performance of modern electronic music. The solution suggested here is to demonstrate to the audience the processes at work, connecting the agency and intentionality of the performing musician to the agency of the machine systems. It is further suggested that this may be done by innovative use of pseudo-live elements, which are defined and explored in Section "Liveness".

VISUAL AGENCY

For most of us, sight is our primary sense, and the effect that visual input has on our listening experience is profound, especially in live musical performance. Probably the most dramatic evidence of this is the McGurk effect: the linguistic phenomenon where if a video of someone speaking has the audio replaced with a different sound (for example, "ba" instead of "fa"), the brain will ignore the new sound and actually reinvent the original (Mcgurk and Macdonald, 1976).

As discussed by Carlson, there are multiple semantic and cultural contexts of the term performance (2004). Given that (as mentioned in the introduction), it is perhaps more useful to consider at what point something *stops* being musical performance, rather than trying to pin musical performance down to a single specific definition.

A useful starting point is to take the expansion of the definition of musical performance offered by Carr et al. in the *Encyclopaedia Britannica* (2017): "In Western music, performance is most commonly viewed as an interpretive art".

It follows then that to be able to interpret a song or piece of music, there is a requirement for a degree of skill and control over the instruments used. While d'Escrivan makes an interesting case for a future without the cultural requirement of skill in music, suggesting that computer games culture causes a break in the requirement for input effort in sound output (2006), the importance of authenticity in performance is demonstrated, in both Rock and DJ culture, most notably in the controversy surrounding button pushers and press play sets (Cochrane, 2012).

Accordingly, the degree of visual agency and intentionality determines both the uniqueness of that interpretation as well as the ability to achieve effective musical expression and communication with an audience. The suggestion here is that a musical event stops being a performance when it has *no* significant element demonstrating visual agency and intentionality.

Godlovitch makes a case for a definition of musical performance based on a distinction between primary and indirect causation: "when primary causation is absent the notion of performance undergoes major change. When based on indirect causation a distinctive synthetic form of art-making emerges" (1998, p. 97).

The omission of the word performance in this "distinctive synthetic form of art-making" is key. In some extreme examples this is a valid argument, for example, when popular electronic musicians The Orb appeared on the television show *Top of the Pops* in 1992, the ambient electronic duo played chess while their single played. While this makes an interesting statement about electronic musical performance, there is no actual musical performance taking place.

For the vast majority of musical performances that incorporate electronic elements (referred to as "electronic musical performances" from this point), Godlovitch's definition of musical performance is far too reductive and reactionary for contemporary popular electronic music. A more useful analysis of musical performance is to take a wider view of visual agency in musical performance and consider analyzing what indirect causation brings to performance rather than rejecting it outright. Three categories of causality are explored here, starting with the concepts of direct causation (what Godlovitch calls primary causation) and indirect causation (this is discussed further in Section "Direct and Indirect Causation"). The additional, and the less orthodox, concept of virtual causation is proposed to both validate and extend the idea of indirect causation, as well as allowing analysis of the technological elements that play a significant part in electronic musical performance (discussed further in Section "Virtual Causation").

Direct and Indirect Causation

For the sake of a consistent approach, instead of using the term primary causation, the term direct causation is used here. Direct causation occurs when the musician is the main cause of the pitch, duration, dynamics and timbre of each note or percussion hit. (This process creates the live elements discussed in Section "Live Elements") This ingrained musical paradigm has been in place for centuries due to the mechanical necessities of acoustic musical instruments.

Indirect causation occurs when the musician is causing the music to be created, but without the complete control of an acoustic instrument. This process creates the pseudo-live elements discussed in Section "Pseudo-live Musical Elements". According to Godlovitch, this occurs when there is "significant procedural remove from the final effect" (1998, p. 99), and he continues: "Because computer processes must intercede between the user and the result, these processes are indirect avenues to the result from the perspective of the agent" (1998, p. 99).

From the perspective of a traditional guild approach, this is a clear threat to the old order. Godlovitch again sums up his resistance to computer interference in music with a painting analogy that sums up the commonplace fear and misunderstanding of digital tools: "Computers produce code, not colour" (1998, p. 100).

While this resistance is understandable from the perspective of avoiding the faking of musical performance, what is often referred to as "press play sets" (Cochrane, 2012), technology has had a long history of enabling musicians to play more complex music while removing finer control, for example, a violin compared to a piano, compared in turn to a synthesizer. The inclusion of digital technology as a tool for modern musical performance is inevitable, and while opportunity for fakery is rife, there is far more to be gained by engaging with the vast creative possibilities that come from interaction between human and machine agents rather than discounting digital technology outright.

While the degree of causation relates to the amount of visual agency, to dismiss any computer intercession from a concept of performance is to deny a vast array of meaningful live experiences between audiences and musicians. A common example of indirect causation in a performance environment would be the triggering and manipulation of samples that is typical of electronic dance music (EDM) performances. While there is the safe haven of the press play set, the techniques of electronic musical performance can very quickly become complicated and sophisticated enough to warrant the term "virtuosos" in the case of artists such as Tim Exile and Madeo.

The issue of whether this counts as performance can no longer be doubted. In terms of popular culture semantics, the term "performance" is the term widely used for electronic music events. Instead then of an approach that says: "*if* indirect causation is present then the event does not count as true musical performance", a more useful approach would be to ask: "*how much* direct causation is needed to make the event count as true musical performance?"

Indirect causation is vital for electronic music, and how this combines and interacts with direct causality clearly requires further exploration. A consideration of both indirect causality and the proliferation of electronic visual elements allows the possibility of other factors that may inform visual agency in electronic musical performance, most specifically the concept of virtual causation.

Virtual Causation

Theatrical performance, in contrast to musical performance, is based largely on the concept of creating a character. While the (non-dramatic) act of performing music does not necessarily require performance in character, there is a musical performance personae, where the distinction between the real person and the musical personae is generally far more ambiguous. For example, Auslander (2004) suggests in summary that in the case of David Bowie's iconic performances as Ziggy Stardust: David Jones can be considered the person, David Bowie the persona, and Ziggy Stardust the character.

It is also observed by both Auslander (2006) and Sanden (2013) that the concept of performing personae is integral to musical performance and that this concept of performing personae relates as much to the audience and social environment as the performer.

Sanden (2013) continues his exploration of personae in his categorization of "virtual liveness", discussed in the context of both the performing personae and a technology-obsessed cyberculture. Wider examples of the successful virtual assistants such as Apple's Siri and Google Assistant show a general willingness to engage with a technological network. There is a level of anthropomorphic imagination typical to human nature accompanying this engagement that allows machines to be attributed human and therefore artistic characteristics.

This willingness to accept a virtual artistic environment results in a situation of virtual causation, where the listener actively engages in incorporating the perceived technological processes into the performance scenario, so that these processes of human and computer interaction are integral to the performance.

Where musicians and visual artists engage with virtual causation to the point of creating virtual performing characters, then the machine agency is made explicit and is given clear artistic direction, or a pseudo-intentionality. The most obvious mainstream example of this would be the four core cartoon characters of the successful virtual band and pop art project Gorillaz. Gorillaz' early performances typically consisted of real musicians playing instruments, who are visible, but generally in the background. These musical parts are then portrayed as being played by the cartoon characters on video screens. The cartoon characters are of course entirely fictitious, but the triple platinum success of Gorillaz' first album (BPI, 2017) serves as a proof of concept that audiences are happy to engage with a fictional universe and incorporate that with a live performance. The musical restriction of this technological live scenario is significant, however, as to play along with the video of the characters, the musicians are having to synchronize to a fixed click track, thus considerably reducing the liveness. This process, both at the point of the early Gorillaz performances and up to now has been the only reliable method of video characterization. However, with advances in both processing power and the creation of more flexible tools for live performers, there is significant value in exploring virtual causation in electronic music in order to bring new means of expression to musical performances.

LIVENESS

Auslander interprets the concept of liveness in the context of dominant mediatized cultural forms and summarizes this dominant form as television; however, this term is out of date (2008). A more accurate title to describe the broadcasts watched over the variety of devices available (including televisions as well as tablets, smartphones and computers) and used here is "broadcast recordings". While the purpose of most recordings of performances is to make that recorded performance (or mediatized performance) available to broadcast, it is worth making this explicit.

Both Auslander (2008) and Sanden (2013) discuss the dialectical tension between the cultural forms of the original live performance and the significantly dominant mediatized forms. Currently, live performance exists in the shadow of broadcast recordings due to the ubiquitous technology of westernized culture, but as noted by a wide range of academics from Auslander (2008) and Sanden (2013) to Emmerson (2007) and Wurtzler (1992), to put the two concepts in binary opposition is far too reductive. Performance cannot be removed from the recording, otherwise there is nothing to record (and leaves only sequencing), and for the performance to be successful in popular culture terms, it must be recorded to be distributed.

This tension highlights the importance of liveness in modern performance.

Live performance fulfills a vital cultural role in establishing the musical skill and the authenticity of a musician as a live performer. This is clearly the case in Rock culture, as discussed by Auslander (2008) and Frith (1998), among others.

Although this is less of an issue to fans of Pop music, the Milli Vanilli controversy of the late 1980s demonstrated that the common practise of lip-syncing to a recorded song was culturally acceptable, but pretending to sing the vocal part by lip-syncing over a part recorded by *someone else* was not (*New York Times*, 1991). The fallout of this included more than 26 lawsuits, the returning of a Grammy award and guaranteed notoriety in popular music history.

In DJ culture, performance is also important, especially in hip-hop or rap music, as there is a requirement for either live rapping or a degree of turntablism (or both), but live musical performance in the traditional sense is less relevant where the DJ is more of a curator than performer. The labeling of "live performance" for someone who plays pre-recorded music is particularly difficult. In this case, it may be worth seeing the DJ club night in a different light: Ferreira describes "performance" in this context as the interaction between the "machine sound" and "human movement", with the DJ as mediator (2008).

Liveness is considered here as the ideal quality of live musical performance. It is the quality of musical performance where the musical elements are created at that moment in time, with musicians as the agents who directly cause the pitch, duration, dynamics and timbre of each note or percussion hit (as discussed in Section "Direct and Indirect Causation").

In some instances, especially in EDM performances, recordings (or to use the music technology term, "phrase samples") and fixed, pre-programmed sequences are so heavily used that for many musical events there is not enough direct causality to be able to consider them as live performance at all. To quote the EDM producer Deadmau5 from his own blog: "we all hit play. it's no secret. when it comes to 'live' performance of EDM . . . that's about the most it seems you can do anyway. It's not about performance art" (DJ Tech Tools, 2012).

A Model of Musical Performance Liveness

A performance-centric (rather than audience-centric) model of musical performance liveness is presented here. This model is based around three categories of performance elements—live, non-live and pseudo-live elements—and illustrates these categories for elements that are both musical and visual.

Live Elements

Figure 7.1 shows the live elements on the left of the diagram. Live elements are created at that moment in time, that is: demonstrating direct causation. Standard examples of live musical elements are traditional instruments played live, such as the violin and piano.

The traditional live visual elements of a performance are the actual musicians playing the music. However, a modern alternative or addition to this are generative visuals: programs such as Reaktor or Max/Jitter that create graphics in real time in reaction to live musical stimulus, typically

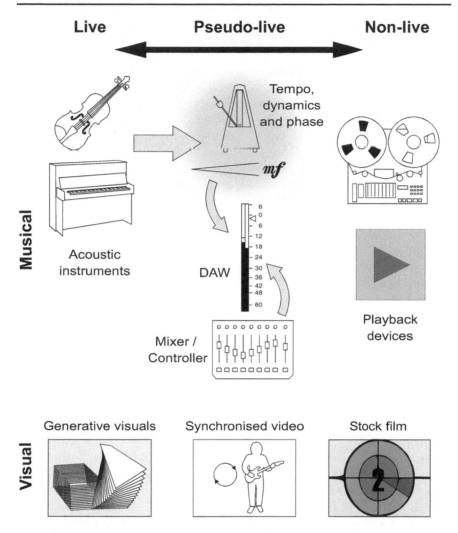

Figure 7.1 A Model of Musical Performance Liveness

in shifting patterns. (The graphic in Figure 7.1 is a combination of modified public domain patterns taken from magmavisuals.com, used to represent generative visuals.)

Non-live Elements

Figure 7.1 shows the non-live elements on the right of the diagram, and aside from the time of triggering, these elements have no real-time causal relationship with the performance, and as such they represent the opposite end on a gradient of liveness. Typically, non-live musical elements are either audio that is produced prior to the performance (otherwise known as a backing track) or a phrase sample/loop that is captured at an earlier point in the performance and is played back in a fixed form.

Standard examples of non-live musical elements are an audio player or a computer (represented here as a reel-to-reel tape player and a play button).

The effect of playing to a backing track is that musicians are forced to follow the same version of the music, removing most of the opportunities for spontaneous musical expression, a very large factor in liveness. The degree to which musicians use backing tracks is a common contentious aspect of ensuring authenticity in popular musical performance (Sherwin, 2013).

Non-live visual elements are typically a static backdrop image or stock film footage that plays independently and is not synchronized to the music after having pressed play. Apart from these, another common method where synchronized visuals are necessary is to have musicians synchronize to the visuals by following a click track, for example, the live scenario used by the virtual band and art/pop project Gorillaz (as discussed in Section "Virtual Causation"). This is essentially the same method as synchronizing with backing tracks and brings with it the same restrictions to liveness.

Pseudo-live Musical Elements

In the middle of Figure 7.1 is the pseudo-live area, which consists of music created prior to that moment, i.e. non-live elements, but the important characteristic of pseudo-live elements is that they are manipulated live, by performers in real time, and as such demonstrate indirect causation (as discussed in Section "Virtual Causation").

Croft offers two useful categories of liveness for this situation: procedural liveness and aesthetic liveness. He defines procedural liveness as "the material fact that live sound is being transformed in real time" (2007, p. 61) and offers aesthetic liveness as a continuation of procedural liveness where "aesthetically meaningful differences in the input sound are mapped to aesthetically meaningful differences in the output sound" (2007, p. 61). A key part of this analysis is that the mapping between performer action and computer response is meaningful, rather than a gimmicky use of new technology simply because it exists.

The amount of liveness that a pseudo-live element has then is dependent on how much manipulation is employed *and* how musically effective it is.

There are two different types of manipulation in the creation of pseudo-live musical elements:

- Direct manipulation, where the pseudo-live musical element is manipulated *as well* as triggered directly by the musicians, typically using a sound-processing device such as a mixer or a controller that may have any arrangement of pads, keys, buttons sliders and knobs.
- Indirect manipulation, where elements of the live performance itself causes manipulation of the pseudo-live element. This could be performance audio, MIDI or gestural data.

Traditionally, a phrase sample/loop that is captured at an earlier point in the performance will be played back as it is, achieving a static repetition. (The technique employed by popular musicians such as Ed Sheeran and KT Tunstall.) This can be stylistically desirable, but currently the alternatives to this scenario are scarce and rarely used, hence the need for exploration of pseudo-live techniques.

Methods for manipulating pseudo-live musical elements generally fall into two different categories, either musical or production.

Musical manipulation would be any combination of tempo, dynamics and phase (i.e. the timing of each note or hit relative to the beat, also referred to as groove) manipulation.

The production aspects that can be manipulated are the same aspects that are available to manipulate in the recording studio, but these need to be much simpler and able to be used in real time. The main production aspects are balance, effects and processes.

Balance manipulation is most commonly achieved with some form of mixer or controller. As more and more music is performed using studio technology, specifically DAWs, the mixer is often reinterpreted as a live creative device or even an instrument in its own right. For example, Matt Cox, the MIDI tech for the EDM duo The Chemical Brothers, describes their use of the Mackie 32:8 mixing desk for live performance:

> It's like an instrument almost, the Mackie. They throw faders up, unmute stuff—they know where everything is exactly. Some of the filters in the racks are fed off direct outs, so if one of them thinks, "That Electrix filter would be great across that synth," he can just grab the direct out jack, put it in the direct out of the synth that he wants to get at, then he's away in the rack twiddling the filter. They've got so used to that interface in front of them that it's almost second nature. They can have an idea and put it into practice in an instant.
>
> (Greeves, 2011)

Electronic music is particularly dependent on the effects and processes and the changes of their parameters. The more these are manipulated live, then the more liveness that set will have, but a visually engaging demonstration of musician and or machine agency is currently lacking.

Pseudo-live Visual Elements

Pseudo-live visual elements are visual elements that are not caused directly by the musical activities, but are manipulated by the music in real time, for example, in a most basic form would be pre-produced video that is synchronized live to musicians.

Pseudo-live visual elements can represent a vital performance element that would otherwise be lost, for example, a vocal part or phrase sampled instrument. This key issue here is that as more technology becomes a part of live musical performance, there is a risk of less visual agency and therefore less liveness as expression and communication of important musical processes are lost to the audience.

One way to counter this problem would be to create visual representations that demonstrate the link between the intentionality of the musician and the machine processes, reconnecting the musician and the audience. Currently, the visualization standard is the use of abstract shapes and patterns, which while interesting do not increase audience engagement or express musical interpretation. An alternative to the abstract audio visualizations would be to bring some characterization to the missing musical elements, in a way that is like the performances of the virtual band Gorillaz, as discussed in the section above titled "Visual Causation". Vitally though, for these elements to be pseudo-live rather than non-live, they would need to synchronize to the real musicians and ideally reflect the intention and interaction of the musicians and machine processes at play. This is opposed to the current convention in popular music where key visuals are employed, that is, requiring musicians to synchronize with a fixed backing track and or click track.

At its most basic (and potentially most reliable implementation), pseudo-live visual elements would consist of video loops that synchronize with the music by varying the frame rate (represented in Figure 7.1 as a cartoon graphic of a guitarist and a loop symbol). Other methods would be to have character visuals that are generated live using software such as Processing or Max Jitter. These visuals could either be sequenced to follow musicians using software such as Ableton Live or triggered by performers. These methods could improve the liveness of a performance and re-connect the music to an otherwise distanced audience.

CONCLUSION

With the continued proliferation of electronic elements in popular music and a wider cultural engagement (or even obsession) in technology, there has never been a more exciting time to create music, especially electronic music. However, the main point raised here is that the tools and techniques that afford electronic music have introduced barriers to musical expression and communication, as the cultural conventions of bodily expression and "touch" (Peters, 2013) are potentially absent in electronic music performance.

These issues need to be explored and understood better, as this chapter barely scratches the surface of such an ignored subject. Because of this (and rather than attempting to nostalgically undo creative innovation to return to a purely physical definition of performance), it is perhaps more useful to explore the differences among live, non-live and pseudo-live musical elements and to critically analyze what these differences do both *to* and *for* performance. In this way, pseudo-live musical and visual elements could allow engagement with the creative possibilities of virtual causality, effectively communicating and expressing musical intention in innovative new ways.

Next steps for this research include the continuing development of the author's software for live musical performance, which can be found at www.reactivebacking.com, and the establishment of a best practice of live performance techniques in the context of the issues explored in this chapter. This will take the form of a book, published as a part of the Perspectives on Music Production series, called *Music Technology in Live Performance: Tools, Techniques and Liveness*, which is planned to be available in 2020.

REFERENCES

Auslander, P. (2004). Performance analysis and popular music: A manifesto. *Contemporary Theatre Review*, 14(1), pp. 1–13. doi:10.1080/1026716032000128674.

Auslander, P. (2006). Musical personae. *TDR/The Drama Review*, 50(1), pp. 100–119. doi:10.1162/dram.2006.50.1.100.

Auslander, P. (2008). *Liveness: Performance in a mediatized culture*. 2nd ed. London and New York: Routledge.

BPI. (n.d.). *British album certifications, Gorillaz, BPI*. Available at: www.bpi.co.uk/bpi-awards/ [Accessed 22 Oct. 2017].

Carlson, M. A. (2004). *Performance: A critical introduction*. London: Routledge.

Carr, B. A., Thomas, J. P. and Foss, L. (n.d.). Musical performance. *Encyclopedia Britannica*. Available at: www.britannica.com/art/musical-performance [Accessed 2 Oct. 2017].

Cochrane, G. (2012). Do superstar DJs just press "go" on their live shows? *BBC Newsbeat*, 20 Nov. Available at: www.bbc.co.uk/newsbeat/article/20218681/do-superstar-djs-just-press-go-on-their-live-shows [Accessed 9 Oct. 2017].

Croft, J. (2007). Theses on liveness. *Organised Sound*, 12(1), pp. 59–66. doi:10.1017/S1355771807001604.

d'Escrivan, J. (2006). To sing the body electric: Instruments and effort in the performance of electronic music. *Contemporary Music Review*, 25(1–2), pp. 183–191.

DJ TechTools (2012). Is deadmau5 right? The "We All Hit Play" debate. 2 July. Available at: http://djtechtools.com/2012/07/02/is-deadmau5-right-the-we-all-hit-play-debate/ [Accessed 5 Sept. 2017].

Emmerson, S. (2007). *Living electronic music*. 1st ed. Aldershot, Hants and Burlington, VT: Routledge.

Ferreira, P. P. (2008). When sound meets movement: Performance in electronic dance music. *Leonardo Music Journal*, 18, pp. 17–20.

Frith, S. (1998). *Performing rites: On the value of popular music*. Cambridge, MA: Harvard University Press.

Godlovitch, S. (1998). *Musical performance: A philosophical study*. London and New York: Routledge.

Greeves, D. (2011). *Matt Cox: MIDI tech for the chemical brothers*. Available at: www.soundonsound.com/people/matt-cox-midi-tech-chemical-brothers [Accessed 8 June 2017].

Lyon, E. (1992). The absent image in electronic music. In: R. Altman, ed., *Sound theory sound practice*. 1st ed. New York: Routledge, pp. 623–641.

Mcgurk, H. and Macdonald, J. (1976). Hearing lips and seeing voices. *Nature*, 264(5588), pp. 746–748. doi:10.1038/264746a0.

Peters, D. (2013). Touch: Real, apparent, and absent. In: D. Peters, G. Eckel and A. Dorschel, eds., *Bodily expression in electronic music*. 1st ed. London: Routledge, pp. 17–34.

Pinch, T. and Trocco, F. (2002). *Analog days: The invention and impact of the Moog synthesizer*. Cambridge, MA: Harvard University Press.

Sanden, P. (2013). *Liveness in modern music: Musicians, technology, and the perception of performance*. 1st ed. New York: Routledge.

Sherwin, A. (2013). Musician calls for big bands to come clean on secret backing tracks. *The Independent*. Available at: www.independent.co.uk/arts-entertainment/music/news/musician-calls-for-big-bands-to-come-clean-on-secret-backing-tracks-8773412.html [Accessed 5 Sept. 2017].

The New York Times (1991). Judge rejects Milli Vanilli refund plan. 13 Aug. Available at: www.nytimes.com/1991/08/13/arts/judge-rejects-milli-vanilli-refund-plan.html [Accessed 5 June 2017].

Wurtzler, S. (1992). "She Sang Live, But the Microphone Was Turned Off": The live, the recorded and the subject of representation. In: R. Altman, ed., *Sound theory sound practice*. 1st ed. New York: Routledge.

8

Liveness and Interactivity in Popular Music

Si Waite

INTRODUCTION

This chapter describes and compares four approaches to the creation and use of audio-visual interactive systems for the live performance of popular music. These approaches were identified through a practice-led process in which the underlying aim was to create a portfolio of work for live performance that demonstrated a high degree of liveness and interactivity. The four approaches were:

- Controller-based interactive-generative tools
- Multi-tool systems for guitar and vocal performance
- Systems controlled by typing the lyrics of a piece
- Systems based on real-world metaphors

Following a review of current issues in the live performance of electronic music, recent theories of liveness and interactivity are discussed. Four aspects of liveness are identified from these theories that form the basis for the comparisons between the four approaches listed. Other work making use of similar approaches is then reviewed, before a discussion on the use of interactive systems in popular music. After an overview of the practice-led methodology used in the project, the four approaches to creating interactive systems are discussed in detail, with reflections on the level of interactivity and the suitability of each type of system for popular music.

Comparisons among the approaches reveal that systems based on real-world metaphors demonstrated the highest levels of liveness. This approach enabled the system to have a presence in space and time distinct from the human performer (spatio-temporal liveness), to reveal the causes of sounds (corporeal liveness), demonstrate interactivity with the human performer (interactive liveness) and reveal the ideas and creative processes behind the composition (aesthetic liveness). The metaphor approach was also well-suited to popular music while allowing significant levels of interactivity during composition and live performance.

CONTEMPORARY ISSUES IN LIVE PERFORMANCE

The well-documented 'problem' in the field of live electronic music performance is that performer gestures do not necessarily correlate to sonic gestures (Fels et al., 2002). Unlike traditional instruments, a performer can instigate the loudest of sounds of infinite duration with an imperceptible gesture. As it is well accepted that audiences are generally motivated to seek out the causes of the sounds they hear, this could potentially lead to a lack of engagement with the performance (Emmerson, 2007). The removal of traditional performance values through the use of pre-recorded media may also detract from the audience's perception of liveness (Auslander, 2008). Despite arguments that for some audiences, the lack of gestures and the use of mediatized material may not be a problem at all (Bown et al., 2014), there is a growing body of work considering live electronic music performance from the audience perspective (Reeves et al., 2005). Recent work has explored the impact of understanding how digital systems work on an audience's enjoyment of a performance (Bin et al., 2016) and has explored strategies for achieving this in the field of interactive audio-visual works (Correia et al., 2017).

DEFINITIONS AND BACKGROUND

Liveness

The binary distinction as to whether a performance is 'live' or 'not live' has largely been abandoned in favor of a continuum: liveness (Auslander, 2008). Several theories of liveness have been proposed. Sidney Fels and colleagues put forward the concept of 'transparency', that relates to the effectiveness of the mapping of gestural input to sonic output (Fels et al., 2002). Philip Auslander discusses how the use of mediatized material in live performance reduces liveness by detracting from traditional performance values (Auslander, 2008). John Croft distinguishes between 'procedural' liveness, in which it is simply true that events are being generated in real time, and 'aesthetic' liveness, where perceptibly meaningful performer inputs are mapped to perceptibly meaningful outputs (Croft, 2007). This relates to the idea of the 'technological sublime', where tools become aesthetic objects in their own right (Demers, 2010). Paul Sanden suggests that the overall perception of liveness arises through a network of different aspects of liveness (Sanden, 2013). For Sanden, liveness can be broken down into:

- Spatio-temporal liveness: an agent's presence in time and space
- Liveness of fidelity: how faithful the performance is to its initial version
- Liveness of spontaneity: the degree to which the performance is fixed
- Corporeal liveness: the strength of the link between a sound and its observable cause
- Interactive liveness: the extent by which music arises from interactions between agents

• Virtual liveness: the extent by which liveness is perceived in mediatized material

Sanden's theory is flexible enough to deal with a wide variety of approaches to contemporary live performance, such as those involving the presentation of fixed material and those where human performers are not present in the same physical space as the audience.

Interactivity

Songwriters and composers are influenced by the tools they use. Rather than exert complete mastery over them, composers exploit the affordances and constraints of the instruments and devices that they work with (Prior, 2009). When working with interactive systems, true interactivity can be said to occur when there is a mutually influential relationship (Noble, 2009) in which both the human performer and the system demonstrate agency over the music being performed (Emmerson, 2007). When the system is afforded significant agency over a piece through the relinquishing of control by the composer/performer, it can be said to be 'generative' (Collins, 2008). Generative systems may demonstrate agency in real-time performance and/or during the composition process. These types of agency have been termed 'performative agency' and 'memetic agency', respectively (Bown et al., 2009).

Allowing the system significant performative agency results in each performance of a work being different. Rather than simply seeking to replicate an idealized version (often the recorded version), unique versions of a work are created in each performance. This sits well with wider ideas of indeterminacy, whereby a piece of music may take on different meanings by different people, or by the same person finding different meanings on repeated listens (Jeongwon and Song, 2002). Allowing the system memetic agency during the composition process results in unexpected system behaviors influencing the fixed aspects of the composition (Waite, 2016). Both the performative and memetic agency of generative systems can bring considerable advantages to the creative process in terms of developing an artist's idiolect through the disruption of familiar creative patterns (Waite, 2014).

Related Work

Interactive-generative processes are used in mobile apps such as *Bloom* (Opal Limited, 2017), where the user instantiates a series of audio-visual objects that interact with one another. Similar processes can be found in generative sequencing tools such as *Push Pong* (Towers, 2014), in which step sequencer cursors collide and bounce off one another to create unpredictable results.

In terms of popular music performance featuring live instruments, systems tend to use live audio recording and playback in a highly controlled way to enable multi-timbral performance of fixed compositions. Such systems may also involve the use of score-following techniques to allow for more variation in the backing track that is possible with standard looping tools. Video recording and playback may be synchronized with the audio recordings in order to reinforce audience understanding and their perception of liveness (Marchini et al., 2017).

A more recent development in the live performance of musical material with lyrics is 'live-writing' (Lee and Essl, 2017), in which text is projected to the audience as it is typed by the performer. Particular letters, words and the rhythm of the typing can be linked to a variety of musical processes and visual effects. This serves to present the act of writing as a real-time process, like that of singing (Waite, 2015).

Real-world metaphors, such as masses and springs and particle systems, have been used as a design strategy to facilitate liveness. Real-world objects provide a shared point of understanding for composers, performers and audience (Johnston, 2013), whether they are grounded in intuitive, physical behaviors or draw on well-known cultural links (Waite, 2016).

Use of Interactive Systems in Popular Music

The use of interactive systems where there is significant agency by the system is not common in popular music. While there are some notable exceptions, such as *Algorave* and *Musebots*, these tend to be limited to blues, jazz and electronic dance music (Bown et al., 2015), which allow for more open forms in performance than song-based music does. As well as the need to adhere to stricter structures than more improvisatory forms (Marchini et al., 2017), there may also be a pressure to reproduce an idealized recording (Cascone, 2002) and therefore minimize risk (Kirn, 2012).

There is also a danger that popular music audiences, who are so used to the playback of audio and video recordings during live performance, may think that the output of an interactive system is actually a fixed backing track. This perception can be so strong that it can persist even when they are told by the performer that the system is live (Biles, 2013). Performing with generative systems that demonstrate a high degree of liveness while adhering to the conventions of popular, song-based music is therefore a significant challenge.

METHOD AND METHODOLOGY

This study is an example of practice-led research (Nelson, 2013). Detailed journals were kept during the composition and rehearsal of the pieces and following public performances. The four approaches arose during the course of the practice, rather than being established at the start. All of the approaches feature significant levels of system agency and a minimum of pre-recorded material, meaning that each performance would be a unique event, with no single idealized version of the work. However, certain structural elements of the compositions were retained in order that they could be identified over repeated performances.

Four types of liveness were then selected to compare the four approaches. These were chosen to incorporate aspects of Sanden's (2013) and Croft's (2007) theories that relate to the live performance of popular music. The four types were as follows:

- Does the system have a presence in time and space? (Temporal-spatial liveness)
- Does the system reveal how it is creating sounds? (Corporeal liveness)
- Is the interaction between human performer and the system evident? (Interactive liveness)
- Is the system's audio-visual output related to the aesthetics of the piece? (Aesthetic liveness)

The systems were realized in *Max* (Cycling '74, 2014) and *Max for Live* (Ableton, 2013). *Max* allows for rapid prototyping of ideas, while *Max for Live* facilitates the creative process by allowing easy access to transport, routing, audio and software instruments in *Live* (Ableton, 2014). *Processing* (Reas and Fry, 2015) was also used for some of the system visuals. As a well-documented C-like programming language aimed at artists, it offers significant efficiency advantages over graphical programming languages for some tasks.

In terms of hardware, a constraint was imposed that the systems should require minimal additional hardware. There are several advantages to this, including minimal additional learning, rehearsal and performance demands on the performer; lower costs; and ease of setup and portability (Richards, 2006).

DISCUSSION OF APPROACHES AND COMPOSITIONS

Interactive-Generative Tool: *Rows, Columns, Collisions* (2012)

The first approach involved creating a complex sequencing tool, with a Novation Launchpad as its user interface. The system's visual outputs, which mirror the behavior of the Launchpad, are shown in Figure 8.1. The functionality of the tool is as follows:

- Pulses moving horizontally across the Launchpad (red squares) are able to create MIDI events themselves and/or act as a playback cursor for a step sequencer, colliding with user-activated squares (small, solid, yellow squares) to create MIDI events (large, hollow, yellow squares).
- Pulses moving vertically across the Launchpad (green squares) could modulate the behavior of the horizontal pulses, create MIDI continuous controller data and/or act as a playback cursor for a step sequencer in the same way as above.
- When horizontal and vertical pulses collide (large, hollow, white squares), further MIDI events are generated.

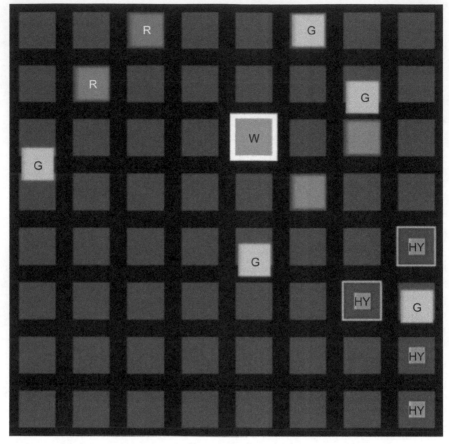

Figure 8.1 System Visuals for *Rows, Columns, Collisions*

Because the speed and direction of each pulse can be set independently, the system's behavior is highly complex. Therefore, while the system is fully under the performer's control, its output is unpredictable and can be considered generative.

The system visuals were designed to highlight the workings of the system to the audience in ways that were not possible on the Launchpad. For example, the motion of the green squares can be set to move gradually rather than in discrete steps in order to represent continuous modulation. Performer actions and collisions were highlighted using additional colors and shapes.

While the system has been used to create and perform an experimental piece of music (National Trevor, 2012), the system has also been used in a popular music context (National Trevor, 2014a). However, this piece involves significantly greater use of pre-sequenced material as opposed to individual note events, which may risk detracting from the overall liveness.

Tool-Led Guitar and Vocal: *Willow* (2014)

The second approach represents an attempt to use interactive systems in a more traditional vocal and guitar performance context (NationalTrevor, 2014c). A suite of devices was created to augment the live guitar performance, including loopers, granular synthesizers, probability-based rhythm generators and spectral processors. These were controlled by real-time performance data from the guitar signal and a mounted accelerometer. A score-following system was implemented in order to modify the behavior of the devices according to the song section.

The system visuals in Figure 8.2 were designed to reveal the system's behavior in terms of inputs, processing and outputs. The sliders at the top show the real-time performance data, with the text box in the top right displaying the current song section. The top three and bottom left boxes represent the looping devices. The bottom central and right boxes represent the outputs of the granular synthesizer and spectral effects, respectively.

While the system was successful in that it could reliably augment the human performance of a fairly traditional song, the system's agency was limited in terms of shaping the composition or performance. A further weakness was the system's visual output. The solution shown in Figure 8.2 is simply an abstracted version of the software itself, which functions primarily as a visual interface for the performer rather than significantly facilitating audience understanding of the system or contributing to the aesthetics of the piece.

Figure 8.2 Studio Performance of *Willow* Showing Performer and System Visuals

Typing Systems: *Kafka-Esque* (2014)

The development of typing systems was a response to the observation that the lyrics of songs are often difficult to follow in the live performance of popular music, particularly when the material is unfamiliar to an audience. Several aspects of a popular music performance compete for attention, such as vocal ability, instrumental ability, physical presence and the presence of system visuals. In *Kafka-Esque* (NationalTrevor, 2014b), typing replaces singing. To reinforce this comparison, samples of sung syllables are triggered to match the sounds of the words being typed, and there is no delete function. Letters are projected to the audience as they are typed, putting the lyrics at the center of the performance (see Figure 8.3).

A further aspect of the piece is the exploration of the musicality of typing gestures. There are strong links between QWERTY keyboards and musical instruments, and daily use of the keyboard results in a degree of virtuosity (Waite, 2015). The use of the keyboard as an input device also satisfies the research criteria of minimal additional hardware. Typing gestures were linked to rhythm capture, generative melody generation and score following, which controlled visual effects. Further interactivity was achieved through a feedback loop between the visuals and synthesizer timbres.

Due to its rhythmic ambiguity and the foregrounding of unpredictable, generative processes, *Kafka-Esque* cannot truly be described as popular music. However, it does feature the live performance of composed lyrics, and makes use of fixed structures. In the near addition, the system could be adapted for compositions featuring a clear pulse and foregrounded melodic motifs.

Figure 8.3 Studio Performance of *Kafka-Esque* Showing Performer and System Visuals

Metaphor Systems: *Church Belles* (2015) and *Piece for Tape* (2017)

The fourth approach marked a second return to guitar and vocal performance. Instead of creating multiple tools, the system was built around a single audio-visual metaphor. As well as facilitating liveness by providing a shared point of understanding for the composer/performer and the audience, the metaphor suggests intuitive possibilities for the system's audio and visual outputs and for how these could be mapped to the live guitar and vocal. By introducing the metaphor at an early stage in the compositional process, the metaphor can be poetic as well as functional and be embedded in the song's themes.

Two pieces were composed using this approach. In *Church Belles* (NationalTrevor, 2015), ten virtual bells (see Figure 8.4) were created in Max, which ring in response to specific notes being played on the guitar, as detected by signal analysis software (Jam Origin, 2017). The force with which the bells ring is determined by the detected velocity of the guitar note. As well as producing a bell sound, each strike of the bell's clapper against the body pitch shifts the live vocal to create complementary backing vocal layers.

As in *Willow*, a score-following system was implemented. When the system detects a change in the guitar part, the constraints on the bells' hinges are removed, enabling them to rotate through 360 degrees. At the same time, the strike messages are connected to a synthesizer in addition to the bell sounds. These new behaviors create a shift in the metaphor

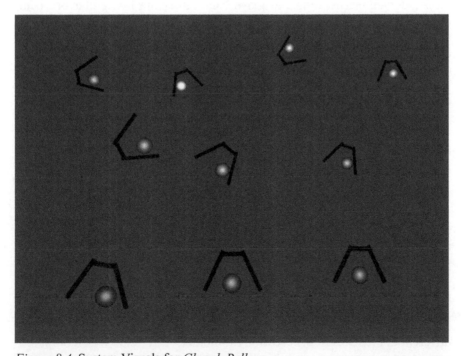

Figure 8.4 System Visuals for *Church Belles*

Figure 8.5 System Visuals for *Piece for Tape*

through a resemblance to air raid sirens. The contrasts and connections between church bells and air raid sirens became a central theme of the piece's lyrics.

Piece for Tape (NationalTrevor, 2017) makes use of the cassette tape as its central metaphor. Each cassette represents an instance of a real-time looper device that records and plays back either the live guitar or vocal (see Figure 8.5). The stopping and starting of recording and playback is accompanied by the playback of samples of actual cassette players, which creates a percussive layer. Again, a score-following mechanism is used to control the behavior of each looper device at particular sections of the song.

As with *Church Belles*, the memetic agency of the system was significant. The selection of the cassette as the central metaphor influenced the inclusion of nostalgia and non-linear time as central lyrical themes. The system also demonstrates significant performative agency. The behavior and output of the loopers is unpredictable, which, coupled with the reflexive aspects of the system, facilitates improvisation during performance.

Like the interactive-generative tool and the typing systems, the metaphor approach allowed the systems to demonstrate significant agency, thereby fulfilling the need for the systems to be truly interactive. They also allow for the composition and performance of music that can most certainly be identified as 'popular', and facilitate the system-building process by suggesting mappings and outputs that are intuitive for both performer and audience.

FINDINGS

Following several performances of each of the aforementioned pieces, the different approaches to the creation of interactive systems were compared according to how successful they were in terms of the four aspects of liveness identified in Section "Definitions and Background". Table 8.1 shows

Table 8.1 Comparison of approaches in terms of demonstrating aspects of liveness

	Interactive-generative tool	Tool-led guitar and vocal	Typing	Metaphor
Spatio-temporal	✓✗	✓	✓✗	✓
Corporeal	✓	✓✗	✓✗	✓
Interactive	✓✗	✓✗	✓✗	✓
Aesthetic	✗	✗	✓	✓

the results. A cross (x) indicates not at all successful, a tick (checkmark) and a cross (x) indicates partially successful, and a tick (checkmark) indicates very successful.

Spatio-Temporal Liveness

In terms of spatio-temporal liveness, all systems were at least partially successful through the use of visuals to endow them with a visible presence. The tool-led guitar and vocal and metaphor approaches were deemed more successful as the system exists and functions separately from the human performer. In the interactive-generative tool and typing approaches, the human performer plays on the system rather than with it, meaning that there is less separation. Here, the system can be considered to be more of a generative tool or instrument rather than a co-performer.

Corporeal Liveness

In terms of corporeal liveness, most of the systems offer clear visual clues to the causes of sonic events. In *Rows, Columns, Collisions*, each sonic event can be linked to the movement of squares or collisions between them. Linking the brightness of the squares to the volume controls provided further clues. However, when used with more highly mediatized material (e.g. triggering drum loops rather than piano notes), the corporeal liveness would be reduced. Also successful is the metaphor approach, which features intuitive causal links based on a shared understanding of familiar, real-world objects. *Willow* and *Kafka-Esque* demonstrate a lower degree of corporeal liveness due to multiple layers of audio being controlled by several simultaneous processes. The attempt to provide visual clues to all of these layers risked information overload in *Willow*, while *Kafka-Esque* avoids this by only providing cues for some of these layers and relying on other aspects of liveness to engage the audience.

All of the pieces composed with the systems begin simply and gradually build, with the intention that the audience can establish causal links for each new layer. Again, this was most successful in the metaphor systems, as the use of a single metaphor limits the number of different simultaneous processes that can be produced by the system. In the case of *Church Belles*, compositional development is achieved through changing the behavior of the simulated objects (the bells become air raid sirens). In *Piece for Tape*, interest is built through gradually adding more instances of the metaphor that intermittently record and play back different parts of the live guitar and vocal at varying speed, direction and duration.

Interactive Liveness

All of the pieces demonstrate interactive liveness to some extent. However, because there is less separation between the human performer and computer system in the interactive-generative tool and typing approaches, inferring interactivity between two separate entities could be more difficult. In terms of demonstrating perceivable interaction between human and system processes, the challenge is to demonstrate that the human and machine performers are in some kind of dialogue, rather than simply allowing algorithmic processes to play out or perform a fixed instrumental or vocal part that is unaffected by the system's real-time behavior. *Willow* is perhaps the least successful in this regard, with the system mostly being limited to following the human performer and providing ornamentation. In parts of both *Rows, Columns, Collisions* and *Kafka-Esque*, the autonomy of the system enables the human performer to stop and listen (and demonstrate to the audience that they are listening), before responding. In *Church Belles*, the interaction can be heard through the real-time pitch-shifting of the live vocal. In *Piece for Tape*, there is a section where the performer improvises with the system in a conversational style.

Aesthetic Liveness

There was considerable variation between systems in terms of aesthetic liveness. In *Rows, Columns, Collisions*, the system was successfully represented in the visuals. However, because the piece was essentially about the system, there was little to reveal in terms of deeper meaning. The visuals of *Willow* were effectively the user interface, and a much higher level of design would be required to offer any significant aesthetic value. *Kafka-Esque* demonstrates significant aesthetic liveness through the careful selection of sounds and images that relate directly to the meaning of the work. The metaphor systems are also successful in this regard, as the metaphor determines the functioning, the look and the sound of the system as well as being intrinsic to the meaning of the song. In both the typing and the metaphor approaches, aesthetic liveness offers the audience an understanding not just of the system but also of the composition itself.

CONCLUSIONS AND RECOMMENDATIONS

This portfolio of work has demonstrated how interactive systems can be used for the live performance of popular music. The pieces and performances produced complement similar work due to their applicability to popular music (Bown et al., 2015), levels of interactivity (Bown et al., 2009) and/or overall liveness (Sanden, 2013; Croft, 2007). The portfolio also demonstrates how the use of interactive systems can develop an artist's range of expression (Waite, 2014) and result in each performance of the work being a unique version (Jeongwon and Song, 2002).

Of the four approaches discussed, the metaphor approach was the most successful overall against the identified liveness criteria. The interactive-generative tools and typing approaches were not as successful due to the performer controlling them directly rather than performing alongside them. In the tool-led guitar and vocal approach, the visuals did not successfully reveal the system behavior in a simple and coherent way, nor did they communicate any of the themes behind the piece itself.

The metaphor approach was also the most useful in terms of creating pieces that could be identified as popular music while allowing the interactive system significant agency. The interactive-generative tools and typing approaches resulted in more experimental material, while the tool-led guitar and vocal approach involved system behaviors that were controlled by instrumental input rather than generative processes. It should also be noted that although the metaphor approach was highly applicable to popular music, the durations of the pieces were significantly extended beyond the usual three to four minutes in order to allow space for system agency and to communicate liveness by gradually revealing how the system works to the audience.

The downside to the metaphor approach is that the simulation of real-world objects and their behaviors is not applicable to every compositional situation. While the tool-led guitar and vocal approach of *Willow* offers more flexibility while maintaining the applicability to popular music, demonstrating liveness is more challenging. Future work with this type of system could examine a more nuanced approach in terms of the aesthetics of the system visuals and audio outputs that, like *Kafka-Esque*, are unique to the work being performed and link closely to underlying themes. More use could be made of gradual increases in complexity (as in *Rows, Columns, Collisions* and *Piece for Tape*) or dynamic mappings (as in *Church Belles*).

Though practice-led methodology (Nelson, 2013) was integral to the identification, implementation and evaluation of the four approaches, further work could involve audience studies. Though feedback from peers and audiences was actively sought as part of the creative process, additional insights could potentially be gained by comparing the responses of audiences to the different approaches by using similar methodologies to Bin et al. (2016) and Correia et al. (2017).

Finally, it is worth looking again at why liveness is important. Through the use of a broad concept of liveness (Sanden, 2013; Croft, 2007), this study and the accompanying collection of pieces aim to show that interactive systems can offer more to composers than new modes of expression and more to audiences than an understanding of technological processes. In keeping with current discussions on aura (Cascone, 2002) and indeterminacy (Jeongwon and Song, 2002), interactive systems allow performers to create a unique version of an enduring work in each performance, while enabling audiences to more deeply understand the ideas behind the composition itself.

REFERENCES

Ableton. (2013). *Ableton live*. Available at: www.ableton.com/en/live/what-is-live/ [Accessed 12 Jan. 2018].

Ableton. (2014). *Max for live*. Available at: www.ableton.com/en/live/max-for-live/ [Accessed 12 Jan. 2018].

Auslander, P. (2008). *Liveness: Performance in a mediatized culture*. 2nd ed. London: Routledge.

Biles, J. (2013). Performing with technology: Lessons learned from the Gen-Jam project. In: *Musical metacreation: Papers from the 2013 AIIDE workshop* [online]. Boston, MA. Available at: www.researchgate.net/publication/272181877_Performing_with_Technology_Lessons_Learned_from_the_GenJam_Project [Accessed 12 Jan. 2018].

Bin, S. A., Bryan-Kinns, N. and McPherson, A. (2016). Skip the pre-concert demo: How technical familiarity and musical style affect audience response. In: *Proceedings of NIME 2016* [online]. Brisbane, Australia. Available at: www.academia.edu/26910541/Skip_the_Pre-Concert_Demo_How_Technical_Familiarity_and_Musical_Style_Affect_Audience_Response [Accessed 26 June 2017].

Bown, O., Bell, R. and Parkinson, A. (2014). Examining the perception of liveness and activity in laptop music: Listeners' inference about what the performer is doing from the audio alone. In: *Proceedings of NIME 2014* [online]. London. Available at: http://nime2014.org/proceedings/papers/538_paper.pdf [Accessed 10 June 2014].

Bown, O., Carey, B. and Eigenfeldt, A. (2015). Manifesto for a musebot ensemble: A platform for live interactive performance between multiple autonomous musical agents. In: *Proceedings of the 21st international symposium on electronic art* [online]. Vancouver, Canada. Available at: www.researchgate.net/profile/Arne_Eigenfeldt/publication/281441670_Manifesto_for_a_Musebot_Ensemble_A_platform_for_live_interactive_performance_between_multiple_autonomous_musical_agents/links/55e71ee708ae65b638994fbf.pdf [Accessed 11 Jan. 2016].

Bown, O., Eldridge, A. and McCormack, J. (2009). Understanding interaction in contemporary digital music: From instruments to behavioural objects. *Organised Sound*, 14(2), pp. 188–196.

Cascone, K. (2002). Laptop music—counterfeiting aura in the age of infinite reproduction. *Parachute Contemporary Art*, (107), pp. 52–60.

Collins, N. (2008). The analysis of generative music programs. *Organised Sound*, 13(3), pp. 237–248.

Correia, N. N., Castro, D. and Tanaka, A. (2017). The role of live visuals in audience understanding of electronic music performances. In: *Proceedings of audio mostly* [online]. Goldsmiths, University of London. Available at: http://research.gold.ac.uk/20878/ [Accessed 6 Sept. 2017].

Croft, J. (2007). Theses on liveness. *Organised Sound*, 12(1), pp. 59–66.

Cycling '74. (2014). *Max software tools for media | cycling '74* [online]. Available at: https://cycling74.com/products/max/ [Accessed 12 Jan. 2018].

Demers, J. (2010). *Listening through the noise: The aesthetics of experimental electronic music*. Oxford: Oxford University Press.

Emmerson, S. (2007). *Living electronic music*. Aldershot: Ashgate Publishing, Ltd.

Fels, S., Gadd, A. and Mulder, A. (2002). Mapping transparency through metaphor: Towards more expressive musical instruments. *Organised Sound*, 7(2), pp. 109–126.

Jam Origin. (2017). *MIDI guitar* [online]. Available at: www.jamorigin.com/ [Accessed 12 Jan. 2018].

Jeongwon, J. and Song, H. S. (2002). Roland Barthes' "Text" and aleatoric music: Is "The Birth of the Reader" the birth of the listener? *Muzikologija*, 2, pp. 263–281.

Johnston, A. (2013). Fluid simulation as full body audio-visual instrument. In: *Proceedings of NIME 2013*. Daejeon, South Korea, pp. 132–135.

Kirn, P. (2012). Deadmau5, honest about his own press-play sets, misses out on "Scene". *CDM Create Digital Music* [online]. Available at: http://cdm. link/2012/06/deadmau5-honest-about-his-own-press-play-sets-misses-out-on-scene/ [Accessed 6 Sept. 2017].

Lee, S. W. and Essl, G. (2017). Live writing: Gloomy streets. In: *Proceedings of the 2017 CHI conference extended abstracts on human factors in computing systems*. CHI EA '17 [online]. New York: ACM, pp. 1387–1392. Available at: http://doi.acm.org/10.1145/3027063.3052545 [Accessed 15 July 2017].

Marchini, M., Pachet, F. and Carré, B. (2017). Rethinking reflexive looper for structured pop music. In: *Proceedings of NIME 2017* [online]. Copenhagen, Denmark, pp. 139–144. Available at: www.nime.org/proceedings/2017/ nime2017_paper0027.pdf [Accessed 9 June 2017].

National Trevor. (2012). *Rows, columns, collisions* [online video]. Available at: https://vimeo.com/54929941 [Accessed 25 Aug. 2017].

National Trevor. (2014a). *I begin where you end* [online]. Available at: https:// soundcloud.com/national-trevor/ibeginwhereyouend [Accessed 12 Jan. 2018].

National Trevor. (2014b). *Kafka-Esque (Live studio version)* [online video]. 11 July. Available at: www.youtube.com/watch?v=PIHl85fR068 [Accessed 25 Aug. 2017].

National Trevor. (2014c). *Willow (Live studio version)* [online video]. 9 July. Available at: www.youtube.com/watch?v=OloPjS0Q06U [Accessed 25 Aug. 2017].

National Trevor. (2015). *Church Belles* [online video]. Available at: www.youtube.com/watch?v=mDGnFQj1MUw&feature=youtu.be [Accessed 25 Aug. 2017].

National Trevor. (2017). *Piece for tape* [online video]. 1 May. Available at: www. youtube.com/watch?v=3Lne3h2Qhms [Accessed 25 Aug. 2017].

Nelson, R. (2013). *Practice as research in the arts: Principles, protocols, pedagogies, resistances*. Basingstoke: Palgrave Macmillan.

Noble, J. (2009). *Programming interactivity: A designer's guide to processing, arduino, and openFrameworks*. Cambridge: O'Reilly Media.

Opal Limited. (2017). *Bloom*. Available at: https://itunes.apple.com/us/app/ bloom/id292792586?mt=8 [Accessed 21 July 2017].

Prior, N. (2009). Software sequencers and cyborg singers: Popular music in the digital hypermodern. *New Formations*, 66(1), pp. 81–99.

Reas, C. and Fry, B. (2015). *Processing*. Available at: https://processing.org/ [Accessed 12 Jan. 2018].

Reeves, S., Benford, S., O'Malley, C. and Fraser, M. (2005). Designing the spectator experience. In: *Proceedings of the SIGCHI conference on human factors in computing systems*. CHI '05 [online]. New York: ACM, pp. 741–750. Available at: http://doi.acm.org/10.1145/1054972.1055074 [Accessed 7 Dec. 2015].

Richards, J. (2006). 32Kg: Performance systems for a post-digital age. In: *Proceedings of NIME 2006* [online]. Paris, France, pp. 283–287. Available at: www.nime.org/proceedings/2006/nime2006_283.pdf [Accessed 24 Apr. 2014].

Sanden, P. (2013). *Liveness in modern music: Musicians, technology, and the perception of performance*. New York: Routledge.

Towers, M. (2014). *Push pong* [online]. Available at: https://marktowers.net/2014/08/20/push-pong/ [Accessed 12 Jan. 2018].

Waite, S. (2014). Sensation and control: Indeterminate approaches in popular music. *Leonardo Music Journal*, 24, pp. 78–79.

Waite, S. (2015). Reimagining the computer keyboard as a musical interface. In: *Proceedings of NIME 2015* [online]. Baton Rouge, pp. 168–169. Available at: www.nime.org/proceedings/2015/nime2015_193.pdf [Accessed 15 Dec. 2016].

Waite, S. (2016). Church Belles: An interactive system and composition using real-world metaphors. In: *Proceedings of NIME 2016* [online]. Brisbane, Australia, pp. 265–270. Available at: www.nime.org/proceedings/2016/nime2016_paper0052.pdf [Accessed 15 Dec. 2016].

9

How Algorithmic Composition Prompts Innovative Movement Design in Full-Body Taiko Drumming

Stu Lambert

INTRODUCTION

In the composition for taiko drums *Times of High Water (ToHW)*, a simple arithmetical rule varies the relative positions of a drummer's hands on five strike points across two drums, while playing a two-beat rhythm, alternating right and left hands. The rhythms are not sonically innovative, but the demands of playing algorithmically designed rhythm patterns, in a form that uses the whole body in striking a drum, prompted innovative movements to solve positioning problems.

Beyond merely managing to strike as directed by the algorithm, the intention is to explore taiko's technical standards and aesthetic framework in a discipline of the body. These are called *kata*—shape or form. It was considering kata that generated most movement innovation, more than playability alone.

Algorithms give near-instant variations. The process of learning to articulate each rhythm is differently demanding as it lacks the progressive familiarization with the task that human composing allows. In *ToHW*, very simple algorithms produced over 20 rhythmic melodies. Three or four have not been used due to difficulty of articulation, but most of them are very pleasing and characterful.

In the realm of the mind then, the rhythms must be memorable, until memory goes into the body. The spirit must find them musically satisfying. To be more than an exercise, they must sequence and combine pleasingly—in general and in a taiko style.

The sound the melodies made and the sweeping progression of the movement conjured association with tides, which influenced design of the vertical element of movement to use the forms of waves, backwashes and tidal changes.

The majority of scholarly work in English on taiko concerns its history and culture (e.g. Bender, 2012; Hennessy, 2005). There is some documenting of different regional playing styles, but little has been written about the strategies and values of performance. This work intends to strengthen discussion of these, which usually happens in rehearsals, workshops and master classes, more commonly than in the scholarly domain.

BACKGROUND—THE MUSIC THAT MADE THE MOVEMENT

The terms used for the strike points are taken from points of the compass, from the drummer's point of view (Figure 9.1).

ToHW uses three kinds of rhythmic melodies in which the sticks, *bachi*, traverse the drums to strike five points across the two drums—the head of the first drum and the front rim, then the front rim of the left drum, its head and the rim at the rear.

The melodies are all played Right, Left, Right, Left continuously in a swung, 'heartbeat' rhythm, known to taiko players as *Shichisan* or Don Go.

Usually, taiko drumming is transmitted in vocal syllables, called *kuchi shoga*—'mouth song'. Kuchi shoga sings the 'melody'—the sound that will play—and sometimes indicates which hand will play. It does not give position—where the strike will happen. Kuchi shoga mainly directs strikes and melodies on one drum.

The two-drum setup requires a usable way to refer to the strike *positions*, rather than their sounds. Using the compass, with the drummer at North, gave a position indicator that is easy to identify, remember and notate (see Table 9.1). The extreme right strike is named East, through to West at the far left.

This method is only useful in step-time learning, not for real-time performance. This too makes it different to kuchi shoga which is performed in real time and gives a good indication of how the drum performance will sound.

Eighteen different parts have been recorded on video, though most are unpolished, for teaching purposes, and are private. Some developing performances of the melodies can be seen and heard in Lambert 2018a.

The 'lead' bachi traverses the strike points immediately, in order from SE. The 'following' bachi plays a variable number of strikes on E before traversing.

The three types of melody are a single measure, which makes four variations of melodies, a double measure, which makes eight usable melodies, and a single measure of three eighth notes per beat to make six melodies.

MUSIC AND MOVEMENT

Composing by predetermined rules means that the rhythms do not refer to the body's abilities to play them. While this can happen from scoring on paper or composing on a DAW, taiko's strong use of space for the body in performance—a taiko player at a drum uses a performance frame of about a 2.5-meter cube—amplifies both the challenges of drumming with

Table 9.1 Two-Measure Melody With One Repeat on East

1	2	3	4	5	6	7	8	9	10	11	12	13	14	15	16
R	L	R	L	R	L	R	L	R	L	R	L	R	L	R	L
E	SE	E	S	SE	SW	S	W	SW	SW	W	S	SW	SE	S	E

arbitrary movements: the challenge of articulation and the challenge of aesthetics. The discourse of movement in taiko is kata.

Kata means form or shape, or way of doing. The term is used in karate, taiko, management studies and coaching (Systems2Win), and computer coding (CodeKata, 2014). It often refers to form in movement—floor exercises in a martial arts move, types of gesture in drumming.

In taiko, 'strict' kata proposes that drummers playing the same part will, in any instant, occupy space in exactly the same way from the soles of their feet to the crowns of their heads and the tips of their fingers (allowing for the differences in their bodies). At any moment the shapes the drummers make should be identical (see Figure 9.2). Press the pause button on a video clip and all angles of the bachi in the hands, the angles of the arms and the positioning of the body between earth and sky should be as one for those playing the same part. Similarly, a single drummer repeating a move should move identically each time. This is both because there is a best way to make the move and because schools, or styles, within a discipline will select, specify and develop moves. Each taiko group has its kata; some place more value on funkiness than precision, some are very athletic, some are meticulously traditional and so on.

Having mastered the rhythms to the point of having each bachi in the right place at the right time and beginning to move the sequences into muscle memory, the aesthetic challenges given by the arbitrary strike positions could be addressed.

The moves between the rims of the drums, SE to S and S to SE, give a challenge because the drummer can strike them without needing to lift. The strikes can be made by moving the arm mainly horizontally, between the rims. This means moving the body sideways, which is done in taiko, in styles where the drum head is vertical to the drummer, (e.g. Miyake style), so that the energy needed is lateral. In Times of High Water's 'beta' style, the drum head is (almost) horizontal, so the body should move vertically downwards. This is slightly less true for rim strikes, which are executed around 30 degrees off vertical, away from the center of the drum. The horizontal movement here does not 'feel like taiko'—the strike is not being delivered as well as possible.

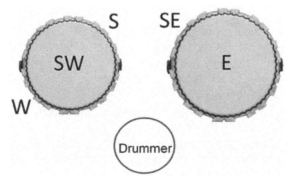

Figure 9.1 Strike Points on Two Drums, Named as Compass Points

The visual aesthetic effect of lateral movement is to flatten the look of the performance by restricting vertical movement in the performance frame. The frame is approximately bounded by the curves described by the bachi when the arms move forward, fully extended, from the height of the drum, outwards to nearly vertical with the hands above the shoulders. The horizontal dimension of the frame might be extended by shifting weight left or right while remaining balanced and, more rarely, the frame can be enlarged by a step or a jump. The horizontal movement in this style does not look like taiko—it occupies very little of the frame in either dimension. (Styles on a slanted or vertical drum head strongly occupy the horizontal dimension.)

For strikes to the drum heads, sideways movement on flat drums produces glancing strikes. These lack some power and also sound less full—rather like playing a stringed instrument near its bridge. It is also more likely that the strike point will vary more in distance from the center of the drum than vertical strikes, which alters the tone. The horizontal movement here does not sound like taiko.

Thus, what *ToHW* needs is a technique which enables a vertical drop while working across the horizontal space from E to W on the two drums. This is similar to a technique taught in a drill composed by Oliver Kirby of Kagemusha Taiko, though this drill separates the move into a vertical move followed by a horizontal move, so that the player is only managing one dimension at a time. To play SE, S in *ToHW* using that method, the drummer strikes SE, raises the arm above SE to the top of the frame in use, moves sideways to above S and drops to strike S. This does answer all of the reservations above, and each strike is optimal, but moving the arm in a series of right angles was not aesthetically pleasing for this piece.

Following the strike with a curving sweep up to the top of the frame in use above the next strike point improved the vertical drop. These shapes, characterized as similar to cursive 'j' and a curved 'i', then as like shapes of waves, then applied similarly for many other strikes, apart from E, which is always down, and some strikes returning from W towards E in the two-measure melodies.

Working against a resistance gives more precision to the arrival at the top of the frame in use: to finish a SE beat in the direction of the next beat on S gives too much energy and an unpleasing line to the top of the lift. Curving away, towards stage right, and back in as the arm rises controls the energy and gives a more airy top to the kata, like a wave crest.

These strategies are not the product of personal design preference or the aesthetic layer of kata. They come from the kata for the body and are felt to be as imperative as standards of construction are to architecture—though there may be some subjectivity in deciding what is best for the body. It is the intention to obey the body imperative that precipitates innovation. Merely executing the rhythmic melodies would not require it.

The curving movements give new challenges: the kata becomes multidimensional, compared to any linear movement including diagonal movement. Within a solo performance, it is harder to be consistent because one is moving in two dimensions at once. In a group, there is greater chance of difference between the kata of the drummers due to their shades of interpretation of the shape of the curve. This is a crucial consideration: like dance or karate, if it looks ragged it isn't adequate.

This difficulty is acknowledged in Times of High Water and, rather than being designed out, is deemed an opportunity to study multidimensional kata, still aspiring to the standard of kata described earlier. This is gained by an understanding of the curve's ideal shape, generated by the ideal shape of the body that is making it and shared understanding of that among the drummers.

In two-measure melodies with less than four strikes on E before the right bachi follows, the hands come close together as the left hand returns while the right hand is still 'outbound', as in beats 8–11 in Figure 9.2, using a stroke like a tennis backhand with the right hand on W, rising to the top of frame to strike SW (lack of height into W is not very important as it is a rim strike), which has not been seen in taiko to date.

Gaining height to meet objective standards in taiko also soothes a subjective sense of intentional, visual and tonal compromise in a multi-drum setup in taiko. Perhaps because the huge majority of taiko has one drummer at one drum, seeing a player with two or three drums makes a shift in feeling about the performance. The performance starts to be drumming, 'pakashon', not taiko as Shawn Bender describes in *Taiko Boom* (Bender, 2012): "Hitting the top of the drum quickly with the top of the hands (fingers) and the top of the body (arms and shoulders) is emblematic of 'pakashon', a term used generically to refer to Western-style drumming". Executing a perfect strike has become subject to moving to the right place, whereas on a single drum, that place is almost a constant. A close

Figure 9.2 Sun Lotus Taiko Showing Good Kata (Lambert, 2018b). Image supplied by author.

comparison is seeing a guitarist play a double-neck electric guitar. It instantly changes an expectation of what the intention of the performance will be, how the performance will look and particularly what movement it will contain, compared to a conventional solid guitar.

The change may also shape expectations of what the music will be. Seeing a multi-drum setup in taiko may lead to an expectation of a performance which places less value on kata than on technique and, because of the need for lateral movement discussed above, strikes will come more from the top of the body than from the core—pakashon. The taiko performer begins to become 'just a drummer' like any other drummer, as the unity between the player and the drum is impaired.

The kata design of Times of High Water occupies the vertical dimension of the performance frame and refers strongly to the root principles of taiko, while working over two drums and extended strike points.

As well as kata, a further measurement of movement in taiko is 'Ma'. Rolling Thunder's *Taiko Resource* glossary defines *ma* as:

> The space between two events (two notes or beats on the drum, etc.). Somewhat equivalent to a rest in Western notation, but with a deeper connotation than mere absence of sound. Ma is just as important as the notes that surround it, giving shape and contrast to the sounds that we hear.
>
> (Rolling Thunder, 2000)

The example of a musical rest makes ma concerned with time in this definition and the quote also refers to sound, in absence, as a relevant quality.

The usage in taiko playing and discussions is not sonic but visual: it is used of stillness at the end of a movement, a holding of position for emphasis. This corresponds well with WAWAZA's example of ma:

> when Japanese are taught to bow in early age, they are told to make a deliberate pause at the end of the bow before they come back up—as to make sure there is enough MA in their bow for it to have meaning and look respectful.

WAWAZA's discussion of ma also refers to the space between the notes in music and to negative space as an art concept.

Ma's emphatic moment of stillness tests the kata of the group: by pressing the pause button in real life, they allow the audience to appraise how close the group's shared understanding of the kata is. (The group cannot usually appraise this, as only one or two players at most will be in their field of vision.)

Gunter Nitschke, in *Kyoto Journal*, translates ma as 'place', wishing to get away from the rendering of ma as 'imaginary space' (Nitchke, 1988). Nitschke instances the use of ma for defining physical area, such as the number of tatami mats used to measure the size of a room in a Japanese house. This harmonizes with the use of ma in Nichiren Buddhist practice: the Butsuma, or

Buddha-space, is the room in the house where devotion is performed. This is not a negative space; it is a defined and framed space with a purpose.

The term *ma*, in taiko, could be used to refer to the performance frame. It is the place in which kata happens. It is a three-dimensional space, as is Nitschke's interpretation, though the majority of taiko performance is seen in the 2D frame. Wanting to preserve and extend the idea of the 'examined moment' from other interpretations, one may think of ma as the place that kata has taken the drummer to at any time, as in the pause-button analogy.

Returning to the original definition of ma as concerning time, the time gap between moments of place is supplied by the Buddhist concept of *ichinen sanzen*—"3,000 realms in a single moment of life. This gives a dynamic mix of 3,000 named types of effect observable in any instant. I can intuitively attend to these effects at any moment when drumming, created in the individual, in other beings and in the physical world" (Soka Gakkai International, 2012). The unit of time in ma, then, becomes *ichinen*—a single moment, the present moment attended to mindfully (Anupadin, 2014). Though many of the 3,000 possible life-states may not change moment-to-moment while drumming, the framework is rewarding, encompassing one's life-state when executing a strike, a song or a show, as well as effects on listeners—voluntary or involuntary, evident or latent, and the interactions with the acoustic and physical environment in a form which is frequently played outdoors.

THREE THOUSAND REALMS IN A SINGLE MOMENT OF LIFE (ICHINEN SANZEN)

The link with Buddhism was born in the author's own practice and observation of the similarities of development and values between taiko and Nichiren Buddhism in Japan. Bender notes the extensive links between

Ten Worlds		Ten Factors	Three Realms
Buddhahood		Appearance	
Boddhisatva		Nature	Self
Realization		Entity	
Learning	Mutual	Power	Living Beings
Heaven/rapture	Possession	Influence	
Humanity/tranquility	of the Ten	Internal cause	Environment
Anger	Worlds	Relation	
Animality		Latent effect	
Hunger		Manifest effect	
Hell		Consistency from beginning to end	

Figure 9.3 Ichinen Sanzen

taiko and Buddhism and gives a specific instance of a link to Nichiren Buddhism (Bender, 2012). These links are also present in taiko's diaspora to the US West Coast (Nen Daiko, *inter alia*).

MUSIC SHAPED BY MOVEMENT

Times of High Water uses a high proportion of rim strikes, compared to skin strikes. Bachi make a pitched sound on rims—they are like big claves. (On strikes to the drum head, the bachi don't sound detectably.) Changing the place on the bachi that strikes the rim changes the pitch produced. There are three 'agreeable' pitches available from a bachi—High, Mid and Low—that can be used systematically in composition. To emphasize this, an amplification of a technique from a teaching of Shoji Kameda keeps the hand as open as possible when the bachi strikes, ideally leaving the bachi momentarily in free space at the moment of the strike and catching it. This allows the bachi to sound strongly.

ToHW thus has a complexity of pitches for a piece on two drums, as the rim strikes are carried out on each drum, with left and right bachi, on front and rear rim (SE, S and W) and returning towards the East with a different kata, all of which can vary the strike point on the bachi and thus the pitch.

Making the melody mostly on the rims is unusual in taiko. Pitching rim strikes has not been discussed in taiko practice or literature, nor has it been evident in shows and videos. Taiko drummers probably do attend to it, intuitively and consciously, but developing pitched strikes more consistently into the fabric of composition is innovative.

APPLICATION BEYOND TAIKO

Striving for beauty of form in *Times of High Water* has developed precision and grace of arm gestures, in free space, in two dimensions, in music. This extended study of kata could suggest taiko as a beneficial complementary discipline for musicians using gesture to control electronic and virtual instruments, such as the Mi-Mu gloves used by Chagall van den Berg at the University of Westminster and at TEDx-AmsterdamWomen 2016 to control sounds and sound effects in a computer.

While there has been much written about adding performance values to electronic music through gesture—Bob Ostertag wrote: "I think most musicians working with electronics are probably not very satisfied with the state of electronic music today, and the crucial missing element is the body" (Ostertag, 2002)—the preoccupation has been with interface design or the meaning of gesture (e.g. Bahn et al., 2001), which explores: "the design of the gestural controller itself; the choices of sounds, synthesis and digital signal processing methods for a particular performance and the integration of new sonic display systems in the performance feedback loop". Little has been written about physically beneficial technique or about the formalization of gesture as a means of control, so that there is a language of gesture *as* control.

At the Innovation in Music Conference 2017, Jenn Kirby, in her presentation The Performance of Electronic Music, spoke of aches and pains from making gestures in the air to simulate playing an instrument, with sensors and virtual instruments creating sound. This is unsurprising, as the support of the wood and the resistance of the wire that are intrinsic to the experience of playing, say, a guitar or violin, with non-harmful technique are missing, as are the centuries of development of playing techniques that are sustainable for the body.

Taiko first asks of movement how it can be done with best use of the body. Most work is done by the lower body and the core, unlike most Western playing's concentration on the weaker parts such as fingers, elbows, biceps and shoulders. All kata is resolved by bringing the body to rest, in balance. Precise replicability of a movement in free space is as essential to taiko as precise replicability of movement on a keyboard or fretboard is to music. This replicability is presumably required when using controllers in free space, but performers may lack a supporting body discipline for it. Once 'objective' kata is achieved, there is the opportunity to design gesture to achieve aesthetic or communication goals and to create a language of gesture as control so that performers may use a new controller with the ease with which they might switch from guitar to banjo.

CONCLUSION

In this chapter, innovations in movement for taiko on two drums are the physical outcomes. Multi-drum setups are becoming more popular in taiko, with some risk of losing or changing the definition of the form. By fully occupying *taiko-ma*, a space in which taiko is done, the critique of 'pakashon' in a two-drum setup is answered. The impetus for innovation is ascribed to largely objective considerations of form when adding a new movement idea; innovation is not done by choice, but by need to meet the aesthetic.

The added strike point of W, the algorithms of composition and the kata of performance have combined to make a piece that has a high level of pitched information from one drummer, compared to what's usual taiko.

The chapter makes statements of definition of kata in taiko, formalizing general ideas expressed in taiko discourse. The definition of ma is developed and recast as, or returned to, a measure of occupied space rather than of time or of negative space. A notional unit of time that sets points for the appraisal of kata (in ma) is taken from Buddhist thinking and ascribed 3,000 types of effect as an appraisal framework.

Times of High Water was not intended to innovate; it was intended to summon *ki*, spiritual energy or joy, in drummers and audiences. Innovating has slowed development of the work but has enriched its performance values to an extent unimaginable when the melodies of the algorithms were first played.

The next development for *ToHW* is to decide when the melodies are to be the melodic part or the rhythmic part in the piece. The melodies have not been called rhythms because the same beat plays all the time, though pitch variation makes them tuneful. This defines these melodies as baseline rhythms, *ji-uchi*, that give the pulse over which melodies are usually played and don't use much of the performance frame. Their kata, though, gives the melodies strong visual performance values, like conventional top-lines.

There is then the question of whether to use only algorithms. So far, this has produced pleasing polyrhythms but lacks dynamic variation. An algorithm that dictates where a rest occurs might give this variation. The alternative is to compose in response to the melodies, writing sound and movement for the tidal metaphor as well.

REFERENCES

Anupadin. (2014). *The search for the meaning of life.* Available at: https://anupadin.com/tag/ichinen-sanzen/ [Accessed 15 Jan. 2018].

Bahn, C., Hahn, T. and Trueman, D. (2001). Physicality and feedback: A focus on the body in the performance of electronic music. *International Computer Association,* 2001.

Bender, S. (2012). *Taiko boom.* Berkeley: University of California Press, p. 126.

CodeKata. (2014). Available at: http://codekata.com/ [Accessed Oct. 2017].

Hennessy, S. (2005). Taiko Southwest: Developing a "New" musical tradition in English schools. *International Journal of Music Education,* 23(3), pp. 217–226.

Lambert, S. (2018a). *Times of high water (playlist).* Available at: www.youtube.com/playlist?list=PLAHG0t8zQVcdgIuZJa9DCYyuW4MhGBRf2 [Accessed 30 May 2018].

Lambert, S. (2018b). Sun Lotus Taiko showing kata.

Nen Daiko. *Buddhism and Taiko.* Available at: http://nendaiko.weebly.com/buddhism-and-taiko.html [Accessed Oct. 2017].

Nitchke, G. (1988). Ma: Place, space, void. *Kyoto Journal,* 8. Available at: www.kyotojournal.org/the-journal/culture-arts/ma-place-space-void/ [Accessed 10 Oct. 2017].

Ostertag, B. (2002). Human bodies, computer music. *Leonardo Music Journal,* 12, pp. 11–14.

Rolling Thunder. (2000). *Taiko glossary.* Available at: www.taiko.com/taiko_resource/history/glossary.html [Accessed 11 Jan. 2018].

Soka Gakkai International. (2012). *SGI quarterly.* Available at: www.sgi.org/about-us/buddhism-in-daily-life/three-thousand-realms-in-a-single-moment-of-life.html [Accessed 25 Oct. 2017].

Systems2Win. *Kata Lean coaching cycle*. Available at: www.systems2win.com/
LK/lean/kata.htm [Accessed 10 Jan. 2018].

Wawaza. *When less is more: Concept of Japanese "MA"*. Available at: https://
wawaza.com/pages/when-less-is-more-the-concept-of-japanese-ma.html
[Accessed 12 Jan. 2018].

Part Two

Production

Production

10

Collective Creativity

A 'Service' Model of Commercial Pop Music Production at PWL in the 1980s

Paul Thompson and Phil Harding

The following study explores the creative co-production workflow system of Mike Stock, Matt Aitken and Pete Waterman (SAW) at Pete Waterman Ltd. (PWL) Studios during the 1980s, focusing on the agency and influence of Pete Waterman within the process as the production team worked within the creative system of pop music making.

INTRODUCTION

Pop music's crucial characteristic is that "it is music produced commercially for profit, as a matter of enterprise not art" (Frith, 2001, p. 94). Because it is so overtly commercially driven, pop music is typically viewed as derivative, rather than innovative, following stylistic and musical trends of the time. In musical and economic terms, pop music is "defined by its general accessibility, its commercial orientation, an emphasis on memorable hooks or choruses, and a lyrical preoccupation with romantic love as a theme" (Shuker, 2001, p. 96). Simon Frith argues that pop music "is about providing popular tunes and clichés in which to express commonplace feelings—love, loss, and jealousy" (2001, p. 96). In commercial pop music, the artist takes center stage and the songwriters and producers, who are often driving the pop production process, remain behind the scenes. However, because "much of pop is regarded as disposable, for the moment, dance music" (Shuker, 2001, p. 201), there has been very limited attention given to the scholarly study of pop music production from the 1980s onwards with only a handful of academic studies in this area (e.g. O'Hare, 2009).

In his introduction to *The Art of Record Production: An Introductory Reader for a New Academic Field* (Frith and Zagorski-Thomas, 2012), Frith proposed that producers in pop and dance music genres have a significantly different role from music producers in other music genres such as rock. In commercial rock music, the producer is often a single person who typically has creative, financial and logistic control over the production process and acts as the intermediary between the artist, the engineer and the record company. Often, in rock music, the artist or band has written the song to be recorded, and it is the producer's job to help them present it in the best way possible. Their central role in the process is to guide the production, elicit a performance from the artist or musicians and provide support, guidance and knowledge throughout. A prominent difference between rock and pop music is that pop producers are often part of a larger production team involving direct and ongoing participation with songwriters, programmers, musicians, artists, management and record company representatives. Pop music songwriting and production teams are therefore more frequently part of a larger 'creative collective' (Hennion, 1990) in creating a musical product.

One of the most prominent and commercially successful production teams of the 1980s were Mike Stock, Matt Aitken and Pete Waterman (SAW), who adopted a complete in-house co-production approach (songwriting, recording and production) and employed artists to simply sing on the records. SAW's embrace of the label 'Hit Factory' and their continued chart successes led to criticisms of manufactured-style music that was the result of a production line (Napier-Bell, 2001). However, as Peter O'Hare argues: "the fact was that this means of production was highly efficient and commercially successful" (2009, online). The following chapter presents this creative production workflow system at Pete Waterman Ltd. (PWL) Studios during the 1980s, focusing on the agency and influence of Pete Waterman within the production team. Through analysis of the creative system of pop music making, a pop music 'service' model has been developed, which highlights collectivist rather than individualist thinking, and underlines Pete Waterman's agency at various stages of the commercial pop songwriting and production process.

METHODOLOGY

This study draws upon a series of semi-structured interviews with programmer and producer at PWL Studios Ian Curnow, artist manager for Pet Shop Boys during the 1980s Tom Watkins, pop journalist Matthew Lindsey and songwriter John McLaughlin. Interviews were conducted from May 2014 to December 2015 and focused upon the practices and processes within PWL from 1985 to 1992. The interviews were recorded using a Dictaphone and later transcribed and analyzed for common themes, ideas and observations. As one of the authors of this chapter, Phil Harding began working at PWL Studios as chief engineer and producer from 1985 and was involved in recording and mixing most of the artists who worked at PWL Studios through to 1992. These artists included Dead Or Alive, Brilliant, Mel & Kim, Rick Astley, Bananarama, Kylie Minogue, Jason Donovan and Pet Shop Boys. Harding's autoethnographic data such as personal diaries, alongside press articles, sound recordings, sketches and information collected for the book *PWL from the Factory Floor* (Harding, 2010) have all been used to support some of the themes and observations highlighted in the interviews. For ease of presentation, the majority of responses from interviewees have been summarized and integrated into the main body of the text.

CREATIVITY IN CONTEXT

Romantic ideas of creativity view creative individuals and creative products as somehow set apart from a particular cultural tradition, but recent research has shown that new products have a firm grounding in what has come before. Tradition therefore is a necessary part of creative activity. In order to create something new, it is first necessary to acquire knowledge of the old. However, the creative process is more complex than this, and contemporary research in this area has shown that creativity is the result of convoluted and dynamic interaction among biological, psychological, cultural and social factors. The 'systems model of creativity' proposed by Csikszentmihalyi (1988, 1996, 1999) underlines this dynamic process and proposes three necessary and interrelated parts: (1) a set of symbolic rules, practices and guidelines called a 'domain', (2) an 'individual' who brings something unique into that domain and (3) a 'field' of specialists or experts who recognize and substantiate that novelty (Csikszentmihalyi, 1996, p. 6).

The interactions among the domain, the field and the individual, are ongoing, overlapping and don't have distinct stages or a specific start or end. This is because the relationships between each of the elements are: "dynamic links of circular causality" (Csikszentmihalyi, 1988, p. 329). In other words, the elements of domain, field and person both influence, and are influenced by, each other. For creativity to occur then, there must exist a domain, which contains a set of symbol systems, a body of knowledge and culture of practice. An individual must acquire a comprehension of that domain and produce something that has an element of novelty. This novelty is then evaluated and validated, by the field.

A domain is made up of a "set of symbolic rules and procedures" (Csikszentmihalyi, 1996, pp. 27–28). Domains are engrained within the culture or symbolic knowledge of a particular group, institution, society or humanity more broadly. In order for individuals to be creative, a working knowledge of the domain is essential, particularly because "creative products are firmly based on what came before. True originality evolves as the individual goes beyond what others had done before" (Weisberg in Sternberg, 1988, p. 173). In order to produce something new, the creative individual must first gain knowledge and understanding of previous creative works. This domain-specific knowledge "serves to provide the background so that the individual can begin to work in an area and also serves to provide ways in which to modify early products that are not satisfactory" (ibid).

As well as internalizing knowledge of the domain, the individual must also learn the rules that govern the selection of creative work by the field (Csikszentmihalyi, 1996, p. 47). Therefore, the individual and their knowledge of the domain are important but not sufficient on their own. For creativity to occur, the field must provide the right conditions for creative contributions, which include "training, expectations, resources, recognition, hope, opportunity and reward. Some of these are direct responsibilities of the field, others depend on the broader social system" (Csikszentmihalyi, 1996, p. 330). In addition to providing the right conditions for creativity to occur, it is also the function of the field to judge the creativity of an idea or product. However, it is important to note that "no judgment ever occurs in a vacuum . . . those who hold the knowledge are also important contributors to the system as they have the background to make those necessary judgments" (McIntyre, 2012, p. 151). The field is therefore not a socially isolated group; the field is both a contributor to, and user of, the domain, and the field decides whether or not an idea or product should be valued or implemented into the domain (Csikszentmihalyi and Wolfe, 2000, p. 81). Creativity therefore occurs at the intersection of interacting elements and for an idea or product to be creative, it must be valuable to a particular group of people (the field), original and implemented into the cultural matrix or symbol system (the domain).

During pop music production, an agent (which could be a single person or the production team) must draw from the domain in order to choose a fitting selection of ingredients from this body of knowledge and symbol system. This selection of ingredients is then presented to the field—the social organization that recognizes, uses and alters the domain—for evaluation (Csikszentmihalyi, 1996). This process can be seen in action

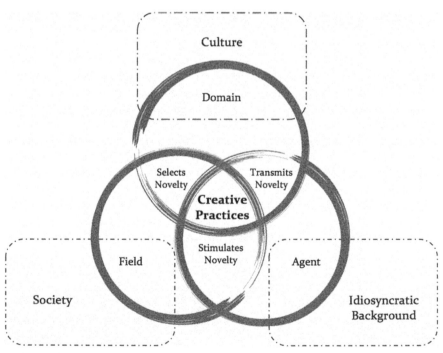

Figure 10.1 Revised Systems Model of Creativity Incorporating Creative Practice (Kerrigan, 2013, p. 114)

when a pop record is released to the public and the field, which includes the popular music press, audiences, musicians etc., decide on the record's novelty or creativity. The systems model of creativity can also be viewed in action during the making of the pop record by scaling the model to a group level and contextualizing the domain and field so that they apply to the specific context (Kerrigan, 2013; Thompson, 2016). The interaction among the system's elements can then be observed in action as the production team collaborates during the tasks of pop music production such as songwriting. The creative system can therefore be viewed as scalable, which "applies equally well at the individual level and also at the group, organizational, institutional or sociocultural level" (McIntyre, 2013, p. 91) (see Figure 10.1).

THE CREATIVE SYSTEM OF POP MUSIC PRODUCTION

Group creativity within pop music production can be studied by first re-contextualizing the elements of the creative system so that they apply directly to pop music production. The following section first introduces the domain and then field of pop music and finally the agents involved.

The Domain of Pop Music

In order to make a pop record, creative agents need to access the domain. The domain holds the symbolic rules, practices and guidelines involved in manufactured pop, and it can be divided into four key areas: musical, technical, cultural and industrial.

The musical part of the domain of pop music is vast, but at its center is the contemporary Western popular song, its structure, form, lyrical themes and settings, and the various ways instrumentation can create its arrangement. Pop producers also require a working knowledge of chords, chord structures, melody and harmony. Domain knowledge also extends to playing and programming both traditional and electronic instruments, as Mike Stock explains:

> Matt [Aitken] and I throughout all of our success *were* the band, we performed on all our records. We were songwriters, we were a band, we were musicians—we were the artists, really and they [the credited artists such as Kylie Minogue] were the guest singers. Nobody's really given us credit for that.
>
> (in Egan, 2004, p. 283)

Technical areas of the domain of pop music include the vast array of recording and music technologies. From the late 1970s and early 1980s, electronic instruments have allowed pop producers the potential to program everything on a pop record including drums, bass, synthesized keyboards, samplers and so on—all driven via MIDI from a master keyboard. Technical knowledge of electronic instruments, computers, samplers and music software is needed, in addition to a working knowledge of microphones, mixing consoles, room acoustics and monitoring equipment in order to record the artist's vocals.

The domain of pop music includes socio-cultural aspects such as how to collaborate with artists, programmers, songwriters and management; learning the lexis, terminology and various ways of communicating to each of them. The socio-cultural elements of the domain also include cultivating relationships among the various personnel involved in the production and maintaining a relaxing or comfortable atmosphere in which to work. Pop producers must learn how and when to elicit a performance from an artist. For example, the vocal sessions for music recordings can be the most delicate part of the process; setting the right studio atmosphere, ensuring that the technology is working efficiently so as not to disrupt the creative flow, consistently encouraging the artist—regardless of tuning and timing errors. It is an ongoing hand-holding, confidence-inducing, ego-massaging process where the producer's role is to 'service' the artist in order to capture the best vocal performance possible for the song.

Pop producers also require contextual knowledge of the pop music industry as it has progressed since the 1960s and 1970s, as well as the current practices and methodologies. The domain therefore also includes commercial aspects of pop music production and, although creative agents involved in manufactured pop music can initially survive on a minimal knowledge of the commercial workings of the music industry, a fundamental knowledge of songwriter and producer business deals are useful. For example, the inner workings of record companies, how they pay royalties for sales, how they market artists and how to promote songs are the types of knowledge that are needed once commercial projects are initiated and completed. A working knowledge of how producers are remunerated before, during or after making a record is also part of the commercial aspect of the domain of pop music. These include but are not limited to producer deals, producer/manager deals and producer/songwriter publisher deals, knowledge of Performing Rights Society (PRS), Phonographic Performance Limited (PPL), social media promotion and digital distribution. Some knowledge of these financial mechanisms is essential in order to recoup costs and receive an income from pop music production.

Finally, included in the domain are all the pop records that have been accepted by the field in the past. The field may add a pop record to the domain for its economic success or its cultural or musical novelty. Pierre Bourdieu refers to these accepted cultural products as "the field of works", and these records illustrate the cultural boundaries of form within pop music production as they include "techniques and codes of production" (McIntyre, 2012, p. 75). In other words, pop music records that have been released characterize the structure, form and constraints of pop music to a creative agent. Creative agents involved in specific tasks in the pop production process will use different parts of the domain. For example, songwriters may primarily use the musical part of the domain or engineers may predominantly use the technical processing part of the domain. In either case, internalizing relevant parts of the domain is necessary in order for creative agents to contribute to the co-production process and carry out their creative tasks.

The Field of Pop Music

The creative contributions of agents within pop music production occur within the broader system of commercial pop music—the field of pop music. The field "includes all those who can affect the structure of the domain" (Csikszentmihalyi, 1988, p. 330) and is the social group that comprehends, uses and modifies the content of the domain of pop music. This includes artists, engineers, songwriters, programmers, record producers, artist management, record label representatives, the pop music press (TV, Internet and radio), audiences and social media commentators. Each area of the field offers its own criteria for selection of particular works over others, and creative agents must internalize the various criteria and mechanisms for selection of successful pop records.

Bourdieu labels fields as cultural arenas of contestation (Bourdieu, 1993) in which there is an ongoing struggle for dominance. Within this struggle, various types of capital are employed: cultural, economic and symbolic. Bourdieu identifies fields as places for "production, circulation, and appropriation of goods, services, knowledge, or status, and the competitive positions held by actors in their struggle to accumulate and monopolize these different kinds of capital" (Swartz, 1997, p. 117). Fields are then "structured spaces that are organized around specific types of capital or combinations of capital" (Ibid). The field consists of "a complex network of experts with varying expertise, status, and power" (Sawyer, 2006, p. 124), and the field of pop music is where records are produced, outputted, considered, validated or rejected. Everyone involved in the co-production of a pop record are also members of the field. They, too, are agents who can alter the content and structure of the domain of pop music. The field is therefore omnipresent during the pop production process, and it is of equal importance to the domain and the agent to the creative process (Csikszentmihalyi, 1996, p. 330). The dynamic system of causality also explains how the field can impact the way in which "musicians work. . . [and] the technological means through which music is recorded, broadcast, circulated, and the aesthetic form and meaning of popular music" (Swiss et al., 1998, p. 103). The selection criteria of the field of pop music is therefore "important in shaping the content and form of the musical product" (Robinson et al., 1991, p. 238).

The Agents of Pop Music Production

The final element within the creative system of pop music production is the 'agent', which in this case is the production team. The production team often includes various members with different types of expertise and domain knowledge. A typical production team has a team leader with programmers, musicians, a lyricist and a top liner, and each team member has a specific responsibility. The songwriting process may also be split into tasks, where a person could be responsible for the song's melody (the top liner), the song's lyrics (lyricist) and the song's musical arrangement and instrumentation (the programmer). The programmer, for example, "is responsible for choosing all the sounds, including drums, bass percussion, keyboards, strings, horns and other instruments . . . using these to create an arrangement in the style of music involved" (Hannan, 2003, p. 88).

Agency Within the Creative System of Pop Music Production

Although it illustrates the three central components of domain, field and agent, the systems model does not depict the ways in which agency is enacted during a creative task. Each role within the production process has varying degrees of creative agency that relates to the power relationships that operate within the specific context and in pop music production more generally. Each person involved has a certain amount of economic, symbolic or cultural power. Pierre Bourdieu argues that the ability to exercise power is directly related to an agent's accumulation of different types of capital. The different forms of capital include:

> economic capital, which is immediately and directly convertible into money and may be institutionalised in the form of property rights; cultural capital, which is convertible, on certain conditions, into economic capital and may be institutionalised in the form of educational qualifications; and social capital, made up of social obligations (connections), which is convertible, in certain conditions, into economic capital.
>
> (Bourdieu, 1986, p. 243)

In PWL's case during the 1980s, Pete Waterman had the most agency within the final decision-making process and unquestionably undertook the role of team leader. When he formed SAW (Stock Aitken and Waterman), Pete Waterman could have been described as having the most cultural capital and symbolic capital (Bourdieu, 1984), owing to his experience and knowledge of previously overseeing many hit records with his former business partner and producer, Peter Collins. In other words, he had accumulated notable cultural, economic and symbolic capital.

PETE WATERMAN'S CAPITAL

As team leader of SAW, Pete Waterman had accumulated the most cultural, symbolic and economic capital (Bourdieu, 1984). Waterman began his career in the 1960s as a DJ playing Tamla Motown and Northern Soul records at Northern England dance clubs such as the Mecca Dance Halls. After spending some time in the early 1970s as a record promoter, he was offered a job in A&R at Magnet Records working with acts such as Susan

Cadogen and the JALN Band. Waterman later became senior A&R at MCA Music publishers while also forming Loose End Productions with producer Peter Collins. Waterman was fully immersed in commercial pop music culture from the mid-1970s through to the early 1980s, which contributed to his accumulation of cultural capital.

By 1984 Loose End Productions had achieved multiple hits with Waterman as team leader and Peter Collins as producer. Waterman became recognized by the UK music industry as a distinguishable figure who could deliver a commercially successful record for an artist. Loose End Productions' successful artists from 1978 to 1983 included Matchbox, The Lambrettas, Musical Youth, Tracy Allman, Nik Kershaw and Matt Bianco. Although Waterman was never credited as the producer or co-producer, he often decided which artists Collins would work with, the overall direction of the production, and assisted Collins at the 'pre-mix' stage in offering his view of the record's commercial viability, its suitability for the record company and how to meet market expectations. This successful string of hit singles and albums contributed to Waterman's accumulation of cultural and symbolic capital. This was also a good grounding in the art of music production and provided the conditions for Waterman to form the SAW production team in 1984.

The success of Loose End Productions also provided Waterman with the economic capital to purchase the latest synthesizers (PPG Wave and The Fairlight System, for instance) and audio equipment while SAW were based at The Marquee Studios. Later, the commercial success of the SAW-produced "Spin Me Round" single by Dead Or Alive allowed Waterman to approach his bank and secure the required loans to purchase PWL Studios in South London. These developments were crucial to Waterman's accumulation of economic capital.

Pete Waterman used his successful cultural and symbolic capital in persuading record companies that his new SAW production team and the PWL studio team could provide hit records for their artists. Waterman used his cultural and symbolic capital to arrange visits to record companies so he could ascertain which artists were being signed, what their marketing budget would be and what type of production sound they required. As success grew for SAW from 1985 onwards, this process lead to a reversal of roles in which record company A&R personnel would consult Waterman on which acts SAW was developing. Waterman was both the instigator and provider of the work for SAW at PWL Studios and was well positioned to meet the creative requirements of a client and translate them to the individuals within his production team. This ability to interpret, liaise and mediate was a central aspect of Waterman's role as team leader at PWL, and he enabled a state of 'flow' (Csikszentmihalyi, 2002) throughout the whole studio team from 1985 onward. Also his accumulation and deployment of cultural, symbolic and economic capital was employed in financing the studios for the production team to work in, sourcing the commercial work that would fulfill the running costs of the studios and paying members of the production team.

Jimmy Cauty's PWL studio sketch (Harding, 2010, p. 68) is an interpretation of the daily workings of the PWL Borough Studio during 1985. It is unlikely that all of these people involved in the 'mock' session would be in the room at the same time, but Cauty is depicting how he saw the events across one or two days in the studio in a single scene, whilst recording there with his band, Brilliant, in 1985. Each production team member would be present in the studio at various times throughout the process, but Cauty illustrates everybody together at one time, and places them all onto the flight deck of the *Starship USS Enterprise*, from the television series *Star Trek*. By placing Pete Waterman in the captain's chair, Cauty is suggesting that Waterman is the main arbiter of the creative process, and he has the ultimate decision, not the recording artist. This is not to say that SAW overlooked the song or the artist, on the contrary as Jason Donovan explains: "SAW were fantastic at understanding their artists' character and writing a song fashioned to suit that person exclusively" (Donovan, 2015). Fundamentally though, the artist's views were secondary to Waterman's command of the field of agents around him. Pete Burns, the front person of Dead or Alive, once stated that: "Waterman had Woolworths ears" (Burns, 2015), highlighting Waterman's judgment on which records the UK public would buy in Woolworths on a Saturday morning the day after their release. The depth of PWL's relationship with Woolworths was highlighted during the autumn of 1988. Woolworths contacted Pete Waterman, stating that press rumors of a Kylie Minogue and Jason Donovan Christmas duet single had resulted in a large number of customer orders of the record. However, Kylie and Jason had not yet recorded a duet, so Waterman took advantage of this privileged commercial knowledge and team-led the production of "Especially For You" on November 28, 1988, which reached number 2 in the UK charts by Christmas 1988 and peaked at number 1 in January 1989.

Placing Waterman at the center of the picture in the captain's chair emphasizes Waterman's status within the process, using his extensive knowledge of pop music and culture to inform his decisions and exercise his agency to ease the "tension between commerce and creativity" (Negus, 1992, p. 153). Waterman could say with confidence what would, and what would not, work on a final product leaving the studio, and every member of the production team would trust Waterman on such decisions. If at the point of the final mix Waterman was still not happy with something, be it musical or technical, then a team member would have to re-work it until the 'captain of the ship' was happy. One could describe the production team working on these PWL projects as 'in service' to Pete Waterman—whether they realized or acknowledged it. Certainly all of the agents working in the recording studios at PWL were fully engaged in what Keith Sawyer calls 'Group Creativity' (Sawyer, 2003).

Waterman's cultural capital was therefore deployed within the evaluation process, and his knowledge of the domain and the mechanisms and criteria for selection formed part of this evaluation. This is because "the influence of the market—what will sell—is important in shaping the content and form of the musical product" (Williams in Robinson et al., 1991, p. 238). As the pop music market showed its enthusiasm for PWL records, those involved in the process (and principally Waterman) grew in their self-confidence. This resulted in the emergence of a PWL house style (or distinctive sound), in which tried-and-tested production methods and sounds were re-used, which further increased Waterman's symbolic capital as a hit-maker. In this way, the PWL production team created a unique sound through a process of their own internal re-evaluation. The creative system can therefore be seen operating inside the recording studio on a group level in which they drew from a limited area of the domain to rearrange it in a new way and then internally evaluate their contributions with Waterman as the lead authority within each process. On a group level then, the domain becomes a microdomain and the social organization inside the recording studio become a microfield, accepting or rejecting new creative ideas, as shown in Figure 10.2.

Waterman began the pop music co-production process as the person with the most amount of agency. He gathered other agents with specialist

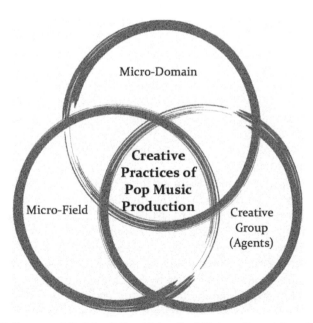

Figure 10.2 Systems Model of Creativity Scaled to a Group Level During the Pop Music Production Process

musical and technical knowledge of the domain and steered them to deliver a creative product that would be successful within the broader cultural domain of commercial pop music. When these pop records succeeded, the overall credit was given to Waterman, which enhanced his cultural and symbolic capital. This also incensed the other primary agents involved in the co-production process—Stock and Aitken—whose contributions were often overlooked or under-represented.

SIMPLIFIED SERVICE MODEL IN ACTION

The daily workflow priorities for the creative studio team were shown on a display board situated just outside the control room of the main PWL Borough Studio. This board could contain up to six projects that were in production by SAW. These projects were deliberated each morning, and Waterman would decide on the work schedule and targets for that day. Upon moving into the PWL Studio in The Vineyard, London SE1 in early 1985, Waterman may well have stated to those present on the first day that he intended to create a UK equivalent to Detroit's famous Tamla Motown organization of the 1960s. This equivalent was in the production workflow and in-house co-production processes that eventually lead to a PWL 'signature sound' (Zagorski-Thomas, 2014). Music journalist Matthew Lindsay supports the idea of an SAW production formula and musical framework as follows:

> To me, that was like taking House [music] and Hi-Nrg [music] and wedding it to an early Tamla Motown songwriting formula. If you listen to the early Supremes songs you've got the PWL song formula and wedding that with Hi-Nrg was a genius combination. What was interesting about PWL to me was that they were capable of doing so many different sounding records from Princess's "Say I'm You're Number One", a really good Soul/R&B record, why didn't they make more records like that? Also "Roadblock" by SAW was a really good Soul/R&B record.
>
> (Lindsay, 2014, personal interview)

The production process at PWL during the 1980s can be depicted as a flow diagram (Figure 10.3) that models the involvement of Waterman at various stages of making a pop record. Most notable are the points at which Pete Waterman engaged. Nothing began without Pete Waterman's agreement; Waterman would choose the song to record or would have some involvement in writing a new song, which would typically be the song title. The process would then be led by Stock and Aitken, who composed a minimal backing track that included drums, bass, keyboards and a rough arrangement of the song. Vocals, including lead and backing vocals, were then recorded and arranged. The production team then completed the rest of the record in sympathy with the vocal, principally because this is considered to be the most important element of a pop recording. Waterman

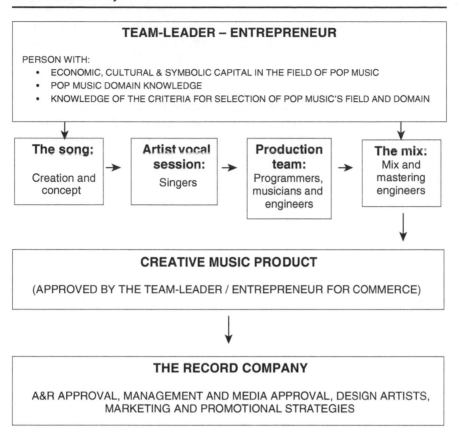

Figure 10.3 Simplified Service Model of Commercial Pop Music Production at PWL in the 1980s

was not involved in these two stages (vocal recording and production) and entrusted these parts of the process to the rest of the production team.

Waterman returned during the mixing stage and, after listening to a mix, offered his opinion on elements that required revision, which sometimes involved song arrangements and song structures. On occasion this involved revisiting previous stages to amend or add particular musical parts. In rare instances, it involved starting the process again in order to react to new musical tastes, trends or market conditions. For example, Mel and Kim's single "System" (1986) had already been completed to the creative music product stage but, prior to its release, Chicago House music had been introduced into the London club scene. A last-minute decision was taken by Waterman to persuade the record company to stop pressing the single, and the production team began working on a fresh new song entitled "Showing Out" (1986), which incorporated the aesthetics and style of Chicago House. "System" (1986) was released as the B-side to the A-side "Showing Out" (1986).

Waterman deployed his various types of capital during the pop production process. Nothing would leave the studio without his approval or his involvement, and his involvement at the beginning and end of the studio-based process ensured that he had a key influence. It would be true to say therefore that each member of the PWL studio team by the late 1980s had become "an agent who is conditioned through creative practices" (Kerrigan, 2013).

Once Waterman had approved the mix, the musical product was finalized and presented to the record company. Every Monday morning there would be a meeting between Waterman, PWL's managing director David Howells, PWL's label director Tilly Rutherford and record promoters Robert Lemon and Ron McCreight (Sharp End Promotions). They decided among themselves if the record was ready to be released and whether or not to include the involvement of design, marketing and promotion departments of the record company. The simplified service model for PWL is shown in Figure 10.3.

Within each of the production stages of the service model, the creative system can be seen operating on a group level with different numbers and types of agents involved at each stage. For example, Waterman would be involved in the songwriting stage to the extent of writing the song title, but the task of completing the song, including the song structure, chords, melody (top line) and lyrics, was completed by Stock and Aitken. Stock and Aitken drew from the domain of pop music, rearranged it in a new way and then evaluated their contributions, both internally and collectively, in order to create the final song. In this way, Stock and Aitken drew from a microdomain of pop music and formed a microfield during the songwriting process, accepting or rejecting creative ideas to produce the finished song, as shown in Figure 10.4.

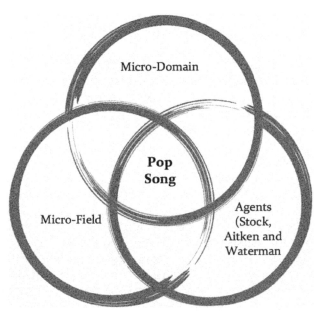

Figure 10.4 Systems Model of Creativity Scaled to a Group Level During the Pop Songwriting Process

CONCLUSION

Pop music's focus is typically on the artist, but there are often numerous personnel working behind the scenes who make up the production team, which include songwriters, programmers, musicians, lyricists, top liners and a team leader. One of the most prominent and commercially successful pop production teams of the 1980s were Mike Stock, Matt Aitken and Pete Waterman (SAW), who adopted a complete in-house production approach (songwriting, recording and production) and employed a team of engineers, programmers, musicians and artists in creating a musical product.

This study has explored the creative production workflow system of SAW at Pete Waterman Ltd. (PWL) Studios during the 1980s, focusing on the agency and influence of Pete Waterman within the process as the production team worked within the creative system of pop music making. In setting up PWL studios in the 1980s, Waterman had accumulated enough cultural, economic and symbolic capital to undertake the role of team leader. As team leader, Waterman's involvement was both at the beginning and the end of the pop production process. In other words, the pop production process in the recording studio began and ended with Waterman and, as the person with the most agency, he had the final say on whether the product was sufficient.

In the production stages between Waterman's participation, the creative system can be seen operating on a group level with different numbers and types of agents involved at each stage. Here, other members of the production team drew from the domain of pop music, rearranged it in a new way and then evaluated their contributions in relation to the group's, Waterman's and the broader field's mechanisms and criteria for selection. This formed a scaled system of creativity on a group level consisting of a microdomain of pop music and a microfield made up of those involved in accepting or rejecting creative ideas to produce the finished song.

To date there has been limited academic enquiry into the production of pop music from the 1980s onwards, and in seeking to redress this imbalance, further work is needed to explore the various facets of pop music production and co-production, the creative systems involved and the ways in which pop music producers work to create a musical product.

BIBLIOGRAPHY

Bourdieu, P. (1984). *Distinction*. Oxford: Routledge.

Bourdieu, P. (1986). The forms of capital. In: J. Richardson, ed., *Handbook of theory and research for the sociology of education*. New York: Greenwood, pp. 241–258.

Bourdieu, P. (1993). *Field of cultural production* (Edited by R. Johnson). New York: Columbia University Press.

Burns, P. (2015). *#1's of the 1980's*. London: ITV [Accessed 10 Nov. 2015].

Csikszentmihalyi, M. (1988). Society, culture and person: A systems view of creativity. In: R. J. Sternberg, ed., *The nature of creativity: Contemporary psychological perspectives*. New York: Cambridge University Press, pp. 325–329.

Csikszentmihalyi, M. (1996). *Creativity: Flow and the psychology of discovery and invention*. New York: Harper Collins.

Csikszentmihalyi, M. (1999). Implications of a systems perspective for the study of creativity. In: R. J. Sternberg, ed., *Handbook of creativity*. Cambridge: Cambridge University Press, pp. 313–335.

Csikszentmihalyi, M. (2002). *Flow: The classic work on how to achieve happiness*. 2nd ed. New York: Harper Collins.

Csikszentmihalyi, M. and Wolfe, R. (2000). New conceptions and research approaches to creativity: Implications for a systems perspective of creativity in education. In: K. A. Heller, et al., eds., *International Handbook of Giftedness and Talent*. 2nd ed. Oxford: Elsevier, pp. 81–93.

Donovan, J. (2015). *BBC radio 5 interview*. Broadcast, 31 Dec.

Egan, S. (2004). *The guys who wrote 'Em: Songwriting geniuses of rock and pop*. London: Askin Publishing.

Frith, S. (2001). Pop music. In: S. Frith, W. Straw and J. Street, eds., *The Cambridge companion to pop and rock*. Cambridge: Cambridge University Press, pp. 93–108.

Frith, S. and Zagorski-Thomas, S. (2012). *The art of record production: An introductory reader for a new academic field*. Farnham: Ashgate Press.

Hannan, M. (2003). *Australian guide to careers in music*. Sydney: University of New South Wales Press.

Harding, P. (2010). *PWL from the factory floor*. London: Cherry Red Books.

Hennion, A. (1990). The production of success: An anti-musicology of the pop song. In: S. Frith and A. Goodwin, eds., *On record: Rock, pop and the written word*. London: Routledge, pp. 185–206.

Kerrigan, S. (2013). Accommodating creative documentary practice within a revised systems model of creativity. *Journal of Media Practice*, 14(2), pp. 111–127.

McIntyre, P. (2012). *Creativity and cultural production: Issues for media practice*. Basingstoke, UK: Palgrave Macmillan.

McIntyre, P. (2013). Creativity as a system in action. In K. Thomas and J. Chan, eds., *Handbook of research on creativity*. Cheltenham: Edward Elgar, pp. 84–97.

Napier-Bell, S. (2001). *Black vinyl white powder*. London: Ebury Press.

Negus, K. (1992). *Producing pop: Culture and conflict in the popular music industry*. London: Oxford University Press.

O'Hare, P. (2009). Undervalued Stock: Britain's most successful chart producer and his economy of production. *Journal on the Art of Record Production*, (4) (Nov.) Available from: http://arpjournal.com/undervalued-stock-britain%E2%80%99s-most-successful-chart-producer-and-his-economy-of-production/ [Accessed Jan. 2018].

Robinson, D. C., Buck, E. and Cuthbert, M. (1991). *Music at the margins*. London: Sage.

Sawyer, K. (2003). *Group creativity: Music, theatre, collaboration*. New York: Taylor & Francis.

Sawyer, K. (2006). *Explaining creativity: The science of human innovation.* Oxford: Oxford University Press.

Shuker, R. (2001). *Understanding popular music culture.* 2nd ed. London: Routledge.

Swartz, D. (1997). *Culture and power: The sociology of Pierre Bourdieu.* Chicago: University of Chicago Press.

Swiss, T., Herman, A. and Sloop, J. M. (1998). *Mapping the beat: Popular music and contemporary theory.* Malden, MA and Oxford: Blackwell Publishers.

Thompson, P. (2016). Scalability of the creative system inside the recording studio. In: P. McIntyre, J. Fulton and E. Paton, eds., *The creative system in action: Understanding cultural production and practice.* Basingstoke: Palgrave Macmillan, pp. 74–86.

Weisberg, R. W. (1988). Problem solving and creativity. In: R. J. Sternberg, ed., *The nature of creativity: Contemporary psychological perspectives.* Cambridge: Cambridge University Press, pp. 148–176.

Zagorski-Thomas, S. (2014). *The musicology of record production.* Cambridge: Cambridge University Press.

DISCOGRAPHY

Dead or Alive. (1985). *Spin Me Round* [vinyl 7" single]. UK: Epic.

Kylie and Jason. (1988). *Especially For You* [CD single]. UK: PWL Records.

Mel and Kim. (1986a). *System* [vinyl 7" single]. UK: Supreme Records.

Mel and Kim. (1986b). *Showing Out* [vinyl 7" single]. UK: Supreme Records.

Princess. (1985). *Say I'm Your Number One* [vinyl 7" single]. UK: Supreme Records.

Stock, Aitken and Waterman. (1987). *Roadblock* [vinyl 7" single]. UK: PWL Records.

11

Mix and Persona

Analyzing Rejected Mixes

Dan Sanders

INTRODUCTION

A mix may be defined simply as the relationship between multi-track elements in terms of relative loudness, spectral balance, stereo positioning and spatial and effect processing, created in order to realize a stereo master recording. But the process of highly skilled, professional-standard mix engineering exhibits a seeming contradiction: although it is arguably the most technically complex aspect of the process of production, the task is oriented to achieve the most emotive and qualitative aims of a project.

Mix engineers exist in a world of subjective demands and targets, where numerous stakeholders are entitled to contribute to the process without providing clear technical direction. Any member of a wider production team (including artists, producers, managers and A&R representatives) may legitimately make demands such as "more energy", "more angst" or "less Disco" from the mix engineer. These "technically or artistically naïve" requests (Izhaki, 2012, p. 23) are not unreasonable; the craft of the mix engineer is to develop frameworks and strategies to achieve these extrinsic goals to the satisfaction of the team in addition to his or her intrinsic, artistic aims.

The central question of this investigation is how to examine any relationship between the mix presentation of a piece of recorded music and the persona of the artist. Professional understanding of this key concept is relatively easy to find: Adam Brown was staff engineer for London's Olympic studio for 11 years from 1994. He recalls assisting on a mix of The Verve's 1996 single "Bittersweet Symphony" by a Grammy-winning mix engineer, which was rejected by the band as being unrepresentative of them:

> The mix was what you'd expect it to be. It was clean and dynamic and polished. The band listened to it in the studio with the mix engineer and Richard Ashcroft explained that it was a brilliant mix, but entirely wrong for the band. Nobody got upset, because at that point the direction of the band became clearer and this allowed Chris Potter [who

subsequently mixed the final version] to understand how to progress. The band needed a looser, messier sound to suit their indie image.

(Brown, 2015)

To address the question of how artistic persona can be reflected or furthered in the mix of a recording, it would be fruitful to search for differing mixes of the same multi-track material and analyze these with reference to documented reports, reviews and recollections of the participants where available. Consideration will focus on performer identity as either soloist or group member and how mix presentation can imply commercial or alternative genres.

EXPLORING MIX ANALYSIS AND PERSONA

The key technique to be employed in this investigation is listening. The complexity of listening is well documented both academically and professionally; many ways to extract meaning from audio combine in analysis of a mix of discrete sources.

Although specifics vary, many writers broadly agree on different "modes" of listening and "aural awareness". Influential throughout are Schaeffer's (1966) four modes, which might be characterized as: hearing without listening; listening to identify; selective listening; and listening to understand. Others expand and refine these concepts, describing for example, "listening-in-search", actively seeking information from sound (Truax, 2001, pp. 21–24); "contextual listening", allowing for interpretation through cultural experience (Norman, 1996, pp. 2–9); and significantly for this investigation, "technological listening", where technical and technological details can be inferred (Smalley, 1997, p. 109). A concept emerges of an axis of different functions and " 'levels' of listening attention" (Truax, 2001, p. 21). The ability to shift aural attention intentionally, famously recognized as the "cocktail party effect" (Cherry, 1953, p. 976), is a key part of the process of both mixing and mix analysis. A potential invalidity of listening itself, rather than the mode, can be illustrated by the Edison "Tone Tests", where listeners failed to discern live performance from phonograph recordings in the early 20th century (Olive, 2010).

The aural differentiation of minute detail in addition to consideration of the effect of the wider ensemble may be illustrated best with a visual metaphor: focus and zoom. A listener's attention can move continuously between tight, selective focus and diffused unified hearing. The linking of the visual and aural experience has support in both the academic and professional fields: mix visualization is a significant framework for analysis whereby the sound elements of an audio recording can be represented graphically or (more commonly) mentally in two- or three-dimensional visual space. This form of analysis relies on a number of the various modes of listening, often in conjunction. Dockwray and Moore (2010, p. 181) discuss the notion of "the sound-box", which has parallels to Smalley's "spectral space" (1997, p. 122), in particular the assertion that a hypothetical third dimension unaddressed by stereo, that of height, can be assigned to frequency. Gibson (2008, p. 22) posits a long-established recording industry method for representing these visualizations as an aid to practitioners wishing to analyze or replicate extant works. Gibbs and Dack (2008, p. 176) consider the complexity of space and place in recorded music when multiple recorded room acoustics are present in a recorded artifact, as is common in Rock and Pop. Izhaki (2012, p. 440) advises that consideration should be given to how differing acoustics interact rather than seeking to avoid multiple spaces entirely; thus sound-stage visualization does not need to describe a real space and may describe a far more complex form than nature would allow.

With the sound-stage concept in mind, opinion that the relative placement in visual space of individual mix elements can affect a listener's perception of the music is relatively easy to find in non-academic professional-facing texts. Both Mixerman (2014, p. 57) and Izhaki (2012, pp. 61, 68) note that prominence is connected to a sense of importance, and Owsinski (2014, p. 121) implores engineers to "find the most important element and emphasize it", offering points of advice more subtle than simply raising level. This is a key element of analysis for the connection between mix and persona. One area relevant to this investigation is analyzing examples where there is opinion that a mix or balance has inaccurately reflected the persona of an artist or band, for example, when a band is considered to be under-represented at the expense of its singer or vice versa.

Another area linking mix to persona is illustrated by acknowledgement in both professional and academic material of a concept of complex emotional response to music, although often this is without detail or explicit consideration of technique. For example, Gibson (2008, pp. 248–253) discusses "chills" and "magic" without defining either, advising practitioners to attempt to formulate a personal method of creating these effects. Similarly, Nusbaum and Silvia (2011, pp. 199–204) utilize concepts of personality research and music preference to examine physical emotional response, focusing more on the listener than the audio. Regardless of the absence of detailed deconstruction of these concepts, professionals frequently state their aim of working towards emotional and subjective goals. The Motown mix engineer Russ Terrana (Chin, 2010, para. 31 of 35) states his aim to allow a listener to "feel the emotions change"; emotions logically form a large part of the persona of an artist, and evidence of opinion on the appropriateness of these in mixes is a strong candidate for investigation here.

Artistic persona as a concept crosses many disciplines. Van Winden (2014, pp. 11–17) outlines some of the key contributors from the fields of anthropology, social psychology and performance studies among others. It is beyond the scope of this work to stray too far away from the popular musicological viewpoint, so to look at a connection with a mix is to focus on an artist's public persona. Van Winden demonstrates that the concept was being considered in the early part of the 20th century by thinkers such as Mauss and Jung; more recent work by Auslander and Frith looks at the realities of "performance persona" and "star personality", respectively.

Auslander (2004, pp. 6–7) enumerates three different layers of persona (that contained in a song lyric; the public persona of the performer; and a "real person"), refining Frith's concepts of "star personality" and "song personality" (1996, p. 212) and pointing out that the delineation of the "real person" is often less than clear due to the inference of the viewer or listener rarely being shaped without passing through the filter of various media. This has parallels with Goffman's notion of the (true) self "as a performed character" (1959, p. 245), always shaped by its audience. These layers form a useful framework to use to analyze a mix, as feasibly, a recording could contain all three as much as a live performance. Variation in mix outputs, which potentially may lead to rejection, could derive from differing perceptions of persona, or representation of one layer when another is required.

Evans and Hesmondhalgh (2005, p. 19) argue that the persona is a "distinctive image of a person built up from the sum of their mediated appearances". Mediated appearances must include audio recordings, suggesting that persona need not only be represented in music, but the music could also contribute to perception. It could be fruitful to investigate to what extent audio can act similarly to visual media in this context, but an overlapping area research would be that of interplay between visual image and audio artifacts. Both Auslander (2004, p. 12) and Donze (2011, p. 50) refer to visual imagery as part of the overall persona, noting connections between sexualized dress and live performance, for example. It follows that relevant performances are likely to be presented in a consistent manner within a recorded artifact. Stylized or exaggerated performance may well be able to be shown to map to placement of mix elements. Broadening the scope from the visual, parallels to wider cultural representations (political stance, lifestyle descriptors and so on) of a band or solo artist might also be shown to be represented within the presentation of a mix.

As hinted earlier, one important limitation must be noted with regard to persona: in a sophisticated, commercial industry which is frequently fiercely aware of its own perception, there must always be an acknowledgement that mediated information may have passed through the filter of the complex world of public relations, which historically has a very different relationship with the need for actuality than other areas of society. This provides a significant, but useful restriction to what can be learned, but also highlights that an exploration into the mechanisms which form a listener's understanding of the artist are beyond the scope of this project.

Thus, this investigation will hypothesize that there is a connection between mix and persona and consider how the area might be further investigated. Different listening strategies, informed by the modes and considerations discussed here, will be applied—listening attentively to elements of performance, cultural references and technological processing but also acknowledging overall impact in a less attentive manner.

ANALYSIS OF REJECTED MIXES WITH SUBSEQUENT REPLACEMENTS

On occasion, a decision may be reached by any or all of a wider production team that a mix is not "working" in some way and should be revised, restarted or abandoned. Infrequently, this decision is taken after an entire album has been presented as a candidate for release. In the case of an album which becomes successful and is eventually deemed by its record label to be noteworthy enough to support a commemorative re-release on an opportune anniversary, the rejected version becomes an available choice to add value to an expanded release, which will be typically referred to as a "deluxe" or "super-deluxe" edition.

There can be a wide variation in the intended purpose of unreleased mixes; some may be sketches, demonstrations, works-in-progress or so-called overnight mixes intended to be reflected upon between mixing sessions. Of most interest to this investigation are genuine candidates for release which were rejected in favor of subsequently successful alternatives created with no significant re-recording. A potential reason for initial rejection is if the music is considered to misrepresent the artist in some way; in these cases analysis of these mixes, attempting to ascertain the aims of the definitive version and considering press and public response to the re-release may throw some light on any relationship between artist persona and mix technique.

Thus, the pool of available material which satisfies all relevant criteria is not enormous, but nonetheless potential candidates include the "Detroit" mixes of Marvin Gaye's *What's Going On* album (1970–71) and Bernard Edwards and Nile Rodgers of Chic's original mixes of *diana* [sic] by Diana Ross (1980, re-released in 2010), both on the Motown label; the UK version of *The Holy Bible* by Manic Street Preachers (1994, re-released in 2004) on Epic; and Butch Vig's original "Devonshire" mixes of Nirvana's *Nevermind* (1991, re-released in 2011) on Geffen. On examination, Gaye's original mixes are interesting, but booklet notes state that some were embellished with additional recording (Edmonds, 2001, para. 13 of 20) and so may be seen more as a progression of the material than a straightforward mix. A relatively smaller amount of documentation for *The Holy Bible*, mixed again for the US market to the approval of the band (Izhaki, 2012, p. 7), has led me to focus on *diana* and *Nevermind*.

In order to progress, I have assumed equity between versions where, in reality, many variables are active. In dealing with subtle detail, it is possible that post-mix mastering and characteristics of any playback system will affect perception. Although each "deluxe edition" is mastered as a whole, it is not clear how equitable technical treatments have been between versions of the album. Where an album is in two versions on one CD, volume levels are likely to be similar, but variables such as what studio or mixing desk was used to prepare each version will significantly affect outputs. More structural issues, such as comparing previously mastered audio with a modern, non-contemporaneous mastering of the unreleased work, for example, may create further variation. With the impact on a listener in mind, I have only compared between album versions within each "deluxe" edition (avoiding "original" or other releases) and consistently used one system of powered studio monitors in one room. Naturally, this playback system cannot be the same as the systems used to create and review a subject recording, but a professional mix engineer's aim to deliver a balanced sound across all playback systems provides some counteraction against the issue.

The original mixes of Diana Ross' 1980 album *diana* were felt to be more representative of the commercial "performance" persona of the producers than was appropriate for an artist of Ross's stature:

> [Nile] Rodgers and [Bernard] Edwards [of Chic, producing] were accused of trying to make their sound shine and Motown's first lady look poor by comparison, so Ross and long-time engineer Russ Terrana pulled the album and remixed it at the eleventh hour.
>
> (Easlea, 2003, para. 4 of 8)

> It seemed like a Chic album with a Diana Ross voice. It wasn't a Diana Ross album.
>
> (Russ Terrana, in Chin, 2010, para. 30 of 35)

That this view was held by both the artist and her record label is reinforced on the official archive discovery website of Motown's owner, Universal Music, which states that Ross herself was unhappy with the results of the sessions:

> [Ross] gave them specific remixing instructions; they made slight changes and suggested that if she still didn't like them, she could get them remixed herself. Ross did so, reworking the whole album with . . . Terrana to downplay the funk element and make her voice more prominent.
>
> (uDiscover, 2014, para. 4 of 5)

However, other reasons are outlined by Nile Rodgers:

> [With] Diana Ross, we were working on [the *diana*] album, and somehow we offended her even though we were trying to say in the most delicate way possible that she was singing just a little bit flat. She walked out of the studio. I never saw her for another two months.
>
> (Nile Rodgers, interviewed in Alleyne, 2007, answer 20 of 38)

Whereas both aspects are likely to be valid but symptomatic of a complex breakdown in trust within the wider production team, it is clear that in this situation any evaluation of the music as being "unrepresentative" is simultaneously both a commercial and artistic concern; here there was unity from both the artist and the label, sides which elsewhere may be opposed. To interpret Terrana's point that it "wasn't a Diana Ross album", the story being told by the producers did not reflect the persona of Diana Ross in the way the artist and label understood it.

Applying sound-stage visualization to the recordings, it becomes clear that the differences between the mixes are more subtle and complex than simply concerning the relative level of the lead vocal. Indeed, on "Have Fun Again" and "Now That You're Gone", Ross is proportionally quieter in the Terrana mixes. Much like a band avoiding drawing attention away from a revered solo artist on a stage, mix engineers can consider hierarchical placement: "Placing the vocal prominently in a mix establishes its importance. This is the primary method of harnessing a listener's focus" (Mixerman, 2014, p. 57). To this end a number of the final mixes reduce the prominence of Rodgers' rhythmic guitar playing, either by lowering its level relative to the vocal ("Tenderness", "I'm Coming Out", "Give Up" and "Friend to Friend", although the opposite is true in "Inside Out") or by moving it from center stereo position ("I'm Coming Out", "Have Fun (Again)"). Allowing Ross to be audibly the most important element in the music is consistent with both stated intent and ultimate effect of the final mixes.

Terrana generally corrects excessive reverb on Ross's vocal (for example, "Inside Out", "Friend to Friend"). As sounds in proximity exhibit a lower proportion of reflected, reverberant sound, this results in Ross appearing nearer to the front of the sound-stage, making her more intimate with the listener without necessarily affecting vocal level. In places there is a reduced focus on low-mid vocal frequencies, producing a thinner, lighter sound, moving Ross to a less crowded part of the sound-stage; the difference in vocal tone between versions of "Friend to Friend" is stark, for example.

Curiously, Edward's bass guitar is not sidelined like the rhythm guitar. In some tracks ("Tenderness", "I'm Coming Out") it appears louder. In terms of frequency, one dimension of the 3D sound-stage concept, a bass guitar could be said to compete less with the vocal than a lead guitar. It could also be argued that the traditional persona of a bass player is less oriented towards demanding stage attention and so might provide less of a challenge to a lead singer. Terrana allows each groove to be led, but not overpowered, by the bass.

Greater dynamic control overall is exhibited in the Terrana mixes. Referring to the original mixes and explaining that the source material contained enough "excitement", he states:

> The dynamics seemed limited. . . . I tried to create the dynamics that seemed missing, so you could feel the emotions change and hear the subtleties already in the music.
>
> (Chin, 2010, para. 29–31 of 35)

Terrana's mixes display more compression deployed to focus on a rhythmic sound's more percussive attributes, subtly reducing legato or sustain. This opens up space in the sound-stage without necessarily reducing the impact of the instrumentation; the instruments continue to do their job, but proportionally more vocal (for example, on "I'm Coming Out") can be heard. Dynamic control is key to Terrana's balancing of the rhythm section, too. Techniques other than compression would have been available in 1980 (perhaps drum replacement or synthesis using a triggered drum machine or synthesizer), but the relationship between kick and snare drum, more consistent on "Have Fun Again" and "I'm Coming Out", is characterized by a louder, more present kick drum that is perhaps more indicative of a disco groove. To generalize about the difference between the mixes, Terrana seems to have concentrated on establishing a two-layer hierarchy—a "groove" (or rhythmic unit) supporting the significantly more important vocal. In comparison, Chic's mixes could be characterized as portraying only one layer—a more egalitarian band.

It is important to consider whether Terrana and Ross's versions are true mixes as opposed to embellished "remixes" or developed productions. It is entirely feasible that the final mixes used only the existing multi-track material almost exclusively. Although the arrangement of "Friend to Friend" appears quite different, this could be the result of cutting the grand piano and delayed guitar from the mix and reinstating synth pads, which typically may have been present throughout to provide pitch reference to vocalists, rather than re-recording. Elsewhere, alternate vocal takes were clearly utilized. The "Inside Out" lyric "The time has come for me to break out of this shell" (Edwards and Rodgers, 1980a), at 1'27" in the Terrana mix and 1'46" in the Chic mix, is phrased and pitched differently and gives some credence to Rodgers' description of Ross's singing as being sub-standard. Similarly, in "Have Fun (Again)" around 2'52" in both mixes, the phrase "Money won't be enough, when the going gets tough it's rough" (Edwards and Rodgers, 1980b) is phrased wholly differently. These alternatives could reasonably come from the same session; Rodgers explains (Chin, 2010, para. 25 of 35) that their working method was to let the artist learn the song in the recording session to promote a sense of live performance. Ross, who Rodgers recalls had never worked that way before, would likely have recorded a number of different interpretations of each line. However, an inability (or lack of desire) to select the best vocal takes may be indicative of the damaged relationship between artist and producer.

Easlea's (2003, para. 4 of 8) reference to "the eleventh hour" may suggest that no further recording was made and that the Terrana mixes used only extant multi-track material, enhanced by razor-blade editing of the master to shorten some sections and vari-speed to increase the tempo ("I'm Coming Out" and "My Old Piano" exemplify both). There is no suggestion that lead vocal re-recording without Rodgers and Edwards took place, although sleeve notes state that "some harmonies and percussion" were added after the initial mix (Diana Ross, 2010, album credits, Chic mix). Edwards and Rodgers are listed as sole producers, and it is likely that any extra recording without the consent of the producers would be a professional discourtesy worthy of being hidden at the least, if not enormously offensive.

Terrana, with Ross (some pressings carried the credit "Mixed by Russ Terrana and Diana Ross", Diana Ross, 2010, album credits, original released mix) was able to proceed with only Ross's persona in mind, unencumbered by being required to simultaneously represent Chic. The "performance persona" (Auslander, 2004, pp. 6–7) of Diana Ross as a talented individual had developed steadily since first finding success with The Supremes in 1964, becoming "Diana Ross and The Supremes" in 1967 before going solo in 1970 (Myers, 2015, section 3 of 10). Opinion is easy to find of popular acceptance for the notion of her cultivation of "diva" status (Percy, 2014, para. 1 of 8); in this reading, her performance persona relies on isolation and being the recipient of focused attention. Chic was well established by the time of recording *diana* (Huey [n.d.], para. 3 of 6), and both Rodgers and Edwards are recognized as distinctive instrumentalists (Gold, 2009, para. 1 of 15; Gress, 2015, para. 5 of 12). Thus, in bringing Rodgers and Edwards in to produce Ross, Motown imported another considerable performance persona based not only upon being significant musicians, performers, composers and producers but also of being successful industry practitioners. It is perhaps unsurprising that there was an issue in accommodating both personae.

The situation with Nirvana's 1991 album *Nevermind* exhibits similarities with *diana*, but here the driving force for change is far less clear and the relevance of the rejected mixes arguably less. Nirvana's second album was due to be mixed by its producer Butch Vig, but time constraints led the band to select Andy Wallace for the task (Azerrad, 1993, p. 179).

There are clear differences between Vig's earlier "Devonshire" mixes and Wallace's final, successful work. For example, Wallace's first chorus of "Smells Like Teen Spirit" places Kurt Cobain's lead vocal more prominently, requiring more delay and reverb to blend it into the mix. Drums and guitar are balanced together more (Vig's guitar level sounds low by comparison, favoring clarity of the drums). A similar blending of the guitars and drums is apparent in "Come As You Are", as is a less-marked representation of the vocal with slightly less reverb and a slightly fuller spectral balance. "In Bloom" continues the pattern, but perhaps focuses more on refining the guitar's sense of space in the mix to be more musically distinct, allowing the vocal more space. To over-simplify the difference between the mixes, it could be said that Wallace presents Cobain more as a frontman where Vig presents a more unified band. This is a similar type of development as observed in the *diana* mixes, though the reason for the change in creative direction here is less documented. Krist Novoselic (bass) and Dave Grohl (drums) are arguably still represented more equitably than Chic's performers were in the final *diana* mixes, but there are fewer sources to share the sound-stage here.

The band's reaction to the success of the album is indicative of their performance persona at the time. Unusually, the band let it be known that they were not happy with Wallace's mix, Cobain famously saying, "I'm embarrassed by it now. It's closer to a Motley Crue [sic] record than it is a punk rock record" (Azerrad, 1993, p. 180), referring to an earlier, commercially focused sound. Significantly, opinion that Vig's mixes were better does not seem to be present, so it is possible to understand this as an attempt to distance the band from its own success. Novoselic identifies this, saying, "We tried to have a fine line between being commercial and sounding alternative" (Azerrad, 1993, p. 180); it was significant to the band at the time to be perceived as an authentic part of alternative, non-mainstream culture to support their lyrical and visual imagery. An album certified four times Platinum for four million sales in the US just over a year after release (RIAA, 2015, database result) could certainly be argued to be part of the mainstream. An examination of credibility and authenticity is beyond the scope of this investigation, but as facets of persona, in this case they are significant.

Media reaction to the Vig mixes varies by taste, but many accept a polarized argument that Wallace's mixes equate to success and Vig's mixes equate to authenticity, a position which seems logical in terms of marketing the re-release. The BBC review complains that "Wallace's heavy-handed mixing of *Nevermind* wiped clean some of its rawness" (Diver, 2011, para. 4 of 5); *The Telegraph* describes Vig's mixes as "muddier and more in keeping with the punk aesthetic of Nirvana's 1989 debut, *Bleach*" (McCormick, 2011, para. 5 of 9). Despite this, these points are possibly moot: Vig's mixes were not necessarily intended for release and may simply be snapshots of work-in-progress. Reasons identified for co-opting Wallace vary; in addition to time pressure, Vig himself described to *Rolling Stone* (Vozick-Levinson, 2011, para. 7 of 11) how interference by the band while working together hampered his mixes. A greater focus on his role as producer could be achieved in the remaining time if the mixes were finalized elsewhere.

If Vig's mixes were not finished to release standards, the most significance may be in consideration of one specific layer of persona. If Vig's mixes can be characterized as delivering a more balanced representation of the band and Wallace's more focused on Cobain as lead singer, it is likely that Vig's mix reflects his experience as producer working alongside the trio and his understanding of their relationship: the band's "real person" persona as understood by Vig (and by extension, by the band itself), explaining some of the subsequent ill feeling towards the Wallace mixes as portrayals by an outsider. Perhaps tellingly, Vig's positive working relationship with Dave Grohl has continued on the Foo Fighters' albums *Wasting Light* (2011) and *Sonic Highways* (2014).

The differences between the *diana* and *Nevermind* alternate versions illustrate the complexity of attempting comparison; analysis becomes fairer if both versions were genuine release candidates. In searching for a connection between artistic persona and mix presentation, there is some value in both cases, *diana* being the most convincing, although without explicit confirmation that the Chic mixes were presented as being for release in exactly the same way as Rodgers and Edwards' previous productions had been, caveats are required. The impact of the use of different recorded takes of performances between mixes should not be overlooked. There is likely to be much more information to be discovered regarding external or commercial creative influence or direction, and it would be beneficial to find the opinions and influence of creative and marketing stakeholders within the record labels at the time, as this is generally less documented.

It is entirely possible that no two genuinely equal release candidates of any one album will be found, but, regardless of the intended use of a mix, evidence that the aims of engineers are supported by listeners is encouraging, even if the process to achieve those aims is opaque.

CONCLUSION

It is clear that there is scope for further investigation here. In searching for links to persona when comparing rejected mixes with successful releases, the key area of focus may be importance or, to visualize, the artist's position with regard to the rest of the sound-stage.

Exploring the hypothesis has shown that any link between mix and persona is complex and subject to numerous influences. In the case of Diana Ross and Chic, personal relationships may have been the most significant player. In the case of Nirvana and Andy Wallace, concerns regarding authenticity in the face of success may have over-ridden a purely audio-based appraisal. Looking for more evidence of intent from engineers and outcomes from listeners could test the idea further.

The conclusions of any one listener's process, as in this case, will be affected by the understanding and degree of familiarity with the subject. As objectivity can be compromised and results cannot be generalized, a logical next step would be testing the responses of individual listeners to the two sets of mixes to examine if there is any consensus, though careful construction of the testing environment and consideration of the musical sophistication and level of listening skills present in the subjects would be required.

Similarly, the small number of subjects under investigation here shows that ultimately, only a much wider choice of test candidates will allow the research to develop towards reliability. The manufacturing of test examples may be required, a task which may be made easier as occasionally professional-quality multi-tracks are made available legitimately by an artist (both Björk and Trent Reznor have done this publicly in recent years, for example). Due professional process in mixing and a desire to avoid introducing other variables will be required to get the best experimental results. Another option would be to work closely with industry-level professionals to further test the theory. It is clear that a more primary-research-based focus, moving towards practitioners and listeners, will drive this investigation further.

REFERENCES

Alleyne, M. (2007). Interview with Nile Rodgers. *Journal on the Art of Record Production*, 2. Available at: http://arpjournal.com/interview-with-nile-rodgers/ [Accessed 8 Aug. 2015].

Auslander, P. (2004). Performance analysis and popular music: A manifesto. *Contemporary Theatre Review*, 1(14), pp. 1–13.

Azerrad, M. (1993). *Come as you are: The story of Nirvana*. London: Virgin.

Brown, A. (2015). Private correspondence [email].

Cherry, E. C. (1953). Some experiments on the recognition of speech, with one and with two ears. *The Journal of the Acoustical Society of America*, 5(25), pp. 975–979.

Chin, B. (2010). Diana: Inside out and round and round. Booklet notes in *diana* [sic]. Deluxe ed. Motown, Universal 0600753279458.

Diver, M. (2011). Nevermind (20th Anniversary Boxset [sic]) review. *BBC Music*. Available at: www.bbc.co.uk/music/reviews/9pj9 [Accessed 10 Aug. 2015].

Dockwray, R. and Moore, A. F. (2010). Configuring the sound-box 1965–1972. *Popular Music*, 2(29), pp. 181–197. Available at: http://dx.doi.org/10.1017/S0261143010000024.

Donze, P. L. (2011). Popular music, identity, and sexualization: A latent class analysis of artist types. *Poetics*, 39, pp. 44–63. Available at: http://dx.doi.org/10.1016/j.poetic.2010.11.002.

Easlea, D. (2003). Diana Ross: Diana—deluxe edition review. *BBC Music*. Available at: www.bbc.co.uk/music/reviews/wm4h [Accessed 8 Aug. 2015].

Edmonds, B. (2001). A revolution in sound & spirit: The making of what's going on. Booklet notes in *What's going on*. Deluxe ed. Motown, Universal 0600753279557.

Edwards, B. and Rodgers, N. (1980a). *Have fun again*. New York: Chic Music, Inc.

Edwards, B. and Rodgers, N. (1980b). *Inside out*. New York: Chic Music, Inc.

Evans, J. and Hesmondhalgh, D. (2005). *Understanding media: Inside celebrity*. Maidenhead: Open University Press.

Frith, S. (1996). *Performing rites: Evaluating popular music*. Oxford: Oxford University Press.

Gaye, M. (2010). *What's going on*. Deluxe ed. Motown, Universal, 0600753279557.

Gibbs, T. and Dack, J. (2008). A sense of place: A sense of space. In: M. Doğantan-Dack, ed., *Recorded music: Philosophical and critical reflections*. London: Middlesex University Press, pp. 172–186.

Gibson, D. (2008). *The art of mixing: A visual guide to recording, engineering and production*. 2nd ed. Boston, MA: Course Technology.

Goffman, E. (1959). *The presentation of self in everyday life*. London: Penguin.

Gold, J. (2009). Nile Rodgers. *Guitar Player*. Available at: www.guitarplayer.com/artists/1013/nile-rodgers/16959 [Accessed 11 Aug. 2015].

Gress, J. (2015). Funk it up like Nile Rodgers. *Guitar Player*. Available at: www.guitarplayer.com/artist-lessons/1026/funk-it-up-like-nile-rodgers/50349 [Accessed 10 Aug. 2015].

Huey, S. (n.d.). Chic artist biography. *AllMusic*. Available at: www.allmusic.com/artist/chic-mn0000092942/biography [Accessed 10 Aug. 2015].

Izhaki, R. (2012). *Mixing audio: Concepts practices and tools*. 2nd ed. Oxford: Focal Press.

Manic Street Preachers (2004). *The holy bible*. 10th Anniversary ed. Sony/Epic, 518872 3.

McCormick, N. (2011). Nirvana: Nevermind (20th Anniversary Edition), CD review. *The Telegraph*. Available at: https://www.telegraph.co.uk/culture/music/cdreviews/8781291/Nirvana-Nevermind-20th-Anniversary-Edition-CD-review.html [Accessed 21 Feb. 2019].

Mixerman. (2014). *Zen and the art of mixing*. Revised ed. Milwaukee: Hal Leonard.

Myers, J. (2015). 10 girl group stars who went it alone. *Official Charts*. Available at: www.officialcharts.com/chart-news/10-girl-group-stars-who-went-it-alone__8768/ [Accessed 28 Aug. 2015].

Nirvana. (2011). *Nevermind*. Super Deluxe ed. DGC/Universal, B0015885–00.

Norman, K. (1996). Real-world music as composed listening. *Contemporary Music Review*, 1(15), pp. 1–27. Available at: http://dx.doi.org/10.1080/07494469608629686.

Nusbaum, E. C. and Silvia, P. J. (2011). Shivers and timbres: Personality and the experience of chills from music. *Social Psychological and Personality Science*, 2(2), pp. 199–204.

Olive, S. (2010). Why live-versus-recorded listening tests don't work. *Audio Musings*. Available at: http://seanolive.blogspot.co.uk/2010/07/why-live-versus-recorded-listening.html [Accessed 29 Aug. 2015].

Owsinski, B. (2014). *The mix engineer's handbook*. 3rd ed. Boston, MA: Course Technology.

Percy, C. (2014). In honour of Diana Ross, can we reclaim the word "Diva"? *The Guardian*, 26 Mar. Available at: www.theguardian.com/women-in-leadership/2014/mar/26/honour-diana-ross-reclaim-diva. [Accessed 28 Aug. 2015].

RIAA. (2015). *Recording industry association of America gold and platinum searchable database*. Available at: www.riaa.com/goldandplatinumdata.php [Accessed 20 Aug. 2015].

Ross, D. (2010). *Diana*. Deluxe ed. Motown, Universal, 0600753279458.

Schaeffer, P. (1966). *Traité des Objets Musicaux*. Paris: Seuil.

Smalley, D. (1997). Spectromorphology: Explaining sound-shapes. *Organised Sound*, 2(2), pp. 107–126. Available at: https://dx.doi.org/10.1017/S1355771897009059.

Truax, B. (2001). *Acoustic communication*. 2nd ed. Westport, CT: Ablex.

uDiscover. (2014). When Diana Ross got a chic makeover. *uDiscover*. Available at: www.udiscovermusic.com/diana-ross-got-chic-makeover [Accessed 6 Aug. 2015].

van Winden, J. (2014). Thoretical framework: Persona, performance, method. In: *Destabilising Critique: Personae in Between [sic] Self and Enactment*, pp. 11–21. Available at: https://jessevanwinden.files.wordpress.com/2014/12/jesse-van-winden-destabilizing-critique-theoretical-framework-persona-performance-method.pdf [Accessed 25 May 2015].

Vozick-Levinson, S. (2011). Inside the 20th anniversary reissue of "Nevermind". *Rolling Stone*. Available at: www.rollingstone.com/music/news/inside-the-20th-anniversary-reissue-of-nevermind-20110823 [Accessed 10 Aug. 2015].

12

Mixing Beyond the Box

Analyzing Contemporary Mixing Practice

Alex Stevenson

INTRODUCTION/CONTEXT

Despite their initial design as a simple audio recording and editing system, in the last 30 or so years since their introduction, Digital Audio Workstations (DAWs) have gradually replaced the majority of equipment found in traditional analog recording studios; from microphone pre-amps, signal routing matrices and dynamic range compressors, all the way through to equalizers, time-based effect devices and summing mixers. There is strong evidence to suggest that these digital technologies have been overwhelmingly successful in replacing their analog counterparts, with virtual plugins for all stages of the signal chain available within a holistic software environment conveniently available on the user's desktop, laptop or now even tablet or smartphone. Furthermore, the design of the Graphical User Interface (GUI) of the vast majority of popular DAWs throughout their history has been firmly rooted in the traditional design concepts of large-format analog consoles with the intention of easing users' transitioning from the 'old' format to the 'new' (Bell et al., 2015).

This relatively recent transition from analog to the dominance of digital audio recording technology has coincided with the demise of commercial music recording and production budgets and concurrently the closure of a substantial number of large professional recording studios; these changing conditions have led to significant shifts in the working methods and practices of recording and mixing engineers, with many having to undertake some if not all of their work in smaller, often home-based studio facilities using larger commercial studios only when a specific recording or production task, such as tracking drums, grand piano or ensembles of musicians, is required and, crucially, where the allocated budget allows for this additional expenditure. It has therefore become common practice for engineers to move projects between different studios throughout the recording stage of a production. Additionally, during the mixing stage, engineers are often required to perform recalls, tweaks and edits to mixes in response to client feedback. Both of these practices are significantly simplified by an 'in-the-box' workflow, as it provides an "integrated environment . . . that offers a greater level of dynamism and the ultimate flexibility" (Paterson,

2016, p. 79). Despite these obvious benefits, however, many professional mixing engineers still favor the use of specific analog processors and tools, such as dynamic range compressors, equalizers or analog summing mixers. This apparent conflict between the ease of portability and recall of an in-the-box workflow and the perceived sonic benefits of analog processing has required mixing engineers to develop their own strategies which allow them to incorporate their favorite analog processing into an in-the-box workflow. This is commonly referred to as hybrid mixing—an attempt to "embrace both analog and digital" (Massy, quoted in *Studio Science: Sylvia Massy on Hybrid Mixing*, 2017).

With the demise of large commercial studios, opportunities for traditional apprenticeship training into recording engineering and production roles have become extremely limited. It has therefore become far more common for those with an interest in this area of work to instead enroll in a college or university program in subjects such as Audio Engineering, Music Production or Music Technology (Davis and Parker, 2013). These programs often give students access to professional recording studios and experienced teaching staff who can pass on their knowledge of recording and production practices. Within these programs, it is common to find the application of a teaching model whereby students will be transitioned from predominantly analog recording systems on to digital and in-the-box systems. Initially, students will be introduced to a traditional analog console and taught signal routing and monitoring entirely in the analog domain. This develops students' understanding of signal flow, especially when teaching recording techniques and practices, with the DAW often being used as simply a recording and playback system (akin to a tape machine), with all the signal routing and processing undertaken on the analog console, often in conjunction with external processors and effect units routed via a patch bay. In many ways, this emulates recording practices in the early 1990s, where Pro Tools systems began to replace tape machines and other digital recording systems. Students will then often transition from this analog workflow to an exclusively in-the-box workflow inside a DAW, often in a computer lab or, depending on the available facilities, a digital or in-the-box recording/mixing studio. Initially, this approach seems logical as it allows students to transfer their knowledge of analog systems into the architecture of a DAW, but it is common for these two systems, approaches or workflows to be treated as discrete, binary or mutually exclusive. This issue is further exacerbated by the absence of any specific discussion of hybrid mixing strategies or practices in many of the key mixing textbooks used in support of these programs (e.g. Hodgson, 2010; Izhaki, 2017; Owsinski, 2014; Savage, 2014), which instead tend to tackle analog and digital workflows in isolation. Despite the fact that these authors assume that any use of analog processing is likely to be in a hybrid system, this lack of discussion about the specific issues related to hybrid systems is somewhat problematic, as it does not encourage students to fully explore the range of possibilities available in hybrid analog-digital systems to discover the potential benefits of incorporating analog processing into modern in-the-box workflows.

METHODOLOGY

This chapter aims to explore the factors that impact on mixing engineers, highlight some emerging techniques and practices, and finally consider how these could be implemented into audio education programs to better equip future mixing engineers in their practice and support the knowledge transfer between the current and future generations of mixing engineers. The data collection for this project utilized a combination of face-to-face, telephone and email semi-structured interviews with 13 practicing mixing engineers. Although the number of participants is limited, they were selected as far as was possible to cover a diverse spectrum of age, experience and gender to allow for the findings to be broadly representative. All of the participants were based in the UK, although a number have either moved here from another country or have experience working internationally. These interviews were also supplemented by additional observations of professional mixing sessions, workshops and guest lectures from some of the participants at Leeds Beckett University in the period between 2015 and 2017. In response to these observations, following from Braun and Clarke's guidance (2006), thematic analysis of the interview data was employed to help draw out some common themes and practices which are presented in this chapter.

The first section of this chapter will discuss the broader emerging themes and issues that impact on the work of the participants. The second section will then highlight specific techniques and practices employed by the participants to address these issues. Finally, the third section will consider the possible implications of this knowledge in curriculum design in educational institutions.

EMERGING THEMES

All of the participants acknowledged significant changes to their working practices over recent years. Many of the comments made were surprisingly consistent despite the participants working in different spheres of the industry, from those predominantly working with independent, self-funding artists to those working on major-label commercial records. Even the participants who were relatively new to the industry in the early stages of their professional careers seemed as aware of these changes as those with decades of industry experience.

Budgets

The impact of reduced budgets became one of the most significant emerging themes in the participant comments. The reduction in recording and production budgets seemed to be the overarching factor feeding into all of the comments, which ranged from general comments recognizing diminishing budgets to those related to changing working practices.

Steph Marziano:[1] For the most part, the budgets are increasingly smaller.
Mick Glossop:[2] You tailor to the budget, that's the big thing; trying to find out what the budget is, is hard enough; nobody wants to tell you . . .

my clients were by and large self-funding artists or independent labels
and that kind of stuff, so yeah, low budgets, or no budgets, or very
small budgets.

Marco Pasquariello:[3] Sometimes I'll mix on a desk, but only if there's
budget/the track is up the board and sounding great, and we want to
commit. . . . This happens rarely now because of budget and time
restraints.

Al Lawson:[4] I wouldn't feel comfortable saying to someone, "yeah, pay
£500 for the studio, then pay me, and it will all be great".

Simon Gogerly:[5] One of the main reasons for changing to the new model
was client requests that I mix in my own studio to reduce costs.

Small and Home Studios

As exemplified by Simon Gogerly's comment in the previous section, budget limitations and the closure of the majority of large commercial recording studios have led to most mixing engineers changing their main place of
work from large commercial studios to significantly smaller, often home
studios or mixing rooms. This theme was highlighted by both participants
working on major record label projects, with access to commercial recording studios, and also with participants who might be considered emerging
engineers working with independent artists.

Steph Marziano: I've got my own studio so I tend to mix out of there.

Mick Glossop: We've had to be flexible and move with the times, build
our own [home] studio. . . . I also wanted it to be built to the same standards as a professional studio, so I didn't feel like I was maybe having
to apologize to clients that I couldn't do this . . . even though it's small
by comparison [with commercial studios].

Simon Gogerly: I moved over to this [hybrid] way of working about
9 years ago when I opened my own studio.

Aubrey Whitfield:[6] I was basically producing my own material, so I'd
invested in my home studio. . . . I trust the sound there, you know I'm
used to it, and its flexibility.

Al Lawson: My studio is . . . in a basement office thing, it's all kind of
music studios in the basement. Just a simple space . . . I listen at a level
that would irritate a neighbor in a flat, so I couldn't mix anything at
home.

Studio Fees

During the discussions on using their own studios for professional work, participants highlighted a range of issues around how this affected their income.
Concerns were raised around charging a separate fee for the studio time and
the mixing engineer's time, with a range of different approaches emerging. These differing approaches and opinions reflect the changing nature
of the work and, perhaps more significantly, a change in who is paying the
fee. With a move towards more artists self-funding their own releases, or

working with smaller independent labels with limited production budgets, compromises are being made on where to prioritize the spending.

Phil Harding:[7] With your own studio, beyond your overall production budget you have to think about, "what am I going to charge per day?", because the business affairs at the label will want to know, "well, what is that going to cost us?".

Mick Glossop: I tend to add a daily [studio] rate, even though it's really small, because that acts as an incentive to get the job finished.

Al Lawson: I generally don't charge extra for the studio, I just charge my day rate, so it's the same if we go somewhere else, or we go to my studio. . . . If you [the client] want a fancy studio, pay for one, and if you can't pay for one, then don't complain about where we are.

Aubrey Whitfield: When you obviously work with independent artists who aren't on a record label, their budgets aren't like big record labels. . . . I always tell my artists, "if you've got the budget, hire a professional mix engineer to mix this track", but a lot of them don't listen.

Mix Revisions

Another emerging theme, recognized as a key driver in the changing working practices for the participants, was client requests for mix revisions. Although this practice has been common for some time, participants' comments highlighted a change in client expectations over recent years in relation to both the number of recalls and expectations regarding any associated costs.

Phil Harding: The clients are very savvy on the technology, and they know that it's hard for someone to justify saying, "it's going to take me half a day or a day . . . to set up"; they know that that doesn't exist anymore. . . . Clients are . . . continually asking for more adjustments.

Simon Humphrey:[8] The dynamic nowadays is either the record company or the client are essentially paying your wages . . . so do you mix for yourself? Do you mix for them? That's the big debate now, isn't it? You mix something for yourself, and the client goes, "well, I don't like it, change it".

Steph Marziano: You have labels asking for recalls every 5 minutes.

Mick Glossop: If they [the client] drift on asking for more and more revisions, it's going to cost them something.

Marco Pasquariello: My clients often want changes quickly and affordably.

Al Lawson: Even if [the mix] is great on that day, [the client might say], "oh, actually can we just change this?" or "that's not quite how I wanted it", and you've got to be prepared for that situation.

Furthermore, some of the participants commented on this becoming an issue in how they approach their work, suggesting a range of different strategies for overcoming the issue. These ranged from writing it into their contract, to getting the artists to sign off on the mix in the listening sessions, to negotiation strategies to convince the client to accept the mix, to downright refusal to recall:

Phil Harding: I think the studios and producers have been bullied out of what was an industry standard in the 80s and 90s . . . where our lawyers would make sure in our production contracts that we are only delivering one further mix beyond the initial mix. Your first mix is your listening mix, the second mix, you take on board comments and so on, and the third mix you start charging an extra fee. . . . One could justify at least half-a-day's studio time, if not a whole day.

Steph Marziano: Most people, with a couple of tweaks, are happy with the mix. Sometimes you have to, like, kind of convince them that actually it was really good on mix 2 as opposed to mix 4; just sort of getting in their heads a bit, but I've only really had to say once, maybe twice, "hey, listen, it's kind of getting a bit too much", but you know, it's their tune so they are going to be as precious as they can about it.

Mick Glossop: The way I try and play it is to mix on my own and then get [the client] in to listen to stuff, because then you get the feedback.

Simon Humphrey: At some point you get a couple of clients who say, "we want a mix with you because you did this, and you did that, and you were there in the good old days, and we really want you to recreate that", and you get sold this idea that people want that, and you try to sit them down and say, "Do you understand what this means? Do you want analog mixing like it was in the 80s?" And they say "yeah, everything sounded great in the 80s", and then you do it and find out that's not what they want because they're schooled in the modern demands. I had a couple of very bad experiences that way, in terms of saying, "No, I can't just tweak this in two minutes".

MIXING STRATEGIES AND TECHNIQUES

In light of the emerging themes discussed in the previous section, all of the participants interviewed seemed to be at some point on the journey towards mixing exclusively 'in the box'; however, a majority of participants were still incorporating some elements of analog, and sometimes digital, outboard processing into their workflow. During the interview process, participants revealed a wide range of strategies and techniques that they either have previously used or still use to enable them to work with their preferred tools while still being able to cope with their clients' expectations. These strategies and techniques will be discussed in the following section, which will begin by first highlighting techniques which predominantly rely on the traditional model of mixing on an analog console with minor adjustments to incorporate a DAW, and will be followed by discussions of practices which successively transition more and more towards an entirely in-the-box workflow.

DAW as a Tape Machine

Digital audio workstations were initially designed to replace tape machines as a recording and playback device with the substantial benefits of digital editing. Despite the vast development of DAWs to enable audio processing and summing allowing complete audio mixing inside the box, some of the participants highlighted a preference for mixing using this more traditional approach. These approaches mostly rely on using an analog console along with analog and digital outboard processing and effect units in conjunction with a DAW which fundamentally acts as a tape recorder. In these workflows, practitioners highlighted using the DAW only for basic sub-mixing tasks, such as mixing numerous individual tom and overhead drum microphones to a single stereo output—a practice many engineers would do during a tracking session to a 16- or 24-track tape machine. Significantly, however, they also highlighted strategies to aid with recall, such as using the DAW to store track levels by setting all the faders on the console to a constant level, a technique also demonstrated by producer/engineer Sylvia Massey (*Studio Science: Sylvia Massy on Hybrid Mixing*, 2017).

Ken Scott:[9] I use Pro Tools to submix and then bring it all out on the desk.

Simon Gogerly: I do a lot of submixing in Pro Tools.

Phil Harding: What I would call a nice compromise digital/analog mix, . . . is that you take maybe 24/48, as many outputs as you've got on Pro Tools, and you run it through an analog desk. The last time I did that was five years ago in Los Angeles mixing a Cliff Richard album. The compromise then in that type of mixing was that one ends up using Pro Tools as a tape machine, so there is no processing in Pro Tools, but what I chose to do, should we need to recall . . . was to have all the faders on the SSL at 0[dB], so Pro Tools was not just a tape machine, it was my balance.

A Preference for Analog Consoles

The preference for mixing on an analog console was highlighted by the majority of participants, but the vast majority had moved away from this approach, predominantly due to practical reasons such as reduced budgets, limited space and the need to instantly recall sessions.

Marco Pasquariello: I haven't been mixing professionally for longer than 10 years really, but where I started out, I was a bit of a purist and mixed mainly analog. This included the use of the desk automation/recalls/full comprehensive stems. It sounded great. Sometimes I'll mix on a desk [now], but only if there's budget, the track is up the board and sounding great, and we want to commit.

Steph Marziano: Obviously, if I had a massive room with an SSL, I'd be mad not to use it, but you know . . . I'd probably always go in the box now, which I never thought I'd say.

Al Lawson: It's been so long since I've actually mixed on an SSL or something.

Richard Formby:[10] I think I might quite like to go back to using a desk. . . . I'd probably consider putting a good desk in [my studio] . . . you know it's much better on a desk.

Participants therefore highlighted that mixing on an analog console was becoming more and more of a niche practice, which would only be suitable in very specific circumstances.

Phil Harding: Looking forward, I think it will take some pretty unique situations for people [to] be totally analog mixing in the future.

Simon Humphrey: [Chairworks Studio] ended up getting this great [SSL] console, and once you've got it, you've more or less got to embrace it, so you are committed, whether you like it or not . . . and once other studios fall away you . . . it evolves into more of a unique place, it stands out by being different, so you kind of promote that.

Printing

While discussing their preference for using analog consoles, many of the participants highlighted strategies to allow them to use aspects of a console signal path while maintaining a digital workflow which facilitates the ability to easily recall and tweak mixes. Although some of these strategies took advantage of manufactured technological solutions (such as Clint Murphy's use of the SSL Duality's 'digitally controlled analog' feature, which allows fader and processor values to be stored and recalled via a combination of stand-alone software and DAW automation), many of the participants had developed their own unique strategies, many of which included printing channels or stems from the console back to the DAW. This allowed the participants to incorporate their existing practice and toolset, but would speed up the ability to recall and edit mixes at a later date.

Simon Humphrey: People . . . think that mixing is processing and summing at the same time. . . . The only way I could get it to work was to completely divorce the two processes. So, you process, you re-print; basically stem mixing . . . and then you sum, or you mix the re-processed, re-printed analog. . . . This is what you have to do to get it to work.

Simon Gogerly: I will print mix stems into Pro Tools, a process I started doing quite some time before moving to hybrid mixing.

Steph Marziano: I mixed the Idris [Elba] album sort of like a half [analog] and half [digital] way. . . . I did two days on an SSL desk, where I put stems through, and . . . put EQ and compression and recorded them back into Pro Tools. . . . I did all tweaks and stuff [in the studio], and then the last day was just adjustments in the box.

Interestingly, many participants who favored the use of stereo bus compression in the mixing would state that they would leave this process off of any stem-mixes printed back into the DAW, which was mostly due to issues around the ability to accurately recall a mix from stems (an issue discussed later in this chapter), which would be exacerbated by the use of stereo bus compression on individual printed stems.

Marco Pasquariello: If I'm mixing [on a] desk, I'll do stems back into Pro Tools but will leave the 2 bus processing off generally—then recall that at a later date if needs be—or recreate it in the box if a mix tweak is needed.

Al Lawson: [I] would print the stems without the mix bus processing (compressor/EQ), and then if [I] recalled from stems, run the tweaked balance back through the mix bus gear.

In addition to printing channel and stem-based processing, many of the participants also discussed the practice of printing effect tracks back into the DAW to allow them to recall and tweak mixes at a later date without needing access to specific outboard effect units. Some participants, however, justified this practice for more creative rather than pragmatic reasons.

Richard Formby: I print all of my analog effects back into the session and treat them as I would any other audio track.

Marco Pasquariello: I often use external FX such as spring reverbs/tape delays etc. Also analog reverbs or real spaces reamped; these will get printed back into the session.

Al Lawson: I might reamp [guitar tracks] through my amps or send something to a tape delay and print it to the session. . . . I print tape delays. . . . I can put on any pair of headphones and go "oh, well, I've turned up that bit, and I've turned that bit down, or I've turned the reverb up on that bit" . . . and can kind of judge . . . knowing that what else I'm hearing is right. I can make those little tweaks without being in a studio.

Simon Gogerly: I will now print individual effects tracks into Pro Tools as well so that my recall process is faster and simpler.

During the interviews, however, some participants clearly stated their preference for not printing channels, stems or even effect tracks back into the DAW as part of their workflow. Many suggested this was due to the fact that they now had alternative methods to recall the mix.

Mick Glossop: Well, actually . . . I don't tend to print anything back in, I just work on the [principle of] "well, I can recall it". I suppose if something breaks down . . . I mean I've got a spring reverb there, if that decided it wasn't going to work anymore, I'd get it fixed wouldn't I, I mean it can't be that difficult!

Al Lawson: Any stems that I do, I would be doing as deliverables to the client; I wouldn't be doing them so that I can then work from them, because I'm mixing in the box.

Although many practitioners discussed using these strategies at some point in the recent past, many disclosed that they have either significantly adapted their approach, with some going as far as stating that they would no longer choose to work in this way. Although the reasons given for these shifts were somewhat varied among the participants, the overarching themes of client expectations, budgets and time were the most common.

Simon Humphrey: [Printing tracks/stems became] way too complicated for me and clients. You have to have so many ins and outs to do it . . . you need good converters . . . you need to understand analog workflow in terms of noise. . . . Most people come with 90 tracks of audio now, so you are basically trebling that overnight, so it makes it very, very hard. I've kind of really reigned that back in; I stripped out a lot of that. . . . I decided that . . . if you want EQ and compression they sound better analog . . . whereas everything else is digital, so reverb, delays, all the automation, all of that all has to go digital, throw all that into the computer, give the computer as much work as it can do, that it is good at.

Steph Marziano: To be honest, I don't know if I'd do [stem printing] now. . . . [The Idris Elba album] was I guess two or three years ago. . . . I'm so used to mixing in the box since then, there are just things that you can't really do outside of the box as well. I just genuinely don't think I could imagine doing a mix any other way.

Aubrey Whitfield: [Printing] sounds like a right headache to me, doing all this bouncing out tracks and bouncing back in. Busy producers like me don't have the time for fiddling with all that.

Static Outboard

Another technique utilized by a number of participants was the use of static outboard settings. This is the practice of using an outboard processor, or effect unit, and using the same settings that don't change. Different participants highlighted their use of static settings between either different tracks (channels) within a mix, across similar tracks (e.g. lead vocals) on different songs if mixing an album, or even using the same settings across all of their mixing projects, utilizing the 'sweet spot' of the device. In this approach, the level of signal going to, and returning from, the outboard device is controlled via the level controls inside the DAW, and is therefore saved within the session and instantly recallable.

Mick Glossop: I use analog gear, but if I'm mixing an album of ten songs then I'll set up certain chains for the vocal or the drums or whatever, using analog gear, and it will stay the same for every track, and if I have to change anything for a track, I'll do it in the box.

Clint Murphy:[11] I'll generally always use the same settings [on the 1176] and adjust the level from Pro Tools until it sounds right. [This] is just a technique that's worked for me. . . . I think a lot of people are doing this actually when they are using the same room.

Simon Humphrey: There are people, Chris Lord Alge for instance, who was such an analog freak in the 80s, SSL, analog, the way it was, and he championed that to such a point that I thought "he's my hero", cause he talks about things in a way that I understood. He talks about an 1176. . ., "you listen for the sweet spot, when you've got it you know . . . that's what you do", and I think, "yeah, I know that, why don't people know that?".

Analog Summing

Although many of the participants were working mostly in the box, a number of these also cited their preference for the use of analog summing through either an analog console or a dedicated summing amplifier. Participants discussed the sonic differences between analog and digital summing as being a key driver for this approach, as well as highlighting techniques to overcome the potential issues of using analog summing within a digital workflow.

Mick Glossop: I've got a summing box, which I built myself for £35. . . . It's just passive mixing, and then you lose about 25 dB of gain, so I put it back through two SSL mic-pres which are pretty quiet, and that brings it, bangs it back, up to a decent level and then it goes to the SSL compressor. So I might feed stuff out and not bring it back into Pro Tools, and just put it into the summing box, cause it's all on the patch bay, it's pretty easy to do, so yeah, I'll construct some kind of system using the analog gear, and then it will stay like that.

Richard Formby: I've mixed two albums through this [Alice 828 analog broadcast desk] . . . . I really like the limiters that are built in to it. . . . It's eight channels, so I'm doing it really basically, yeah, I mean the last one I did I either had . . . mono drums and bass, all the guitars [on channels 3 and 4], all the keyboards [on channels 5 and 6] and all the vocals [on channels 7 and 8] . . . it summed together quite nicely.

Gary Bromham:[12] I really like the SSL Matrix . . . it is a summing mixer . . . a clean summing mixer which kind of troubles me, but actually for me the best part about that is that it's got a floating insert matrix, so in other words you can patch analog stuff across channels.

Simon Gogerly: [I] sum the resulting stem mixes on an analog console.

Al Lawson: A few times I've summed on an API 1608.

Providing Stems

An issue associated with the use of analog summing mixers, which proved to be somewhat controversial, was the topic of mix stems; not in using stems as part of the mixing process, which was common among the majority of participants, but in relation to the requirement to provide mix stems to clients along with the main mix file. Some participants discussed technical issues with the very concept of even using mix stems, with a common opinion expressed that it is impossible to recall and recreate an entire mix, even entirely in the digital domain, from summing the component mix stems, as the result will always sound different. Some participants, however, were more concerned with the moral, and potentially legal, issues with providing the stems to labels without any clear stipulations about how they might be used in the future. Other participants, however, regularly provided mix stems as part of their work and seemed to have no issues with this practice.

Phil Harding: What is either psychologically expected or is now contractually written by the labels, is an expectancy to deliver stems . . . even if something has been totally mixed analog, [the record company will ask] "supply us the mix stems". So you do a drums stem, and [if] someone says, "I don't like the drum sound", . . . it can't be changed from within a stem.

Mick Glossop: I don't do stems, in fact I haven't been asked to do stems yet. I think that's something [the MPG] are dealing with. . . . I haven't done anything with a major label for some time, and I think it seems to be more major labels. It's in the contract, and then there'll be some kind of studio manager person at the label whose job it is to collect the files, the assets as it were, and they'll have their list of stuff, and they've got to tick them off, "have you got the stems . . . the mixes . . . the instrumental . . . the multi-tracks?" . . . The theory is that you can recreate the mix, for a start that's a myth anyway; in my opinion, it's never going to happen exactly the same way. . . . It's [also] not like the record company has even said what they are going to use them for, which is another issue . . . a lot of artists are using the stems in live performances [and] no one is getting paid for that. We're getting paid to make a record.

Al Lawson: Nowadays I only print stems as a deliverable to get paid. They are required by most labels . . . in terms of delivering what I think people should want, rather than what I'm being paid for. I come from [a time when] assisting people we'd do stems, do all the mix versions, we'd do a vocal up, an instrumental, an a capella, and a TV mix, so I generally do all of those, because that is what a professional service is.

Stereo Bus Processing

The preference for using stereo bus processing was very common among the participants, even with those who are predominantly otherwise working entirely in the box, with many participants declaring their use of analog outboard devices, such as compressors and equalizers for this purpose. These participants also highlighted strategies they had developed to allow them to be able to easily recall mixes while incorporating these analog processors in their mix process.

Marco Pasquariello: I'll often use an analog mix bus chain in addition to a digital master chain in Pro Tools; this will usually be some kind of harmonic enhancer/EQ/Compressor . . . 2 bus recalls tend to be quite easy because it's just a couple of pieces of gear to recall. I tend to take photos of them and put those photos on a dropbox folder relating to the project.

Mick Glossop: [The mix] goes to the SSL compressor. . . . I make notes, but it's just a couple of sheets of paper. I've got outline drawings here. . . . it's all downloadable now, so I just tick all that off, stick it in the folder, so . . . I can recreate everything.

Many other participants cited the importance of using in-the-box bus processing in their workflow. Despite many expressing a preference for the use of analog equipment for this process, there was a recognition among these participants that this would not be financially or practically viable for them.

Aubrey Whitfield: The one [plugin] that has absolutely saved my productions is the SSL compressor on my master bus . . . it just makes the records sound . . . professional and glues everything together. . . . I really want to buy these amazing compressors, which I've got plugin versions of . . . but at like five grand a go, they are so expensive . . . and I think I can get pretty much the same effect by buying a £200 plugin . . . there is no need, it's not a gap for me. I've got all the tools that I need in the box to be able to produce the mixes that I need. It's all serving a purpose at the moment.

Steph Marziano: I tend to put a lot of that stuff on the master bus. The SSL compressor [plugin] is pretty much identical to the SSL on the desk, so I tend to just use that. . . . I sometimes play with getting . . . like that thing with the master chain, and getting a couple of [outboard] compressors, and always having that setup on my [output] 1 and 2, but the only reason I don't is purely for the fact that if I'm somewhere else and get mix recalls. . . [and] I don't feel like it's enough of a gain.

Al Lawson: My preference would be to have like a hardware mix bus insert chain, like a hardware EQ and compressor over the mix, and maybe a couple of things as features or whatever, like a lead vocal compressor or EQ or something. . . . I love gear, so I think that would be nice from a sound point of view and an ergonomics point of view, but the reality of having to be back in my studio to do every single thing is just like. . . "I've got to go back there?".

Plugins

The perceived improvement over recent years of the sonic characteristics of digital mixing, and more specifically plugins, has proved to be a significant factor in participants choosing to transition to mixing almost exclusively in the box. A significant number of the participants specifically mentioned Universal Audio's UAD plugin platform as being a fundamental factor in their transition to working in the box. These plugins seemed to be favored by many participants due to not only their perceived sonic qualities but also their relationship with key classic audio equipment manufacturers, with the UAD platform exclusively hosting many re-creations of classic analog and digital hardware which the participants were familiar with. This familiarity with the tools has enabled them to easily adapt their existing workflow from the analog domain into an entirely digital environment in the box without sacrificing the sonic characteristics they desire, while getting all the benefits of an entirely digital workflow.

Aubrey Whitfield: I use the SSL [bus compressor] and that API [channel strip] plugin from Universal [Audio] . . . their plugins are just amazing. They are by far better than waves. I've got loads of waves stuff that is really good as well, but they're the two that I would use on every production without fail.

Steph Marziano: I use a lot of UAD plugins. . . . It's pretty much the thing that [let us] have professional mixes with being in the box, for me at least.

Phil Harding: Universal Audio, okay, it doesn't cover everything, but it covers tons of classics. . . . Clients come to my team asking for a retro 80s sound, and all the 80s retro hardware is on there, and very well R&D'd. My only criticism is that some things can be too clean, like the Lexicon 224, but the DBX 160 compressors sounds great, the Pultec sounds great, the Roland equipment, the Dimension D and the Space Echo all sound great . . . most of what I would call the premiership mixers out there worldwide are all invested in Universal Audio. . . . It was Universal Audio that finally did it for Haydn [Bendall], when finally in the box he felt that, regardless of the fader thing, that he had enough equivalent to hardware outboard plugins that sounded quality to him, finally he could commit to being in the box.

DAW Control

The preference for using control surfaces rather than a mouse or trackpad was a topic that very much divided opinion among the participants. Interestingly, many participants who had a significant number of years of experience using large-format analog consoles in their early careers, who one might assume would prefer using physical faders over a mouse, expressed the biggest reservations around using control surfaces, and often seemed much happier using just a mouse.

Phil Harding: In my experience, engineers of my years of experience and age, we don't seem to adapt well to control surfaces.

Gary Bromham: I'll be honest with you, I mix most of my stuff with a mouse these days; actually, I use a trackball.

Hadyn Bendall:[13] I prefer to use a mouse.
Simon Gogerly: I still prefer to use a mouse. . . . I had a built-in control
surface on my last console (SSL AWS) but rarely used the facility.

Although in the minority, some participants were keen advocates of using
control surfaces and embracing some of the more recent technology such
as touchscreen controls, from companies such as Avid and SlateMT, with
others suggesting they might incorporate them into their workflow in the
near future.

Mick Glossop: [The Raven MTi] is not bad. I mean it's HUI, which is
a drag . . . so four characters per track . . . but it's configurable, and
you've got a load of buttons along the bottom, [however] it's a little
bit slow at times. . . . I [previously had] two Artist Mixes, and that
[Artist Control], and had them side by side, but the software issues . . .
it didn't seem to be supported that well. . . [but] the touchscreen is
amazing, and it's so configurable. I use both of them [The Avid Artist
Control and Slate Raven MTi] all of the time.
Gary Bromham: I really like the idea of just having solo, mutes and pan
pots and faders, I like the idea of that . . . but I don't really need anything else, I really don't. So actually I really like the SSL Matrix.
Aubrey Whitfield: I've got arriving in the next couple of days . . . a
Mackie control . . . to be honest I've always wanted some kind of mixing desk, because I'm one of these people who likes to feel the faders,
and not push up volumes on a mouse. I'm used to doing it now, but
I don't know, it seems more intuitive doing it with faders.

Analog Tracking

In recognition of the move towards an entirely digital mixing process for
many of the participants, many cited the increased importance of capturing the sonic characteristics of analog equipment during the tracking process, both in their own work as recording engineers and producers and in
the work of others whose recordings they are mixing.

Al Lawson: If I'm mixing something I've recorded then I've EQed,
Compressed, distorted to 'tape' and recorded room mics etc., so [dur-
ing mixing] I'm not usually trying to inject a lot of attitude into the
tracks.
Steph Marziano: Most of the projects I tend to record myself, and I record
in a good studio, so I tend to use loads of outboard on the recording.
Gary Bromham: I think there is far more focus now on recording. I think
recording is way more important than it used to be; it didn't used to
be as important when you were using an analog console, because you
could kind of mess things up much easier in the analog domain, and
analog distortion is much kinder than digital distortion is . . . mixing
starts on day one of recording.

IMPLICATIONS FOR CURRICULUM DESIGN

In recent years, audio manufacturers have recognized the emerging preference for hybrid mixing workflows, and as discussed by Reilly (2017), they have released a range of products to allow users to combine aspects of analog signal routing and processing with the digital audio workstation's total recall functions, ranging from digitally controlled analog devices to dedicated analog summing mixers, and even hybrid microphones. Although many educational institutions make every effort to upgrade their facilities to match current working practices, it is common for many institutions to focus their investments in completely in-the-box digital systems for mixing studios, commonly with control surfaces, while maintaining traditional large-format analog consoles and outboard in their recording/ tracking studios. While this allows students to master "how established, often analog-based, practices translate in the digital domain" (Bromham, 2016, p. 248), focusing on analog and digital as discrete practices does not fully allow for students to study "old skills . . . in the context of new ones to inform future work practices" (Fazekas, cited in Bromham, 2016, p. 246). It is therefore of paramount importance for academic institutions to engage with current practitioners, for example, through guest lectures and workshops. Furthermore, it is imperative for academics to critically analyze these practices to inform approaches to audio production teaching that equip students with the understanding required to build on existing practices and be ready to adapt to the ever-changing world of music production. In the United Kingdom, Professional, Statutory and Regulatory Bodies (PSRBs), such as JAMES and UK Music's Music Academic Partnerships in the UK, reinforce the link between industry and education, giving academia access to experienced professionals to input to curriculum design. Explicitly introducing students to these different approaches used by professional mixing engineers, such as those highlighted in this chapter, and encouraging them to experiment with and adapt these techniques during their studies will go some way toward addressing this issue. This will encourage the mixing engineers of the future to develop their own creative strategies while building on the foundations of existing practice. Furthermore, the strategies and techniques discussed in this chapter have been developed by mixing engineers in order to maintain efficiency to meet real-world expectations of clients, whereas the majority of student work within educational institutions is based on longer-term project deadlines based around academic calendars and semesters. Therefore, it is important for academic institutions to offer students the opportunity to apply their acquired learning from what Davis and Parker term their 'student toolbox' (2013, p. 4) in some kind of real-world project with short deadlines, as it is only in these situations that students are able to observe the benefits of these efficient workflows. As students reflecting on working on short-term projects state,

> you have to keep the flow of the sessions. . . . When we do modules at uni, the flow doesn't necessarily matter. . . . You can spend hours

and hours and hours learning how to use microphones [and] compression . . . it means nothing until you are in with someone working on a genuine product.

(Davis and Parker, 2013, p. 4)

NOTES

1. Interview with Author, 2017.
2. Interview with Author, 2017.
3. Interview with Author, 2016.
4. Interview with Author, 2016.
5. Interview with Author, 2016.
6. Interview with Author, 2017.
7. Interview with Author, 2016.
8. Interview with Author, 2016.
9. Interview with Author, 2016.
10. Interview with Author, 2017.
11. Interview with Author, 2016.
12. Interview with Author, 2017.
13. Interview with Author, 2016.

BIBLIOGRAPHY

Bell, A., Hein, E. and Ratcliff, J. (2015). Beyond skeuomorphism: The evolution of music production software user interface metaphors. *Journal of the Art of Record Production*, 9. Available at: http://arpjournal.com/beyond-skeuomorphism-the-evolution-of-music-production-software-user-interface-metaphors-2/ [Accessed 30 Nov. 2016].

Braun, V. and Clarke, V. (2006). Using thematic analysis in psychology. *Qualitative Research in Psychology*, 3(2), pp. 77–101.

Bromham, G. (2016). How can academic practice inform mix-craft? In: R. Hepworth-Sawyer and J. Hodgson, eds., *Mixing music*. Perspective on Music Production. Oxon: Taylor & Francis, pp. 245–256.

Davis, R. and Parker, S. (2013). Collaboration, creativity, and communities of practice: Music technology courses as a gateway to the industry. In: *Audio engineering society conference: 50th international conference: Audio education*. Audio Engineering Society. Available at: www.aes.org/e-lib/browse.cfm?elib=16851.

Hodgson, J. (2010). *Understanding records: A field guide to recording practice*. New York: Bloomsbury Publishing.

Izhaki, R. (2017). *Mixing audio: Concepts, practices, and tools*. Oxon: Taylor & Francis.

Owsinski, B. (2014). *The mixing engineer's handbook* [Electronic resource]. Australia: Course Technology.

Paterson, J. (2016). Mixing in the box. In: R. Hepworth-Sawyer and J. Hodgson, eds., *Mixing music*. Perspective on Music Production. Oxon: Taylor & Francis, pp. 77–93.

Reilly, B. (2017). *Hybrid mixing techniques: Leveraging the best of in the box and out of the box environments*. Available at: https://vintageking.com/blog/2017/06/hybrid-mixing/ [accessed 11 Feb. 2018].

Savage, S. (2014). *Mixing and mastering in the box: The guide to making great mixes and final masters on your computer*. Oxford: Oxford University Press.

Studio science: Sylvia Massy on hybrid mixing. (2017). Hollywood. Available at: www.youtube.com/watch?v=T1PmKtP1WPE [Accessed 18 Jan. 2018].

13

Optimizing Vocal Clarity in the Mix

Kirsten Hermes

It is generally agreed that vocal clarity is one of the most important quality parameters of popular music mixes (Ronen, 2015; Hermes et al., 2017, pp. 10–11). However, there is currently no formally proven basis of what makes recorded singing clear, nor a generally agreed-on definition of vocal clarity. Two computational predictors exist for the spectral clarity of a range of isolated sound sources, including piano, guitar, strings and one vocal take (Hermes et al., 2017). It might be possible to develop similar predictors specifically for vocal clarity in mixes if this depends on a set of clearly defined, generalizable acoustic parameters. Findings in this area would be useful as they could, for example, inform assistive, artificially intelligent mix tools for vocals or help aspiring mix engineers further their skills. In order to assess vocal clarity further, it appears beneficial to draw upon the knowledge of both scientists and creative practitioners in a cross-disciplinary approach. A broad overview of current findings on vocal clarity can be found in the first section. The aforementioned computational predictors of single sound spectral clarity (Hermes et al., 2017) are subsequently tested in the context of a vocal mix in the second section in order to assess whether they offer a useful starting point for measuring vocal clarity in mixes. This is initially done as an autoethnographic study, but feedback is also sought from other audio professionals in a short, indicative pilot listening test. Based on this assessment, suggestions for further research are presented.

VOCAL CLARITY AS A SCIENTIFIC AND CREATIVE CONCEPT

The current section is an overview of existing findings on vocal clarity in music mixes. Findings stemming from scientific studies will be presented alongside creative concepts. The importance of vocal clarity is apparent from the fact that accomplished producers and mix engineers often mention this as one of their key goals in the mix process. Ken Kessie strived to create a "clear, bright vocal" (Clark, 2011, p. 202), and Kevin Shirley aims for a "proud", "present and clear" vocal sound (Clark, 2011, pp. 209–211).

Jeff Strong (2009, pp. 249–276) states that a good vocal mix is character-ized by clarity, presence, brightness and fullness, without "muddiness" or "sibilance", and Bazil (2008, p. 2) stresses the importance of "clarity" and "separation" of important sound sources in mixes. Bob Clearmountain, as quoted by Clark (2011, pp. 193–197), aims to avoid vocal harshness.

An interdisciplinary approach is likely to lead to a nuanced and broad understanding of vocal clarity. While academic literature is usually sup-ported by thorough scientific research, sources aimed at creative practitio-ners provide more information about the vocal mix techniques employed by producers and mix engineers. Therefore, both academic and creative sources are included in the following literature review. Since no scientific literature was found specifically for recorded vocal clarity, several related areas will be investigated. This will include research on general sonic clarity and the two predictors of spectral clarity mentioned earlier. Fur-thermore, the literature review will touch on vocal timbre parameters and overall sound quality, as quality, timbre and clarity are likely all related. These scientific findings, in combination with the opinions of creative practitioners, are likely to yield useful starting points for further investi-gating vocal clarity.

Sonic clarity has been an important focus across audio research areas, including clarity in concert halls (e.g. Beranek, 1996), speech intelligi-bility (e.g. IEC standard 60268–16 2011) and timbral clarity (Solomon, 1959, pp. 492–497). Other research areas that are directly related to clar-ity include loudness (Fletcher and Munson, 1933, pp. 82–108), masking (Moore, 2012, pp. 67–32) and auditory scene analysis (Bregman, 1990). According to ANSI/ASA S1.1 (2013), "masking is the process by which the threshold of audibility for one sound is raised by the presence of another, masking sound". Auditory scene analysis (ASA) is the process of forming mental representations of individual sound sources from the summed waveforms that reach the ears (Bregman, 1990). Findings stem-ming from all of the aforementioned research areas are summarized in the following.

Both Solomon (1959) and Hermes et al. (2017) have shown that in the context of timbre, an increased high to low frequency ratio correlates positively with clarity change. At the same time, strong resonances can reduce clarity in music mixes (Hermes et al., 2017, pp. 214–224). Mask-ing influences the ability to hear and identify sounds (Hafezi and Reiss, 2015, pp. 312–323) and is related to spectral overlap, the phases of over-tones, the timing of onsets, the presence of modulation and reverb in target and masker (Moore, 2012, pp. 67–132) and their loudness and loca-tion (Zarouchas and Mourjopoulos, 2011, pp. 187–200). Aichinger et al. (2011) state that masking reduces the clarity of sounds in music mixes and should be avoided depending on the sound source: "for instruments in a polyphonic melody or voices in a choir it may be worth-while to create a merging sound. . . . For soloing instruments it may be worth striving for a disunited sound". Since vocals tend to play a leading role in mixes, it is likely that they require separation. In ASA, the fusion or separation of sonic elements depends on their amplitudes, similarities in their timbres, harmonic relations, binaural frequency matches, spatial features and their temporal behavior (Bregman, 1990, pp. 641–697). In concert halls, reverb, echoes, noise, tonal distortion and sympathetic ringing tones reduce clarity (Beranek, 1996, pp. 477–481). The ratio between early and late-arriving sound energy, as measured by, for example, the C80 and C50, and the early decay time (ISO 3382-1, 2009, pp. 13–19) are common clarity measures. Loudness is directly linked with audibility and therefore clarity, and depends on a sound's intensity, frequency spectrum (Fletcher and Munson, 1933), perceived location and binaural relations (e.g. Moore and Glasberg, 2007, pp. 1604–1612). Temporal factors such as the duration, envelope, amplitude modulation and the phases of partials also influence loudness (Moore, 2012, pp. 133–168). Speech intelligibility, as measured by the speech transmission index (IEC 60268-16, 2011), is impacted by the presence of noise and distortion and the audibility of important cues in the high-mid frequency area (Fry, 1979, pp. 129–143).

All of the aforementioned factors are either spectral, spatial, temporal or intensity related and can be manipulated in the mix process: spectral parameters can be influenced by spectral equalization (EQ); spatial parameters can be influenced by tools such as panning or reverb; the intensity of sounds can be altered through volume and compression; and all parameters can be changed over time through, for example, automation. Following the aforementioned literature review, spectral parameters appear to be especially important across all areas of research. It has been possible to establish two predictors of the spectral clarity of single sounds in music mixes (Hermes et al., 2017), that is, the harmonic centroid (HC, a measure of a sound's power distribution over frequency) and mid-range spectral peakiness (related to sharp peaks in the frequency spectrum). The HC is a weighted mean, indicating the harmonic at which the center mass of energy of a spectrum is situated. It is defined in Equation 1.

$$HC = \frac{\sum_{k=0}^{K-1} f(k) X(k)}{F \cdot \sum_{k=0}^{K-1} X(k)} \tag{1}$$

$X(k)$ is the magnitude of frequency bin number k, $f(k)$ is the center frequency (Hz) of k, K is the number of bins output from a discrete Fourier transform of the sound, and F is the sound's median fundamental frequency. This is defined as the pitch directly in the middle between the highest and lowest note played (Hermes et al., 2017, pp. 157–159). The HC needs to be raised above around 1.5 harmonics before clarity increases. Mid-range spectral peakiness is calculated by measuring the height of sharp peaks situated towards the middle of stimulus long-term average spectra (LTAS). An existing computational model fits a curve to the stimulus long-term average spectrum (LTAS) such that potentially unpleasant-sounding peaks lie above it, while the remaining frequency areas lie below it (Hermes et al., 2017, pp. 157–159). By comparing the model output pre- and post-EQ, changes in mid-range spectral peakiness can be estimated. More information about this topic, including a download link for the computational model (MATLAB), can be found in Hermes et al. (2017, p. 288). Most naturally occurring sounds have spectra that fall with increasing frequency and, therefore, clarity can usually be increased in the mix by applying low-Q EQ to boost the less-audible higher-frequency regions (raising the HC). However, it is important that no timbrally unpleasant peaks or resonances are introduced or increased in this process (Hermes et al., 2017, pp. 214–225).

While no existing study specifically focuses on vocal mix clarity, timbral characteristics of singing voices, as well as overall singing voice quality, have previously been investigated. Several areas of the spectrum appear to be important for vocal quality, but authors come to somewhat differing conclusions as to where exactly these areas are. This may be due to the specific context of each study. Zwan and Kostek (2008, pp. 710–723) state that the formants around 2.5–3.5 kHz are important for detecting different voice types and to measure overall quality. Winckel and Krause (1976) presents 3 kHz as being particularly important for detecting a singer's level of training. Omori et al. (1996, pp. 228–235) mention both 2–4 kHz and 0–2 kHz as important areas for measuring voice quality in classical singing, albeit only for the vowel "A". Most of the mentioned areas appear to lie close to the frequency area that the ear is most sensitive to (Fletcher and Munson, 1933). Similarly to quality, timbral attributes also depend on the energy in specific frequency areas. An "edgier" vocal sound is related to increased energy around 2–4 kHz and 8–10 kHz, according to Brixen et al. (2012). According to Hamlen (1991, pp. 729–733), vocal richness and depth is related to more harmonic content above 1.9 kHz. Parameters that relate to song key and the harmonic structure seem to have an impact on vocal quality (e.g. Williams, 2015; Zwan and Kostek, 2008, pp. 710–723), alongside the vocal level, pitch and musical articulation (Zwan and Kostek, 2008, pp. 710–723), but this has only been tested for classical singing. Furthermore, vocal timbre also depends on the noise-to-harmonic ratio (Feijoo and Hernández, 1990, pp. 324–334; Eskenazi, 1990, pp. 298–306; Farner et al., 2009).

Overall, it appears that vocal quality and timbral parameters strongly depend on spectral factors: the amount of energy in specific frequency areas, the level of voice harmonics and formants, factors relating to the song key, noise and tonal components are all important. Since timbre, quality and clarity are likely to be related, it would be useful to investigate whether similar parameters may influence vocal clarity.

When mixing engineers and producers describe how a specific vocal sound is achieved, they usually quote the equipment used, their process chain, their creative influences and anecdotes surrounding their creative process. For example, in order to achieve the vocal sound for The 1975's single "Chocolate", Mike Crossey wanted to "capture the energy you have at [a young] age" (Tingen, 2013). To do this, he used a Neumann KM84 "boosted super-bright" in an *Avalon 737* channel strip, going into a *Thermionic Phoenix* compressor with a long attack and short release. For the mix, he used several equalizers and compressors. When mixing "When Love Takes Over" (David Guetta), Veronica Ferraro worked towards an "angelic" vocal sound by splitting the vocal onto three tracks with up to five differing plugins each, including compression, EQ and deEssing (Tingen, 2010). Fabian Marasciullo uses EQ to "get rid of frequencies that bug him" (Flo Rida and T-Pain) for vocal mixes (Tingen, 2008).

In a telephone interview, Grammy award-winning producer and mixer Simon Gogerly (e.g. Paloma Faith, Gwen Stefani, Shirley Bassey) explains that EQ is his first step towards achieving vocal clarity, followed by several compressors, one of which is used for parallel compression. Sometimes a second EQ is used later on. For the EQ, after applying a high-pass filter, he largely focuses on five frequency areas, i.e. 200 Hz ("warmth"), 300–500 Hz ("mud area"), 800 Hz ("nasal area"), 3 kHz ("harshness") and 8 kHz upwards ("brightness"). For each area, Gogerly either removes or adds energy, a process guided by the natural harmonic structure and "character" of the vocal. Often, the EQ is either automated or multiband compression is used to account for spectral variation throughout the track. Compression is mainly used to reduce the variation in loudness throughout the track. Saturation may also be added. and occasionally a De-Esser, although Gogerly prefers to automate EQ and level to correct sibilance.

Since clarity is an important vocal mix parameter and since EQ and compression are virtually always applied, it is likely that these processes influence vocal clarity. Since EQ is used to alter a sound's spectrum, vocal clarity once again appears to heavily depend on spectral factors. Intensity over time, as altered by compression, may be important additionally.

It is possible that the acoustic parameters of vocal clarity are in part context dependent and depend on, for example, the style of the song, fashion or taste of the artist. Simon Gogerly stressed in the interview that the vocal characteristics of each singer need to be enhanced, and similarly, producer and remixer Rick Snoman (2009, p. 364) stresses that in dance music, the specific style of the mixing engineer makes the mix special: "All engineers will approach a mix in their own distinctive way—it's their music and they know how they want it to sound". While Pestana and Reiss (2014) state that lead vocals are often set to the same loudness as the rest of the mix, King et al. (2012) point out that engineers can have differing preferences for vocal loudness. Zagorski-Thomas (2007, pp. 189–207) notes that the sound of recordings has changed as a result of technological innovations and describes the exaggeration of intimacy in vocal tracks as a fashion trend. Hence, the optimum level of clarity may also be context dependent. Despite this, it appears that spectra are important and that relative changes in vocal clarity can, at least to some degree, be measured. Therefore, it appears useful to test the previously introduced predictors of spectral clarity in music mixes (HC and mid-range spectral peakiness) on a vocal mix.

TESTING THE CLARITY PREDICTORS IN THE MIX PROCESS

In order to gain an initial understanding as to whether the previously established predictors of single sound spectral clarity in music mixes may also work for vocals and in order to identify further important vocal clarity factors, an autoethnographic study was carried out in combination with an indicative pilot listening test. As mentioned earlier, an interdisciplinary approach is likely to provide useful findings. Through a combination of autoethnography and data gathered from listening test participants, the author was able to include her own observations while simultaneously ensuring that conclusions would remain objective.

Since most of the references for vocals were related to overall quality, rather than clarity specifically, vocal quality was also investigated. In this context, it was attempted to establish whether clarity and quality seem to be affected by similar factors. The focus is on spectral factors, as adjusted by EQ, since spectral factors appeared to be especially important throughout the literature. Another aim of the pilot study was to test a potential setup for a larger-scale listening test that could be used in a future in-depth study.

As an electronic artist (*Nyokeë*), the author has been writing, producing, performing, mixing and mastering original songs for approximately ten years. In cases such as this, where the entire creative process is undertaken by just one person, external, objective feedback on the mix process can be useful. Therefore, the clarity metric was tested informally on a vocal recording from an electronica music production, as follows. The vocal was mixed first without and then with the help of the metrics, readjusting EQ settings accordingly. Additionally, a version without EQ was created, resulting in three spectrally differing versions that were compared in terms of either clarity or quality by a listening panel as part of a short, indicative combined ranking and verbal elicitation task. The results of the listening test were analyzed alongside a self-reflection. In the following, the creation of the different vocal mix versions, as well as the pilot test setup, are summarized.

The program item was a 30-second vocal recording featured in the author's electronica production "Serendipity", with a range of pitches spanning an octave and a third—a typical range for modern pop music. The vocal was recorded into *Apple Logic Pro X* in an acoustically untreated home studio, using an *SE2200* microphone and an *SSL* channel strip. According to informal listening, the recording lacked clarity and had unpleasant spectral peaks, which made it a useful starting point for testing the predictors.

The vocal was first mixed by the author using EQ, compression and deEssing, since these processing tools are typically used for vocal mixes (*version 1*). Although the predictors were not directly used for this version, their previous knowledge influenced the process: low frequencies were cut and high frequencies were boosted in order to raise the HC, and it was attempted to reduce the unpleasant resonances with some success. However, there still appeared to be room for improvement in terms of clarity.

Another version was created which was identical to version 1, with the exception that all EQ was removed (*default version*). The HC in version 1 was higher than in the default version by around ten harmonics. Version 1 had a higher degree of mid-range spectral peakiness than the default version, hence another version was created with the help of the predictors (*version 2*). The HC of this version was increased by another six harmonics compared to version 1. Using the computational model of mid-range spectral peakiness mentioned in the previous section (Hermes et al., 2017, pp. 214–225), it was attempted to eliminate peakiness entirely. Figure 13.1 shows an example of the curve being fitted to the version 1 stimulus.

The EQ in version 1 was altered until the LTAS of the vocal did not have any more mid-range spectral peakiness (*version 2*), which required additional EQ and a large number of fine adjustments. The result of this is shown in Figure 13.2.

All vocal versions were exported as 44.1 kHz, 24-bit. wav audio files. While the presence of the other sounds in the mix is likely to have an impact on clarity, small spectral variations are more likely to be audible in the isolated vocal. Hence, the vocals were exported both in the finished mix, as well as in isolation. The resulting set of stimuli is listed as follows:

- Isolated vocal mixed by the author (version 1)
- Isolated vocal with no EQ (default version)
- Isolated vocal with clarity metric informed EQ (version 2)
- Version 1 in the mix
- Default version in the mix
- Version 2 in the mix

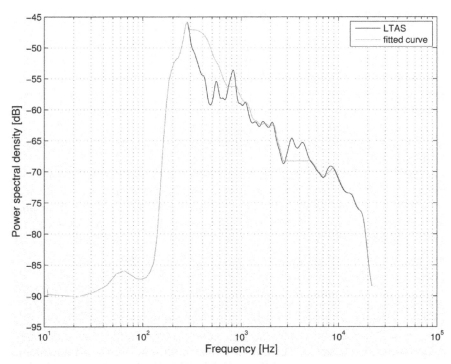

Figure 13.1 The computational model of mid-range spectral peakiness is applied to the version 1 vocal stimulus. Potentially unpleasant-sounding spectral peaks lie above the curve.

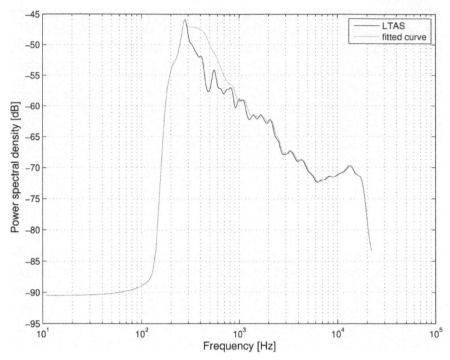

Figure 13.2 The computational model of mid-range spectral peakiness is applied to the version 2 vocal stimulus. The entire LTAS now lies below the curve.

As loudness is likely to influence clarity (Hermes et al., 2017, pp. 63–73), the isolated vocals were loudness matched in LUFS using *Adobe Audition*. In the pilot listening test, listeners rated each vocal version against each other version in each group (isolated and in the mix), in terms of either clarity or quality in a graphics user interface (GUI). No definition was given for clarity because, although there is no commonly accepted definition for clarity, a common understanding appears to exist of what clarity means (Hermes et al., 2017, p. 102). In a previous study (Hermes et al., 2017, pp. 182–202), clarity definitions were elicited from participants in the context of an EQ clarity adjustment task. In many definitions, clarity was linked to naturalness or expectations of how an instrument should sound (i.e. the "readability" of an instrument's characteristics). Therefore, in a later experiment, the clarity definition "the extent to which all the important components of a sound's natural timbre to be heard" was used. However, informal conversations with listeners revealed that sometimes, sounds could be clear when not all timbral detail was audible and, conversely, some sounds with audible detail could also be unclear. This appears to indicate that while the previous definition might offer useful insights into what clarity may be, a listener's internalized concept of clarity is still more accurate. The author has carried out six listening tests on spectral clarity without providing any clarity definition, and all yielded useful results (Hermes et al., 2017). Quality was defined as listener preference in this case, as it is likely that preference is related to clarity: clarity appears to be related to how an instrument "should sound", as mentioned earlier.

As concentration levels drop when listening tests are too long, for now, it was decided to present no more than six GUI pages with four stimulus pairs being tested for clarity and two for quality, as shown as follows:

- Default version and version 1 in isolation—clarity
- Default version and version 2 in isolation—quality
- Version 1 and version 2 in isolation—clarity
- Default version and version 1 in the mix—quality
- Default version and version 2 in the mix—clarity
- Version 1 and version 2 in the mix—clarity

Additional attributes that are also affected by spectral balance (fullness, warmth and brightness) were included but not tested to account for dumping bias (Hermes et al., 2017, pp. 99–104). Brightness seems to be related to high- and low-frequency balance (Brookes and Williams, 2010); warmth is negatively correlated with brightness (Brookes and Williams, 2010); and fullness can be affected by low-frequency spectral fluctuations (Alluri and Toiviainen, 2010, pp. 223–242). On each page, listeners could play back two versions by pressing buttons labeled 'A' and 'B' as often as required after pressing 'play', and a 'stop' button could be used to reset the stimuli to their beginning position. The stimulus presentation order was randomized overall and on each page to avoid sequential bias. Clarity and quality ratings were recorded as slider positions, which indicated if and by how much one of the stimuli was clearer or preferred. Listeners were also asked to enter text into a box, in order to indicate which factors may have led to any clarity and quality differences between the stimuli, and to suggest how clarity and quality might be further improved. Figure 13.3 shows an example of a test page.

The chosen listening panel comprised ten male and female students and sound professionals at the University of Westminster, as ten is the minimum recommended number of listening test participants (Bech and Zacharov, 2006, pp. 112–118). The participants were experienced in critical listening and in verbalizing sensations of timbre and did not report having any hearing damage. Following a familiarization stage, listeners sat the test individually and listened to the stimuli via high-quality studio headphones in an acoustically isolated space.

In the following, the listening test results are presented alongside a self-reflection. Listeners took on average 15 minutes for the test. In the author's opinion, both equalized versions (versions 1 and 2) were notably clearer than (and preferred to) the default version (no EQ), both in isolation and in the mix. The unequalized version sounded somewhat dull and lifeless, and words were less intelligible in the mix. In the mix, presence and separation from the other instruments had improved, and the overall sonic quality seemed much more similar to that of other, similar vocal mixes.

The prerequisites for using parametric statistical methods were not quite fulfilled by the listening test data, and a more extensive, fully factorial test

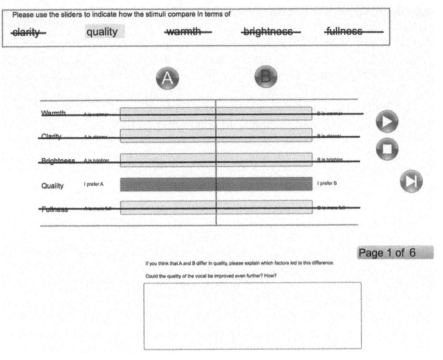

Figure 13.3 An Example of a Test Page

(based on IoSR, 2016)

would need to be carried out to infer generalizable conclusions. Therefore, the listener data are presented in simple, non-parametric box plots, which make it possible to see the direction of clarity change between stimuli and whether this is significant. In the box plots, the box is the interquartile range and the whiskers include the rest of the data except from outliers (crosses). The notches are the confidence intervals. The notch is centered on the median (line in the middle) and extends to $\pm 1.58 \dfrac{IQR}{\sqrt{N}}$, where N is the sample size and IQR is the interquartile range. The limits that define outliers (crosses) are defined as $Q_1 - 1.5\,IQR$ and $Q_2 + 1.5\,IQR$, where Q_1 is the 25th percentile and Q_2 is the 75th percentile. Where the notch lies completely outside the center line, it can be assumed that the direction of the clarity and quality ratings is statistically significant.

For single sounds, version 1 was rated clearer than the default version, but this was not quite statistically significant (Figure 13.4). Similarly, version 2 was preferred (quality) to the default version but this was also not quite statistically significant (Figure 13.5). These findings seem to be in line with the results of the self-assessment. The listeners seemed to mostly agree with some exceptions, which may have been due to the fact that some judged the vocal as an isolated solo vocal, whereas others imagined it in the context of the rest of the mix. This was confirmed by some informal conversations with the test subjects.

Version 1 clearer

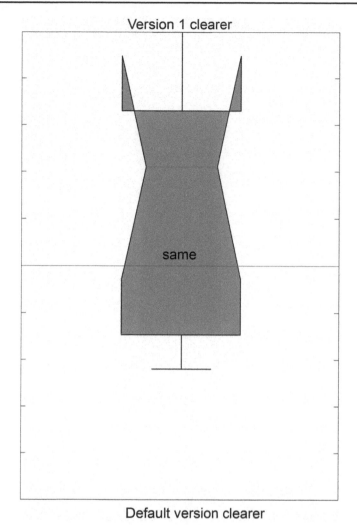

same

Default version clearer

Figure 13.4 Extent to Which Version 1 Was Rated Clearer Than the Default Version

Apart from one outlier, all test subjects strongly agreed that in the mix, version 2 was much clearer than the default version (Figure 13.6). This strong agreement shows that listeners seemed to share the same idea of what clarity may be for vocals in the mix and once again confirms that for the solo vocal, some listeners may have anticipated the presence of the rest of the mix, but others did not. Version 1 was preferred to the default version but again, this was not quite statistically significant (Figure 13.7). It is possible that quality, as a higher-level parameter, depends not only on clarity but also on taste and other timbral attributes and may therefore be more difficult to measure. Given that versions 1 and 2 sounded very similar in the mix, it is likely that the version 1 would have also been rated clearer.

According to informal listening, there was little difference between versions 1 and 2, especially after the vocal was placed in the mix, and this was

Version 2 preferred

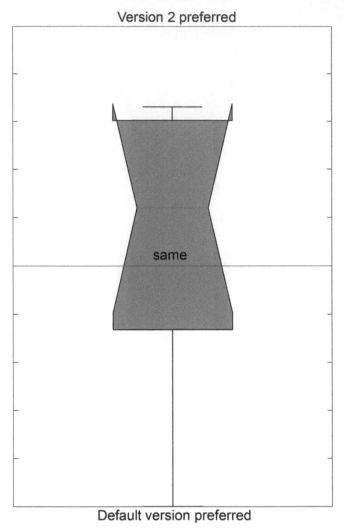

Default version preferred

Figure 13.5 Extent to Which Version 2 Was Preferred to the Default Version

also confirmed in informal conversations with the participants. It is highly likely that the author's prior knowledge of the predictors had influenced the initial mix. Despite being more heavily processed, version 2 sounded somewhat smoother, which seemed to improve both clarity and quality slightly. However, there still seemed to be room for improvement. Listening test subjects were in disagreement about vocal clarity in the mix: some judged version 1 to be clearer, and others version 2 (Figure 13.8). For the isolated vocal, most listeners in fact judged version 1 to be clearer, which may be due to an overly processed sound in version 2, resulting from the additional EQ (Figure 13.9).

Overall, there was an improvement in both clarity and quality for verions 1 and 2 over the default version, in the mix and in isolation. Since

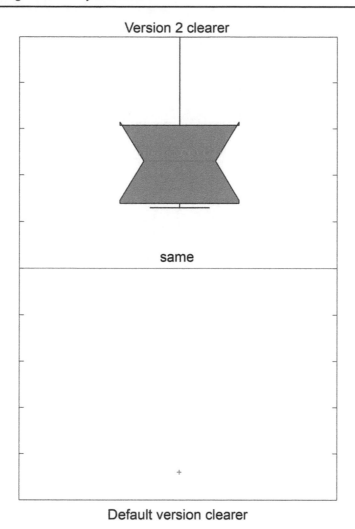

Version 2 clearer

same

Default version clearer

Figure 13.6 Extent to Which Version 2 Was Rated Clearer Than the Default Version in the Mix

both equalized versions had an increased harmonic centroid compared to the default version, this predictor appears to be relevant for recorded vocals. Mid-range spectral peakiness was greatest for version 1 and smallest for version 2; however, listeners disagreed on which of the versions had greater clarity or quality. Therefore, mid-range spectral peakiness appears to be a weak predictor of vocal clarity in this case. It is possible that version 2 sounded overprocessed due to the extensive EQ treatment, impacting negatively on clarity and quality. The qualitative data provided by listeners appears to confirm this assumption, as detailed as follows.

When asked which factors may have led to differences in clarity and quality, listeners most often mentioned the relative amount of energy in specific frequency areas, most likely due to the fact that the stimuli

Version 1 preferred

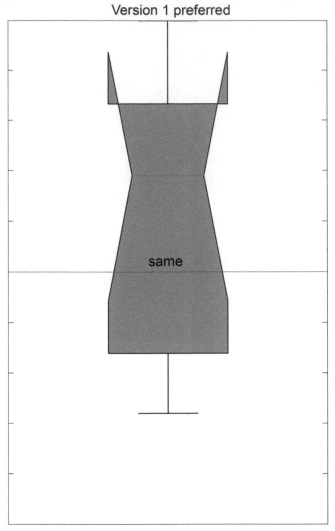

Default version preferred

Figure 13.7 Extent to Which Version 1 Was Preferred to the Default Version in the Mix

differed spectrally (22 instances, 18 for clarity and 6 for quality). The specific areas mentioned differed among test subjects, however. Since most vocal mixes feature compression, as established in the previous section, the impact of intensity changes over time seems to have an impact on vocal clarity and quality and, therefore, next most often, compression and dynamics were mentioned (16 instances, nine for clarity and seven for quality). The impact on other timbral attributes was mentioned 14 times (brightness was mentioned seven times, warmth was mentioned three times for quality, breath/air was mentioned four times for clarity and body

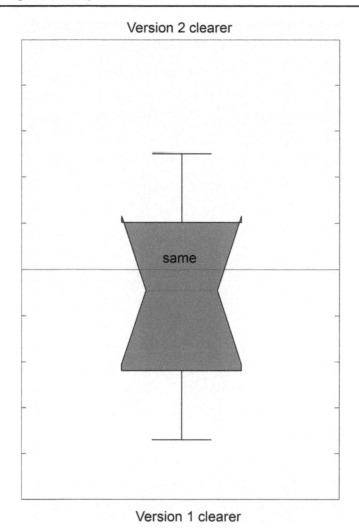

Figure 13.8 Comparing Clarity for Versions 1 and 2 in the Mix

was mentioned once for clarity). Sibilance and deEssing were mentioned five times (four times for clarity), and one subject mentioned "ringing" in the context of clarity. Some of the parameters that may influence clarity and quality merely appeared to be synonyms of clarity—i.e. presence and cutting through (four instances) and "human", "close", "intimate" or "upfront" (clarity); and similarly, "presence" and "cutting through" (quality). Some subjects commented on the clarity of the sung words, mentioning pronunciation, intelligibility and clarity of lyrics and sentences (four instances, two for clarity and two for quality). Noise, artifacts and distortion were mentioned five times, and three subjects described the stimuli as sounding overprocessed. Overall, the elicited factors were similar for clarity and quality.

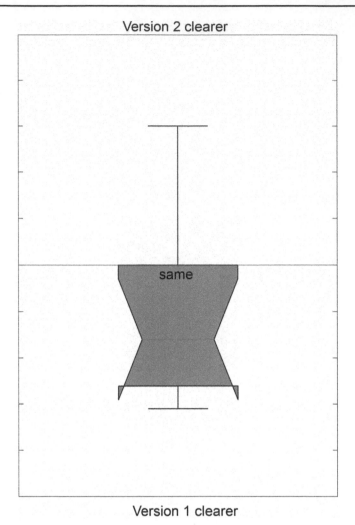

Figure 13.9 Comparing Clarity for Versions 1 and 2 in Isolation

DISCUSSION

For most stimulus pairs, listeners appeared to agree on clarity and quality judgments, but this was more evident for clarity. Quality may be a high-level concept influenced not only by clarity (since both seem to depend on similar factors), but also other perceptual attributes and personal taste, and may be more difficult to measure for this reason. Therefore, it would be useful to fully understand clarity before further investigating overall quality. An increased HC appears to correspond to increased clarity and, therefore, still appears to be a useful predictor that should be included in further research. Changes in mid-range spectral peakiness did not seem to correlate with changes in clarity and quality, which may have been due to the fact that the additional EQ in version 2 led to an unpleasant, overprocessed sound. Drawing on the existing literature, as well as the factors mentioned by test subjects, additional factors may need to be considered. The following additional factors were mentioned both by test participants and in previous studies. The relative amount of energy in different frequency areas appears to be important, but it would need to be assessed in further detail which spectral areas contribute to clarity in which way. Compression appears to have an important impact on vocal clarity, which may be due to its impact on the spectrum, as well as the resulting changes in intensity over time. Additionally, the ratio between tonal and noise components appears to be important and unwanted distortion and artifacts, or an over-processed, "fatiguing" sound caused by EQ or compression overuse may also be related to this. In order to improve confidence in these results and to find out more about the exact relationship between these parameters and clarity, it would be useful to carry out a large-scale listening test where listeners rate vocal clarity for stimuli that vary in all of these factors. Since the listening test setup appeared to deliver usable results without taking too long, a similar setup could be used in the future. When vocals are tested for clarity in isolation, it should be clarified whether the vocals should be imagined as part of a mix or as alone-standing solo recording, since this distinction appears to have led to listener disagreement in the pilot study.

CONCLUSION

The importance of vocal clarity in music mixes is widely agreed on, but currently there is no formally proven basis of the factors that this depends on. While certain aspects of vocal clarity may be context dependent, it appears that acoustic parameters exist that all clear vocals have in common.

Sonic clarity has been an important focus across audio research areas, including clarity in concert halls, speech intelligibility, timbral clarity, masking, loudness and auditory scene analysis, and it was established that overall, spectral parameters are especially important for clarity but spatial, temporal and intensity-related factors also matter. Two predictors of the spectral clarity of single sounds exist for music mixes—i.e. the harmonic centroid (a measure of a sound's power distribution over frequency) and mid-range spectral peakiness (related to sharp peaks in the frequency spectrum).

While no existing study specifically focuses on the clarity of recorded singing, other timbral characteristics of singing voices, as well as overall singing voice quality, have previously been investigated. Nearly all of the parameters determining vocal timbre and quality are spectrum related and, therefore, it is possible that the two existing predictors of single sound spectral clarity also apply to vocal clarity and quality. This was tested on a vocal recording in an electronica music production in a combined pilot listening test and self-reflection.

It was concluded that the HC appears to be a useful predictor for vocal clarity and possibly also quality, although quality seems to be more difficult to measure. Mid-range spectral peakiness did not seem to correlate well with changes in clarity and quality, which may have been due to the stimuli sounding overprocessed where additional EQ had been applied to reduce peakiness. Additional clarity factors were established from the literature and pilot study. These are the relative amount of energy in different frequency areas, intensity over time, as altered by compression, the ratio between tonal and noise components, and distortion and artifacts. It would be useful to carry out a large-scale listening test where listeners rate vocal clarity for stimuli that vary in all of these factors. This might improve confidence in the results and make it possible to find out more about the exact relationship between these parameters and clarity. Lastly, it is possible that the perception of clarity is in part subjective and dependent on listener preference. Follow-on research should establish to what extent this is the case.

REFERENCES

Aichinger, P., et al. (2011). Describing the transparency of mixdowns: The masked-to-unmasked-ratio. In: *Audio engineering society 130th convention*. London, UK. 13–16 May.

Alluri, V. and Toiviainen, P. (2010). Exploring perceptual and acoustical correlates of polyphonic timbre. *Music Perception: An Interdisciplinary Journal*, 27(3), pp. 223–242.

ANSI/ASA S1.1. (2013). *American national standards. Acoustical terminology S1*. New York: American National Standards Institute.

Bazil, E. (2008). *Sound mixing: Tips and tricks*. Norfolk: PC Publishing.

Bech, S. and Zacharov, N. (2006). *Perceptual audio evaluation: Theory, method and application*. Hoboken: John Wiley & Sons.

Beranek, L. L. (1996). *Concert and opera halls: How they sound*. Acoustical Society of America, American Institute of Physics. New York: Woodbury.

Bregman, A. S. (1990). *Auditory scene analysis: The perceptual organization of sound*. Cambridge: Bradford Books, MIT Press.

Brixen, E. B., et al. (2012). On acoustic detection of vocal modes. In: *Audio engineering society 132nd convention*. Budapest, Hungary, 26–19 Apr.

Brookes, T. and Williams, D. (2010). Perceptually-motivated audio morphing: Warmth. In: *Audio engineering society 128th convention*. London, UK, 22–25 May.

Clark, R. (2011). *Mixing, recording, and producing techniques of the pros*. 2nd ed. Boston: Cengage Learning.

Eskenazi, L. et al. (1990). Acoustic correlates of vocal quality. *Journal of Speech, Language and Hearing Research*, 33(2), pp. 298–306.

Farner, S., et al. (2009). Natural transformation of type and nature of the voice for extending vocal repertoire in high-fidelity applications. In: *Audio engineering society 35th international conference: Audio for games*. London, UK, 11–13 Feb.

Feijoo, S. and Hernández, C. (1990). Short-term stability measures for the evaluation of vocal quality. *Journal of Speech, Language and Hearing Research*, 33(2), pp. 324–334.

Fletcher, H. and Munson, W. A. (1933). Loudness, its definition, measurement and calculation. *Bell Systems Technical Journal*, 5, pp. 82–108.

Fry, D. B. (1979). *The physics of speech*. Cambridge: Cambridge University Press.

Hafezi, S. and Reiss, J. (2015). Autonomous multitrack equalization based on masking reduction. *Journal of the Audio Engineering Society*, 63(5), pp. 312–323.

Hamlen, W. A. (1991). Superstardom in popular music: Empirical evidence. *Review of Economics and Statistics*, 73(4), pp. 729–733.

Hermes, K., et al. (2017). *Towards measuring music mix quality: The factors contributing to the spectral clarity of single sounds*. PhD thesis, University of Surrey, Guildford, UK.

IEC 60268–16. (2011). *Objective rating of speech intelligibility by speech transmission index*. 4th ed. Geneva: International Electrotechnical Commission.

Institute of Sound Recording IoSR. (2016). *A Max/MSP patcher for MUSHRA listening tests*. Available at: http://iosr.uk/software [Accessed 10 Jan. 2018].

ISO 3382–1. (2009). *Acoustics—measurement of room acoustic parameters. Part 1: performance spaces*. Geneva: International Organization for Standardization.

King, R., et al. (2012). Consistency of balance preferences in three musical genres. In: *Audio engineering society 133rd convention*. San Francisco, USA, 17–20 Oct.

Moore, B. C. J. (2012). *An introduction to the psychology of hearing*. Bingley: Emerald.

Moore, B. C. J. and Glasberg, B. R. (2007). Modeling binaural loudness. *Journal of the Acoustical Society of America*, 121(3), pp. 1604–1612.

Omori, K. et al. (1996). Singing power ratio: Quantitative evaluation of singing voice quality. *Journal of the Voice*, 10(3), pp. 228–235.

Pestana, P. D. and Reiss, J. D. (2014). Intelligent audio production strategies informed by best practices. In: *Audio engineering society 53rd international conference: Semantic audio*. London, UK. 26–29 Jan.

Ronen, Y. (2015). Vocal clarity in the mix: Techniques to improve the intelligibility of vocals. In: *Audio engineering society 139th convention*. New York, USA. 17–20 Oct.

Snoman, R. (2009). *Dance music manual: Tools, toys, and techniques*. Oxford: Elsevier.

Solomon, L. N. (1959). Search for physical correlates to psychological dimensions of sounds. *Journal of the Acoustical Society of America*, 31(4), pp. 492–497.

Strong, J. (2009). *Home recording for musicians for dummies*. 3rd ed. Indiana: John Wiley & Sons.

Tingen, P. (2010). Secrets of the mix engineers: Veronica Ferraro. *Sound on Sound*. Jan. Available at: www.soundonsound.com/people/secrets-mix-engineers-veronica-ferraro [Accessed 10 Jan. 2018].

Tingen, P. (2008). Secrets of the mix engineers: Fabian Marasciullo. *Sound on Sound*. Aug. Available at: www.soundonsound.com/techniques/secrets-mix-engineers-fabian-marasciullo [Accessed 10 Jan. 2018].

Tingen, P. (2013). Secrets of the mix engineers: Mike Crossey. *Sound on Sound*, Dec. Available at: www.soundonsound.com/people/secrets-mix-engineers-mike-crossey [Accessed 10 Jan. 2018].

Williams, D. (2015). Affective potential in vocal production. In: *Audio engineering society 139th convention*. New York, 17–20 Oct.

Winckel, F. and Krause, M. (1976). A quick test method for the diagnosis of speakers' and singers' voices under stress. In: *Audio engineering society 53rd convention*. Zurich, Switzerland, 2–5 Mar.

Zagorski-Thomas, S. (2007). The musicology of record production. *Twentieth-Century Music*, 4(2), pp. 189–207.

Zarouchas, T. and Mourjopoulos, J. (2011). Perceptually motivated signal-dependent processing for sound reproduction in reverberant rooms. *Journal of the Audio Engineering Society*, 59(4), pp. 187–200.

Zwan, P. and Kostek, B. (2008). System for automatic singing voice recognition. *Journal of the Audio Engineering Society*, 56(9), pp. 710–723.

14

Plugging In

Exploring Innovation in Plugin Design and Utilization

Andrew Bourbon

INTRODUCTION

Audio plugins have become the sound-processing hub for the increasingly dominant in-the-box (ITB) mixing environment. As the digital audio workstation (DAW) has become the primary environment for the task of music mixing, the breadth and complexity of audio processing tools has grown considerably, covering an ever-expanding range of tools to solve specific problems since the introduction of Waves Q10 in 1992 as the first available third-party plugin.

As plugins have developed, designers have focused their development on several key areas. Emulation of classic hardware has been one of these key areas, with developers continuing to find new techniques for the measurement of classic hardware and the re-creation of that hardware in the digital environment. The market hunger for these tools is clear, with manufacturers continuing to develop for this apparently saturated market sector. In addition to the innovations leading to a higher-quality, more faithful emulation of the hardware, there have been further examples of innovation, allowing functionality that would not necessarily have been available to the hardware user, offering greater flexibility in audio processing and also tacking some of the issues inherent in exploring exact emulations of hardware in a DAW environment where gain staging limitations and contemporary delivery levels are significantly different from those faced in the analog domain at the time of the development of the emulated hardware.

While the innovation is clear in the development of the technologies required to create faithful emulations of complex hardware, there is also an argument that emulation lacks innovation in the development of new control paradigms. The question of emulation and innovation will be discusses in more detail, exploring the potential user base for plugins and the apparent hunger for emulated plugins among the music production community.

One recent development in plugin design inspiration has been to look away from the specific hardware, and instead to the human user associated with record production. Companies such as Waves have developed signature series plugins, which have embraced the more traditional emulation approach, and also explored new interfaces for control that embrace the particular sonic signature associated with a particular engineer. These interfaces have moved away from the traditional interfaces associated with hardware, instead often moving towards a simple interface offering direct control of a semantic descriptor associated with a particular engineer. The celebrity engineer element of plugin innovation has also led to the development of a number of endorsed preset libraries for tools, again designed to give the end user access to the sound of their favorite engineer and record.

Innovation in plugin design can also be found in the interfaces presented to the user, with developers providing new ways of visualizing and controlling the parameters offered by a particular process. In some cases, the aim of the interface is to simplify the controls, providing an interface that makes it as easy as possible for the end user to interact with their sound, without necessarily having a complex understanding of the process behind the interface. Other interfaces provide an entirely different experience to the end user, allowing configuration of every element of the process undertaken in microscopic detail, requiring significant prior knowledge of the audio process in order to take full control of the interface provided. Other manufacturers such as Fabfilter create expert areas in their plugins, offering greater levels of control to users who wish to take greater control of their signal processing through contemporary interfaces that offer significant user feedback through visualization of signal processing.

The final area of innovation that this chapter will explore is in the development of tools for non-traditional audio processing. There has been significant technological innovation in processes such as tuning, spectral noise reduction and manipulation and creative processing, with many of these tools offering functionality that has only become available as users have move into the digital domain and processing power and technology have increased. Companies such as Izotope have created a range of innovative tools exploring visualization, mixing, mastering and noise reduction through the use of technologies such as machine learning. These tools are clearly at the height of innovation in music production, offering users new approaches to both traditional production processes and also to tackle problems that would traditionally be incredibly difficult, or indeed impossible, to fix.

CLASSIC HARDWARE EMULATION: THE SEARCH FOR REALISM

When exploring the plugin portfolio of the major plugin manufacturers, there is a prevalence of emulated hardware tools. Exploring the list of dynamics processors that form the catalog for the Universal Audio UAD

DSP platform reveals a list of compressors that include the following processing tools:

API 2500	Precision Bus Compressor
Manley Vari-Mu	Empirical Labs EL8 Distressor
Fairchild Collection	Elysia Mpressor
Teletronix LA-2A Collection	Teletronix LA-3A
1176 Classic Limiter Collection	Vertigo Sound VSC-2
SSL 4000G	Elysia Alpha
Neve 33609	dbx 160
Tube Tech CL1B	Valley People Dyna-mite
Summit Audio TLA-100A	

The list of compressors here represents an exhaustive list of classic and contemporary studio dynamics processors, all of which have been meticulously emulated by Universal Audio and their partners since their inception as a programming company in 1999. Competitors such as Plugin-Alliance, Softube, Slate Digital and Waves all feature similar product lists, with either direct emulations of much of the hardware noted here or tools inspired by these classics but without the official licensing to allow use of the specific model name.

The demand for high-quality emulations is clear, with only one of the plugins named representing an original design, with the Precision Bus Compressor conceived with the specific aim of offering clean gain reduction from a voltage-controlled amplifier (VCA) style compressor. The 1176 is established as a studio standard (Massy, 2016) and is one of the most emulated compressors as a plugin, with all the major manufacturers offering emulations and native versions of these compressors available in most major DAWs, and is famous for its use of a field effect transistor (FET) in the gain control circuit (Case, 2007). The 1176 is a particularly colorful compressor, offering very fast attack times, with a real sense of "hair" added to the sound when in compression and a sense of "air" provided as the release speed is increased. Commonly used for vocal processing, bass processing when harmonic enhancement is desirable, and also on drums, Universal Audio has revisited the 1176 in their 1176 limiter collection, adding new improved emulations to the already existing legacy version. The main focus of the upgraded emulation has been to model the complete circuit path of the compressor, including the transformers and amplifiers in the input and output circuitry. Through measurement and subjective auditioning of the compressor, the increase in harmonic complexity is clear, with resultant increases in harmonic content both in compression and out of compression, confirming the enhanced impact of the complete modeled circuits on the sound. Universal Audio has also provided three different editions of the 1176, with a Rev A, Rev E and AE versions of the compressor available to the end user having been through the component modeling process.

The three modeled versions provide three different sets of sonic characteristics, with each having different impact on the sounds that they are processing. The Rev A, for example, exhibits higher distortion, and also has a different release profile from the Rev E due to the variation in the program-dependent release behavior.

Recent innovation in creating realistic emulation of hardware performance has seen even more detailed modeling taking place. The world-famous SSL 4000 mixing console has been modeled by a number of manufacturers, with companies such as Waves, Universal Audio and Softube modeling the channel drive either as part of the channel strip or as additional components, allowing the user to build a full SSL channel strip in the digital environment. Massey describes the process of "creaming the mic pres", in order to bring excitement into recordings through the dynamic compression and distortion associated with various vintage circuits. Universal Audio have also modeled the performance of the dbx voltage VCA found in the channel output section, as well as adding the optional Jensen transformer on the microphone input, offering the user the full SSL channel experience within the DAW with full control over gain staging, as found in the hardware and the tonal options to explore the classic techniques associated with creative abuse of the channel strip.

Brainworx have been responsible for further innovation in their emulations of contemporary channel strips. Through their process of Tolerance Modeling Technology (TMT), Brainworx have measured the component tolerances in every piece of the channel strip, and then realized those variations in the console, with significant and subtle variations present throughout the 72 TMT channels. This process has been competed for three different consoles, with a Neve VXS and SSL E and G series all measured and emulated. Control is provided to the user, allowing either random allocation of channels or specific choices to be made, with phase variation and stated value changes in the channel strip offering changes to the presentation of audio, particularly when applied to stereo material. The impact of this randomization can be significant, resulting in perceived changes often considered desirable by end users, with increased image "width", "depth" and "punch" experienced by the listener.

Component circuit modeling has been a significant innovation in the recent history of plugin design. The TMT technology and component modeling used by other manufacturers has seen a significant increase in the performance of emulations. Digital audio consoles primarily designed for live sound such as the Midas Pro series have plugins built into the console developed from modeling their own hardware, built specifically for the purpose of digital modeling, and later released by Klark Technik as the Square One Dynamics processor in order to offer the best possible real-time performance to engineers. Waves and Universal Audio are also incorporating their plugins into dedicated digital signal processing (DSP) units to be integrated into the digital console workflow with a minimum of latency. Improvements in reinforcement systems are seeing users focusing further on the signal processing engaged in live mixing, turning to the emulated processors commonly used in the studio environment to re-create the signature sounds from records in real-time live performance.

It is important to note the influence of convolution technology on plugin design, and in particular on the development of reverb processing in the DAW. Products such as Altiverb offer engineers an extensive collection of classic reverb units and real spaces, all captured through convolution technology. Universal Audio is using combinations of algorithmic delay networks and convolution in their BX spring reverb in order to faithfully re-create the unique sound of the AKG spring.

Perhaps the most innovative use of a convolution approach is being pioneered by Acustica audio, who are using dynamic convolution to create emulations of a range of classic processing tools. Though the DSP load of this technology on the host machine is significant, and there are some associated complications in the real-time manipulation of parameters, the sonic quality afforded by this technology is impressive and represents an exciting innovation in emulation methodology.

REALITY RULES: SKEUOMORPHISM, NOSTALGIA AND PLACE

As the DAW has become the primary interface for the creation and mixing of music, there is now a generation of engineers who, rather than moving from the analog studio into the digital domain, have their entire studio experience in a computer-based environment. Despite not having had access to hardware tools, many young engineers do have an appreciation of classic hardware tools as explored in the digital domain. As already discussed, much of the innovation in plugin emulation has been focused on creating ever more accurate models of hardware. In the digital domain, the changes in the process of capturing, processing and generating sound have led to challenges for those exploring these emulated tools without the experience of using these tools in an analog domain. It is possible when working in the modern DAW to almost completely ignore traditional gain staging approaches, driving channels beyond the levels achievable using vintage recording equipment and simply turning down the master fader to avoid output clipping. Floating-point processing has led to internal clipping simply not being an issue; as long as the signal at the output is not clipping the digital-to-analog converter, the signal will remain clear of distortion. As music has become louder at distribution, young engineers are creating pre-mastered mixes that are already at levels above those that would have been traditionally provided pre-mastering. Many of the emulated hardware tools will react aggressively to this signal level, with tools such as Fairchild compressors, for example, in significant gain reduction with the threshold set at a high level and the input gain low; manufacturer-driven solutions to this contemporary gain staging issue will be discussed later in this chapter.

The proliferation of tape processing tools, classic console processing and vintage hardware in the digital domain has also influenced the sense of nostalgia for the tools in the hardware environment, and in turn is feeding the constant increase in emulated tools in the DAW. This is particularly interesting with engineers who have only experienced music production in a digital environment, yet still feel a sense of nostalgia towards analog technology that there is no physical experience of using. The incorporation of arguably undesirable elements such as noise have added to the digital emulation experience. Bell et al. (2015) state in regards to Slate VTM: "The VTM makes digital *look* vintage, but more importantly, it makes digital *sound* vintage, going so far as to offer optional tape hiss. This is 'technostalgia' at its apogee, and it extends our definition of skeuomorphism beyond visual to aural". The impact of tech nostalgia has been further enhanced by the significant rebirth of vinyl as a format, with new challenge faced by those cutting vinyl from the aggressive, bright and loud masters generated in contemporary popular music. The integration of classic processing in the contemporary DAW environment provides a significant challenge to educators and enthusiasts looking to embrace these tools without the limitations enforced by the analog environment.

The influence of nostalgia on plugin design innovation is not limited to the emulation of classic tools, but also with the places in which those tools could be found. Waves, for example, offer a series of *Abbey Road*–inspired plugins, Universal Audio offer access to an emulation of the Ocean Way recording rooms, and Eventide offer Tverb, providing the Hansa studio reverb made famous during the recording of David Bowie's *Heroes*.

All of the emulated places and hardware tools have also seen significant development in the look of the plugin, embracing literal re-creations of the hardware visually. This skeuomorphic approach provides the user with an interface that affords the same control as the real hardware, with the quality of the visual impact of the interface an important part of the user experience. Slate Digital plugins, for example, exhibit different levels of rack rash and general damage each time a plugin is instantiated. Though this clearly has no effect on the sound of the plugin, it is clearly important to the end user and as such represents an important innovation in plugin design.

ALTERNATIVE REALITY: ADDING FUNCTIONALITY IN EMULATIONS

Though the focus in emulation innovation has been in creating ever more accurate emulations of hardware, manufacturers have also been adding functionality beyond that found in the original units. The tolerance modeling technology found in the Brainworx console plugins, for example, can be varied randomly, providing variations not originally enacted by the user. The impact of this random processing is a change in image size and depth, creating according to Brainworx, a "hyper realistic mixed on a big console sound for the first time in the box". There is also user control over other features, including a total harmonic distortion (THD) control adding variable color, configurable per track. Wet dry functionality is added to the dynamics, as well as a secondary release time to customize the release of the compressor and avoid audible pumping. Middle and side monitoring functionality is provided in the plugin, with additional gain staging control to that of the original console to manage noise floor and color through the plugin itself. This concept of hyper-reality, where the user gets a level of control beyond the hardware original is a key innovation, both in the plugin market and also in the hardware market. The SSL G Bus Compressor is available as a DIY project and has also been created by a number of other manufacturers, often adding high pass filtering in the sidechain to avoid the compressor reacting to low end and compromising the lower octaves that have become so important in contemporary popular music production. Similarly, the option to modify the sidechain stereo linking is provided in a number of hardware tools, with the ability to use a summed sidechain or true stereo detection, resulting in either perceived width or central punch, depending on the mode selected. Plugin compressors are innovating with even more sidechain modification potential, which will be explored later in this chapter when exploring contemporary non-emulated interfaces.

Another recent innovation in plugin design has seen the ability for the end user to modify the calibration of the plugin itself. The levels of signals inside the DAW can be incredibly variable, depending on engineer approach, genre and positioning of the audio processing in the audio chain. By default, the Fairchild 660 plugin is calibrated so that 0 dBFS will result in what would be approximately +20 dBu into the hardware unit. At this input level, there would be significant coloration in the audio circuit, and the unit would be subject to high levels of gain reduction. In order to control this, it is possible to modify the reference level from the default 16 dB, with a range of variation in calibration from 4 to 28 dB. This creates a range of control, allowing subtle coloration and control over small amounts of compression that is useful in bus processing and mastering to more aggressive channel and parallel bus usage. It is also of note that Universal Audio have added wet dry control and sidechain filtering to the parameters traditionally offered in a Fairchild compressor, providing yet more innovative control in addition to the default accurate emulation.

Tape processing has become more prevalent in emulation, with significant steps being made by manufacturers in the emulation of magnetic performance. The range of control offered in tape plugins varies significantly, depending on the provenance of the tape tool being employed. Universal Audio, for example, have created faithful emulations of the Ampex ATR-102 and Studer A800 tape machines, and have also created a simplified tape machine with a focus on ease of use and for real-time processing at the point of capture using their own DSP interfaces. The two emulated tape machines both offer full calibration control, with the ability to modify noise, sync and repro equalization and level. The manual provides the recommended calibration for the four tape formulations provided within the emulation. It is possible to vary these parameters, allowing a level of flexibility in tape processing that would not have been viable to instantly compare on an analog machine. It is important to note that the headroom control found on the latest dynamic processors is not available in this tape emulation, meaning that the end user has to be very aware of the gain staging within their production depending on how the tape is to be driven and where the tape sits in the processing chain. As has already been discussed in this chapter, there is a generation of young engineers who have only experienced audio processing in a digital environment, and who have knowledge of classic processing gained only through the use of emulation tools. With emulations such as the Studer A800, a knowledge of gain staging related to traditional studio practice is desirable in order to maximize the potential of the tape in the classical sense. It is also true to say that one of the great innovations in contemporary production comes from what would be seen in traditional approaches as a creative misuse (Keep, 2005) of tools such as tape, where traditional approaches would not afford the processing opportunities offered in the DAW.

As well as specific emulated tape machines, manufacturers such as U-He have provided an interface that appears to the user as a tape machine, but offers control over a greater range of parameters without reference to classic machines and tape formulas. Two formulas are offered, simply described as vintage and modern. Tape speed control is provided, but unusually is continuously variable to allow specific dialing in of tape response. A pre-emphasis control is also provided and offers control over transient response and tone, and a compander section allows exploration of noise reduction technology as used and misused in the traditional analog studio both to add dynamic range and famously in the processing of backing vocals to create a sense of "air" and "height" using Dolby A processing. Further innovation is provided through control of parameters of tape processing often considered to have negative connotations. It is possible to manipulate noise performance, wow and flutter and a range of advanced features such as asperity, crosstalk and bias. It is also possible to manipulate the emulated physical characteristics of the tape machine. The captured frequency response is significantly impacted by the size of the reproduction head gap, with this physical characteristic fixed in an analog tape machine. The head gap width on a mastering machine, for example, will tend to be very small to allow high-frequency extension. Companies such as U-He allow manipulation of this physical feature, along with other characteristics such as bias and the tape bump, giving the user exceptional levels of control over their audio signal both in terms of frequency response and distortion characteristics. Through manipulating these parameters, users are able to re-create their desired tape machine characteristics to facilitate more traditional tape processing with enhanced control, and also engage in creative processing not easily achieved with an analog machine.

PICK-AND-MIX PROCESSING: BUILD YOUR OWN TONE

Tape processing provides clear examples of manufacturers creating a pick-and-mix approach to plugin design, with users able to combine elements of different emulated units into a single signal path. Similar functionality can be found in EQ units, where the bands can each feature different EQ curves to represent different emulations in different frequency bands. It is possible in a single interface to select a Neve filter, with mid-range bands from an SSL and the top end from an APL 550, for example. This process has been taken further by Waves, who as part of their Manny Marroquin signature series provide an EQ that features the preferred hardware EQ choices for specific frequency ranges as employed by Manny Marroquin. The influence of mix engineers on plugin design will be discussed in more detail in the following section.

As well as creating emulations that are faithful to the original machines, it is clear that extreme control over traditional processes sits at the heart of plugin innovation and use in contemporary processing, with the majority of modifications made to support use in a contemporary production environment. Soundtoys, for example, use generic interfaces that contain processing elements inspired by hardware. Decapitator, for example, is a distortion plugin that offers multiple distortion circuits from devices, including the germanium stage of a Neve preamp and the tube distortion found in the Thermionic Culture Culture Vulture. Drive and EQ controls allow the characteristics of the distortion to be controlled, with an additional "punish" control driving the distortion to relatively extreme levels. The high-order distortion created by this approach has become more prevalent in contemporary production as creative abuse has seen high levels of distortion produced and captured with high fidelity. There are also a number of commercially available products such as Dada Life The Sausage Fattener embracing the heavily compressed and distorted aesthetic associated with particular musical genres and specific artists with a greatly reduced set of interface controls designed for instant musical gratification.

MIX ENGINEER INSPIRED PLUGIN DESIGN: SEMANTIC DESCRIPTORS

As well as being inspired by classic hardware tools, plugin manufacturers such as Waves have also embraced contemporary mix engineers in their plugin design approach. The Waves signature series tools embrace a range of engineers, including Chris Lord Alge (CLA), Manny Marroquin, Andrew Scheps, Jack Joseph Puig (JJP), Eddie Kramer and numerous other engineers associated with contemporary pop and rock music. These signature series tools in combination with online resources such as Mix with the Masters, Slate Audio Legends and Pensado's Place have all led to mix and recording engineers becoming recognized names, and their sonic signatures becoming known by engineers, producers, musicians and consumers.

The CLA Classic Compressors and JJP Analogue Legends are examples of Waves signature plugin bundles that focus on emulation of classic hardware associated with the engineers, and have continued the tradition of hardware emulation already discussed in this chapter, only this time bringing the sonic personality of the engineer into consideration. Perhaps a more interesting innovation has taken place in the creation of the signature series plugins, which rather than providing the tools found in the studios of the engineers focus more on the sonic impact of the processing undertaken by these engineers.

The Waves CLA signature tools target specific elements within the mix through a series of faders, each focusing on a particular characteristic of the target sound; it has controls for bass, treble, compression, reverb, delay and pitch. Each of these faders has three options for enhancement, with the use of some semantic descriptors for each of the processes. The treble control, for example, offers options such as "bite", "top" and "roof" rather than citing specific frequencies or tools that would be used by CLA to add a sense of "bite" or "roof" to a vocal. The compress fader offers options such as "push", "spank" and "wall", all of which conjure strong images of how the sound may well be manipulated by the fader. This use of semantic descriptor is an interesting development in plugin design, and is certainly something that could be expected to develop as manufacturers run out of tools to emulate and are forced to find new methods of innovation in order to maintain the commercial success of their plugins.

The Waves CLA Drums module interface is very similar to the Vocals interface, with the most significant difference being the choice of drums to be processed using the dial located on the left of the plugin. The other significant difference is the removal of the pitch stereo widening tool in favor of a gate, which has a simple hard or soft setting with a fader to control the impact of the gate on the target sound. Very similar descriptors are used, with some variations in the reverb choices, again to represent the process undertaken by CLA in his own mix approach, as documented in detail in both Mix with the Masters (*Mix With The Masters*, 2018) and Slate Audio Legends (*Audio Legends – Slate Digital*, 2018) resources.

The chosen descriptors found in these tools very much align with the sonic signatures of a CLA mix, providing a user with direct tools that bring the immediacy and impact of a CLA mix to their own production, or indeed the impact of the various engineers who have signature processing available. The ability to mix and match different signature plugins also has exciting potential in utilization, with engineers able to use the audio fingerprint or 'auroprint' as an influence for an engineer finding their own sound through semantic descriptors rather than targeting a single engineer tone or set of classic emulated hardware tools.

Within the signature series of plugins, there are also tools that explore alternative approaches to audio processing. As is suggested in the name of the Waves 'Scheps Parallel Particles' plugin, this tool offers a set of present parallel processing approaches that are designed to add thickness, air, bite and sub, with only the sub processing offering control beyond a simple send. The Andrew Scheps mix approach is discussed in depth in Mix with the Masters (*Mix With The Masters*, 2018), with his mix template available to registered users. This template shows the reliance on parallel processing in his mix system, and this plugin provides a tool that provides the desired processing results in a simplified interface, again driven by semantic descriptors without requirement for knowledge of the process, only the imagination to use the tools to create the sound desired by the end user.

The Manny Marroquin bundle is the most traditional of the processing bundles available from the Waves Signature Series. The EQ, for example, provides a set of EQ filters that represent the tools regularly used by Manny Marroquin in his mix process. The reverb captures settings from the reverbs used in his sessions, and the delay is a relatively common setup but with some additional processing options including distortion and reverb within the delay plugin itself. The Marroquin Signature Series very much represents the mix approach taken by Manny Marroquin, affording a great deal of user control while still embracing the associated processing approach.

Semantic descriptors have become an important feature in the analysis of mixes, leading to the development of tools for automatic mixing that use instrument tags and semantic descriptors to drive the automated mixing process (De Man and Reiss, 2013). Though not the focus of this chapter, feature extraction and knowledge-engineered automatic mixing solutions are becoming increasingly prevalent in plugin design. Equalization, panning and dynamic processing are all undertaken by automatic mixing tools, with companies such as Sound Radix offering tools for automatic adjustment of phase of elements within a mix that could not be achieved without the algorithmic analysis undertaken by the plugin.

It is also noteworthy that there has been a proliferation of preset libraries provided by engineers for more traditional processing tools. As well as providing an effective endorsement for the processing plugin in question, these presets also provide a useful starting point for both experienced and novice engineers. The market demand for presets is also a significant contributor to the proliferation of preset lists, with the reliance on presets in synthesis, for example, well established through the development of complex synthesis technology in the 1980s (Théberge, 1997) and still present in today's plugins. As discussed in this chapter, one of the challenges faced by plugin developers is around issues of gain staging and headroom, with presets often likely to be influenced by the gain staging of the user themselves, and as such will still potentially require modification in order to achieve the desired outcome.

CONTEMPORARY INTERFACES

The Waves Signature Series plugins represent a new approach to plugin innovation, embracing a simplified approach to processing driven by signature sounds of engineers and via an interface that has taken a step away from traditional hardware interfaces. Plugin manufacturers such as Fulfiller and Soundtoys have taken the decision to move away entirely from the traditional interfaces associated with hardware, and instead have focused on creating digital tools specifically to provide the maximum control and feedback to the user in the DAW environment. DAWs have offered a range of native processing, with the traditional DAW EQ and dynamics strip representative of the typical digital channel strip also found in the majority of digital consoles. In the case of dynamics, the classic digital channel strip demonstrates a great deal of information, showing the compression curve, giving clear indication as to the status of the compression and providing easy access to parameters such as ratio, attack and release. In contrast, the basic dynamics - processing strip found in Pro Tools offers very little in the way of tonal control.

The Fabfilter Pro C-2 interface, which although has some similarities with the traditional digital processing found natively in that DAW, offers significantly more detailed control and feedback. As well as offering tonal choices via the semantic descriptors found in the mode control, Pro C-2 provides detailed metering, giving visual feedback on the action of the compressor. The knee of the compressor can be manipulated using a high-resolution slider, the gain reduction range can be limited, and wet dry controls are provided. There are also a series of advanced control options in the expert area, allowing sidechain detection linking control, and multi-point filtering of the sidechain signal, offering the end user significant control over the action of the compressor and the impact on the resulting audio. It is also interesting to analyze the range of preset options available in this compressor, and indeed the potential for users to study the settings that are used to create these presets in further enhancing their ability to create their own approaches.

DMG Audio Compassion is another tool that has been developed with the aim of putting full control of the audio processing in the hands of the end user, offering a wide variety of control parameters that require an in-depth understanding of compression and compressors. The manual describes every parameter in detail, and provides insight into compression for even the most experienced engineers. DMG Equilibrium is similar to Compassion in that it places its focus in providing full user control, with the interface setup dictated by a configuration wizard when instantiating the plugin. Fabfilter and DMG Audio are examples of two companies who are developing tools that place control and metering of audio signal at the heart of their design, with Fabfilter focusing on a work-friendly user interface and DMG providing maximum control and technical detail. For the experienced engineer who enjoys dialing in their tools, these innovative plugins offer huge processing potential, with the quality of processing on offer matching the quality of the interface.

RESTORATION AND TUNING

Audio restoration and tuning has become increasingly important in the development of contemporary workflows and sounds. Paterson (2017) discusses the use of auto tune in contemporary production, with the process of auto tune being used beyond simple correction and instead becoming a detectable sonic signature among the general populace. There has been significant innovation in the processes of detection and control of tuning, with Celemony Melodyne offering the ability to re-voice chords taken from a single recording, manipulate formats in vocal recordings and to facilitate complex sound design through manipulation of an existing signal.

Izotope have developed a series of DAW plugins and stand-alone applications, designed to support mixing, mastering and spectral restoration and manipulation. The noise reduction control offered by Izotope has become industry leading, and has been developed through techniques of machine learning to identify and manipulate noise, with impact from the removal of unwanted noise on music recordings to the addition of location ambience to ADR in postproduction.

HYBRID APPROACHES AND HARDWARE INTEGRATION

Recent innovation has seen an increase in the development of audio products that provide a hybrid approach to processing. The Universal Audio Apollo interfaces, for example, feature the Unison microphone preamp technology, where the analog characteristics of the preamp work alongside a DSP-based plugin to provide a deeper level of integration and realism in the emulation of classic studio hardware. Townsend Labs have created a hybrid microphone system, which uses a specific microphone in combination with a plugin to model the complex characteristics of classic microphones, including the previously not emulated off-axis response and proximity effect.

Other companies are exploring the benefits of recall by offering plugin-based control and recall of hardware settings. Total recall is identified as one of the core advantages of mixing in the box (Paterson, 2017), with plugin control also offering complex automation opportunity not traditionally offered by hardware recall systems. Tegeler Audio provide a range of hardware tools including reverb, compression, equalization and full channel strips with full DAW integration, with Bettermaker going a step further in offering hardware that can only be controlled via plugin interface.

Hardware control over digital re-creations of classic hardware has provided challenges to manufacturers, with many building on the HUI protocol to create generic tools for the control of the DAW environment. In recent years, Eucon has seen a number of innovations, allowing manufacturers to create mappings for complex control using Eucon-enabled devices for direct parameter control. One of the difficulties faced by users in exploring these hardware control interfaces is in their generic layout, with the hardware failing to emulate the layout that has been so specifically re-created in the plugin interface. Softube have developed an innovative solution to this problem, providing a hardware channel strip that is capable of supporting a range of classic consoles in a layout that offers familiarity and a clear link to the emulated channels. Though workflow integration across multiple platforms remains a challenge, the development of dedicated hardware for the control of digital channel strips marks an interesting and powerful development for engineers for whom consistency of workflow and layout are important factors.

THE FUTURE: INNOVATION, EMULATION AND RE-EDUCATION

The impact of innovation in plugin design since 1982 has played a significant role in defining the DAW as the primary environment for music creation and mixing. There is clear potential for the development of new interfaces, but these rely on developments to take place in the entire control paradigm explored by the DAW. New techniques have seen hardware emulation improve to the point where many professional engineers feel entirely comfortable replicating their traditional mix approach while mixing ITB. Although we have seen further developments, particularly in the creation of digital tools that offer a level of control and processing potential that would previously have been unavailable, a number of questions still need to be answered as to the future developments of plugins.

As the quality of emulation reaches a plateau, and the range of desirable tools available to be emulated reduces, designers are faced with new challenges in satisfying the market demand for improvements and innovation. Despite the developments in semantic-driven processing and innovative interfaces, the majority of focus in the industry still falls on emulation and addition of features to provide flexibility and control around defined sonic signatures. Strachan (2017) discusses concepts of democratization of technology, but the addition of features to classic processing tools is arguably more likely to see a gentrification of such processing, with the required knowledge of not just the specific tool but also the advanced control and calibration of that tool required to successfully replicate the mix approaches established in popular music production practice. With companies such as Slate Digital developing engineer-led education programs, perhaps the next innovation is not in interface design or emulation quality, but in the methodologies explored in educating the market in the use of plugins and their broader utilization in mix and production practice.

BIBLIOGRAPHY

Audio Legends – Slate Digital (2018). Available at: https://slatedigital.zendesk. com/hc/en-us/categories/115001623068-Audio-Legends (Accessed: 1 April 2019).

Bell, A., Hein, E. and Ratcliffe, J. (2015). Beyond skeuomorphism: The evolution of music production software user interface metaphors. *Journal on the Art of Record Production*, 9.

Case, A. (2007). *Sound FX: Unlocking the creative potential of recording studio effects*. New York: Focal Press.

De Man, B. and Reiss, J. (2013). A knowledge-engineered autonomous mixing system. In: *Proceedings of the 135th audio engineering society conference*, New York, 17–20 Oct.

DMG Audio. (n.d.). *Compassion manual*. Available at: https://dmgaudio.com/dl/ DMGAudio_Compassion_Manual.pdf [Accessed 14 Jan. 2018].

Fabfilter Software Instruments. (n.d.). *Pro.C2*. Available at: www.fabfilter.com/ products/pro-c-2-compressor-plug-in [Accessed 14 Jan. 2018].

Massy, S. (2016). *Recording unhinged.* Milwaukee: Hal Leonard.

Milner, G. (2009). *Perfecting sound forever: An aural history of recorded music.* New York: Faber and Faber.

Mix With The Masters (2018). Available at: https://mixwiththemasters.com/node (Accessed: 1 April 2019).

Paterson, J. (2017). Mixing in the box. In: R. Hepworth Sawyer and J. Hodgson eds., *Mixing music.* New York: Routledge.

Schmidt Horning, S. (2013). *Chasing sound: Technology, culture & the art of studio recording from Edison to the LP.* Baltimore: The John Hopkins University Press.

Strachan, R. (2017). *Sonic technologies. Popular music, digital culture and the creative process.* New York: Bloomsbury Academic.

Théberge, P. (1997). *Any sound you can imagine.* Hannover: Wesleyan University Press.

Universal Audio. (n.d.). *UAD plug-ins.* Available at: www.uaudio.com/uad-plugins.html [Accessed 14 Jan. 2018].

Waves. (n.d.). *Signature series bundles.* Available at: https://dmgaudio.com/gallery_all.php [Accessed 14 Jan. 2018].

15

Mixing and Recording a Small Orchestral Ensemble to Create a Large Orchestral Sound

Jenna Doyle

INTRODUCTION

As Bennett quite accurately states within his guidelines for musicians on computer orchestration (Bennett, 2009):

> Film composers regularly use large sample libraries to produce mockups of their score. . . . You may not want or need to replace the synthetic orchestra or you might want to use it in conjunction with real players and instruments.
>
> (p. 1)

His advice from 2009 is still very relevant today within the music industry and is particularly pertinent to orchestral music created for music for media (films, television, online video, games etc.). The technique of combining an orchestral stem (of either a solo instrument or a group of instruments from the same family) with a Virtual Studio Technology (VST) stem to enhance the perception of a piece of purely acoustic orchestral audio is common practice among composers, producers and studio engineers. As Pejrolo and DeRosa suggest (Pejrolo and DeRosa, 2017), this is particularly true for recording projects with a restrictive music budget (p. 7.2.4).

The simple layering of 'real' and 'fake' (VST) stems often involves a process of simple volume manipulation and matched digital effects (FX) to blend the two sources together, giving the impression of one captured performance. Although this technique may work for many composers or producers who lack time for more detailed mixes, the question still remains—can a method be developed to blend 'virtual instruments' with real recordings to create a truly authentic symphonic, orchestral recording?

The intention of this chapter is to document practice-based research, in which a method was developed over a number of months (through trial and error) within a controlled studio environment, aiming to create 'real-with-fake' orchestral recordings which could truly fool the 'inexperienced' (non-musician) and 'experienced' (professional musician) listener into believing that they were hearing real recordings of an authentic 90-piece symphony orchestra. There are varying recommendations for the numbers of players for each section of a symphony orchestra, with Chew et al. (2010) (p. 16, 76, 126, 176) recommending the following as a guideline (see Table 15.1).

The trials and methods were explored in Crown Lane Studios, a carbon-neutral recording studio based in Morden, London. The studio had not undertaken a project of this size before, with regard to the number of tracks involved (24 in total) and combined number of instrumental players and vocal performers (40 in total). Nonetheless, a commitment from studio owner and engineer John Merriman to completing the project, and exploring the creative questions involved, ensured the time and flexibility that would be needed to undertake the research.

The piece itself, the heart of the entire project, is a 120-minute-long musical called *The Battle of Boat*, composed by long-time collaborator Ethan Lewis Maltby, with myself as lyricist. The show was written for 14 instrumental players and 27 young performers/vocalists, with an orchestral palette throughout. There is no woodwind within the arrangements—a purely creative choice. The show was premiered in August 2016 at The Rose Theatre (Kingston, London) by The National Youth Music Theatre (NYMT), an organization which "offers exceptional opportunities in pre-professional, musical theatre training for talented young people aged 10 to 23 years" (NYMT, 2013; https://nymt.org.uk/).

In October 2016, a Crowdfunder campaign was set up to raise the necessary financial backing to create an original cast recording, featuring all of the performers from the London shows. The final product was promised to contributors as a double-CD stereo album. The commitment to creating this final CD product had a significant influence on the decision to focus on stereo recordings only. The obvious potential for 5.1 mixing,

Table 15.1 Recommended Configuration for a Symphony Orchestra

Number	Instrument/ Grouping	Number	Instrument/ Grouping
16	1st Violins	3	Clarinets
14	2nd Violins	3	Bassoons
12	Violas	4	Horns
10	Cellos	3	Trumpets
8	Double Bass	3	Trombones
2	Flutes	1	Tuba
1	Piccolo	3	Percussionists
2	Oboes	2	Harps
1	English Horn	2	Keys (piano)
		90	Total Players

Key: ▪ Strings ▪ Woodwinds ▪ Brass ▪ Keys & Percussion

(Note: Some instruments have been combined to form one grouping, e.g. bassoons and contrabassoon)

particularly with regard to the experience within a cinematic setting, is an area that I do intend to explore in future.

The following methodology explores factors and variables which directly contributed to and formed a key part of the overall workflow for the final recording and mixing process: microphone choice, sequence of live recording, post-production FX (focusing on convolution reverb and panning), choice and configuration of VST-powered orchestral sample libraries and placement of the live players within a digitally constructed live 'space', using the aforementioned FX.

Finally, qualitative methods of evaluation will be employed to assess the 'success' of the method, i.e. do both experienced and inexperienced listeners perceive the recording to be:

1. a captured recording of a real 90-piece symphony orchestra (omitting the woodwind entirely equates to a 78-piece orchestra, using the Chew et al. recommended model in Table 15.1)?
2. without impression of any synth or VST-based enhancement?

The qualitative assessment will take the form of personal interviews with professional musicians (with experience in working with orchestral music). Additionally, a written questionnaire will be conducted with a group of musicians and non-musicians (who lack significant music experience) to assess their reaction to the piece in relation to the two questions.

BACKGROUND AND RELATED WORK

One important aspect of the piece at the center of this research, *The Battle of Boat*, is its identity within the context of musical theatre as a genre. My collaborator, Ethan Lewis Maltby, and I did not approach the piece as one to be written for a smaller orchestral ensemble, which theatre producers frequently favor in order to make musical productions more cost-effective, as suggested by Adler (2004, p. 53). In terms of orchestration, the piece was written with a symphonic orchestra in mind (minus woodwind). These arrangements were then scaled back to the available number of players for the NYMT show. We wished to create a 'cinematic musical'—one which blended the form and structure of a two-act musical, but with a sonic identity and compositional approach more in-keeping with a film score. In this way, we concluded that references to technology and innovations in recording for film scores were more appropriate to the project than those utilized for musical theatre cast recordings.

The 'cinematic' component of our 'cinematic musical' term is therefore a very important aspect of the research. This is, of course, influenced by the expectations that cinema-goers have with regard to film music. Larsen (2005) rightly points out that "after the breakthrough of the sound film, film music was for a long time synonymous with symphonic orchestral music; today, film music can be everything under the sun" (p. 7). Therefore, to assume that a cinematic sound equates to a symphonic, orchestral sound could be presumptive. However, it cannot be ignored that a large number of modern composers working on large-budget, commercial films are still relying on the orchestra as a key part of the sound for their score. The popular electro group Daft Punk combined their signature synth-based approach with a 90-piece symphony orchestra on their score for *Tron: Legacy* (2010). In an interview with *The Guardian* (Michaels, 2010), the French DJ duo outlined the timeless quality of an orchestral film score:

> We knew from the start that there was no way we were going to do this [soundtrack] with two synthesisers and a drum machine. . . . A cello was there 400 years ago and will still be here in 400 years. But synthesisers that were invented 20 years ago will probably be gone in the next 20.

So while the definition of film music may be in constant flux, for the purposes of the term 'cinematic music' utilized throughout this chapter, the cinematic component equates to an authentic, non-augmented symphonic, orchestral sound.

METHODOLOGY

The recording process began on November 11, 2016 at Crown Lane Studios, and the final mastered recordings (24 separate tracks in total) were delivered on June 30, 2017.

The following time line (see Table 15.2) gives an overview of the process from start to finish.

Project Initiation

The team involved in this trial process included myself, co-writer Ethan Lewis Maltby and studio engineer John Merriman. Maltby is a classically trained percussionist and has had a 20-year career in composition for theatre and film, while Merriman (also trained classically) has been working with live players and virtual instruments since 2003 in his studio. I have been producing music for television and the commercial pop music market since 2007, often working to limited budgets (which allows for a minimal number of live players). Together, we formed a creative team that was confident in our perception of an authentic orchestral sound, and decisions were almost always unanimously agreed upon.

To reduce the amount of time required to test the live players against the virtual library trial options, it was decided that exported audio stems of VST-powered, MIDI-based excerpts from the show would be bounced before the live players arrived. The players would then be able to play along with the VST audio, giving the creative team flexibility to switch microphones or experiment with volumes or FX. This preparation was a time-consuming, but necessary, part of the experimentation process.

Table 15.2 Simplified Time Line for the Trialling, Recording, Mixing and Mastering Process

Date	Activity
November 11, 2016	Raw MIDI data from the entirely VST-powered demos were exported for each instrument/grouping of the orchestra—(as outlined in Figure 15.1).
November 12, 2016	This MIDI data was exported to audio stems through three chosen VST-orchestral sample libraries (detailed in 4.2, 4.3 and 4.4).
November 19, 2016	First testing session—a solo violin player was invited to play excerpts of the show, overdubbing the equivalent MIDI part. A number of microphones were trialed against each of the three potential VST libraries.
November 25, 2016	Second testing session—testing trumpet (trialing microphones and VST libraries)
December 1, 2016	It was concluded that the base sample library would be EastWest Platinum Orchestra.
January 10– February 1, 2017	Recording sessions for strings
February 13– March 2, 2017	Recording sessions for brass
March 10, 2017	Recording session for keys
March 15–16, 2017	Recording sessions for percussion
April 10–June 15, 2017	Mixing and post-production process
June 29–30, 2017	Mastering final 24 tracks

Three professional-level, orchestral virtual libraries were selected to be part of the trials. These are as follows:

1. EastWest Platinum Symphonic Orchestra by EastWest Sounds
2. Essential Orchestra (Special Edition Volume 1) by Vienna Symphonic Library
3. Albion by Spitfire Audio

There are a large number of virtual orchestral libraries currently available on the market, but budget and time constraints meant that only a few could be tested. Other popular products not included as part of these trials include: Hollywood Orchestra (EastWest Sounds), Cinematic Studio Series, and LA Scoring Strings (Audiobro, strings only), among others. The three chosen libraries were selected for their contrasting qualities, which the creative team hoped would allow for a healthy range of choice

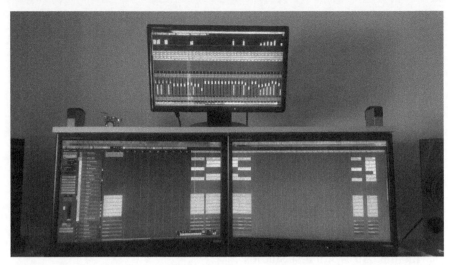

Figure 15.1 Bouncing Audio Tracks From MIDI, Running Through Three Trialled Virtual Libraries

in terms of tone, dryness of recorded capture and microphone placement (a popular feature in high-end orchestral VST sample libraries).

EastWest Platinum Symphonic Orchestra

Recorded by Grammy-nominated classical recording engineer Professor Keith O'Johnson, this library is described by the composer (Jonathan Wright, 2017) as having "a bright, lush, present sound which means it can cut through complex mixes very well", with respected music technology magazine *Sound on Sound* (Stewart, 2009) claiming that the library is "big, bold, naturally reverberant, tailored for the big screen". There is a natural reverb on the samples, and the library is widely considered to have been recorded in Benaroya Hall, Seattle (Lichtman, 2014), although EastWest have not confirmed this outright. The sample library features three microphone positions on each instrument, labeled as 'close', 'far' and 'surround' within the software interface (Killchrono, 2013).

Essential Orchestra—Vienna Symphonic Library

VSL's Essential Orchestra was recorded in a specially designed room, known as 'The Silent Stage' in Ebreichsdorf, Austria (Vienna Symphonic Library, 2007). As (Jackson, 2012) points out in his review for the library, while the provided samples are "exceptionally detailed and nuanced, they are—by design—a bit plain and dry". The dryness of the samples is perceived as a strength by others, however. Poit argues that "the key is to understand VSL's approach towards providing the most basic building blocks and giving the composer a massive toolkit for manipulating the dry samples. Versatility is the key word here" (Poit, n.d.). Within the context of this research project, the creative team thought it wise to include a dry library of its kind—both EastWest Platinum Symphonic Orchestra and Albion include samples which include natural reverb from the space in which they were recorded. The selection of Essential Orchestra as part of the trialing process would allow the creative team to explore the versatility of completely dry samples. Unlike EastWest Platinum Symphonic Orchestra and Albion, Essential Orchestra only provides a singular close mic option for the included samples.

Albion

Sound on Sound observed the following on Spitfire Audio's Albion:

> Adopting a British-is-best approach, the company hired the best UK orchestral players. . . [English Session Orchestra] . . . and recorded them at Air Lyndhurst Studios'.
>
> (Stewart, 2011)

Air Studios' Lyndhurst Hall is well regarded for its "tremendous acoustic properties . . . influenced by a motorized acoustic canopy, which is suspended from the vaulted roof". (Dazeley and Daly, 2014)
The distinctive characteristics of the recording space have become somewhat of a hallmark within orchestral film music, with scores for numerous big-budget Hollywood films being recorded at the studio. The sample library features four microphone positions, including 'close', 'tree' (vintage Neumann M50 mics placed above the conductor's podium), 'outrigger' (widely spaced AKG C40 mics) and 'ambient' (a more distant gallery placement). This library offered the greatest range of microphone placement of all three libraries studied.

All three of the chosen libraries featured high-quality orchestral samples, including strings, brass and percussion, necessary to cater for most areas of the orchestration for *The Battle of Boat*. The harp and piano parts are always used in a solo capacity throughout the piece, so it was decided that these parts would be entirely VST-powered (to dedicate more studio budget towards the larger orchestral sections) and not duplicated with live performers. The EastWest Quantum Leap Pianos' Steinway D was used for piano parts, and EastWest Platinum Orchestra's harp sample was used for harp parts. Real recordings of a grand piano and harp would have been more ideal, but the creative team was confident that these virtual libraries created very realistic (dry) parts which could be integrated into the overall mix very well.

The MIDI parts for the corresponding sections of the orchestra were passed through the relevant samples from all three of the selected libraries. The number of players emulated within each sample patch varied depending on the library. This is currently a limitation of many orchestral virtual libraries currently available; there may be some options to have a smaller number of players within the section, but the larger numbers are limited to a set figure. Table 15.3 compares the number of players per section (most relevant to the sizes needed for a symphonic, orchestral sound).

Table 15.3 Number of Players Per Section Across the Three Selected Orchestral Libraries

Orchestral section	EastWest Platinum Orchestra	VSL Essential Orchestra	Spitfire Audio Albion
1st Violins	18	14	11
2nd Violins	11	(no separate patch)	9
Violas	10	10	7
Cellos	10	8	6
Double Bass	9	6	4
Horns	6	4	4
Trumpets	4	3	3
Trombones	4	3	3 (doubled in octaves—6)
Tuba	3	1	1 (doubled in octaves—2)
Percussionists	Individual samples	Individual samples	Individual samples
Harps	EastWest Platinum Orchestra's harp	EastWest Platinum Orchestra's harp	EastWest Platinum Orchestra's harp
Keys (piano)	Bösendorfer Piano (EastWest Pianos)	Bösendorfer Piano (EastWest Pianos)	Bösendorfer Piano (EastWest Pianos)
Total (minus percussion, harp & keys)	79	74	48

(Soundsonline, 2010) (Vsl.co.at, n.d.) (Stewart, 2011)

The creative team did consider that bounces of singular, or solo, instruments from each section could then be grouped to create a consistent comparison and allow for even more mix control. However, unique expressions of each sample, or 'round-robin' functionality (as named within most virtual libraries), were limited, and we could not guarantee sufficient individuality from each solo player across the three trialled products. We therefore resolved to use the large-section samples from each of the chosen libraries.

Combinations of Virtual Libraries With Microphones

Bounces from the three selected virtual libraries were contrasted against live recordings of solo players, recorded with different microphones, in two dedicated studio 'testing' sessions (see Table 15.2 for time line). Tables 15.4 and 15.5 summarize the findings of these sessions, comparing the various combinations between virtual libraries and microphones.

First Violin Player

Table 15.4 Combinations of Microphone (Recording a 1st Violin Player) With the Three Selected Virtual Libraries

Virtual library/ Microphone	Rode NT2	AKG C414	Sontronics Sigma Ribbon	Neumann TLM-49
EastWest Platinum Orchestra	A very balanced combination	A good overall sound, perhaps a little thin	Mic sounds a little too rich	A pleasant, warm sound
Essential Orchestra	A fairly good combination, though the samples sound quite pure against the live player	A very thin sound, lacking warmth	The samples sound too pure against the live player	A good combination
Albion	A fairly balanced combination, though the samples sound more compressed than other libraries	Similar to the Rode NT2, the strings sound quite processed	The rich sound of this mic works well with the library samples	The warmth of this mic equally works well

Trumpet Player

Following these experimentation sessions evaluating the combinations between virtual libraries and microphones, it was decided by the creative team that the EastWest Platinum Orchestra (EWPO) would be the most suitable orchestral library for the project. This was, in part, due to the complementary relationships with the microphones that were tested, particularly the AKG C414, Rode NT2 and Neumann TLM-49 microphones. The creative team found the samples from EWPO to sound realistic and exhibit versatility in terms of post-production manipulation.

Limitations within the orchestral libraries themselves also became apparent throughout the testing process, which naturally influenced the final library selection. The doubling of notes in octaves within specific sections of the Spitfire Audio Albion library (as listed in Table 15.3) was somewhat problematic, as parts were not always written octaves apart within the piece. This reduced the flexibility and suitability of Albion as the base library overall, and suggests a limitation that composers and producers should be aware of when looking to create score-accurate orchestral demos.

Table 15.5 Combinations of Microphone (Recording a Trumpet Player) With the Three Selected Virtual Libraries

Library/ Microphone	Rode NT2	AKG C414	Sontronics Sigma Ribbon	Neumann TLM-49
EastWest Platinum Orchestra	The combination is a little bright overall	A great, balanced sound	The samples sound quite thin within this combination	A great, balanced sound
Essential Orchestra	A good combination, though the samples sound a little thin	This microphone is too bright against the library samples	The timbres of the virtual and live sounds are quite mismatched	A reasonable combination, though the timbres are not matched very well
Albion	A good combination	The samples sound fairly processed in this combination	This combination is overly dark	An attractive, warm sound

It was also noted that Albion featured patches with comparatively lower numbers of players (see Table 15.3), when compared directly against EWPO and Vienna Symphonic Library's Essential Orchestra (VSL EO). The large-scale orchestral sound that was desired could be more difficult to achieve with this library. One can also observe from Table 14.3 that EWPO features the largest combined number of players (only slightly ahead of VSL EO), so it made sense to the creative team that EWPO showed the highest potential for success in conveying the impression of a symphony orchestra, when combined with live players.

VSL EO samples allowed for flexibility in re-creating accurate arrangements where Albion did not (however, they did not provide the variable mic positions that EWPO and Albion offered), but the creative team felt that the VSL EO ensemble sounds were overly 'pure' in timbre and did not blend particularly well with the live players, even with a combination of microphones. EWPO and Albion samples contained more natural 'noise' within the individual samples themselves, blending more easily with the live recordings captured in the studio environment.

Final Microphone Selections

Once the EWPO library was selected, many of the microphone choices for other instruments in the orchestra were made during the final recording sessions. This involved testing a range of microphones and selecting the best fit against the EWPO bounced parts. Table 15.6 shows the final microphone selections.

Table 15.6 Final Microphone Selections for the Live Players

Live Players (14 total)	Selected Microphone
Violin x 2	Rode NT2
Viola x 1	Rode NT2
Cello x1	Sontronics Sigma Ribbon
Double Bass x 1	AKG D112
Horn x 2	AKG C414
Trumpet x 2	Neumann TLM-49
Trombone x 1	Shure SM57
Percussion x 2	AKG C414 (pairs)—overheads Rode NT2—low tom & bass drum Sontronics STC-1S—cymbals Sontronics DM-1T—toms Shure SM57—snares Neumann TLM-49—glockenspiel
Keyboard x 2 (piano, harp)	Sample triggering—no microphones needed

Between January 10 and March 16, 2017, the live players were recorded at Crown Lane Studios, with the EWPO bounced parts included within the Pro Tools sequence to check for matched timings, performances and potential for an overall complementary mix (see Figure 15.2).

Microphone Placement

During the microphone/virtual library trial sessions, the creative team also experimented with microphone placement. Although EWPO featured very desirable stage and hall placements, which allowed for a very realistic emulation of a physical space, it was found that combining this option with the live players was particularly problematic. As the live players were being recorded within a dry studio space, the very specific Benaroya Hall sound (the recording space used for EWPO) was difficult to emulate. This proved to be true even with the EastWest Quantum Leap Spaces convolution reverb, which features a convolution reverb of this exact hall.

The creative team concluded that the best result would likely be achieved by using a close (or spot) microphone placement for both the virtual instrument stems and the live players. This would achieve the driest possible sound for both elements, which could then be more effectively 'glued' together in the post-production process.

Compression

To avoid an orchestral sound that could be perceived as augmented or enhanced within the post-production process, the creative team opted for

Figure 15.2 Recording Live Players in Crown Lane Studios, Morden

a very light touch with regard to compression across the entire project. As Dowsett (2016) observes on mixing strings:

> Strings are both very dynamic and rich-sounding instruments, which sound very forced with anything more than slight post-production tweaking. In fact, a lot of the time there is no EQ used (other than maybe high-pass filtering). Compression is also used infrequently but sometimes the style/dynamics of the player might result in a spot mic needing to be compressed slightly. Room mics are rarely compressed as the amount of space between the players and the mic(s) helps to even out any level discrepancies.
>
> (p. 13.3.6)

Using the highly regarded Teletronix LA-2A compressor, a low ratio of 1:2 with a medium-low threshold was applied to the live players to apply a subtle, but beneficial, leveling effect on the overall dynamic range. An additional plugin compressor (Pro Tools' Dynamics III Compressor) with a higher 4:1 ratio, together with a higher threshold, prevented sudden peaks in volume from becoming too prominent in the overall mix. A medium attack of 20ms and medium release of 100ms was favored for a subtle application of the compression effect.

Reverb—Placement of the Live Players Within the Mix Using Panning and EastWest Quantum Leap Spaces

As the aim of the project was to record a piece that gave the impression of one-time capture in a physical space, the decision to opt for convolution reverb was a very natural one. The creative team was keen to extend beyond basic panning for the live players and a blanket convolution reverb to 'glue' the real and virtual instruments together.

As well as left-to-right pan positions, we discussed the idea of emulating depth/distance between live players by creating three separate reverb 'zones' within a perceived concert hall space. This would be achieved via three distinct 'send effects' of convolution reverb, giving the impression of varying distances between instrumental sections within the final mix. The construction of these three separate zones will be discussed momentarily. The team did recognize the potential of a 5.1 mix for creating an even more convincing experience for listeners, but given that the final mixes were destined to be released on a CD (in stereo format), a 5.1 mix was not explored. As many composers and musicians are required to create orchestral mock-ups in stereo format, we do believe that this experimentation within a stereo mix made for valuable exploration.

The creative team considered a range of digital convolution reverb products, including Waves IR-1, Audio Ease Altiverb, LiquidSonics Reverberate, SIR Audio Tools SIR2 and EastWest Quantum Leap Spaces. The team settled on using Quantum Leap Spaces (QL Spaces), not only for its high-quality impulses and functionality but also due to the inclusion of an impulse response taken from Benaroya Hall, labeled as 'Northwest Hall' within QL Spaces (Boberg, 2015). This hall was the location in which the chosen virtual library (EWPO) was recorded years previously, so it seemed a logical option to choose a convolution reverb based on the original space in which the samples were recorded.

Both the live and virtual stems were bussed through one of three QL Spaces reverbs (all based on the Northwest Hall impulse response), which were configured to represent the front, middle and rear of the hall. These configurations involved unique settings for the wet and dry signal pots, input signal level and pre-delay. In an effort to mirror section positions within a traditional orchestral setup, the percussion and keys were bussed through the rear configuration, brass through the middle configuration and strings through the front configuration.

The rear configuration had a higher proportion of wet signal, while the front configuration featured a higher proportion of dry signal (with the middle configuration creating a balance between the two). The rear configuration processed a lower input signal than the front configuration, in an effort to create an acoustic illusion of greater distance between the listener and the rear sections of the orchestra. The front configuration featured a stronger input signal to suggest a closer position.

(Weiss, 2010) observes that "pre-delay is a key element in determining the front-to-back relationship of the dry sound and the space it exists in. . . . Generally speaking, the longer the pre-delay, the closer the dry sound". Naturally therefore, the creative team opted for a longer pre-delay for the front configuration and a shorter pre-delay for the rear configuration (with the middle splitting the difference), but these settings contrasted minimally in order to maintain a natural sense of the acoustic space.

Finally, the creative team discovered that panning the live players to their first-chair (or leader) positions within the relevant sections of the orchestra created an effective lead instrument within these sections. Second-chair positions for the trumpet and horn sections were also utilized. The volume balance between the live stems and the virtual stems was automated considerably, as we did not want to resort to harsh compression to maintain a dynamic balance. The panning configuration for the live stems was calculated using Figure 15.3 (based on traditional first- and second-chair positions outlined by Eargle (2003, p. 304)) to outline player positions and Figure 15.4 to calculate the appropriate amount of panning.

As explained by Miles-Huber (2007, p. 35), panning within a stereo mix ranges from 0 (hard left) to 64 (center) to 127 (hard right). In many sequencers, such as Logic and Pro Tools, this can be represented as −64 (hard left) to 0 (center) to +63 (hard right). This representation of panning range is used within Figure 15.4. One can observe that the difference in panning between the string and the brass sections is very subtle, while the live double bass is panned fairly harshly to the right. The live percussion is panned across −40 (left) and +20 (right), depending on the instrument. The overall effect means that there is a very concentrated area of principal 'live' players within the mix, but together with the 'zones' of convolution reverb providing variations in distance, the overall effect was aurally desirable to the creative team (in our aiming of achieving an authentic, symphonic, orchestral sound).

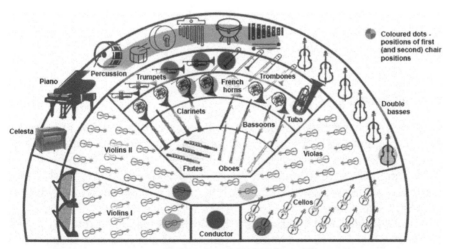

Figure 15.3 First-Chair (and Second-Chair) Positions Within a Large Symphony Orchestra

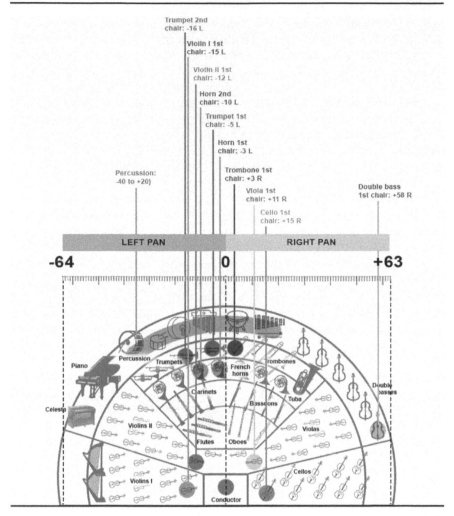

Figure 15.4 Calculated Pan Setting Based on the First- and Second-Chair Positions Outlined in Figure 15.3

EVALUATING SUCCESS

Evaluating the success of the project was carried out via a number of qualitative methods. The creative team was very satisfied with the final product, and a very positive outcome of the project came in the form of a publishing deal for *The Battle of Boat* with Theatrical Rights Worldwide, who were impressed by the quality of the piece and the authenticity of the orchestral recordings.

It was decided that interviews with professional (experienced) musicians, as well as a questionnaire with musicians and (inexperienced) non-musicians, would be an effective way of evaluating perceptions of the piece, with regard to the aims of the research project, as outlined in this chapter.

Qualitative Interviews With Music Professionals

Session Percussionist Paul Townsend

Paul Townsend is a professional session drummer and percussionist, classically trained with a BMus degree from Canterbury Christ Church University. As well as playing as a drummer in West End shows (e.g. *Jersey Boys* and *Whisper House*), Townsend has played in orchestras throughout his career, percussing for the London Youth Philharmonic Orchestra, The Canterbury Orchestra, the Sittingbourne Choral Society and the Folkestone & Hythe Orchestral Society. With a background that includes performances in both orchestral and contemporary ensembles within the context of music theatre, Townsend was an ideal interviewee based on the 'cinematic musical' aspect of the project. He was interviewed (Townsend, 2017) in September 2017, where he listened to two instrumental tracks from the project.

Townsend interpreted the recordings to be indicative of a "symphony [orchestra]—I hear full range brass, pizzicato strings and lots of percussion". He remarked that "the natural quality of the reverb makes it sound like a concert hall", and "only the strings sound synthetically enhanced as I think the pizzicato effect seems unnatural, perhaps it is too loud and powerful in the mix". Despite this observation, Townsend concluded that the recordings "sound very natural and real to me. The mix is very balanced and the dynamic range is huge".

Composer Warren Meyers

Warren Meyers is a professional composer and producer based in Tatsfield, Kent and is published by Peer Music UK. Meyers works on a variety of projects, which range from commercially released dance tracks to music for television. A self-taught musician, he has occasionally worked with large orchestral ensembles (not to symphony size) and solo orchestral session players, but he more often works with virtual instruments where a project requires orchestral instruments. This makes him an ideal candidate as a musician who is very aware of virtual orchestral libraries and recognizes particular traits within faux-orchestral productions. Meyers was interviewed (Warren, 2017) in October 2017 and listened to two instrumental tracks from the project (the same tracks played to Paul Townsend).

Meyers agreed that the recordings had a purely orchestral palette, but rather than using a symphony orchestra, he stated that "it sounds like it's a 16-piece orchestra. I may be wrong but I feel there has been some added artificial elements from a classy plugin of sorts". Meyers suggested that the recorded sound suggested "an open-space recording area" and, while he identified that the recordings may not be truly authentic as real symphonic orchestral pieces, he was complimentary about the production:

> As a professional producer myself, I am aware that these days it's very difficult to tell the difference between a true recording of a symphonic orchestra and a high class plugin which uses sample-based software.

This obviously relies on the user having a full understanding of stage layout, positioning, velocity and feel of the orchestral performance. You have to be fully educated and understand every instrument to create a piece like this, if that were the case.

Qualitative Questionnaire With Musicians and Non-musicians (Students)

Students of Ravensbourne (London) were invited to listen to an instrumental version of the first track from the final CD, "What We Get Up To", and subsequently filled in a questionnaire on the piece. Closely relating to the initial aims outlined in the introduction of this chapter, the questions were designed to assess the degree to which the students felt the track:

1. captured a recording of a real symphony orchestra
2. did not convey an impression of any synth or VST-based enhancement

Questions included:

---

1. Thinking about the track "What We Get Up To", what do you hear in terms of instruments and what kind of piece do you interpret it to be?
2. Does the piece sound like it was recorded by a symphony orchestra or a smaller ensemble? Please elaborate in your answer.
3. If you had to guess, what kind of space would you say the piece is recorded in?
4. Does the piece sound natural to you? Is there anything that sounds out-of-place?
5. Does the piece sound synthetically enhanced in any way? If it does, describe what you mean.

In total, 18 students participated in the questionnaire—ten individuals were students studying music, while the other eight students were studying Sound Design (these students were comfortable and confident discussing their perception of sound, but lacked experience with orchestral instruments and working with musical instruments in general).

Comments from the music students included:

- "This sounds like a symphony orchestra—there is a massive sense of depth of instruments".
- "It sounds pretty natural".
- "I don't think it is a massive orchestra, but it has a few instruments from each section".
- "Symphony orchestra—due to the use of instruments and the amount of instruments you can hear, not a little-sounding ensemble".

- "It's orchestral and sounds quite filmic, but it lacks the size that I normally expect from a film score".
- "[It sounds like it was recorded] in Air Studios?"
- "[It sounds like it was recorded] in a concert hall".
- "It would sound better if the bass was fattened up".
- "The strings could be synthetic, but very well done? I would guess natural instruments".
- "It sounds like a smaller ensemble. . . [there was a] very harsh crispness to it. This is only to trained ears though if it is so".

Comments from the non-music students included:

- "The piece seems to be recorded by a symphony orchestra due to the range of instruments used".
- "The piece sounds natural and I don't think any part is out of place. I think reverb may have been added to the recording".
- "It sounds thinner than most orchestras I hear on movies".
- "[It sounds like it was recorded] on a stage".
- "Symphony orchestra due to the large variation of instruments and strings—sounded like more than one of each instrument".
- "Low end percussion sounded like it was boosted".
- "A symphony orchestra—it's just because I can picture it in an amphitheatre with a symphony orchestra (giving this kind of sound)".
- "It sounds like an orchestra, but maybe less players than normal? I don't know, I think it should be bigger".
- "It does not sound synthetically enhanced. But the thing is, my audio listening is not fine enough to distinguish a synthetic sound from a natural sound".

It was apparent from the results of the questionnaire that the non-music students were more likely to be convinced that the recording was 'natural' (i.e. no synthetic enhancement) and featured a real symphony orchestra. Some of the music students were convinced, but a number of them suggested that the piece was not recorded with a symphony orchestra but with a smaller orchestral ensemble (under 90 players). One music student (whose comment is included) suggested that the string section "could be synthetic, but well done".

CONCLUSION

The results of the qualitative interviews and questionnaire highlighted some interesting variations between musicians and non-musicians and their perception of the recordings. The musicians who were approached, who often come into contact with 'faux' orchestral recordings as part of their work within the music industry, were much more inclined to identify traits which suggested that the piece was not truly authentic (as a piece recorded with a symphony orchestra). They were more likely to suggest that the piece was recorded with a smaller ensemble and not a full 90-piece configuration. Feedback was variable in terms of the authenticity of the instruments themselves, with some musicians suggesting that 'synthetic' elements were apparent and others suggesting that it was completely real and 'natural'. The non-musicians were much more convinced, although the Sound Design students were more likely to suggest some possible augmentation in postproduction (e.g. reverb and 'boosting').

Nearly all individuals who were approached were confident that the recording had been captured in a physical space (e.g. a concert hall or stage) and not in a controlled, dry studio space. The creative team was also very satisfied with the way in which the final mixes maintained the prominence of the vocals while placing the orchestral instruments in a realistic, crafted hall space.

Some of the comments, which make specific reference to films and film scores, raise an interesting question: has the modern orchestral film score altered the perception of recorded orchestral music? Some participants commented that they found the recordings to lack the size or fullness of symphonic, orchestral recordings from film soundtracks. The original creative intention of Ethan Lewis Maltby and myself was indeed to create a 'cinematic musical', with elements borrowed from cinematic film music, so it seems natural that the interviewed individuals would draw parallels with film music (they were not told about this 'cinematic musical' intention when they were interviewed, however). The comments on the underwhelming size of the orchestra (specifically within the context of film music) is interesting as the creative team did not use any synthesizers (e.g. synth bass, lead synth) to enhance the sound in any way. We wanted to create a 'cinematic musical' with the sound of a real, unaltered symphony orchestra.

Meyer (2017) discusses the "synth-augmented orchestra of Hans Zimmer" (p. 52) in his score for Ridley Scott's *Gladiator* (2000), while Audissino (2014) highlights the Oscar-winning composer's "rock music arranged for orchestra" in *The Rock* (1996). Other notable composers currently working on blockbuster films with enhanced orchestral scores include Danny Elfman, James Newton Howard and John Powell. It could be suggested that, with modern orchestral film scores now being so commonly enhanced with the addition of synths, guitars and other post-production effects, the mainstream perception of a purely symphonic, orchestral piece is being slowly being altered.

This raises further questions: with evidently changing perceptions and expectations from orchestral film music, is the pure, unenhanced sound of the symphony orchestra losing its value outside of live performance and classical music? Should composers and producers of orchestral music for media be expected to enhance the size of the orchestra in creating music for film (possibly extending to music for television and games)? In the context of this particular project, should the creative team have enhanced the orchestral sound to match the sonic size of augmented orchestral film scores? In creating a 'cinematic musical', does this augmented, synth-enhanced sound form a key part of a cinematic piece of music? With more period-specific projects, it would be a natural choice to opt for unenhanced orchestral instruments to match the orchestral sound from the era, but should modern, orchestral projects for media lean towards the augmented orchestra?

These questions are interesting ones, which would require more of a socio-musicological research approach, identifying the expectations of creators and consumers in the context of music for media.

As mentioned earlier in this chapter, a particular area of future development for the project is the exploration of 5.1 (or higher) surround sound mixing, now that the commercial responsibilities (i.e. releasing a stereo CD recording) have been met. This would certainly offer the creative team much greater possibilities in terms of creating a sense of space within the orchestral mix. Given that cinematic music is now often enjoyed in surround sound at home or at the cinema, it makes sense that a 'cinematic musical' would endeavor to explore this kind of listening experience.

Since conducting the research, the creative team has also considered the potential of trialing EastWest's Hollywood Strings and Hollywood Brass products within the final method used for the recordings. These products were recorded by well-regarded sound engineer Shawn Murphy, who specializes in recording and mixing large, blockbuster film scores. The potential of these products was brought to my attention during the Innovation in Music 2017 Conference (University of Westminster), where a question posed at a post-presentation (Doyle, 2017) Q&A session highlighted the possible benefits of using these products within the context of this project. Hollywood Strings boasts desirable "full-blown symphony orchestra section sizes (16 1st violins/14 2nd violins/10 violas/10 cellos/7 double basses)" (Soundsonline, 2010) (although EWPO still offers similar large-section sizes) with very strident styles of playing, indicative of the cinematic Hollywood sound. Reviewers of the product are very complimentary about the "rich, full, symphonic sound and an opulent quality that comes from combining great players with a great sound engineer" (Soundsonline, 2010), so the trialing of this library is a definite consideration for the future.

REFERENCES

Adler, S. (2004). *On Broadway: Art and commerce on the Great White Way*. Illinois: Southern Illinois University, p. 53.

Audissino, E. (2014). *John William's film music*. Wisconsin: The University of Wisconsin Press, p. 84.

Bennett, S. (2009). *Computer orchestration: Tips and tricks*. Norfolk: PC Publishing, p. 1.

Boberg. *QL spaces [Archive]—soundsonline forums*. Available at: www.soundsonline-forums.com/archive/index.php/t-50721.html [Accessed 1 Oct. 2017].

Chew, L., DeReiter, D., Doheny, C., Gilbert, C., Greenwood, T. H., Loewy, S., Maples, M. and McQuilkin, J. (2010). *The daily book of classical music*. Irvine: Walter Foster, pp. 16, 76, 126, 176.

Crowdfunder. *The battle of boat—original cast recording*. Available at: www.crowdfunder.co.uk/thebattleofboat/ [Accessed 18 Aug. 2017].

Dazeley, P. and Daly, M. *Ever wondered what goes on at air studios?* Available at: www.kentishtowner.co.uk/2015/01/01/ever-wondered-goes-air-studios/ [Accessed 27 Sept. 2017].

Dowsett, P. (2016). *Audio production tips: Getting the sound right at the source.* New York: Focal Press, p. 13.3.6.

Doyle, J. (2017). *Mixing and recording a small orchestral ensemble to make a large orchestral sound*, Sept. Paper session presented at Innovation in Music Conference 2017. London: University of Westminster.

Eargle, M. (2003). *Handbook of recording engineering.* Norwell, MA: Kluwer Academic Publishers, p. 304.

Gladiator. (2000). [film]. Directed by Ridley Scott. DreamWorks Pictures.

Jackson, G. *Review: Vienna instruments special edition* (Vol. 1). Available at: www.xlr8r.com/gear/2012/08/review-vienna-instruments-special-edition-vol-1 [Accessed 25 Sept. 2017].

Killchrono. *Help mixing EastWest symphonic orchestra.* Available at: http://ocremix.org/community/topic/35931-help-mixing-eastwest-symphonic-orchestra/ [Accessed 4 Sept. 2017].

Larsen, P. (2005). *Film music.* London: Reaktion Books, p. 7.

Lichtman, P. *Hollywood orchestral percussion by EastWest.* Available at: http://soundbytesmag.net/hollywoodorchestralpercussion/ [Accessed 4 Sept. 2017].

Meyer, S. (2017). *Music in epic film: Listening to spectacle.* Oxon: Routledge, p. 52.

Michaels, S. (2010). *Daft Punk recruited 90-piece orchestra for Tron: Legacy soundtrack.* Available at: www.theguardian.com/music/2010/nov/19/daft-punk-orchestra-tron-legacy [Accessed 18 Aug. 2017].

Miles-Huber, D. (2007). *The MIDI manual: A practical guide to MIDI in the project studio.* Oxon: Focal Press, p. 35.

NYMT. *NYMT—National Youth Music Theatre.* Available at: https://nymt.org.uk/ [Accessed 18 Aug. 2017].

Pejrolo, A. and DeRosa, R. (2017). *Acoustic and MIDI orchestration for the contemporary composer: A practical guide to writing and sequencing for the studio orchestra.* 2nd ed. New York: Routledge, p. 7.2.4.

Poit, D. *Vienna special editions and vienna instruments pro.* Available at: https://musiclibraryreport.com/software-reviews/vienna-special-editions-and-vienna-instruments-pro/ [Accessed 25 Sept. 2017].

Stewart, D. *EWQL symphonic orchestra play edition.* Available at: www.soundonsound.com/reviews/ewql-symphonic-orchestra-play-edition [Accessed 3 Sept. 2017].

Stewart, D. *Spitfire audio albion.* Available at: www.soundonsound.com/reviews/spitfire-audio-albion [Accessed 27 Sept. 2017].

Soundsonline.com. *EastWest Hollywood Strings.* Available at: www.soundsonline.com/hollywood-strings [Accessed 24 Oct.].

Soundsonline.com. *Symphonic orchestra.* Available at: www.soundsonline.com/symphonic-orchestra [Accessed 27 Sept. 2017].

The Rock. (1996). [film]. Directed by Michael Bay. Buena Vista Pictures.

Townsend, P. (2017). Personal interview, 5 Sept.

Tron: Legacy (2010). [film]. Directed by Joseph Kosinski. Disney.

Vienna Symphonic Library (2007). *Capturing timeless classics.* Available at: https://www.vsl.co.at/en/AboutUs/Silent_Stage/ [Accessed 20 Feb. 2019].

Vsl.co.at. *Vienna symphonic library*. Available at: www.vsl.co.at/en/Special_Edition_Complete_Bundle/Special_Edition_Vol1 [Accessed 1 Oct. 2017].

Warren, M. (2017). Personal interview, 2 Oct.

Weiss, M. *Mixing with reverb—how to use reverb in a mix*. Available at: https://theproaudiofiles.com/the-importance-of-space-in-a-mix-part-ii/ [Accessed 7 Oct. 2017].

Wright, J. *Orchestral string libraries compared (Updated)*. Available at: www.jonathanwrightmusic.com/orchestral-string-libraries-compared/ [Accessed 3 Sept. 2017].

16

Committing to Tape

Questioning Progress Narratives in Contemporary Studio Production

Joe Watson

INTRODUCTION

In contemporary recording studios, from large professional complexes to the smallest project setup, the Digital Audio Workstation (DAW) is ubiquitous, normative, and is generally presented in historical discussion as the logical successor to tape-based, analog 'precursors'. Furthermore, the DAW is often assumed to supply all of the functionality of a tape-based studio but with added affordances such as vastly improved editing facility. Underlying such assumptions are progress narratives and a succession logic that color and in many cases occlude the ramifications of a move from multi-track tape to an increasingly all-encompassing digital environment.

This chapter presents ongoing practice-based research into recording and production using analog multi-track tape. The author has many years of professional experience engineering, producing and composing using digital technologies and has recently turned his attention to the analog 'forebears' of the DAW in self-production of his third solo pop recording under the moniker Junior Electronics.[1] The practice commits to tape the following questions: given the skeuomorphic[2] nature of the DAW, and the way its design avows indebtedness to the legacy of traditional analog engineering, what insights can be gleaned by engaging with the actual analog equipment itself? As the DAW increasingly swallows up the whole studio (recorder, mixer, outboard, instruments, personnel), what can be learned by the digitally 'literate' producer/composer from a physically situated and primarily analog production process? What are the practical effects on the music produced if a producer used to 'unlimited' tracks is forced to work with, for example, only eight? What are the effects on the production process when editing is restricted to what one can achieve with a razor blade and there is no recourse to versioning (previous saved versions of a project)[3] or undo?

THE PRACTICE[4]

In order to explore these questions, the studio recordings[5] of this practice-based research have been produced under the following rule:

Audio shall not be digitized at any point in the process from source sound to final product (vinyl or other analog format).

No digital instruments, samplers or synthesizers are used: sound sources are analog electronic or acoustic. All recording happens to tape: all tracking is to eight-track half-inch multi-track;[6] this is then mixed to quarter-inch two-track; the mix is then cut to vinyl (or cassette, or other analog format) via a wholly analog mastering process. At no stage will tape tracks be dumped into the DAW for editing (this is a fairly common hybrid procedure, and has much to be said for it, but since one of my primary concerns is the affordance of digital editing, I eschew this approach here); mastering will not include a digital stage for processes such as brickwall limiting (a normative contemporary procedure even for most recordings presented as 'analog'); in other words, there will be no conversion of audio into or out of the digital domain. The idea behind the rule is to see what can be learned about recording, in its analog and digital guises, by the removal of the elements that are, in contemporary practice, normative and ubiquitous: digitized audio and the DAW. Ironically, as an experienced DAW-literate engineer, I have found that removing the DAW altogether is the best way to learn how it shapes and regulates studio practices. This removal places into sharp focus (and forces one to confront) those aspects of recording and production that have become second nature, such as micro-editing[7] and a primarily visual environment, and that simply do not exist in the tape studio.

Limitations on chapter length preclude detailed discussion of the analog/digital distinction, which has a long and complex history (see, for example, Gerard, 2016 [1950]; Haugeland, 1981; Sterne, 2016), but if one adopts the "standard interpretation" (Schonbein, 2014, p. 415) of digital as discrete and analog as continuous, then one can note that even if the recording medium is analog, digital elements are still in evidence throughout the practice discussed here: analog drum machines use digital memory (preset or programmed). Any on/off elements (e.g. buttons and switches) are discrete and can be viewed as digital, such as the digital logic matrix underlying the tape transport that does important work preventing the tape machine entering fast-forward and rewind at the same time! Discrete clocks and triggers have been used extensively for syncing, and may also be viewed as digital. Digital documentation has been used throughout, including the use of the DAW to chart the evolution of the 8-track tape, though this digitized audio is never sent back to tape. Online resources (manuals, forums, etc.) are consulted as needed.

In short, this work is not an attempt at 'analog purity', and has no pretension at being closer to nature/the past/life (see Sterne, 2006). The practice removes one aspect of contemporary recording procedure—digital audio—in the hope of learning something about recording, production and the digital procedure itself. It foregoes the obvious benefits and affordances of the DAW to ask some questions about those benefits and affordances, and to push at the boundary between analog and digital, wondering whether we might just have tipped the balance too far in our wholehearted embrace of digital *everything*.

NOSTALGIA AND AUTHENTICITY

Before moving on to discuss progress narratives, I feel it is important to note that the practice described here is not an exercise in nostalgia or authenticity. The former term, much discussed over the last 20 years, and grounding some recent academic work on studio recording (e.g. Bennett, 2012; Kirby, 2015; McIntyre, 2015; Williams, 2015), implies just the kind of periodizing logic that this project attempts to question: the work presented here does not yearn for the past, does not believe that things were better yesterday, and is not attempting to be retro, vintage, old school or any other term that valorizes times gone by. Current academic work on the studio that draws on narratives of nostalgia often focuses on use of 'vintage' equipment: for example, Williams argues that "an element of nostalgia is present when 20th century machines are used to capture 21st century music" (Williams, 2015, p. 2), and Théberge notes that "the association of multitrack studios with the sound of much classic rock has, in the digital age, resulted in its own form of nostalgia for 'vintage' analogue gear" (Théberge, 2012, p. 81, cited in Kirb, 2015, p. 318). There are, however, scholars such as Samantha Bennett who question such attributions:

> So far, there is very little evidence that recordists and practitioners using vintage technologies or precursors do so due to fashion, trends, nostalgia or sentimentalism. . . . [T]he attribution of vintage technology usage to nostalgia alone is deeply flawed and ignores more important factors such as musical and recording aesthetic intention on the part of the musicians and recordist(s), sonic characteristics of chosen technologies, client expectations as well as time and budget constraints.
>
> (Bennett, 2012, p. 14)

Bennett's reasoned account is a result of extensive conversation with studio practitioners and resists many of the implicit progress narratives that color other contemporary accounts (though her persistent use of the term 'precursor' to refer to analog equipment indicates an evident succession logic). For my own part as practitioner, I can assert that nostalgia is not a motivating factor in the work presented here.

The notion of authenticity has similarly been heavily mined in studies of studio technology and practice (Auner, 2003; Dickinson, 2001; Knowles and Hewitt, 2012), though perhaps without due regard to the problematic and often paradoxical nature of the term. It is, I'll admit, one of the terms I find most difficult when encountered in the mouths or writing of my students, where it is generally used in isolation, as a stand-alone term assumed to have meaning in and of itself: authentic (good), more authentic (better), less authentic (bad), not authentic (very bad). The fact that any idea of authenticity must be *relational*, that the object, process or performance can only be authentic in relation to something else, seems to have been lost somewhere along the way.

Still, the word exists and holds meaning for many. One of the more interesting treatments, of direct relevance to recording practice, is Simon

Zagorski-Thomas' article (2010), which discusses anxieties around "performance authenticity" that result from the affordances of digital micro-editing where "a proportion of editing is done because it is possible rather than necessary—or even desirable" (ibid., p. 206). These issues arose from "an observation several interviewees made", such as:

> Justin Scott . . . sitting with an engineer who would look at the wave forms of his drum performance on a computer screen, at a scale where several centimeters of screen represented only a few milliseconds of audio, and correct inaccuracies in his bass drum timing through this visual representation even though neither of them could hear the difference.
>
> (ibid.)

The anxieties arise from negotiations around the question 'how tight to edit a performance?' given that the DAW allows the whole gamut of responses, from 'leave it as it is, warts and all' to 'let's hard-quantize the entire drum track, and while we're at it let's nudge the overheads forward so the transients line up with the close mics'. Does leave-it-as-it-is lead to 'authentic'? Or does it just sound out of time/tune? Does hard-quantize sound like a robot? Do we need to 'humanize' the quantize value? Should we leave *some* of the original timing in there (personality/feel) but *correct* the more egregious moments? Engineers, producers, composers and performers all have their ways of dealing with this anxiety, and it is a serious topic of negotiation in the studio. One of the results of working on tape, where micro-editing is not possible, is that the question simply never arises. Again, I note that the practice discussed in this chapter is making no claim for authenticity.

PROGRESS NARRATIVES[8] AND SKEUOMORPHIC EMULATION

> Digital audio technology is progressively substituting most devices and tasks traditionally implemented via analogue electronics and components. In this sense, digital audio technology may be considered to represent a progression of practices and techniques known during many past decades in electro-acoustics and audio engineering.
>
> —(Mourjopoulos, 2005, p. 300)

The narrative that positions the DAW as successor to, evolution of and improvement on the tape studio is commonplace and has assumed the position of common sense, such that it is rarely questioned. A short sampling will give a flavor:

> [The DAW] has deeply broadened the process of music creation and reception.
>
> (Savage, 2011, p. viii)

Due to the inherent micro editing and automation possibilities, a DAW exceeds the capabilities of a conventional studio.

(Kirby, 2015, p. 17)

After the DAW became the common recording and mixing environment the traditional studio model rapidly became an anachronism.

(ibid., p. 269)

Digital audio systems are quite elegant, suggesting that they easily surpass more encumbered analog systems.

(Pohlmann, 2005, p. 20)

Computers can quite easily assume the role of a fully-equipped traditional production facility.

(Leider, 2004, p. 46)

The DAW can effectively replace and encapsulate much or all of the functionality present in a traditional console- and outboard-gear-based studio.

(ibid.)

These assertions all come from scholarly enquiry and all display more or less celebratory assumptions of progress; they all assume a logic of 'succession'; several also posit an idea of replacement (of a specific piece of analog technology by a digital version) along lines of continuity (the replacement 'does the same thing').

To be sure, there is no questioning the 'periods of dominance' that these progress narratives lean on: the tape studio was the dominant mode of production from the 1950s to the mid-1990s; the DAW has been the dominant mode since the mid-2000s, with the intervening decade or so one of transition and hybridity. In that these periods coincided with the maturing of the respective technologies there is no doubt. In that these narratives posit inexorable technological and concomitant sociological 'progress', there is much that can and should be questioned, in my opinion. The most commonly cited sociological benefits are those of 'democratization'—bringing the power of production to the masses—and the related ease of use. Paul Théberge's investigation is one of the more insightful accounts:

"Democratization" came to be equated with the availability of consumer technologies and prefabricated music resources. . . . The conflation of democracy, consumption and ease of use is, of course, a hallmark of contemporary capitalism. This ideology continues to dominate the promotional discourses of digital culture (musical and otherwise) in the early twenty-first century.

(Théberge, 2015)

One of the most notable aspects of the ongoing evolution of the DAW is the way it has encouraged more and more of the traditionally performed elements of studio production to be offloaded onto the computational domain. Performance here refers not only to the traditional role of the musician, but also the many and varied activities of engineer and producer. Performative real-world studio activities include: musical performance; calibrating and operating hardware; editing; mixing; conversing with various personnel (musicians, writers, arrangers); and, of course, listening. All of these activities have, to some degree, benefited from digital 'solutions' which automate, appropriate and assimilate the activities into formalized algorithmic representations. These, in general, appear in the guise of skeuomorphs, which offer replacement along the lines of continuity, and promise to remove or ease those aspects of studio procedure that are deemed 'difficult', 'hard to learn' or 'expert'. They are thus presented as beneficial, particularly to the (increasingly prevalent) 'amateur' (home, bedroom, hobbyist) producer. This process, in evidence from the earliest MIDI specification that separated performance from instrument, shows no sign of letting up, and we are now witness to virtual performers, such as Logic X's 'Drummer', as well as automated machine-learning based 'intelligent' mixing and mastering: "Mixing multichannel audio comprises many expert but non-artistic tasks that, once accurately described, can be implemented in software or hardware" (De Man and Riess, 2013, p. 1). Why, though, are these 'tasks' considered 'non-artistic'? For many, the mixing process is one of the most creative and artistic parts of the job of engineering. 'Accurately described' here means formalizable, codable and posits an idea of standardization in music production which many bands, engineers and producers baulk at. That the machine-learning process mines historical databases of pre-existing mixes indicates an irony at odds with the supposed 'innovative' nature of such software solutions: these automated mixes can only ever listen to the past.

This incremental incorporation of the physical, analog and human elements of the recording studio into the DAW seems, with retrospect, to have followed an inexorable logic of convergence. Once audio is added to the micro-processor based MIDI sequencer and the DAW is born, increasing numbers of elements of traditional studio engineering become housed within the computer environment: the multi-track tape becomes multi-channel digital audio; the analog mixer becomes the virtual mixer; the studio 'rack', containing processors and effects units, becomes available as 'plugins' on 'insert' slots on the virtual mixer; the live performance of the mix, with multiple fingers on faders, becomes automation, which allows any element within the DAW to have programmed parameter changes in 'real time'. Then instruments themselves get incorporated: initially this uses on-board sampling technology, but then physical modeling brings effective emulations of the Rhodes electric piano, the Hammond B3 organ, the Hohner Clavinet, for example. Physical modeling also takes in emulation of analog synthesis, known as 'virtual analog' or 'analog modeling': hugely expensive and rare classic synths, such as the Arp 2600 or the Yamaha CS-80, are available in multiple instances, all skeuomorphically rendered as 'authentic' emulations. Even amplifiers get digitized: amp simulators being common ways to 'get that sound', without the tedious need to buy a decent amp, have it played at volume, or set up microphones on the amp to capture the sound. By this point, nearly the entire studio environment can be seen to have been swallowed by the DAW, except for three features of the traditional studio: microphones, loudspeakers and the personnel who operate the system.

The microphone, an analog transducer, will continue to be valued in the studio environment, and no serious commercial studio will be without a selection, but the contemporary business of production does not *need* the microphone: samples, sound libraries, virtual instruments and performers played via MIDI, even Vocaloid software,[9] all mean that everything can be constructed within the DAW itself, without the need to ever record anything. Loudspeakers: it is worth remembering, in the 'digital age', that the final stage a recording goes through before meeting the medium *we* listen in, the air, will *always* be analog. Pro Tools may happily emulate a guitar amp in software, but the actual things we listen to the emulation on stubbornly refuse to be emulated. Personnel: one of the more intriguing contemporary developments of this shift from physical/analog processes to ones instantiated within a computer is the potential to remove the studio personnel themselves: in addition to engineers being replaced by automated artificial intelligence (AI) mixing and mastering, we also see the drummer's old adversary, the drum machine, raised to skeuomorphic heights in the latest incarnation of Logic, with Drummer—a selection of virtual performers, handy little homunculi, each with their own 'personality' and subtle performance differences, always on hand to lend their expertise to your production. In effect, these demon drummers all sit in a little locker in the studio, ready and waiting to perform for you, all free from the pesky foibles that make real drummers potentially difficult to work with: these demons don't turn up late, they don't drink on the job, they don't have squeaky pedals, they don't ask for the drums to be turned up in the mix, they don't speed up towards the end of the song, they don't lock themselves in a cupboard and refuse to come out until *their* song is put on the B side of the single that's obviously going to be a smash . . .[10]

Time and again we see the changes the DAW brings being celebrated as liberating the user, of upscaling affordances, saving the engineer time, allowing them to do more, more quickly, cheaper. We expect such celebration in the marketing hype of DAW manufacturers, but, as noted above, we also see it in much of the scholarly literature addressing the DAW and digital audio. One of the oft-noted key changes is 'vastly improved editing functionality', but even if the icon of the editing tool relies on skeuomorphic reference to physical work done with a razor blade (in Logic, for example, it is represented by a pair of scissors), the affordances of digital editing are of a fundamentally different order to those of tape splicing. This is a difference in kind, not of degree: digital audio editing is a discretization, a coding, a taking out of time, and this removal of sound from its dynamic milieu is neither predicted by nor develops out of analog tape. On multi-track tape it is pretty much impossible to move individual track elements in relation to other tracks, and the only real editing possibilities are global: horizontal stretches of tape (time) can be sliced, removed,

moved and reversed, but there is a limit to the smallest (shortest time) edit possible on tape, and it is of a vastly greater magnitude than the infinitesimal slice that digital audio affords.[11] This near-infinite manipulability of sound, possible even at the level of the individual sample (individual slice, typically one of 44,100 per second) once it has become data within the DAW, is well known and much discussed. What seems to be less readily acknowledged, perhaps because it is hidden behind the skeuomorphs and the progress narratives that posit the DAW as the natural successor to analog forebears, is that this ability, literally, to 'take sound out of time' is a radical, paradigm-shifting break with analog techniques and technologies—it is most definitely *not* a continuation, an upscaling of affordance or a necessarily logical successor.

Although this increasing digitization of the studio does seem to follow its own internal logic, this does not imply that this history is one of progress. Yes, digital audio has become better in quality as computers have become faster and storage has become cheaper. Yes, the studio can now become truly mobile and production can happen anywhere. Yes, the whole process has allowed for more and more people to make more and more music more cheaply and quickly: the monopoly of the professional studio, paid for by the record company, on making 'professional sounding' recording has gone. But does that necessarily equate to progress? Are there things we have lost as we have rushed headlong into this embrace of the computational? As Strachan (2017) notes:

> Computer-based music production enables and demands that the user work directly with captured and generated sounds that are at a remove from the processes and competencies of performance traditionally associated with musicianship. . . . The computer environment needs to be interrogated for the way that it allows, encourages and facilitates the making, processing and manipulation of sound. In other words, the computer environment should not be understood as a neutral way of recording, capturing and presenting sound but as highly influential to the creative process in its design, construction and capability which in turn have a central influence on the sounds and eventual recordings that are produced.
>
> (ibid., p. 7)

How are we to get to grips, in our actual studio practice, with the seemingly inexorable move away from the physical? When even performance, historically one of the fundamental studio procedures, gets offloaded onto the computational domain; when personnel, including, increasingly, the person of the engineer, seem no longer to be necessary for the production of recorded music; when everything is made so much easier, cheaper and more convenient by staying 'in the box'. How do we make decisions about where to draw the line between what we want to play and what we want a machine to play for us; about how tightly to edit; about what we want to have control over and what we are happy to offload onto automated processes?

My own approach has been to remove these questions, by removing the DAW from the process. The following sections detail that process and draw out some of the implications of a primarily analog tape studio system. It will make comment on comparisons with the DAW along the way.

COMMITMENT

> Digital artifacts often do not exhibit commitment to actions; in fact, being able to index at random into the past of our creation through undo/redo and versioning may be the single most important characteristic that separates digital from physical interactions.
>
> —(Klemmer et al., 2006, p. 145)

It is no accident that traditional studio engineers talk of 'committing to tape'. The lack of any kind of undo or versioning means that every move made on tape requires great care and commitment. It is perfectly possible, while recording, to press the wrong button, resulting in days' worth of work being deleted. This forces one to take great care, and to double-check everything before making critical moves. The old carpenter's maxim of 'measure twice, cut once' is apposite here. This added pressure encourages a sharpening of focus which, in many ways, is a useful contrast to the somewhat blasé attitude often found in DAW recording, which may well be sufficient where necessary attention to the task in hand is leavened by foreknowledge of versioning and undo. No such distracted wandering of attention is possible when recording on tape: if you change something it is changed permanently, so you'd better make sure it is a change for the better.

This sense of commitment is also, as discussed later, central to all taped *performance*: the knowledge that what you play is what the listener will hear; the knowledge that DAW-style micro-editing is not an option; the knowledge that previous/better versions of a take may be jettisoned in search of the 'definitive' take. There is also a greater commitment to the post-performance audition of a take: the performer records their take and sits with the engineer to listen and decide if it is good enough or whether it needs to be redone (or where possible, if certain moments will be fixed through dropping-in). Engineer Carl Beatty, who trained on analog equipment and witnessed the transition to DAWs in professional studios, notes:

> [Of] all the things that I see that have changed drastically because of Pro Tools, the biggest one is communication. For the performers on the "other side of the glass" . . . there's very little feedback from the producers, there's very little guidance in terms of performance, because they know they have acres and acres of "real estate" to collect data, and very often the guidance is "do another one". From the world I come from, of limited real estate, 24-track analog recording, that wasn't really part of the process, you had to be listening and making decisions as you go.
>
> (Beatty, 2007)

Insights such as this are a useful antidote to assumptions of isomorphism such as: "In contemporary recording sessions, digital technology mimics that of older analog tape-based processes, so that for the performing musician the experience is nearly indistinguishable" (Williams, 2012, p. 1). Williams' article is noteworthy for the exploration of how the primacy of visual display in the DAW influences and regulates the production process, but fails to critically interrogate deeper implications of the screen, choosing rather to celebrate how the shared screen, in a studio control room, affords participation: "Observing the ease with which an engineer can delete, repeat, stretch time, or literally 'flip it and reverse it', the musician is given access to a staggering number of creative possibilities" (ibid., p. 3). Perhaps, but this also alerts the musician to the fact that there is far less riding on their performance than they might have thought. This is a liberation, of course, an easing of pressure, but it can also be viewed as an abdication of responsibility. The performer in the DAW studio tends to be much less committed to their own performance than the performer in the tape studio.

PERFORMANCE AND EDITING

> The actual process of analogue recording is by necessity far more focused on performance than digital recording.
> —(Kirby, 2015, p. 361)

On tape, if you want something to *sound* like a performance it must *be* a performance. This is not the case in the DAW, where emulation, correction or construction of performance are commonplace. Of course, the nature of performance to tape is different from that of live performance in front of an audience; the very nature of recording moves sound from the evanescent to the repeatable, and all kinds of affordances facilitate the recording of performances that musicians are happy to let out into the world, knowing that they may be listened to again and again. Traditional analog recording offers: re-recording, until you are happy with the take; 'dropping-in' to remove more localized errors; 'comping', which allows the compilation of various takes to select the best bits from each. However, all of these affordances involve *further performance* to correct unacceptable performance. Even comping requires a real-time performance on the part of the engineer to swap between different takes on the multi-track, according to a predetermined script, using mute buttons or faders, and also involves the inevitable loss of quality that comes from internal track bouncing. Comping in a DAW is free both from quality degradation issues and any need to perform.

On tape, editing is intimately related to performance. This occurs on at least two levels: (1) the craftsman-like wielding of razor blade, chinagraph pencil, splicing tape and ears, around the physical medium of magnetic tape is a performative combination of tacit knowledge and hand/ear/eye coordination that is mostly absent from the 'line the scissors up with the transient on the screen' type of editing common in the DAW; and

(2) foreknowledge of the limitations of tape editing influences musical performance: if you make a mistake you know you will either have to live with it or replay it—there are none of the possibilities of localized post-performance editing that exist in the DAW, where timing errors are easily corrected by moving the event relative to the time line, and errors of pitch can be corrected by DSP-based 'solutions'.

Furthermore, on tape the decisions about whether to live with a performance, or attempt to correct it, directly feed into the *sound* of the performance. A take considered 'good' may include errors, but to fix these the previous performance (or parts of it through drop-ins) will usually be wiped. You can always re-record, but you run the risk of losing aspects that may never be as good. When, and how, do you decide the take is 'good enough'? How much time do you have? Does your playing get worse as the part gets 'stale'? The sound of the final take is often a compromise as well as a composite of all of these decisions made throughout the process, and may contain things you would rather were not there, but seem like a better compromise than doing another take. This leads to some recorded elements having 'rough edges' or even obvious mistakes, and puts the performer in a position of a much greater commitment to the performance than one sees in DAW recording, where there is almost always the option of 'fixing' mistakes later, and where you keep every version of a take so you can choose the best bits later.

However, it is important to note that micro-editing is not a *default* in DAW production, though it is the norm in many areas of music production, such as pop music. Where instruments are recorded simultaneously in the same space (and multiple microphones exhibit 'bleed'), micro-editing becomes difficult, if not impossible, and thus many types of music, such as most classical music, and much improvised music, are not subject to the same kind of tendency towards construction and manipulation discussed here.

DECISION-MAKING

The making of judgments and decisions happens constantly during tape recording, and is a much more heavily weighted part of the process than is evident in most DAW recording. The interface and affordances of the DAW positively encourage the deferral of all kinds of decisions, from the compiling of takes once a performer has left the studio, to the oft-heard 'we'll fix it in the mix'. Unlimited track number means that huge amounts of 'speculative' recording can happen, and large banks of tentative 'maybes' are often amassed before definitive arrangement decisions get made. The very tight constraints of multitrack tape recording mean that decision-making is an integral and necessary part of the recording process, and is thus intimately woven into the fabric of the recording in its becoming. This is perhaps the most important concrete result of this foregrounding of decision-making: the *sound* of the final recording is in a direct and contingent relationship with ongoing performative decision-making. This is different from the comparatively minimal impact ongoing decision-making can have on the sound of a DAW recording. Of course, decisions get made in the DAW environment, but they don't necessarily, and in many cases tend not to, happen concurrently with processes of recording.

On tape, many of these decision-making processes are related to the restrictions of track count: bouncing of multiple elements fixes their internal relationship as they will appear in the final mix, both in terms of their relative levels and where they appear in the stereo field. Decisions taken about effects and processing of a track cannot be removed later. Decisions made about and during performance have a similarly pressing vitality and have a direct bearing on how those performances occur. There is a considerable difference between the pressure performers feel when they are wiping everything they previously did to get the best take, and the pressure they are put under when they can offload anxieties about 'best' performances onto a retrospective compiling of 'all the best bits' once the red light has gone out.

Much of this is related to the DAW's ability to take sound 'out of time', remove it from the ongoing temporal flux. This doesn't just affect things in the micro, moment-to-moment level of the temporal flow of the song, but also at the macro level of studio environment and recording process—the encouragement of deferral of decision-making in the DAW impacts performer, engineer and the ongoing evolution of the song in progress. The tight weaving in of decision-making to the fabric of the song in progress on tape will tend to result in a very different-sounding final recording.

CONTINGENCY

> The good craftsman places positive value on contingency and constraint.
> —(Sennett, 2008, p. 262)

On tape, however thoroughly we plan how the track is going to sound, however detailed and prescriptive the demo is, the song in the making *will change*. The becoming of the recorded song will have a relationship to, and dependence on, unpredictable soundings that are an inevitable and not always unwelcome result of performance, constraint and circumstance.

Contingency is the dependence on, and reaction to, the environment that one finds oneself within at a particular juncture—the knot that emerges unforeseen in the carving of a piece of wood. It is the inevitable unplanned (and unplannable for) nature of reality in its becoming; the unpredictable, the error or deviation from the plan, the uncontrolled/uncontrollable. The hyperfluidity of digital audio makes it perfect for exerting total control, eliminating contingency; sound, performance, the song, become programmable, and there is an irony to how often this coding is pressed into service in the construction of an *emulation* of performance.

In the tape studio, contingency is clearly related to decision-making in the construction of the recording, the becoming of the recorded song. Because decisions about arrangement, about sound, and about elements of the final mix cannot be endlessly deferred, there is a sharper focus on how each of the separate elements fit together into the evolving whole. The relationships among different elements assume greater significance than in the DAW because there is minimal possibility for shifting and altering how these elements fit together after they have been *committed* to tape. Editing/altering possibilities still exist, to be sure, but they are of a different order of magnitude than their DAW 'equivalents'. Thus there is a necessary ongoing focus on *listening* to how elements fit together, with a concomitant mental 'mapping forward' of how they will fit into the final mix.

Each element in the growth of the recorded song can be seen as being highly context-specific, much more so than in the 'equivalent' building of a song in a DAW, where there is a tendency, arguably an encouragement, to accumulate elements that will be 'fit together' at some point later in the process. Tape has no truck with this deferral of decision-making—it insists one deals with contingency and context in the moment of making.

LISTENING AND GESTURE

> Audio engineering practices are not reducible to one sense alone. Every widespread form of engineering developed until today has depended on the body for the manipulation of interfaces and on audition through headphone or loudspeaker audition systems. All computer-based audio engineering technologies depend upon the visualization of abstractions of sound and also a visualization of the interface for manipulating sound. However, scholarship on audio engineering has ignored the sensing body for the most part.
>
> —(Bates, 2009, p. 1)

When one removes oneself from the DAW and into the realm of its supposed forebear, it becomes apparent just how hard the DAW tries to ape the moves, processes and space of the analog studio through skeuomorphic emulation within the "square horizon" (Virilio, 1997, p. 90) of the computer screen. But consider a clutch of related assumptions: the assumption of similarity between the physical space of the tape studio and its virtual, logical 'successor'; the assumption that techniques and procedures move smoothly from one domain to the 'next' via a "digital switchover" (Buckley, 2011); the assumption that all that has changed is increased affordance, ease of use, efficiency; all these assumptions are misplaced, for they fail to recognize fundamental differences of modality, gesture and the way the 'sensing body' moves around the space of production.

The analog studio is truly the domain of the ear: listening necessarily has primacy. When we consider the DAW arrange window as skeumorphic emulation of the tape head with tape passing by it, we downplay the important and fundamental change that happens to the engineer/producer when waveforms become visualized. If we look at the tape passing by the head, there is *absolutely nothing to see*, just an expanse of black tape. When we start digitizing audio and presenting it on screen in a manner seemingly analogous to the way tape passes the head, something fundamental happens: we start *looking at sound*. This sounds like an oxymoron, but such language is common in activity around the DAW: 'I can see a silence here we can cut into', 'that master looks a bit over-cooked', 'it looks like you're playing consistently behind the beat'. All such pronouncements depend on a visualization of waveforms (and the grid of the time line in the last example). Sight is used in the analog studio: we look at meters to check levels, we visually line up pieces of tape we're editing, we look to check we haven't record-enabled the wrong track by mistake. But sight carries significantly less weight than listening or gesture. The ocular-centrism of the DAW is manifest in countless ways, such as the visual editing that Zagorski-Thomas (2010) explores, or the 'visualizer' built into many EQ plugins, which I've witnessed students use to make the sound 'look more even': they turn the visualizer on and cut where they see peaks. In some circumstances this can help, in others it definitely does not, but one thing is for sure, it *teaches you nothing* about what an EQ does to sound: as the focus moves from ear to eye, attention is diverted away from the actual aural effect, and we tell ourselves a tale of how we have made the sound 'look better'. I have witnessed many times a similar preference for the modality of sight in my students as they 'listen' to a mix: what they appear to be doing is watching 'blocks of sound' pass by on the screen, and if the screen is not in line with the speakers, they will rather orient themselves towards the screen than the monitors.[12]

Earlier I suggested that the visual modality is of lesser import in the tape studio than gesture, or gesturo-haptics, to use Brian Rotman's term, which folds linguistic and choreographic notions of gesture into modalities of touch and proprioception (Rotman, 2008). In the tape studio, these performative choreographies revolve around both learned embodied routines, such as fader riding and tape splicing, and spontaneously created gestural solutions to engineering problems, such as raising a stringed instrument up to the microphone after it has been plucked.[13] This is a kind of 'manual automation' of later fader riding, an anticipation of how it will sound in the mix. Left as normally played, especially in fairly dense material, the harp in question has a tendency to 'plink' and then immediately disappear. This manual automation allows the tail of the sound to remain audible, and means it doesn't have to be done in the mix where hands are busy with all kinds of other moves. Tape editing is also a good example of the kind of gesturo-haptic/listening choreography where the 'rocking' back and forth of the reels is accompanied by a listening that needs to be expert to hear where the transient is.[14] What I am alluding to here is an embodied engagement with the physical dynamics of the studio; a steering through the complex and resistive materiality (Norman, 2013) of the tape studio environment that will not allow the bending of sound to the imperious will of the producer, but has to *go with* the multiply recursive and co-implicative trajectories and channels of that dynamic, through performative engagement and interaction.

CONCLUSION

The key insights gleaned from this practice-based comparison of tape studio and DAW have primarily been to do with the nature of *performance* in the studio, from the perspective of both musician and engineer. The tape studio *insists* on performance, at multiple levels, and throughout all aspects of production. As well as musicians' performances, which require the highest levels of commitment, the craft of tape studio engineering is also highly performative: from calibration and operation of technology, interpersonal discussion and negotiation, craftsman-like razor-blade editing, through to the final mix. By comparison, the DAW tends to devalue performance, by making micro-editing such a powerful and attractive alternative, by offering such flexible non-linear construction opportunities, by automating traditionally performed engineering tasks such as fader riding, by encouraging an accumulation of material to be assembled later, by endlessly deferring decisions about how the track is eventually going to sound, and through recourse to versioning and undo.

This performance-centric analysis also points up how the DAW relies on skeuomorphic emulation of the tape studio and its physical environment. We have seen how the succession logic that drives this emulation relies on narratives of progress, of increased affordance, of democratization, and how it tells a tale of replacement along lines of continuity, as if the historical function of analog technologies has been to lay the groundwork for their eventual, inevitable incorporation into an all-encompassing digital realm. But this narrative occludes fundamental differences in kind between the tape studio and its supposed descendant. Rather than looking at the DAW as tape studio with added affordances, we should see the two environments as different in fundamental ways, and the practices that happen in these environments as leading to very different-sounding music.

NOTES

1. The first two releases were produced with more conventional DAW technology. See Junior Electronics, 2007, 2012.
2. A skeuomorph is "a design feature that is no longer functional in itself but that refers back to a feature that was functional at an earlier time" (Hayles, 1999, p. 17). A common everyday example is the camera click sound on smartphones, which samples the sound of the shutter opening and closing on a mechanical camera; the smartphone has no such shutter, and the sound, whose actual function is to say 'photo taken', could equally well be a beep. Examples in the DAW often present digital emulations dressed in the clothes of hardware, such as the many software emulations of classic compressors like the LA-2A. See also Bell et al. on the use of skeuomorphs as "analog audio metaphors" in the DAW (2015, p. 1).
3. See Duigan et al. 2010 on versioning in the DAW, and Klemmer et al. 2006 for related concepts in Human-Computer Interaction.
4. The work is documented and discussed in detail at euterprise.com

5. To be released on the Bureau B label as Junior Electronics *Euterprise* EP.
6. The reason why eight-track is used is mainly financial: 16- and 24-track machines are still fairly expensive, and the 2-inch tape they use is very costly now. Certainly, the quality of the multi-track is not analog 'at its best', and the constraints of tape discussed here are made much tighter by using only eight tracks, but most of what is said here applies equally to 24-track recording.
7. Micro-editing refers to one of the key affordances of digital audio and differs from the macro-scale editing possible on tape in two ways: (1) the ability to edit with extreme precision (down to the sample level) that comes from the ability to view the waveform and from the removal of physical constraints; and (2) the ability to dissociate one track from another once it exists in the DAW—on tape you cannot move one multi-track element relative to another.
8. I borrow this term from Tara Rodgers, whose "long-term perspective on the history of synthesized sound resists the linear and coherent progress narratives that characterize many histories of technology and new media" (Rodgers, 2010, p. 9).
9. See www.vocaloid.com/en
10. Reputedly what Roger Taylor of Queen did to get "I'm in Love With My Car" on the B side of "Bohemian Rhapsody" (Irvin, 2007, p. 355). B sides take a 50% split of revenue.
11. With tape editing it is possible to *cut* very small slices (1 mm or less) but, as I can attest from experience, it is more or less impossible to put these fiddly little pieces back together again.
12. This studio-specific term for loudspeakers indicates how monitoring activity in the studio is an aural affair.
13. See euterprise.com
14. See euterprise.com

REFERENCES

Auner, J. (2003). "Sing It for Me": Posthuman ventriloquism in recent popular music. *Journal of the Royal Musical Association*, 128, pp. 98–122.

Bates, E. (2009). Ron's right arm: Tactility, visualization, and the synesthesia of audio engineering. *Journal on the Art of Record Production*, 4.

Beatty, C. (2007). *"We'll Fix It in the Mix": The era of unlimited tracks* [online video]. Available at: www.youtube.com/watch?v=7LDneHRhP-o [Accessed 9 May 2017].

Bell, A., Hein, E. and Ratcliffe, J. (2015). Beyond skeuomorphism: The evolution of music production software user interface metaphors. *Journal on the Art of Record Production*, 9.

Bennett, S. (2012). Endless analogue: Situating vintage technologies in the contemporary recording & production workplace. *Journal on the Art of Record Production*, 7.

Buckley, J. (2011). Believing in the (Analogico-)digital. *Culture Machine*, 12.

De Man, B. and Riess, J. (2013). A semantic approach to autonomous mixing. *Journal on the Art of Record Production*, 8.

Dickinson, K. (2001). "Believe"? Vocoders, digitalised female identity and camp. *Popular Music*, 20, pp. 333–347.

Duigan, M., Noble, J. and Biddle, R. (2010). Abstraction and activity in computer-mediated music production. *Computer Music Journal*, 34(4), pp. 22–33.

Gerard, R. (2016 [1950]). Some of the problems concerning digital notions in the central nervous system. In: C. Pias, ed., *Cybernetics. The Macy conferences 1946–1953. The complete transactions*. Chicago, IL: University of Chicago Press.

Haugeland, J. (1981). Analog and analog. *Philosophical Topics*, 12, pp. 213–225.

Hayles, K. (1999). *How we became Posthuman: Virtual bodies in cybernetics, literature, and informatics*. Chicago, IL: University of Chicago Press.

Irvin, J. (2007). *The Mojo collection: 4th Edition: The ultimate music companion*. Edinburgh: Canongate Books.

Junior Electronics. (2007). *Junior electronics* [LP]. DS45–42. Duophonic Super 45s.

Junior Electronics. (2012). *Musostics* [LP/CD]. bb099. Bureau B.

Kirby, P. (2015). *The evolution and decline of the traditional recording studio*. PhD. University of Liverpool.

Klemmer, S., Hartmann, B. and Takayama, L. (2006). How bodies matter: Five themes for interaction design. In: *DIS '06 proceedings of the 6th conference on designing interactive systems*, New York: ACM, pp. 140–149.

Knowles, J. and Hewitt, D. (2012). Performance recordivity: Studio music in a live context. *Journal on the Art of Record Production*, 6.

Leider, C. (2004). *Digital audio workstation*. New York: McGraw-Hill.

McIntyre, P. (2015). Tradition and innovation in creative studio practice: The use of older gear, processes and ideas in conjunction with digital technologies. *Journal on the Art of Record Production*, 9.

Mourjopoulos, J. (2005). The evolution of digital audio technology. In: *Communication acoustics*. Berlin, Heidelberg: Springer, pp. 299–319.

Norman, S. J. (2013). Contexts of/as resistance. *Contemporary Music Review*, 32, pp. 275–288.

Pohlmann, K. C. (2005). *Principles of digital audio*. 5th ed. New York: McGraw-Hill.

Rodgers, T. (2010). *Synthesizing sound: Metaphor in audio-technical discourse and synthesis history*. PhD. McGill University, Montreal.

Rotman, B. (2008). *Becoming beside ourselves: The alphabet, ghosts, and distributed human being*. Durham: Duke University Press.

Savage, S. (2011). *Bytes and backbeats: Repurposing music in the digital age, tracking pop*. Ann Arbor, MI: University of Michigan Press.

Schonbein, W. (2014). Varieties of analog and digital representation. *Minds and Machines*, 24, pp. 415–438.

Sennett, R. (2008). *The craftsman*. London: Allen Lane.

Sterne, J. (2006). The death and life of digital audio. *Interdisciplinary Science Reviews*, 31, pp. 338–348.

Sterne, J. (2016). Analog. In: *Digital keywords: A vocabulary of information society and culture*. Princeton: Princeton University Press, pp. 31–44.

Strachan, R. (2017). *Sonic technologies: Popular music, digital culture and the creative process*. New York: Bloomsbury Academic.

Théberge, P. (2015). Digitalization. In: *The Routledge reader on the sociology of music*. London: Routledge.

Virilio, P. (1997). *Open sky*. London and New York: Verso.

Williams, A. (2012). Putting it on display: The impact of visual information on control room dynamics. *Journal on the Art of Record Production*, 6.

Williams, A. (2015). Technostalgia and the cry of the lonely recordist. *Journal on the Art of Record Production*, 9.

Zagorski-Thomas, S. (2010). Real and unreal performances: The interaction of recording technology and rock drum kit performance. In: *Musical rhythm in the age of digital reproduction*. London: Routledge, pp. 195–212.

Part Three

Technology

17

Harnessing Ancillary Microgestures in Piano Technique

Implementing Microgestural Control Into an Expressive Keyboard-Based Hyper-Instrument

Niccolò Granieri, James Dooley
and Tychonas Michailidis

INTRODUCTION AND AIMS

Musicians spend a great deal of time practicing their instrument. As a result, they develop a unique set of microgestures that define their sound: their acoustic signature. We consider microgestures as small movements that are part of the pianistic technique, but not necessarily related to the sound production. These movements revolve around the sound-producing gestures and are also called ancillary movements (Cadoz and Wanderley, 2000).

This personal palette of gestures presents unique aspects of piano playing and varies from musician to musician, making a distinctive sound and enabling them to convey their music expressively.

The ability of performers to communicate through their instrument depends on the fluency the performer has with the instrument itself (Tanaka, 2000). Fluency, in this case, is seen as a combination of technical proficiency and expressive charisma. These two elements are themselves dependent on the time spent practicing an instrument and ways of incorporating ancillary movements that are known to convey expressiveness in musical performance (Miranda and Wanderley, 2006).

Concert pianist Xenia Pestova suggests that "the ability to be creative with phrasing, articulation and stylistically acceptable breathing or flexibility are just some of the elements that make for an expressive performance and create a satisfying experience for both the performer and the audience" (Pestova, 2008, p. 68).

The instrumentalist needs to have a sizable kinetic vocabulary to operate an instrument. Thus, the aim of this research is to create a keyboard-based interface allowing pianists to manipulate live audio processing of the piano through microgestural hand and finger control. In this way, the digital instrument will not be seen and treated as a difficulty to be overcome, as described by Rebelo (2006), but rather as an extension of the pianist's known technique. To achieve this, we apply machine-learning algorithms, specifically *Random Forests* classification algorithm, to enable us to accurately identify existing microgestures currently used and performed by the pianist. We free the performer from the challenge of learning any new gestural language or technique that might be required. The learning curve of digital instruments can be the most disruptive and challenging element for a performer since it considerably limits any technical control or freedom (Nicolls, 2011).

Focusing on user-centered and activity-centered interface design approach, we propose a system that interfaces and allows performers to express their creativity and extend it through greater engagement with this innate microgestures in the activity of piano performance. An interface that removes or reduces the steepness of the learning curve when approaching it for the first time can also remove the creative barrier posed by a system designed without the end user in mind (Bullock et al., 2016). The chapter presents a case study investigating an innovative way of extending keyboard interfaces, drawing upon pianists' existing instrumental technique. The goal of this research is to extend the creative possibilities available on keyboard-based interfaces, stimulating the creation of new approaches to build intuitive interfaces for musical expression, as well as exploring new ways of learning and playing digital instruments. The system, pertaining to the augmented instrument class (Newton and Marshall, 2011), offers a creative environment to manipulate live piano sound. Google's Soli alpha sensor, a miniature radar-based technology, was used to detect the pianist's hand movements (Lien et al., 2016). Through machine-learning, we identify specific gestures which we map to the frequency-modulation algorithm parameters. Precisely, the acceleration and energy of the analyzed gesture are mapped and used to control the depth and speed of the vibrato effect.

BACKGROUND

Since the introduction of aftertouch technology in the 1980s, a keyboard feature that allows control of sound parameters through the use of pressure-sensitive keys, there has been a great development in keyboard-based digital instruments. Through the available technology, creators have had the opportunity to enhance features of the instrument by adding several layers of expressive features making effects and modulations possible that are not available to their acoustic counterparts.

Both the *Haken Continuum Fingerboard* (Haken et al., 1992) and *The Rolky Asproyd* (Johnstone, 1985) approached the issue via two different methods. The first approach consisted of a continuous neoprene surface where a classical keyboard was drawn, and the independent tracking of the x-y-z coordinates of up to ten different fingers enabled single-note pitch and amplitude control. The second approach consisted of a transparent surface using light detection to determine the position of each finger and enable single-key pitch modulation. Both of these interfaces had a limited amount of tactile information regarding the location of the fingers and did not manage to provide an intuitive way to enable polyphonic pitch-bending capacity while also allowing effective tuned playing (Lamb and Robertson, 2011). In addition, the Haken Continuum Fingerboard does not have moving keys, while the Rolky Asproyd is not specifically a keyboard-based instrument; instead, it is a touch controller. Both interfaces present the pianist with a level of unfamiliarity that requires adaptation or the learning of new skills.

More recently, the *ROLI Seaboard* (Lamb and Robertson, 2011) and Andrew McPherson's *TouchKeys* (McPherson, 2012) present two innovative keyboard interface developments. The common thread between these two interfaces is that they both require users to alter or adapt their technique to accommodate a new gestural vocabulary built to work with their systems.

The ROLI Seaboard, as described by its creator Roland Lamb, "is a new musical instrument which enables real-time continuous polyphonic control of pitch, amplitude and timbral variation" (Lamb and Robertson, 2011, p. 503). A silicon membrane has been applied, following the traditional keyboard layout, transforming the keyboard into a continuous slate where the fingers' position, pressure and movement can be tracked and mapped to control individual parameters through the provided software.

Similarly, Andrew McPherson's *TouchKeys* coats a standard electronic keyboard, or acoustic piano, with a touch-capacitive sleeve that enables the individual detection of the fingers along the length of the keys, enabling the control of different parameters. Both interfaces take what is known as the pianistic technique and enhance it by implementing individual note pitch-bending capabilities and other sound modulations, all taking information from the pianist's fingers. However, these two interfaces dissemble a familiar pianistic technique into various time-dependent gestures. They extrapolate only the sound-producing gesture, the vertical movement of the finger when pressing a key and build a new set of gestures or technique to control the new sound modulation parameters. While we acknowledge the cutting-edge technology implemented in these innovative interfaces, our research aims to address the steep learning curve that is inherently proposed towards the 'classically' trained performer who already has mastered his or her piano technique.

LOWER DEGREE OF INVASIVENESS

Traditionally, instruments are built and designed to achieve a particular sound; the physical properties of their construction define their timbral identity. For example, the organ or the double bass need to be shaped in the way we know and occupy a certain amount of volume in order to produce their unique tonal qualities. The shape of the acoustic instrument determines the gestural interaction and technique required to play the instruments as well as determining the sonic characteristic and any haptic feedback. Musicians spend years working within these limitations refining their command of the instrument to achieve a desired sonic result, a specific acoustic signature (Chadefaux et al., 2010). The amount of time spent on the instrument itself refining its performative technique is justifiable through the well-established musical culture: one that charts a steady evolution in instrument design and technique. The combination of years of practice and technique development create a unique relationship between the instrumentalist and the instrument.

This concept could also be explained by Heidegger's concept of tools considered 'ready-to-hand' (Dourish, 2004). Acoustic instruments, being embedded into the musical culture, become an embodiment of the sound the performer wants to produce with them, thus falling into Heidegger's category of tool that can become *ready-to-hand*. A *ready-to-hand* tool is one that a user can act through: in this case, the musical instrument becomes an extension of the performer's hands and arms to play music. However, the morphing nature of the musical instrument considered as a tool does not usually apply to digital interfaces, which gestural interaction and timbral identity are not defined by their physical properties. On the contrary, digital instruments are versatile with no fixed properties on how they produce sound. Donald Norman, in a more pragmatic way, defines digital interfaces as problematic because of their nature: "The real problem with the interface is that it is an interface. Interfaces get in the way. I don't want to focus my energies on an interface. I want to focus on the job" (Norman, 1990, p. 210).

The problem with digital interfaces lies in the intrinsic fact that they are interfaces: they interface the user with something else. When we apply this concept to musicianship, an instrument is also an interface, but through years of practice the instrument/interface is no longer disruptive; it becomes a tool. However, a digital interface posed between the musician and the sound produced is an added step that is not present in its everyday practice, thus seen as disruptive, or with a higher degree of invasiveness. Interfacing between the performer-instrument relationship can often become invasive and disruptive from a performer's view. Grandhi et al. (2011) propose the significance of naturalness in interfaces. When we define an interface as unnatural, the definition usually is attributed to the system itself. Instead, we believe that the unnaturalness of a system, or the interaction with it, is the result of bad design and implementation.

A digital interface may be portrayed as poorly designed if it requires performers to relearn a familiar technique. When an interface is built around the designer's idea instead of the user's needs, it often results in fabricating a new type of hybrid performer that combines the creator of the interface, the composer and the performer (Michailidis, 2016). These design-centered, not user-centered, interfaces are not necessarily intuitive to performers other than the creator.

Utilizing a user-centered approach for the development of expressive digital interfaces, our system focuses on the importance of touch-free gesture recognition characterized by a low degree of invasiveness. It is inspired by the work of Dobrian and Koppelman (2006), who highlight the importance of developing systems that allow artists to reach the same level of sophistication achieved with traditional instruments (jazz, classical and so forth). We focus on developing strategies for better mapping and gesture recognition utilizing existing virtuosity and developing new repertoire for piano performances.

RADAR-BASED DETECTION

Here we provide an overview of the capabilities of Google's Soli Alpha Developer Kit (Soli hereafter) sensor, outlining our motives for choosing the device. A thorough technical description of the Soli examining its hardware, software and design is provided by Lien et al. (2016). Soli is capable of using millimeter-wave radar to detect fine-grain and microscopic gestures with modulated pulses emitted at frequencies between 1 and 10 kHz. The strength of a radar-based signal lies in its ability to offer a high temporal resolution, the ability to work through specific materials such as cloth and plastic and to perform independently of environmental lighting conditions (Arner et al., 2017). One significant feature is the highly optimized hardware and software devoted to the prioritization of motion over spatial or static poses. In addition, the compact size makes it an excellent choice for musical purposes that require a low degree of invasiveness from the system.

Other systems are also capable of identifying gestures. This includes color detection from 2D RGB cameras (Erol et al., 2007) to 3D sensing arrays of cameras, such as Microsoft's Kinect (Han et al., 2013). Researchers have developed other means of sensing gestures such as IR technology mainly represented by Leap Motion (Han and Gold, 2014). However, such technologies often lack in precision when aimed for fine-grain gesture detection. Other devices that enable gestural input using radar-like detection are the SideSwipe, which analyzes disturbances of GSM signals (Zhao et al., 2014), and the WiSee that analyzes existing Wi-Fi signals and their perturbances to recognize human gestures (Pu et al., 2013).

The devices mentioned here are unable to capture microgestures with a high level of accuracy. Current devices using radar waves or wireless signals similar to Soli work with lower-frequency bands typically under 5 GHz. Soli uses high-frequency radar of 60 GHz that considerably increases the device's level of accuracy, making it suitable for fine-grain gesture sensing (Wang et al., 2016).

THE SYSTEM

Figure 17.1 shows an overview of the system design and components. Software written in OpenFrameworks manages and visualizes data received from Soli. The *Random Forests* classification algorithm determines whether the gesture is performed or not. This binary outcome is then used as a gate to forward or block the actual data directly mapped to the pitch-shifting algorithm.

Google provides several existing wrappers and examples for Project Soli, including OpenFrameworks, a C++ wrapper specifically designed for creative applications. Nick Gillian's "Random Forests Classification

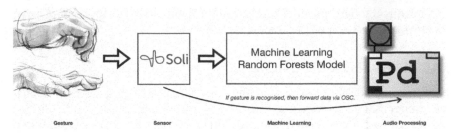

Figure 17.1 Overview of the System Design and Components

Algorithm" from the GRT (Gillian and Paradiso, 2014) was chosen as the initial test algorithm, as it is already implemented as part of the Soli framework, and during the initial prototyping phase of this research, it proved to be a valuable tool due to its ease of use and implementation.

Two core features from the dataset were chosen to control the pitch-shifting algorithm: the energy and the velocity of the gesture analyzed. Through the Open Sound Control (OSC) protocol, the Pure Data (PD) programming environment receives and maps data directly to a pitch-shifting effect: gesture intensity controls the range and amplitude of the effect. The intensity of the effect is also affected by the amount of audio signal incoming from the acoustic instrument, thus giving complete control to the performer regarding the quantity of modulation and volume.

TESTING—INITIAL CASE STUDY

The first prototype of the system was used during a performance at the Beyond Borders conference, held at Birmingham City University in July 2017. The performance enabled us to identify any limitations and constraints and examine potential applications of microgestures of the system before the formal usability test.

The prototype system presented recognized only one gesture: lateral swaying of the hands after the key had been pressed, as shown in Figure 17.2.

We demonstrate the system through a simple piano piece composed in the key of D major exploring the soundscape of the tonal key itself through chords, voicings and different melodic lines superimposed upon one another. The use of the pedal was essential in this piece to create an extended and continuous bedrock of sound that would fill the room with harmonics. It was also aimed to give enough 'room' to the pitch-shifting effect to be heard and noticed. The composition and the performance were tailored to the audience without any musical background to get as much constructive feedback as possible. The piece was divided into three parts, to underline the differences of gestures and gestural nuances in piano playing. During the first part, the pianist used different sizes of wooden sticks, allowing the playing of chords that were otherwise impossible to play. This section underlines the non-expressive elements of performance, by

Figure 17.2 The Lateral Swaying of the Hands After the Key Had Been Pressed. Sequential snapshot of the vibrato gesture.

limiting the abilities of the musician to a mechanic motion: note-on, note-off. By pressing the piano keys with a wooden board instead of the finger, it resulted in a 'binary' and mechanical playing that lacked expression and musicality.

The second part bridges the purely 'binary' playing of the first part, seeing the pianist slowly abandoning the wooden contraptions he had been using until that moment to play, and moving towards a hand-driven exploration of the keys. With the hands on the keyboard, but still performing a binary movement, the system did not activate, and the machine-learning algorithm was not able to recognize any ancillary movements revolving around the piano technique: the playing was still not expressive. This leads us to the third part of the piece, where the pianist makes extensive use of his pianistic technique enhanced by layers of sound modulation.

In the third part, the pianist explores chords, modulating sounds and playing with the sound effect driven by the sensor. The gesture recognition is tailored to the unique hand gestures of the performer. Naturally, the microgestural approach changes depending on the expressive articulations within the score. The piece finishes with a chord struck with one of the sticks from the first part.

The feedback from this initial performance was mostly positive. Mapping the gesture to a frequency-modulation effect gave the illusion to the performer that the acoustic piano could produce a vibrato effect on the notes played. The majority of the audience when asked felt that the gestures produced an organic sound modulation and could not distinguish the different sound sources of the acoustic and electronic textures, even though the speaker was placed directly under the piano. The recognized gesture by the system took place as if it was always there. The pianist mentioned that he was able to control and trigger the vibrato and that the interaction with the system felt natural and non-invasive (Granieri, 2017).[1]

The lateral swaying of the hand together with the vibrato effect turned out to be a really intuitive pair of gesture-modulation to implement. As confirmed later on, one of the pianists from the user testing said "It's helpful to know what a vibrato is so you can try and fit a technique to what you'd imagine it. Or if you would imagine a string player doing vibrato and copy that shape that was kind of what was going through my head". The lateral movement of a hand associated with slight pitch modulation is something that both musicians and audiences can easily relate to the gesture due to the familiarity that vibrato effect has with string instruments.

The performer also found very interesting the control of sound modulations through microgestures, and said "it was very easy to connect with the audience and increase or decrease the amount of the modulation depending on the section of the piece that was being played. This was also due to the piece being very free in its form and composed to accommodate the modulation of the system and the gesture recognized".

USER TESTING

We use an informal formative method for the user testing as described by Martens (2016) to test and assess interaction in a task-based scenario. This method was chosen due to the early stage of the research and the ongoing development of the prototype system. A formal user-testing method including error counting and timed tasks would have been less useful for the further development of the system. Without any previous research as reference, a simple empirical test followed by an interview to gather experiences and impressions from the users on the system was the best approach.

Twelve piano students from the Royal Birmingham Conservatoire, Birmingham City University split equally by gender, participated in the user testing. The tests included students from different stages in their studies varying in age and experience. The musical focus was equally split between classical and jazz trained pianists.

The user tests were conducted in a recording studio using a Yamaha upright piano at Royal Birmingham Conservatoire. An Audio Technica AT4040 cardioid microphone captured the sound, and the effects were emitted via a single Behringer B2031A Active Studio Monitor placed on the studio floor. The microgestures were analyzed using the Soli sensor as described earlier. The system detects the lateral swaying of the hands, as shown in Figure 17.2, and maps the movement to the pitch shifting, which is limited to a maximum of half a tone above and below the note played. The real-time audio analysis allows us to introduce a threshold to avoid unintentional triggering of the system.

Each test lasted approximately 40 minutes per participant. Subjects were briefly interviewed about their pianistic background and current knowledge and experience with electronic music and digital instruments. After the interview and a brief explanation of the system, the users were given 10 minutes to try the system and get comfortable with the effected sound coming from the speaker. There was no dry piano signal coming from the speaker; this choice was made because of the loudness of the piano and the small size of the room. We took advantage of that time to calibrate the system, adjusting to the gesture technique of the pianist. Subjects were then asked to perform a series of simple tasks to assess the precision and reliability of the system. These tasks were the following: play a note, play a chord, play a scale. During the first run all tasks were performed twice and users were asked not to activate the system, while the second time to purposely try to activate it. This was done to make users aware of the threshold and how their gestures may trigger the audio processing. Furthermore, subjects were asked to either perform a piece we provided or perform one from their repertoire.[2] We then asked them to perform the pieces twice with and without the system as a mean of comparison. Two users chose to perform a piece from the provided repertoire, and both were coming from a classical background. The pieces chosen by these two performers were *September Chorale* by Gabriel Jackson and *Bells* by Simon Bainbridge. Finally, they were asked if they were willing to improvise, and then were asked to fill in a User Experience Questionnaire (UEQ) (Schrepp et al., 2017). The UEQ allow us to evaluate the system on its efficiency, perspicuity, dependability as well as aspects of the user experience such as originality and stimulation. Each subject took part in a brief final interview about the experience and the system.

DISCUSSION

The musical background and level of the pianist appeared not to have a significant effect on the result of the test itself. Both classical and jazz pianists were able to perform with the system and commented that they would happily use the system in their performances. During the interview, one user said "this is very diverse, can be applied to classical, jazz, anybody who plays the piano. It can be for anyone". He continued, "it was really interesting to play on a real piano, in its natural form being able to effect sound is not something that is possible without controls and effects" (referring to knobs and effects on his keyboard). The results from the questionnaire were all positive, with higher marks given to the system's attractiveness and hedonic quality, and lower but still positive marks in the pragmatic section, as seen in Figure 17.3. This section concerned the responsiveness and reliability of the system. We anticipated such responses as the system was still in the prototyping stage. During the analysis of interviews, a connection emerged between the piece performed and the feedback given. When they performed one of the proposed pieces, the users tended to be more willing to adapt the composition to their imagination and freely interpret the tempo to accommodate sound modulation through the system. The listed pieces were chosen together with a piano teacher from the Royal Birmingham Conservatoire because of their temporal and rhythmic freedom and long ringing chords. We believe this is something that encourages the pianists to take advantage of the system. When users chose to play a piece from their repertoire, the comments were less encouraging. The users seemed to be less likely to feel the need to add this expressive layer on a consolidated piece that they already knew how to play expressively to convey a certain emotion. This can be related to McNutt's (2004) observations stating that performers need to have a reasonable idea of what sounds they will hear, and in this specific case, what sounds their hands will produce. This link between the pieces and the comments given was also confirmed by the most noticed comment on the system throughout the user testing. All users said that the system was eliminating or at least reducing the learning curve of typical interactive systems, but that the strain had shifted to the ability of predicting and expecting the sound of the instrument. Five users underlined that the hardest element to get accustomed to in the system was not the gestures it involved but the sound of it.

> In this case I heard something I wasn't expecting, before I played I knew how the sound (of the acoustic piano) should have been, and when I played now I was like "wow what is this" because it's something new, and I don't like the sound to be different to what I hear before.
>
> (Classical Pianist 5)

When asked if they had to change their piano technique to take advantage of the system, one user said, "The technique that's needed is the listening, as we say we pedal with our ears. It's really what it's about".

Three out of twelve users pointed out that they would have needed some time to practice the system, to learn what their pianistic gestures would correspond to from a sonic point of view. This is closely tied to the previous statements related to the piece performed during the testing: the fact that the user could not predict what the system would have sounded like meant that the system would have felt invasive from a sonic point of view in contrast to a performance of an already known piece.

The following section of the test was optional and consisted of a short improvisation with the system. This enabled us to assess if within the relatively short time of using the system subjects were able to improvise, and if so, to what extent. This section was aimed mainly towards jazz pianists; however, one classical pianist asked to try and improvise with it. The results had many similarities with the previous part. During the improvisation, users were keen to unexpected sounds and timbres, and were willing to explore the new sonic environment with their technique. When asked to compare the experiences of playing a repertoire piece or improvising, one user said:

> I'd say they were different, I wouldn't say one was better than the other. The theme was less spontaneous, so you knew what was coming up, so I was able to pre-empt. Whereas the improvisation is spontaneous, so I would have to be actively putting it and using it.

Another user said:

> I guess someone could be inspired, and write a piece for it, or someone could use it to aid a performance. Not so sure about pre-existing composition, I am sure that for me if I wrote something I wouldn't want to mess around and perform it in a manner that's adding something that's not in the original scripting of my writing.

Figure 17.3 shows the average values from the 12 users with a breakdown of the different aspects analyzed thanks to the UEQ. On the left is the average of each individual parameter showing the maximum and minimum score on the Y-axis. On the right, the same parameters are grouped under

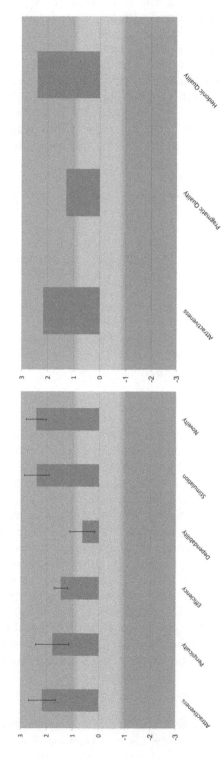

Figure 17.3 On the left, the average of each parameter on a scale from -3 to +3. On the right, the same parameters grouped under three macro categories.

three macro categories, showing a reduced average of the pragmatic qualities, something that we believe is due to the low score around dependability.

With exceptionally high values of 2.4 and 2.17, attractiveness and hedonic quality were the categories that reached the highest score in the test. We believe that the high values were due to the non-invasive character of the system that gave users an additional sonic element with the minimum learning curve. Another less favorable feature of the system was identified in its pragmatic area. While nine users found the system to be innovative and exciting to play with, three users did not feel entirely in control of the system and felt that the system was not responsive enough, resulting in miss-triggering.

The users that felt in control of the system were able to control the triggering of the audio effects in an expressive way through their playing. One user said:

> Yeah I felt mostly in control at some points maybe I was worried I wasn't doing it right. But especially once I got used to it, it felt a lot easier to control. There were a couple of points where I really was thinking If I was performing the gesture correctly, but I don't see it as a long-term issue because I played for a total of 15 minutes.

The comments and feedback, as well as the results of the questionnaire, were expected at this stage of the research. During lab tests, the prototype system was sometimes lacking consistency in providing the data output.

CONCLUSION

From our initial research, the approach we have adopted for developing new interfaces for musical expression has helped to elucidate many factors that musicians face when using digital instruments. By creating interfaces that are non-invasive and build on existing instrumental technique, we can move towards creating less disruptive experiences for performers using technology in performance. We have shown how musicians and pianists, in particular, may benefit from such interfaces. The choice of technologies we have used has allowed us to achieve this. The findings gathered from both the development of the prototype and the usability testing showed positive and encouraging outcomes. The user testing showed how users are keen to adapt and accept such a system which builds upon their existing technique.

With the development of new technologies and devices available, perhaps we need to think about a new communication protocol in instrumental performances that can further explore the potentials presented through microgestures and open new horizons to composers and musician alike.

NOTES

1. The performance can be seen on Vimeo: https://vimeo.com/226180524.
2. Piano pieces provided: *September Chorale* (Jackson, Gabriel), *Nocturne I* (Harrison, Sadie), *Nocturne II* (Harrison, Sadie), *Utrecht Chimes* (Lange, Elena), *Bells* (Bainbridge, Simon), *Yvaropera 5* (Finnissy, Michael).

Repertoire piano pieces: *Nocturne in C-sharp minor*, B.49 (Chopin, Frédéric), *Piano Sonata No. 8 in C minor, Op. 13, Adagio cantabile* (Beethoven, Ludwig van), *Paraphrase de concert sur Rigoletto, S.434* (Liszt, Franz), *Faschingsschwank aus Wien, Op. 26: I. Allegro* (Schumann, Robert), personal arrangement of *When You Wish Upon a Star* (Edwards, Cliff), *Nocturne, Op. 32, Andante sostenuto in B major* (Chopin, Frédéric), *Piano Sonata No. 10, Op. 14, No. 2* (Beethoven, Ludwig van), *Prelude in G-Sharp Minor, Op. 32, No. 12* (Rachmaninoff, Sergei).

REFERENCES

Bernardo, F., Arner, N. and Batchelor P. (2017). O Soli Mio: Exploring millimeter wave radar for musical interaction. In: *Proceedings of the international conference on New Interfaces for Musical Expression* (NIME), pp. 283–286.

Bullock, J., Michailidis, T. and Poyade, M. (2016). Towards a live interface for direct manipulation of spatial audio. In: *Proceedings of the international conference on live interfaces*. Sussex: REFRAME Books.

Cadoz, C. and Wanderley, M. M. (2000). Gesture—music. In: *Trends in gestural control of music*. Paris, France: Ircam—Centre Pompidou, pp. 71–94.

Chadefaux, D., Le Carrou, J. L., Fabre, B., et al. (2010). Experimental study of the plucking of the concert harp. In: *Proceedings of the international symposium on music acoustics*, pp. 1–5.

Dobrian, C. and Koppelman, D. (2006). The "E" in NIME—musical expression with new computer interfaces. In: *Proceedings of the international conference on New Interfaces for Musical Expression* (NIME), pp. 277–282.

Dourish, P. (2004). *Where the action is: The foundations of embodied interaction*. Cambridge: MIT Press.

Erol, A., Bebis, G., Nicolescu, M., et al. (2007). Vision-based hand pose estimation: A review. *Computer Vision and Image Understanding*, 108(1–2), pp. 52–73.

Gillian, N. and Paradiso, J. A. (2014). The gesture recognition toolkit. *Journal of Machine Learning Research*, pp. 3483–3487.

Grandhi, S., Joue, G. and Mittelberg, I. (2011). Understanding naturalness and intuitiveness in gesture production: Insights for touchless gestural interfaces. In: *Proceedings conference on human factors in computing systems is the premier international conference of Human-Computer Interaction* (CHI), pp. 824–828.

Granieri, N. (2017). *Expressing through gesture nuances* [online]. Available at: https://vimeo.com/226180524.

Haken, L., Abdullah, R. and Smart, M. (1992). The continuum—a continuous music keyboard. *International Computer Music Conference* (ICMC), pp. 81–84.

Han, J. and Gold, N. (2014). Lessons learned in exploring the leap motion. In: *Proceedings of the international conference on New Interfaces for Musical Expression* (NIME), pp. 371–374.

Han, J., Shao, L., Xu, D., et al. (2013). Enhanced computer vision with Microsoft Kinect sensor: A review. *IEEE Transactions on Cybernetics*, 43(5), pp. 1318–1334.

Johnstone, E. (1985). The rolky—a poly-touch controller for electronic music. In: *International Computer Music Conference* (ICMC), pp. 291–295.

Lamb, R. and Robertson, A. N. (2011). Seaboard: A new piano keyboard-related interface combining discrete and continuous control. In: *Proceedings of the international conference on new interfaces for musical expression*. Oslo, Norway, pp. 503–506.

Lien, J., Gillian, N., Karagozler, M. E., et al. (2016). Soli: Ubiquitous gesture sensing with millimeter wave radar. *ACM Transactions on Graphics, 35*(4), Article 142, pp. 142:1–142:19.

Martens, D. (2016). *Virtually usable: A review of virtual reality usability evaluation methods*. BSC Thesis. Parsons School of Design.

McNutt, E. (2004). Performing electroacoustic music: A wider view of interactivity. *Organised Sound*, 8(3), pp. 297–304.

McPherson, A. (2012). TouchKeys: Capacitive multi-touch sensing on a physical keyboard. In: *Proceedings of the international conference on New Interfaces for Musical Expression* (NIME), pp. 1–4.

Michailidis, T. (2016). On the hunt for feedback: Vibrotactile feedback in interactive electronic music performances. PhD Thesis. Birmingham City University.

Miranda, E. R. and Wanderley, M. M. (2006). *New digital musical instruments. Control and interaction beyond the keyboard*. Middleton, WI: A-R Editions, Inc.

Newton, D. and Marshall, M. T. (2011). Examining how musicians create augmented musical instruments. In: *Proceedings of the international conference on New Interfaces for Musical Expression* (NIME), pp. 155–160.

Nicolls, S. L. (2011). *Interacting with the piano*. PhD Thesis. Brunel University.

Norman, D. A. (1990). Why interfaces don't work. In: *The art of human-computer interface design*. Addison-Wesley Professional, University of Michigan, pp. 209–219.

Pestova, X. (2008). *Models of interaction in works for piano and live electronics*. PhD Thesis. McGill University.

Pu, Q., Gupta, S., Gollakota, S., et al. (2013). Whole-home gesture recognition using wireless signals. In: *Mobi*. New York: ACM Press, pp. 12–27.

Rebelo, P. (2006). Haptic sensation and instrumental transgression. *Contemporary Music Review*, 25(1–2), pp. 27–35.

Schrepp, M., Hinderks, A. and Thomaschewski, J. (2017). Design and evaluation of a short version of the User Experience Questionnaire (UEQ-S). *International Journal of Interactive Multimedia and Artificial Intelligence*, 4(6), p. 103.

Tanaka, A. (2000). Musical performance practice on sensor-based instruments. *Trends in Gestural Control of Music*, 389–406.

Wang, S., Song, J., Lien, J., et al. (2016). Interacting with soli: Exploring fine-grained dynamic gesture recognition in the radio-frequency spectrum. In: *UIST*, pp. 851–860.

Zhao, C., Chen, K.-Y., Aumi, M. T. I., et al. (2014). SideSwipe. In: *UIST*. New York: ACM Press, pp. 527–534.

18

MAMIC Goes Live

A Music Programming System for Non-specialist Delivery

Mat Dalgleish and Chris Payne

TOWARDS COMPUTATIONAL THINKING IN SCHOOLS

The English national curriculum sets out the subjects taught, the contents of these subjects, targets for achievement, and standards for each subject, in all local authority–maintained primary and secondary schools nationwide. The last six years have seen controversial changes to some subjects, most notably Mathematics and Computing. For instance, the new Mathematics curriculum requires first-year primary students (ages 5–6) to handle and manipulate large numbers. It also requires the earlier introduction of more advanced topics such as fractions and algebra, before they are revisited in more detail in later years (Dominiczak, 2013). The change of ethos in relation to Computing is arguably more radical still: the new curriculum suggests that students should not only become adept users of software (i.e. the Information and Communications Technology [ICT] model) but also understand the fundamental principles that underpin the creation of software artifacts (Andrews, 2014; Curtis, 2013). This ultimately means that students are not only taught how to code, but also encouraged to develop more abstract skills that relate to Computational Thinking.

Following an initial but extensive definition of Computational Thinking by Wing (2006), Selby and Woollard (2014) describe how the term has been contested and a multitude of different definitions proposed. However, there is broad agreement that Computational Thinking relies on multiple stages to solve complex problems (Selby and Woollard, 2014). These stages can be broadly described as follows (Kemp, 2014):

- Decomposition—breaking down problems into manageable components
- Pattern recognition—analysis of identical or similar states in a system so that they can be appropriately categorized, in order to permit the repeated use of a concept
- Abstraction—streamlining a proposed system for reasons of efficiency and stability, by applying only essential concepts
- Algorithm design—designing systems to solve a given problem

For Dehnadi and colleagues (2009), the ability to think computationally is a useful indicator of practical performance. More specifically, they found that those students who are able to develop and consistently apply a mental model of programming were more than twice as likely to pass a practical programming test. For Berry (2013), however, the benefits of Computational Thinking extend beyond the activity of programming or even the discipline of Computing, to provide a kind of toolkit that can be applied to problems in diverse scenarios. To this end, he states that: "Computational thinking provides insights into many areas of the curriculum, and influences work at the cutting edge of a wide range of disciplines" (Berry, 2013). This is notably considered a boon to the modern economy, as well as pedagogically expedient.

TEACHERS AND THEIR DISCONTENTS

Despite the importance placed on the shift from the old ICT model to the current computing curriculum, John (2013) notes that the Graduate Teacher Training Registry (GTTR) saw a 41% decline in graduates training to become ICT secondary school specialists in 2012–13. A 50% decline in the number of computer science teaching applicants was also reported in the same academic year (John, 2013). These issues lead Connell, chair of the Association for Information Technology in Teacher Education (ITTE), to state that:

> [The ITTE is] less confident than the Teaching Agency (TA) that we will have the necessary workforce in place to effectively deliver the new Computing Curriculum for 2014. We believe it will take longer— a view reflected by the BCS itself in its submission to the consultation on the new curriculum.
>
> (Connell, in John, 2013)

There are also concerns from the perspective of existing teachers. Andrews (2014) states that teachers and institutions are skeptical about the ongoing implementation of the new computing curriculum. The broad consensus among many educators is that there is a lack of skills relating to programming specifically within the teacher population. For example, in July 2014, a YouGov survey commissioned by the innovation charity Nesta and Tes (previously the Times Education Supplement) found that 60% of ICT teachers did not feel confident about the new curriculum delivery that was implemented in September 2014. Moreover, over half the surveyed teachers admitted to not seeking any advice on the subject before the start of the academic year. Overall, 67% of teachers surveyed stated that despite the government-funded initiatives, they still did not feel they had adequate support and guidance to aid their subject knowledge and support lesson planning activities (John, 2014).

In an attempt to address preconceptions (including those of teachers), the Year of Code project was established in 2014, with government backing (Year of Code, 2015). Although Year of Code has been praised by some (Cellan-Jones, 2014), teaching groups and academics remained concerned about a lack of necessary skills, prompting a further £3.5 million government training scheme, Computing at School (CAS), relating to computational thinking. The CAS initiative aims to support the new ICT curriculum with training programs and subject resources, as well as instigating reform of awarding body criteria (Computing at School, 2015). In particular, CAS advocates the creation of self-sufficient "Master Teachers" who are trained to deliver professional development sessions to teaching staff across a range of institutions (Humphreys et al., 2014, p. 4). However, while it was anticipated that 400 Master Teachers would be created, Dickens (2016, p. 1) comments that it is unclear how many are currently in post.

LOCATING MUSIC TECHNOLOGY

While increased importance has been placed on technology (and computer technology in particular) elsewhere in the National Curriculum, the state of music technology is less clear. As far back as 1997, Innes (1997) noted that students are reliant on "forward-looking teachers" to implement technological approaches within music. In the two decades since, music technology at school level has continued to be available for study only indirectly, through the inclusion of some of its facets in related subjects, or informally (Gehlhaar, 2002). Indeed, the most recent National Plan for Music Education (NPME) document (Department for Education, 2011) makes it clear that music technology is still considered only as an addition to the music curriculum rather than a subject in its own right. Interestingly, the report includes a number of examples of "appropriate" use of technology within music (Department for Education, 2011). These include:

* sound recordings to aid music listening
* sequencers, loopers, samplers and effects processes for the purpose of musical composition and performance
* online resources and smartphone apps as ways to assist instrumental practice and learning music theory, respectively
* online collaboration
* digital keyboards as a cost-effective replacement for acoustic pianos
* digital audio workstations as a cost-effective replacement for the recording studio

If hardly novel—many of these practices are already established outside of the school domain—the NPME report still asserts that technology in music must be used wisely and not distract from musical or lesson content. At the same time, it also notes that some teachers make good use of music technology, but it is underused by others. Closely related, the Paul Hamlyn Foundation (2014) comments that: "Music technology is not yet sufficiently integrated into school-based music, and many teachers do not capitalize on pupils' confidence and facility with technology".

If the 2011 NPME report suggests that "further work should be under-taken to develop a national plan for the use of technology in the deliv-ery of Music Education—and to ensure that the workforce is up-to-date with latest developments", efforts to date are considered insufficient by some. For instance, as part of a public consultation in February 2013 led by the Secretary of State for Education and intended to implement curricu-lum reforms by September 2014, respondents commented that the role of Music Technology required further definition (Department for Education, 2013).

One particular debate centers around whether Music Technology should be a stand-alone subject. Certainly, a case can be made for the creation of a stand-alone Music Technology subject at school level. For instance, music technology degrees have come to feature prominently in Higher Education (Boehm, 2005), and Music Technology awards are firmly established in Further Education. It could therefore be argued that a specialized Music Technology award at school level might better prepare students for pro-gression onto music technology routes in Further Education and Higher Education. It may also increase recognition and funding for the area, and increased study time might enable students to explore the discipline in greater detail. Nevertheless, a case can also be made for an alternative approach: the spread of music technology into other subjects. For instance, from the perspective of employment, traditional music technology career paths such as studio engineer and producer have suffered significantly with the demise of many large recording studios (Leyshon, 2009) and the decline of recorded music sales. At the same time, new opportunities have emerged elsewhere. These have largely centered around the intersection of music technology with other disciplines, sometimes in combinations that would have seemed unlikely in even the relatively recent past. For example, while Thorley (2005) comments that "there is no such job as a 'music technologist'", music technology specialists make significant con-tributions to fields such as:

- film and television
- theatre
- video games
- virtual environments and augmented reality
- architectural and archaeological acoustics
- product and industrial design
- Human-Computer Interaction (HCI)
- signal processing
- software development

These intersections are closely linked to the flexibility of the computer, as well as its ubiquity. The ability of the home computer to approximate what was previously only achievable using a collection of specialized and expensive studio equipment is well documented (Leyshon, 2009). How-ever, arguably more impactful is the ability of the computer to act as a kind of proxy that is able to leverage skills previously confined to one discipline into new areas—a means of traversing previously largely impermeable disciplinary boundaries. For instance, open-ended programming environments such as Max, Pure Data (Pd) and SuperCollider enable translation between senses: video can become sound, sounds can become rendered 3D graphics, and—aided by low-cost microcontrollers such as the Arduino—either domain can be turned into control signals for a host of physical actuators. Equally important is that these environments recognize that potential users often come from arts backgrounds rather than computer science and make programming concepts accessible to these groups.

As the tendrils of music technology spread into and entangle with other areas, and with Sawyer et al. (2013) stating that music act as a vehicle to expose students to a science, technology, engineering and math (STEM) curriculum, it is useful to consider the notion of transfer learning. Transfer learning relates to the notion that, unconsciously or through concerted reflection, skills developed in one area can transfer to other areas that involve similar processes (Hallam, 2010). For Fleishman, these transfers are an everyday—but vital—occurrence:

> Transfer of learning . . . is pervasive in everyday life, in the developing child and adult. Transfer takes place whenever our existing knowledge, abilities and skills affect the learning or performance of new tasks . . . transfer of learning is seen as fundamental to all learning.
> (Fleishman, in Cormier and Hagman, 1987, p. xi)

Hallam (2010), however, also notes that transfers are not all equal, or equally likely. "Near" transfers are where concepts and performances are closely related. "Far" transfers are between poorly matched contexts and performances. Hallam argues that near transfers are stronger and more likely to occur.

MUSIC AND MATHEMATICS

If music and computing have become closely aligned, it has long been identified that mathematics has a close affinity with music. For instance, as early as 500 BCE, the Ionian Greek philosopher Pythagoras observed the integral relationships between frequencies of musical tones in a consonant interval (Wright, 2009), likely leading to the development of the Greek musical scales (Hawkins, 2012).

If musical rudiments such as pitch and rhythm can be described as fundamentally mathematical, many 20th-century musical works are based on specialized contemporary mathematical concepts. For instance, the musical material of Iannis Xenakis's *Pithoprakta* (1955–56) is rooted in the statistical mechanics of gases, and its title refers to Bernoulli's law of large numbers: namely that, as the number of occurrences of a chance event increases, the more the average outcome approaches a determinate end (Antonopoulos, 2011). Other examples include: (a) the use of the Monte Carlo method of random number generation to make organizational decisions in Hiller and Isaacson's *Illiac Suite* (Sandred et al., 2009); (b) the use of multidimensional crystal algorithms to generate harmonic structures in the work of James Tenney (2008); (c) the use of self-similar structures, particularly melodies, by Tom Johnson (2006); and (d) the use of a Game of Life algorithm to generate the intervals of a triad based on the locations of active cells in Eduardo Miranda's *CAMUS* (Burraston et al., 2004).

As interest in the relationship between mathematics and music continues (Wright, 2009; Loy, 2006; Tymoczko, 2011), there is also some empirical evidence that linking the two disciplines may be of pedagogical benefit for school-age children. For instance, a study by An et al. (2013) involved two student teachers receiving training in relation to the integration of music and mathematics from experts, before incorporating music activities into their usual mathematics classes with 46 first- and third-grade schoolchildren from a range of backgrounds. Activities included composition and performance, delivered as ten 45-minute sessions in total over a period of five weeks. Before the classes began, a series of five Model-Strategy-Application (MSA) tests were administered to every participant, and a further five MSA tests were administered after each interdisciplinary session. An et al. (2013) found that the performance of both first- and third-grade students in all three mathematical areas assessed by the MSA test showed statistically significant improvements after the introduction of the interdisciplinary curriculum.

MAMIC: MUSIC AND MATHEMATICS IN COLLABORATION

The 2011 National Curriculum in England has changed how mathematics and, particularly, computing are taught in schools. However, despite training schemes backed by the government, teachers continue to feel that they are underprepared for the changes. At the same time, the role and position of music technology, itself increasingly computing-centric, remains poorly defined. Responding to these issues, Chris Payne's Music And Math In Collaboration (MAMIC) project is intended to be used by children aged 9–11 years and aims to:

* leverage musical process relating to composition and performance to introduce coding (and visual programming specifically) in an accessible and engaging way
* use musical process to vibrantly introduce mathematical concepts
* reinforce existing mathematical knowledge by making abstract concepts audible and/or visible, and able to be modified in a hands-on way
* facilitate and support "in the wild" delivery by non-specialist teachers
* run on a wide variety of computers, including on the older and budget hardware found in many schools, without requiring permanent installation

The MAMIC project (Payne, 2018) consists of three main elements:

* a library (i.e. extension) for the Pd visual programming environment
* encapsulation of the Pd/MAMIC library combination in a portable, Linux-based live image able to run from a USB pen drive

- tutorial materials intended to aid and support delivery by non-specialist teachers. These consist of a manual, help patches, and step-by-step guides for teachers and students in video and written formats.

Pd (Puckette, 2017) is used as the overarching programming environment as a free and Open Source alternative to Max (Cycling '74, 2017). A visual programming language was chosen in order to minimize the gap between conceptual model and implementational representation, particularly in relation to data flow. Open Source is seen as a boon to enabling reuse or further development by others, and the tutorial materials are also available under a Creative Commons (2017) Attribution Non-Commercial license.

The live Linux image is based on the Ubuntu Studio (2017) distribution, packaged so that non-specialist teachers can boot the system from the pen drive without the need for hard disk installation. It is also designed to be used on the variety of computer systems that can be found in the education system and includes detailed CPU usage-reduction strategies to enable smooth real-time operation on older and budget (i.e. lower specification) hardware. The image also includes shortcuts to SimpleScreenRecorder so that desktop movements and actions can be recorded to various video formats, and VLC Media Player for the use of video playback (e.g. to watch video tutorials or self-capture). Lastly, a dedicated Student Work folder enables students to easily save their work to unused space on the pen drive.

The MAMIC library consists of 75 abstractions that each perform a different compositional or performative task or subtask. The abstractions are categorized and grouped by broad function, leading to the following hierarchy of layers (Figure 18.1):

- External Input
- Internal Sequencing
- Composition
- Sound Generation
- Sound Output

The External Input layer provides a number of abstractions aimed at capturing input from external devices and producing musically useful control data. Controllers currently supported include MIDI and computer keyboards, gamepads and webcams. Multiple input devices can be used simultaneously and external input combined with the internal sequencer.

The Internal Sequencing layer is based around a prebuilt sequencer called Conductor. The Conductor abstraction utilizes an array of 16 steps (colored orange), each able to store a user-determined sequence number between 1 and 16. Activating the Play button causes Conductor to cycle through the steps at the specified tempo (shown in the tempo display). As Conductor arrives at each step, the sequence number stored at that step is sent to its output outlet. This sequence number can then be used to trigger aspects in the composition layer such as scales, chords or one-shot audio samples.

The Composition layer operates primarily at the note level. MAMIC features prebuilt major and minor-scale objects. The Major-scale and

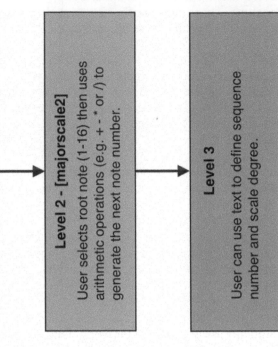

Level 1 - [majorscale1]

User selects root note (1-16) and notes used (1-16).

Level 2 - [majorscale2]

User selects root note (1-16) then uses arithmetic operations (e.g. + - * or /) to generate the next note number.

Level 3

User can use text to define sequence number and scale degree.

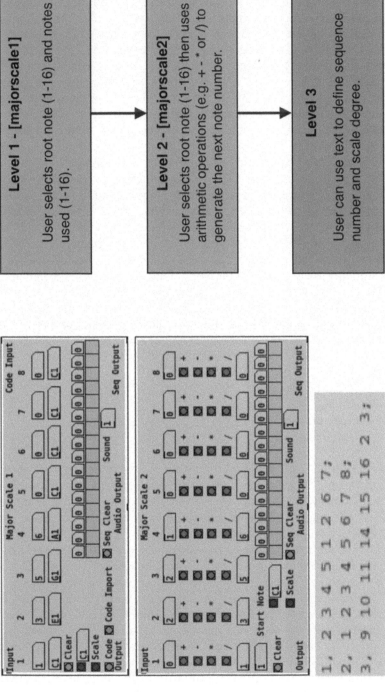

Figure 18.1 An Overview of the MAMIC Library

Minor-scale abstractions contain formulae for whole and half steps. Students then select the tonic and the scale degree to be played. MAMIC also enables students to create diatonic chords (diatonic harmony) in two different ways via the Majorchords and Makechord abstractions.

Majorchords holds and creates all the diatonic chords for the selected major scale. Users can then send sequence numbers (1–8) to select the chord to be played. Roman numeral representations (as commonly used in music theory) are also displayed. Makechord displays three columns of numbers that enable a three note chord to be constructed using the notes of a major or minor scale. When a sequence number (1–16) is sent to Makechord, the programmed chord will then play. Hidden interconnections between objects ensure that Makechord automatically "knows" the scale being used.

The Sound Generation layer contains a modest pallet of simple sound-generation options. These include a number of simplified synthesizers, plus one-shot and looping sample players. While it is considered vital that novice users can very quickly produce basic sounds, some flexibility is provided for more experienced users.

The Sound Output layer contains a simple abstraction that outputs stereo audio.

The more complex or multifunctional abstractions in the MAMIC library implement a system of tiered object granularity, as proposed by Bukvic et al. (2012). This is termed Differentiated Abstraction and means that abstractions are made up of multiple tiers that gradually increase in complexity. Three tiers are implemented:

- Tier 1: involves simple interactions and/or object functionality is automated
- Tier 2: requires a greater degree of interaction and/or interaction across the disciplinary boundaries of music, mathematics and coding
- Tier 3: involves more complex or sophisticated mental operations such as the creation of textual code to set real-time functions or operations

Examples of differentiated levels of abstraction as featured in an early version of the Major-scale abstraction are shown in Figure 18.2.

Higher tiers can be hidden to simplify operation for younger or novice users, or revealed to provide greater freedom or to introduce more sophisticated concepts.

TESTING

Testing of the MAMIC project has been iterative and multi-staged. First, postgraduate music technology students provided expert review of the library at a preliminary stage in its development. All had two or more years of experience with music programming languages and were therefore well placed to provide comment.

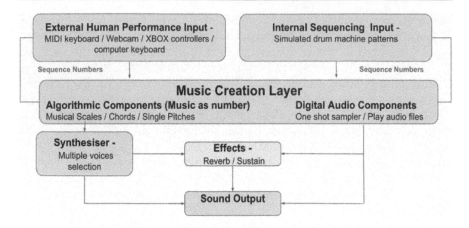

MAMIC System Overview

Figure 18.2 An Example of Differentiated Abstraction Levels in the MAMIC System

After two cycles of this test-and-improve process, a subsequent stage shifted the project out of the lab and into "the wild" (Rogers, 2011). A key premise of in-the-wild studies is that researchers are able to evaluate new technologies as they are really used. In particular, they acknowledge that experimental context can influence results and try to minimize this effect by operating in the participants' usual or "natural" environment (Brown et al., 2011).

To date, one in-the-wild study has been carried out over a period of ten weeks in a Warwickshire (UK) school. A second in-the-wild study is currently underway in another school. To better understand if the MAMIC project can be delivered by non-specialists, the first study was delivered by two student teachers: both held music-related degrees (in production and composition, respectively), but neither had programmed before or studied either mathematics or computing beyond age 16. To introduce them to the project, both student teachers completed an initial 90-minute training session, before a period of self-directed learning and familiarization with query support. The student teachers then used the MAMIC materials to plan and deliver five 2-hour sessions to a class of ten Year 5 children (9–10 years old) over a period of seven weeks. Participants were of mixed ability and from diverse backgrounds. Sessions were playful and open-ended, with participants devising and subsequently exploring personal interests and goals. The sessions took place in the participants' usual classroom environment and, for reasons of manageability, focused on a subset of abstractions that encapsulate the main aspects of the MAMIC library.

Throughout the sessions, Unmoderated Remote Usability Testing (URUT) techniques (Barnum, 2010, pp. 44–45) informed the use of automated video screen capture to document participant activities and outcomes. This focused on how participants manipulated and connected the MAMIC abstractions to create generative music (i.e. the patches created by students), but also captured any student teacher interactions or interventions inside the MAMIC environment. The video screen capture files were retrieved from the USB sticks at the end of every session and archived for later analysis.

Alongside their delivery, the student teachers kept diaries that summarized their sessions. These diary entries provided a narrative outline of each session, together with comments around any positive aspects or points for improvement.

At the end of the seven weeks of delivery, the trainee teachers also evaluated the MAMIC library according to the Heuristic Evaluation model developed by Nielson (1995). The heuristics are general principles that are intended to describe typical properties of usable interfaces. The model consists of the following principles:

- Visibility of system status
- Match between the system and the real world
- User control and freedom
- Consistency and standards
- Error prevention
- Recognition rather than recall
- Flexibility and ease of use
- Aesthetic and minimalist design
- Help users recognize, diagnose and recover from errors
- Help and documentation

For convenience, these heuristics were distilled into an online form with ten questions in total (one question per heuristic). Clarity statements were provided for each question to ensure understanding of its relevance to the project. Each question asked the student teachers to provide a quantitative score (1–5) for the heuristic and then provide rationale for their choice. A final section provided additional space to comment on the quality and appropriateness of the MAMIC documentation and broader experiences of deployment and operation.

INITIAL FINDINGS AND DISCUSSION

Expert review from the postgraduate students tended to focus on technical and experiential aspects. It primarily helped to improve library functionality and usability through interface design and computational optimization. In the latter instance, students informed the reorganization of some abstractions into subpatches so as to enable the extensive use of localized DSP management. These adjustments appreciably aided the ability of the MAMIC library to run effectively on the aging and lower-specification hardware found in the schools.

The diaries of the student teachers in the first in-the-wild study suggest that, by the end of the initial two-hour session, participants had comprehended how to connect and set the abstractions so as to achieve basic functionality. The second session saw participants build their own patches (i.e. collections of interconnected abstractions). This suggests that they had not only become familiar with how to initialize MAMIC abstractions and change their parameters, but also started to conceive and implement their own algorithms. The student teachers adopted a rotation system reminiscent of *Cadavre exquis* (Kochhar-Lingren et al., 2009), so that each participant could experience and contribute to a variety of different algorithms. The diary entry notes that by the end of the session, "most" participants demonstrated confidence in their use of the MAMIC library and how they tackled peer creations. From the third session onwards, participants started to connect and explore the use of hardware controllers such as MIDI keyboards and video game controllers. Participants also developed more fully realized and individualized patches, and there was evidence of peer-to-peer help and problem solving. Interventions by the student teachers involved details of implementation rather than core concepts. They include a reminder of the keyboard shortcut to switch between edit and performance modes in the Pd environment, and how to swap the games controller abstraction between its sampler and sequencer modes. Overall, the student teacher diaries consistently noted that they were pleased with the progress of participants but made relatively little comment about their own experiences.

The Heuristic Evaluations completed by the student teachers at the end of the seven-week delivery period reveal that, despite differences in their backgrounds, both student teachers had largely positive experiences with the MAMIC system overall, with the production graduate rating most aspects of the system more highly than the composition graduate.

In relation to the "match between the system and the real world" and "recognition rather than recall" aspects, the student teachers commented that the MAMIC abstractions are well named and represented; the names of most abstractions and their graphical representation on screen appear to convey their particular functions to the intended users. The exceptions are the more abstract objects. For example, the student teachers identified that the "volume", "load audio sample" and "trigger buttons" abstractions are isolated instances that could be difficult to work with without the use of additional labels. Other aspects relating to recognition rather than recall include quickly understanding the flow from one object to the next. For instance, the student teachers also noted that "students needed little guidance once shown the basics with regards to how to create and link patches".

Responses to the "error prevention" and "help users recognize, diagnose and recover from errors" heuristics are more suggestive of areas for improvement. For instance, the composition graduate reported more system errors than the production graduate and noted that on occasion the "easiest option" was to close the MAMIC system and reload it. However, it was also noted that this was costly in terms of time and is therefore a problematic solution. Additionally, some errors did not produce visible error reports, thereby limiting opportunities for diagnosis and recovery.

In terms of the "aesthetic and minimalist design" heuristic, the student teachers commented that the construction and presentation of the abstractions was largely successful in hiding their inner complexity from participants. However, some participants did occasionally stumble into this interior. This could result in confusion and also caused some issues with source code manipulation (and is therefore also relevant to "error prevention").

"User control and freedom" and "flexibility and ease of use" are arguably more difficult concepts to evaluate. On the one hand, the MAMIC library constrains possibilities compared to the standard, unadorned Pd environment. On the other hand, the MAMIC library still presents an almost inexhaustible number of possible musical topologies. This "guided freedom" is considered a positive aspect by the student teachers, but it is also noted that more guidance or further training would be needed to fully harness these possibilities.

In the additional comments section of the form, both student teachers mentioned the help materials supplied on the MAMIC live USB image: in particular, they found the resources to be very useful as a foundation for their self-study.

The video capture material appears to support many of the comments made by the student teachers. For example, video screen capture from the first session shows that the participants had already discovered how to control the functions of MAMIC abstractions and how to successfully interconnect them. An example of this can be seen in Figure 18.3. However, when abstractions did not act as expected, the MAMIC system was not always able to respond. As a result, the participants or student teacher sometimes had to find the cause of faults on their own. Video capture also provided evidence of audio problems at a system level. Thus, after discussion with the student teachers, shell scripts were used to automate JACK audio server initialization (i.e. how and when the server loads).

While it is more difficult to ascertain how quickly and fully the students developed understanding of the underlying concepts, there is evidence that they have been internalized, at least in some cases. Most notably, some compositions produced by participants do not simply repeat the applications covered in class, but instead use ideas introduced in class as the basis

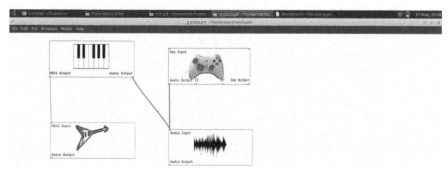

Figure 18.3 A Simple But Functional Example Patch Created by a Participant

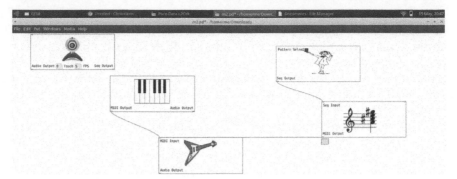

Figure 18.4 A More Sophisticated Patch That Combines Two Different Forms of Input

for improvisation and synthesis. For instance, the participant example in Figure 18.4 combines two different means of input and adds a simple pre-set manager: either a MIDI keyboard or the inbuilt sequencer can be used for input, and sequencer patterns can be stored and recalled. There is also possible evidence of the participant experimenting with webcam control (seemingly discarded) towards the top-left corner of the screen.

CONCLUSION AND FUTURE WORK

Findings from the first in-the-wild study have been promising in terms of how readily users have taken to the MAMIC system and the positive overall experiences of delivery. Issues arising have been primarily of a technical nature and subsequently resolved. A second in-the-wild study is currently underway in another primary school, with delivery over a 14-week period by permanent rather than trainee teachers. It is of particular interest to see how the system fares when delivered by teachers without a musical background (i.e. when the diversity of delivery increases). Indeed, the involvement of participating teachers is considered vital going forward. For instance, teachers at the second school have already asked for the system to be trialed for use with children of reception age (5–6 years old), and for dance mats to be implemented as an additional means of input. The latter takes the MAMIC project towards what Holland et al. (2011) call "whole body interaction in abstract domains". This may extend the project in at least two ways. New matches between physical action and system output may be developed. These could open the system up to new participants (e.g. anyone not able to use current input devices) or help current users to develop new understanding. Additionally, by encouraging users to be physically active, whole-body interaction also has the potential to benefit user health and fitness.

Longer-term work involves deeper exploration of how to situate the MAMIC project relative to the National Curriculum in England. This requires engagement at a variety of scales, from governmental policy, to still largely unconnected subjects, to individual lesson plans.

REFERENCES

An, S., Capraro, M. M. and Tillman, D. (2013). Elementary teachers integrate music activities into regular mathematics lessons: Effects on students' mathematical abilities. *Journal of Learning Through the Arts*, 9(1), pp. 1–21.

Andrews, S. (2014). *What's changing in the computing curriculum?* [onlinc]. [Accessed 14 Dec. 2017]. Available at: www.pcpro.co.uk/features/389875/ whats-changing-in-the-computing-curriculum.

Antonopoulos, A. (2011). Pithoprakta: The historical measures 52–59: New evidence in glissando speed formalization. In: *Xenakis international symposium*. London: Southbank Centre, pp. 1–23.

Barnum, C. (2010). *Usability testing essentials: Ready, set . . . test!* Burlington, MA: Morgan Kaufmann.

Berry, M. (2013). *Computing in the national curriculum: A guide for primary teachers* [online]. Available at: www.computingatschool.org.uk/data/ uploads/CASPrimaryComputing.pdf [Accessed 14 Mar. 2015].

Boehm, C. (2005). The thing about the quotes: "Music Technology" degrees in Britain. In: *International Computer Music Conference (ICMC)*. New Orleans: International Computer Music Association, pp. 682–687.

Brown, B., Reeves, S. and Sherwood, S. (2011). Into the wild: Challenges and opportunities for field trial methods. In: *ACM CHI 2011 conference on human factors in computing systems*. Vancouver: ACM Press, pp. 1657–1666.

Bukvic, I., Baum, L., Layman, B. and Woodard, K. (2012). Granular learning objects for instrument design and collaborative performance in k-12 education. In: *International conference on New Interfaces for Musical Expression (NIME)*. Ann Arbor, MI, pp. 344–346.

Burraston, D., Edmonds, E., Livingstone, D. and Miranda, E. R. (2004). Cellular automata in MIDI based computer music. In: *International computer music conference*. Miami: International Computer Music Association, pp. 71–78.

Cellan-Jones, R. (2014). *Year of code—PR Fiasco or vital mission?* [online]. Available at: www.bbc.co.uk/news/technology-26150717 [Accessed 14 Dec. 2017].

Computing at School. (2015). *About us* [online]. Available at: www.computingatschool.org.uk/index.php?id=about-us [Accessed 21 Dec. 2017].

Cormier, S. M. and Hagman, J. D. (eds.). (1987). *Transfer of learning: Contemporary research and applications*. London: Academic Press.

Creative Commons. (2017). *Attribution-noncommercial 4.0 international (CC BY-NC 4.0)* [online]. Available at: https://creativecommons.org/licenses/ by-nc/4.0/ [Accessed 2 Jan. 2018].

Curtis, S. (2013). *Teaching our children to code: A quiet revolution* [online]. Available at: www.telegraph.co.uk/technology/news/10410036/Teaching-our-children-to-code-a-quiet-revolution.html [Accessed 2 Jan. 2018].

Cycling '74. (2017). *Max* [software]. Available at: https://cycling74.com/products/ max/ [Accessed 2 Jan. 2018].

Dehnadi, S., Bornat, R. and Adams, R. (2009). Meta-analysis of the effect of consistency on success in early learning of programming. In: *Psychology Programming Interested Group (PPIG) annual workshop*. The Open University, Milton Keynes. Available at: www.ppig.org/papers/21st-dehnadi.pdf [Accessed 11 Jan. 2018].

Department for Education. (2011). *The importance of music: A national plan for music education* [online]. Available at: www.gov.uk/government/uploads/system/uploads/attachment_data/file/180973/DFE-00086-2011.pdf [Accessed 16 Dec. 2017].

Department for Education. (2013). *Reform of the national curriculum in England Report of the consultation conducted February-April 2013* [online]. Available at: www.education.gov.uk/consultations/downloadableDocs/NC%20in%20 England%20consultation%20report%20-%20FINAL%20-%20Accessible. pdf [Accessed 16 Dec. 2017].

Dickens, J. (2016). *Government spends £3m in scramble to get 400 "master" computing teachers* [online]. Available at: https://schoolsweek.co.uk/3m-on-and-where-are-all-the-master-teachers/ [Accessed 16 Dec. 2017].

Dominiczak, P. (2013). *Michael Gove: New curriculum will allow my children to compete with the very best* [online]. Available at: www.telegraph. co.uk/education/educationnews/10166020/Michael-Gove-new-curriculum-will-allow-my-children-to-compete-with-the-very-best.html [Accessed 14 Dec. 2017].

Gehlhaar, R. (2002). Music technology and sound art/interactivity: Reality music/ virtual music. In *25th national conference of the Musicological Society of Australia (MSA)*. Newcastle, NSW. Available at: www.gehlhaar.org/x/doc/ musictechnologymsa.doc [Accessed 10 Jan. 2018].

Hallam, S. (2010). The power of music: Its impact on the intellectual, social and personal development of children and young people. *International Journal of Music Education*, 28(3), pp. 269–289.

Hawkins, W. (2012). *Pythagoras, the music of the spheres, and the wolf interval* [online]. Available at: http://philclubcle.org/papers/Hawkins,W20111115.pdf [Accessed 14 Dec. 2017].

Holland, S., Wilkie, K., Bouwer, A., Dalgleish, M. and Mulholland, P. (2011). Whole body interaction in abstract domains. In: D. England, ed., *Whole body interaction*. Human-Computer Interaction Series. London: Springer Verlag, pp. 19–34.

Humphreys, S., Davies, R. and Dorling, M. (2014). *CAS master teacher programme* [online]. Available at: http://community.computingatschool.org.uk/ files/4560/original.pdf [Accessed 22 Dec. 2017].

Innes, K. (1997). Using music technology at key stage 3. In: *British educational research association annual conference*. University of York. Available at: www.leeds.ac.uk/educol/documents/000000329.htm [Accessed 14 Dec. 2017].

John, M. (2013). *Teachers numbers fall but DfE confident on "Computing"* [online]. Available at: www.agent4change.net/policy/curriculum/2029-teacher-numbers-fall-but-dfe-confident-on-computing.html [Accessed 10 Dec. 2017].

John, M. (2014). *More than half teachers "not confident in Computing"* [online]. Available at: www.agent4change.net/policy/curriculum/2264-more-than-half-teachers-not-confident-in-computing.html [Accessed 10 Dec. 2017].

Johnson, T. (2006). Self-similar structures in my music: An inventory. In: *MaMuX seminar*. Paris: IRCAM-Center Pompidou. Available at: http://repmus.ircam.fr/_media/mamux/saisons/saison06-2006-2007/johnson-2006-10-14.pdf [Accessed 10 Jan. 2018].

Kemp, P. (2014). *Computing in the national curriculum: A guide for secondary teachers* [online]. Available at: www.computingatschool.org.uk/data/uploads/cas_secondary.pdf [Accessed 14 Nov. 2017].

Kochhar-Lindgren, K., Schneidermann, D. and Denlinger, T. (2009). *The exquisite corpse: Chance and collaboration in surrealism's parlor game*. Lincoln and London: University of Nebraska Press.

Leyshon, A. (2009). The software slump? Digital music, the democratisation of technology, and the decline of the recording studio sector within the musical economy. *Environment and Planning A*, 41(6), pp. 1309–1331.

Loy, G. (2006). *Musicmathics: The mathematical foundations of music* (Vol. 1). Cambridge, MA: The MIT Press.

Nielson, J. (1995). *10 usability heuristics for user interface design* [online]. Available at: www.nngroup.com/articles/ten-usability-heuristics/ [Accessed 2 Jan. 2018].

Paul Hamlyn Foundation. (2014). *Inspiring music for all: Next steps in innovation, improvement and integration* [online]. Available at: www.phf.org.uk/reader/inspiring-music/key-issues-challenges/ [Accessed 2 Jan. 2018].

Payne, C. (2018). *MAMIC* [online]. Available at: https://github.com/chrispayne1/MAMIC [Accessed 15 Jan. 2018].

Puckette, M. (2017). *Software* [online]. Available at: http://msp.ucsd.edu/software.html [Accessed 2 Jan. 2018].

Rogers, Y. (2011). Interaction design gone wild: Striving for wild theory. *Interactions*, 18(4), pp. 58–62.

Sandred, O., Laurson, M. and Kuuskankare, M. (2009). Revisiting the illiac suite—a rule-based approach to stochastic processes. *Sonic Ideas/Ideas Sonicas*, 2, pp. 42–46.

Sawyer, B., Forsyth, J., O'Connor, T., Bortz, B., Finn, T., Baum, L., Bukvic, I., Knapp, B. and Webster, D. (2013). Form, function and performances in a musical instrument MAKErs camp. In: *44th ACM technical Symposium on Computer Science Education (SIGCSE'13)*. Denver, CO: ACM Press. Available at: http://citeseerx.ist.psu.edu/viewdoc/download?doi=10.1.1.717.2615&rep=rep1&type=pdf [Accessed 20 Dec. 2017].

Selby, C. and Woollard, J. (2014). Refining an understanding of computational thinking. *Technology, Pedagogy and Education*, 2013, pp. 1–23.

Tenney, J. (2008). On "Crystal Growth" in harmonic space (1993–1998). *Contemporary Music Review*, 27(1), pp. 47–56.

Thorley, M. (2005). Music technology education: Who is the customer, the student or the industry. In: *Leeds International Music Technology Conference (LIMTEC)*. Leeds: University of Leeds.

Tymoczko, D. (2011). *A geometry of music: Harmony and counterpoint in the extended common practice.* New York: Oxford University Press.

Ubuntu Studio. (2017). *Ubuntu studio* [online]. Available at: https://ubuntustudio. org/ [Accessed 12 Jan. 2018].

Wing, J. M. (2006). *Computational thinking* [online]. Available at: www.cs.cmu. edu/~15110-s13/Wing06-ct.pdf [Accessed 14 Nov. 2017].

Wright, D. (2009). *Mathematics and music* [online]. Available at: www.math. wustl.edu/~wright/Math109/00Book.pdf [Accessed 19 Dec. 2017].

Year of Code. (2015). *What is year of code?* [online]. Available at: www.yearof-code.org/ [Accessed 17 Dec. 2017].

19

Interaction-Congruence in the Design of Exploratory Sonic Play Instruments With Young People on the Autistic Spectrum

Joe Wright

INTRODUCTION

Through my work as a performer and musical director in inclusive performances, I have found that the preferences for sound and the means of engagement with music shown by our audiences have been very diverse (Street et al., 2015; Griffiths, 2017a). Unconventional sounds or musical communication styles, inspired by my practice in experimental music, have often been helpful in engaging with young people with complex needs, particularly those on the autistic spectrum. These experiences mirror Shaughnessy's observations of inclusive theater that "the children were [also] open to theatre in ways that neuro-typicals [*sic*] are sometimes not" (2012, p. 242). There is room among existing musical resources for these groups—such as the UK government's P Scales (2017), Sounds of Intent (Ockelford, 2013), and Youth Music's Quality Framework (Youth Music and Drake Music, 2015)—for additional materials that would support exploratory or experimental musical play. Such resources might facilitate musical collaboration between adults and neurodiverse audiences in inclusive performances, or simply provide young people with more choices for self-directed musical play. Thus, the research outlined in this chapter aims to explore the following questions:

- How are preferences for music acquired and how does this apply to non-verbal young people on the autistic spectrum?
- Do interaction styles from conventional and non-conventional music making affect the development of instrumental skills in naïve users—a small group of young people on the autistic spectrum?
- What features, if any, of the prototype instruments designed for the pilot study might encourage young people on the autistic spectrum to explore sounds (individually or collaboratively)?

MUSICAL PREFERENCE

The questions of the origins of musical perception and preference are still open for debate, and it is beyond the scope of this chapter to provide a full account of this field. There is, however, growing evidence that musical preferences are culturally and experientially, rather than biologically, acquired (McLachlan et al., 2013; Dellacherie et al., 2011; Kessler et al., 1984; Phillips-Silver and Trainor, 2005). This is reflected in recent studies with typically developing infants and children. Contrary to previous studies, Plantinga and Trehub (2014) found no evidence of innate preference for consonance or dissonance in three- to six-month-old infants. Instead, babies only attended to the more familiar stimulus. Similar results are seen in studies on rhythmic perception. At seven months, babies showed no preference for simple or compound meter. Only prior training, where the subject was bounced in simple or compound meter on the lap of a parent to an unaccented pulse, affected the length of time that infants attended to various rhythmic patterns (Phillips-Silver and Trainor, 2005). Beyond infancy, children typically gain an understanding of harmonic and metric structures relevant to their culture by the age of four (Ockelford, 2013). At six years, this knowledge does not yet affect preferences for sounds outside of that culture; such preferences emerge around the age of nine (Plantinga and Trehub, 2014).

The literature outlined above suggests that for typically developing people, musical structures are typically learned in childhood and become preferences that affect choices for and the perception of music in adulthood. These preferences are learned through exposure to a particular culture.

How might the above theories of socially acquired musical preferences apply to young people on the autistic spectrum? Repetitive or stereotyped behaviors, narrow interests and difficulties with non-verbal communication, social cues, social conventions, language development and social play are typical characteristics of (and diagnostic criteria for) autism spectrum disorder (Autism Speaks, 2013; American Psychiatric Association, 2017; World Health Organisation, 2016; Boucher, 2008). This research project cannot possibly carry out studies into the aesthetic preferences of young people on the autistic spectrum in general. But given the link between autism and the late onset of social and linguistic skills, it cannot be assumed that a non-verbal young person on the autistic spectrum will have the same sonic/musical preferences as a typically developing adult. I suggest that, in some settings, conventional musical structures can risk inhibiting the novice participant and may bias an experienced performer away from auditory events that could go on to form the basis of rich musical experiences. It is for this reason that my study draws on improvised and collaborative forms of experimental music as a source for ideas and for additional means of contextualizing the actions of young people on the autistic spectrum.

COLLABORATIVE AND INSTRUMENTAL PRACTICE IN EXPERIMENTAL MUSIC

Just as experimental practices have been instrumental in successfully reaching neurodiverse audiences in inclusive theater, I believe that musical experimentalism can also be further explored for its potential to facilitate engagement and collaboration. This could build on child-led practices such as sound therapy (Ellis and van Leeuwen, 2000) and intensive inter-action (Hewett et al., 2011), which use the actions of the learner as the basis for joint activities.

The nature of interacting in experimental music forms is necessarily different in this field of musical practice. Modes of interaction that are informed by established musical conventions are often unhelpful or super-fluous in the face of highly unusual approaches to music making. Where conventions do exist, they often arise from close collaboration in small groups and communities. Without the dependencies on established musi-cal techniques or structures, experimental forms of music can have the flexibility to accommodate a breadth of highly individual practices and communication styles, often within a single group.

The improvising trio AMM and more recent community surround-ing Prévost's workshop (2015) serve as good examples of this plurality. Their approach is not actively concerned with pre-existing techniques or structures, but rather with the act described by Cardew as "searching for sounds, and the responses that attach to them" (1971, p. xviii)—a process of exploring sound-making materials along with other people (Prévost, 1995). Seymour Wright (2016) uses the work of AMM as an example of *group learning*, where factors beyond the musical experience of each group member (social context, location, non-musical skills etc.) contribute to the emergence of a unique practice. Indeed, my own musical practice has been shaped by my participation in Prévost's workshop and work with members of its wider community. Though the above examples represent one small area of a much wider field, I have seen first-hand how exper-imental practices can allow diverse groups of musicians to collaborate, regardless of their varying experiences and musical preferences.

As well as nurturing unique collaborative approaches, experimental music forms also afford players the flexibility to develop highly atypical approaches to their instruments or source material. From the late 1990s on, Sachiko M began using only the sinusoidal test tones of her sampling instrument (Jerriel & Fox, 2009). While the repeatability of these tones was adopted by Sachiko M as a consistent part of her musical identity, they also highlight the acoustic differences or discrepancies in equipment or surrounding space from performance to performance, something which can't be controlled (Walls et al., 2003, p. 4; Cox and Warner, 2017, p. 206). Conversely, Nakamura's practice with no-input mixer feedback (Meyer, 2003), the chaotic Blippoo Box synthesizer (Hordijk, 2015) and Gurev-ich's mechatronic instruments (2014) are all examples where control is devolved, at least to some extent, to the instrument—again allowing the performer to work with a degree of uncertainty in his/her music.

In the above examples, musical experimentalism has allowed perform-ers to construct unique instrumental practices and communication styles. For many young people on the autistic spectrum, the social contexts and the concepts behind experimental music may be just as inaccessible as formal or technical expectations found in conventional music forms (if not more so!). However, the collaborative skills and openness to unusual instrument usage in experimental music are very compatible with inclusive performances with neurodiverse audiences. Regrettably, initiatives such as Sound to Music (Sound to Music and Yoshihide, 2013), which brought together experimental musicians and young people with complex needs, are still surprisingly rare.

JACK'S RITUAL

I was fortunate to be involved in another initiative of this kind. *Sound Symphony* brought together an interdisciplinary team of experimental artists for a research project focused on sound in performances for young people on the autistic spectrum (Griffiths, 2017b). Throughout the project, we recorded many examples of students engaging spontaneously with various objects, including spinning bowls, paper strips, bottles of water or glitter and so on. The ritual-like actions of one student in particular (hereafter given the pseudonym 'Jack') stood out to the team.

Jack would spend long periods of time with a Slinky, holding it at one end above his head such that it formed a colorful column, bouncing it from the top downwards (as depicted in Figure 19.1). On closer observation, this was only the surface of a complex ritual. Bouncing would start at a certain pace, which was then maintained steadily. The bottom end of the Slinky would bob up and down as it compressed and expanded in waves corresponding to and interfering with Jack's regular bouncing motions. After a while, Jack's regular bouncing would finish with a more rigorous downward push, resulting in a double bounce of the Slinky on the floor. Jack would then move to a different part of the classroom and repeat the activity. I noticed that despite being repetitious, he was moving to parts of the classroom that had different textured floor surfaces. To me this suggests that a variation in the sensations or sounds caused by the Slinky's contact with the floor was important to Jack's ritual. As well as being intrinsically rich, the movement around the classroom suggested a possible exploratory element to the activity.

I managed to imitate Jack's ritual myself, albeit after some practice. I found that the phase differences between my hand bouncing the Slinky's top end and the rhythm of the Slinky's bottom end hitting the floor quite fascinating: the exact responses of the toy were somewhat unpredictable. When I also attended to the granular sounds coming from the plastic spring itself, the experience was yet richer.

I cannot claim that Jack performed his ritual with the same intent or with a focus on the same stimuli as I had on imitating it. It did, however, require skill to imitate and was certainly far more complex than I had first assumed (I urge you to try it). By viewing Jack's ritual through the lens of my own practice, I at least see that labelling such activities merely as repetitive or stereotyped risks devaluing the potential richness of that activity. Imitating the gestures described above did not bring me closer to understanding Jack's intent or interest, but I did gain a better understanding of the object's behavior and how it reacted to my own gestures. Its semi-incongruent responses (a concept that is discussed below) are a

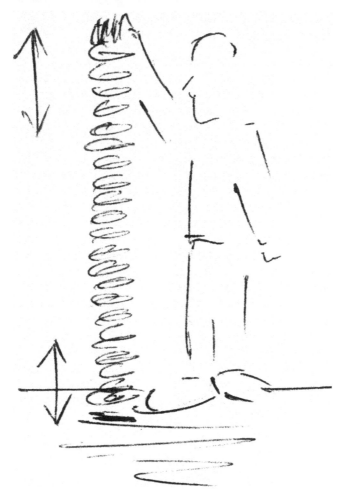

Figure 19.1 Field Sketch of Jack's Activities With a Slinky

feature that was common to many objects that engaged the students we met in the project. Interactions with water bottles, for instance, were frequently observed and felt similar in nature to me. I cannot help but wonder what might have developed if, given more time, I had learned to use the Slinky well enough to collaborate with Jack in his ritual at the school, and am curious as to where this collaboration might have led.

While they may have their differences in their intentions and contexts, the musical work described above and activities such as Jack's ritual involve some degree of play and curiosity. Actions or practices are carried out willingly or spontaneously, for enjoyment, and are at the center of a person's interests or preferences. How, then, can we draw upon knowledge in these areas to create tools that better encourage exploratory play in young people on the autistic spectrum?

INCONGRUENCE, CURIOSITY AND SPONTANEOUS PLAY

The above examples show how chaotic objects or sound-sources can be central to both exploratory musical activities and spontaneous behaviors like Jack's. The conflict between expectation and perception that such objects can cause is central to theories of curiosity and attention. Hebb (1949, pp. 149–151), for instance, proposes the following relationship between perceptual incongruity and curiosity (shown in Figure 19.2): boredom and fear are typical responses to extremely low and high levels of conflict respectively, and moderately incongruent stimuli are optimal for curiosity. Studies with infants show a similar relationship. The *discrepancy hypothesis* (McCall and McGhee, 1977) suggests that in infants, the level of unfamiliarity or discrepancy is correlated with attention in this way. The *Goldilocks effect* found that infants were significantly more likely to look away from highly familiar or highly unfamiliar stimuli (Kidd et al., 2010). These latter two studies do not specifically pertain to young people on the autistic spectrum but *do* involve young people who have yet to develop the preferences and language skills of typically developing adults.

Given that open collaboration without explicit instruction is a goal of this study, it is important to consider the play behaviors of people on the autistic spectrum relative to typically developing individuals. Research on spontaneous play by Libby et al. (1998) found that there were no significant differences in the amounts of basic play and exploratory behaviors of autistic and typically developing children. This study also found, along with others, that symbolic play (also called social or pretend play) in autistic infants is impaired but can be encouraged through structured activities

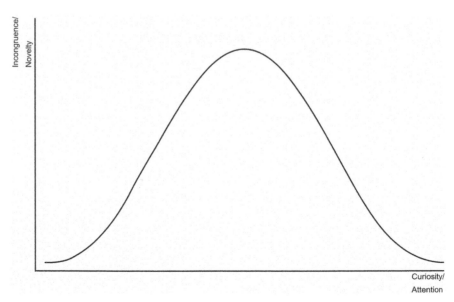

Figure 19.2 Responses to Increasingly Novel or Incongruent Stimuli

and imitation (Libby et al., 1998; Ungerer and Sigman, 1981). The exploration of objects through non-representational play should therefore not be significantly inhibited in young people on the autistic spectrum, though support may need to be given by accompanying adults if the exploratory activity is later used symbolically. Additional support would also be valuable if Lewis and Boucher's *generativity-deficit* hypothesis (1995), that young people on the autistic spectrum may struggle to conceive or initiate spontaneous actions, is correct. Finally, there are multiple studies that stress the importance of the breadth of interest in toys during play and its correlation with later communicative developments, language and skill acquisition (Klintwall et al., 2014; Vivanti et al., 2013; Ungerer and Sigman, 1981; Charman et al., 2000; Kasari et al., 2006). This may mean that exploratory musical play might have some tangible benefits even if it is not compatible with mainstream musical activities.

INITIAL DESIGN IDEAS

From the above theories and studies, important criteria can be gathered that may support exploratory sonic play and any resources that can be designed for it. Prototype instruments or activities requiring prior symbolic knowledge should be avoided. Instead, devices with very clear means of interaction (buttons, for example), or with components (physical or auditory) that can be combined in basic play behaviors are preferable. Conventional instruments give exact responses to exactly repeated gestures. For some, this may impose inaccessible modes of listening and acting that are embedded within some form of pre-defined musical structure. By studying and reproducing the semi-incongruent behavior of objects such as Jack's Slinky, it may be possible to offer additional possibilities for interactions that do not fit conventional (and congruent) musical situations.

AN INTERACTION-CONGRUENCE MODEL FOR TECHNIQUE AND EXPECTATION IN INSTRUMENTAL PERFORMANCE

I have been exploring the use of semi-incongruent instruments in my own practice, through the development of a dynamic feedback system for the saxophone, designed to disrupt conventional (including extended) techniques for playing the instrument. The system (shown in Figure 19.3) is relatively simple, yet can lead to quite complex and unpredictable results, creating an active two-way relationship between performer and instrument: the performer's movements and acoustic tone affect the feedback, and the feedback interferes with the harmonic stability of the acoustic saxophone playing. With this system, the saxophone can be played in unusual configurations similar to those of keyboard, synthesizer or percussion instruments.

The disruption and unusual configurations the system provides often result in complex or unexpected responses that would be difficult to achieve with the saxophone alone. These responses were often the impetus

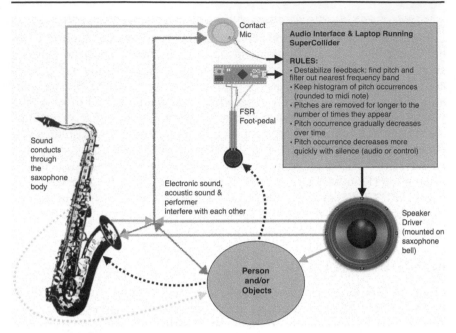

Figure 19.3 Diagram of an Electroacoustic Feedback System for Experimental
Saxophone Playing

for intense sonic explorations in performances. My experiences of this
system and Jack's Slinky developed in a similar fashion. In both cases,
I was able to learn and utilize skills, but the results of my actions were
complex—broadly identifiable but internally chaotic. What follows is an
attempt to describe this type of interaction—from my own perspective—
in terms of expectation, technique, cognition and auditory perception, and
how this differs to a conventional instrumental approach.

For the reasons outlined above, established musical terms will not aid
the research project in describing these interactions. The model shown in
Figure 19.4 describes performer-instrument interactions in three ways: by
the technical demands they require and by the conflict between expected
and perceived sounds produced by them on both general and specific
scales. These forms of conflict—*stream* and *event incongruence*—are
used to distinguish between the expectation of and attention to sounds in
general, and the details therein.

Bregman and Pinker define an auditory stream as "a psychological organization whose function it is to mentally represent the acoustic activity
of a single source over time" (1978, p. 19). The pairing of streams with
individual sound-sources, however, is fallible and can be tricked under
certain conditions. The *many streams illusion* is one such case, exemplified by Parker's later solo saxophone works (illustrated in Figure 19.5). If
they were to be played slowly, the sounding tones in this excerpt (Parker,
2010) would be perceived as a single auditory stream, but the rapid pace
of the emergent tones in Parker's playing give the illusion of multiple

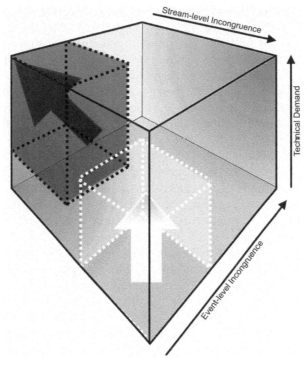

Figure 19.4 A Model of Incongruence and Technique in Performer-Instrument Interactions

Figure 19.5 The Many Streams Illusion in Evan Parker, *Whitstable Solo* (ii), 1'03"-1'07"

streams in different registers. According to this view of auditory streams, the multiple streams perceived in this example would suggest multiple objects. O'Callaghan's (2008) later description of auditory streams does not attempt to ascribe auditory streams directly to single source-objects. Instead, he argues that auditory streams are perceptual objects and that they convey information about events that are happening to a source-object. Such events are not limited to individual streams. In the excerpt shown in Figure 19.5, then, we can think of Parker's interactions with the saxophone as a sequence of events that are experienced through corresponding sets of auditory streams.

The many streams illusion may violate expectations by 'tricking' the listener into perceiving multiple voices from a single performer. The parsing of each stream may also vary between listeners, or change for an individual over time. These complex events are therefore open to multiple hearings, just as Ihde's (2007, p. 188) multistable diagram can be visually interpreted as a protruding or retracting corridor, or that two different aspects can be seen in the duck-rabbit image popularized by Wittgenstein (2010). I suggest that Parker's solo practice is an example of auditory multistability: the long continuous passages of his saxophone solos could be single events, or they could be dense collections of shorter events; the note *E*, written on the bottom line of the stave in Figure 19.5, might be grouped with the middle stream or the bottom stream.

Through performance with the feedback system described above, I have attempted to incorporate these concepts, describing them by technique, the incongruence arising from the experience of an action as a whole (events), and the incongruence caused by the details they convey. To illustrate how the model can represent interactions, let us take another example, shown in Figure 19.6.

In this excerpt (which can be streamed online at (Wright, 2017)), six repetitions of a known saxophone multiphonic are played. The saxophone is first played conventionally, with no additions, and then the feedback system is added. In conventional playing, through practice, I learn to expect the sounds that correspond to physical actions; I can *hear* the sound of the multiphonic in my auditory imagination prior to executing the action. This is perhaps comparable to Ihde's description of inner speech as auditory imagination (2007). Provided the necessary skills are learned, the multiphonic represented in Figure 19.6 can be produced to closely match the expected sound. Under these circumstances, the relationship with the instrument is congruent with expectations. Interactions of this kind can be represented in the darker shaded area in Figure 19.4. Interactions coupled with anticipated sounds are delivered through established techniques, producing expected results. This kind of congruent instrumental practice trends towards maximum replicability and demands technical mastery. Conversely, the addition of the feedback system to the same multiphonic gives rise to a different kind of interaction. Before a note is played with the system, I still have the 'sound' of the fingering in my mind. The semi-chaotic responses of the system, however, ensure that however skillful I may be, the instrument's responses do not seem the same. I may be able to learn and expect an approximate type of response (certain pitches emerge in all three repetitions), but the exact details that emerge within that general response are too difficult to predict. Active streams around certain pitches or registers may be expected, but the change in their emphasis on repeated renditions may allow a performer or listener to experience different aspects of the sound at different times. Exact repetitions through honed technique become irrelevant. In other words, the system is stable at the event-level, yet the auditory streams produced are chaotic, and the results of pre-existing techniques become uncertain. Semi-incongruent interactions of this nature are represented by the lighter shaded area in Figure 19.4.

By aiming to design instruments that function in this region, I hope to produce the moderate incongruence that has been linked to curiosity

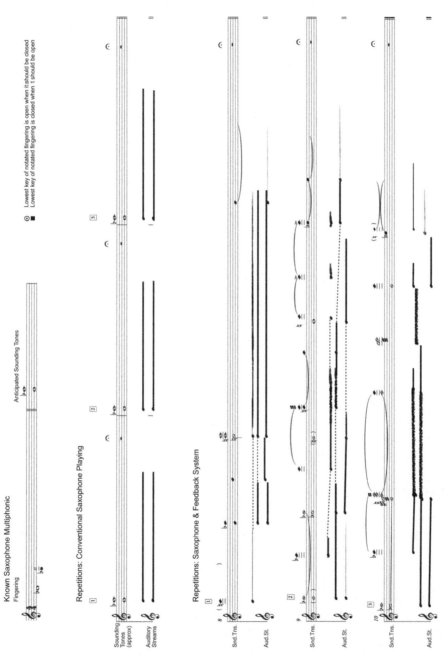

Figure 19.6 Congruent and Semi-Incongruent Responses From a Known Saxophone Multiphonic

and facilitate exploration in sound. If event-level incongruence and technical demands are low, an instrument might remain accessible; causal relationships between actions and sounds can still be understood, and no predefined technique is strictly necessary. The uncertainty over the component streams of an instrument's responses may be moderately surprising or cause these instruments to become multistable. Can instruments with this behavior become part of rituals like Jack's? Might it offer young people on the autistic spectrum another means of interacting with sound, and with other people?

INITIAL DESIGN OF A PROTOTYPE INSTRUMENT

A simple starter interface was devised from features of existing accessible musical instruments with the aim of creating something that was simple to use, needed little or no instruction, and would provide as little distraction from the instruments' sonic behavior as possible. These features and their corresponding instruments are shown in Table 19.1.

Using pliable cubic food containers and the Bela low-latency audio platform (Bela.io, 2016), two starter instruments were constructed according to these requirements (shown in Figure 19.7). This initial prototype was far larger than intended, falling short of the requirements for a small instrument. Regrettably, this was necessary to allow for rapid prototyping, fixes and changes at the early stages of user testing.

The instrument was designed to discriminate between three types of gesture: glancing strikes or taps, small area presses and large area presses. Two versions of the sound-generating software were then developed which allowed the interface to respond to these gestures in two different ways: one suited to a congruent (C) interaction style and one designed for a semi-incongruent (Si) interaction style. As far as possible, the timbre of both versions was restrained to a similar palate of synthesized sine, AM and FM tones, so that the two could be compared for their behavior rather than the sounds they make.

EARLY USER TESTING

This prototype was tested in a special school with seven students (six males, one female; 5–17 years old). All participants have a formal diagnosis of autistic spectrum disorder and are either non-verbal or pre-verbal. A more even gender balance was desired but not possible: of the students at the school who met the above criteria, few were female (which is expected, given that autism is more common in males (Boucher, 2008)). Further issues with permissions and attendance left only one female student that could take part in the project. After two weeks of observations and a week of mock test sessions to help students ease in to a new routine, the instruments were tested over six 10–15 minute sessions. Initially the session structure was identical for each participant. This structure was loosely based on the play research listed above, sessions would take place in an empty space and the object to be tested was placed in the center of

Table 19.1 Design Features Gathered From Existing Accessible Instruments

Design Feature	Expected Benefits	Example Instruments
Embedded Speaker	• Helps to locate sounds with interactions on the instrument	Musii (musiisystems, 2014) Vesball (Nath and Young, 2015)
Cube Shape	• Can be rested on a flat surface - doesn't need to be handheld. • Easy to modify/maintain • Looks the same regardless of orientation	D-Box (Zappi and McPherson, 2014) Skoog (Skoog Music, 2016)
Large Rounded Button	• Simple to operate, little instruction required • Open to different playing styles • May afford poking, pushing, striking etc.	Skoog (*ibid*)
Single Input	• Low-dimensionality may encourage unique or exploratory responses to instruments	D-Box study on dimensionality & appropriation (Zappi and McPherson, 2014b)
Hidden Details	• More able users may discover subtle features that allow for more complex/detailed interactions	Orphion (Trump and Bullock, 2014)
Small Form Factor	• More accessible to younger users and people who struggle to grasp larger objects	Sense Egg (Blatherwick and Cobb, 2015)
Soft Materials	• Safe and tactile	Musii (*ibid.*) NoiseBear (Grierson and Kiefer, 2013) Sense Egg (*ibid.*) Skoog (*ibid.*) Vesball (*ibid.*)
Affordable Materials	• Reduces potential financial barriers to the instruments developed • Allows for the production of more than one prototype for the pilot study	NoiseBear (Grierson and Kiefer, 2013) Vesball (*ibid.*)

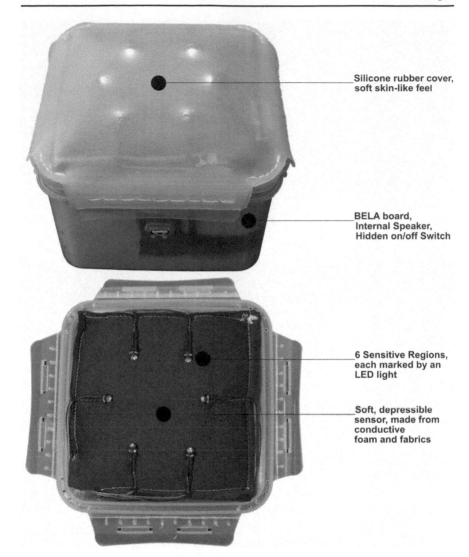

Silicone rubber cover,
soft skin-like feel

BELA board,
Internal Speaker,
Hidden on/off Switch

6 Sensitive Regions,
each marked by an
LED light

Soft, depressible
sensor, made from
conductive
foam and fabrics

Figure 19.7 Simple Prototype Instrument

the room. Participants would be left to interact without guidance from any adults, so that the students' own choices and preferences could be observed. Each week the same reference object (a favorite toy or sound-making object) would be presented for the first five minutes of the session; this would provide a baseline for differences in mood or behavior from week to week. After the reference object, the prototype instrument would be presented in the same way, lasting 10–15 minutes. The behavior of the prototype (congruent/semi-incongruent) and the range of feedback (sound alone, sound and light etc.) would change from week to week. The order in which the behaviors were introduced also differed from pupil to pupil, so that any general trends in the collected data would not be affected by order bias. Sessions would be recorded with a compact HD video recorder placed discreetly in the space. Notes were to be taken immediately after each session and a week later after a review of the video recordings. Analysis of video footage would document the time spent interacting with an instrument, attending to the instrument and not attending to the instrument, for both reference and prototype instruments. Contextual and qualitative documentation of responses were to be used to show evidence of changes in behavior as prototypes changed, and possibly show larger trends occurring in the sessions for all participants.

As the initial test sessions progressed, it became evident that this method—while it had worked for play research—was not entirely suitable for gathering data about the instruments and the potential responses they might provoke. One pupil struggled in the space allocated for the sessions; the last four of her sessions were moved to a smaller, quieter space. Three pupils struggled because of a lack of direction from or interaction with the staff and myself. Later sessions with these students allowed for the use of simple communication, joint attention cues and intensive interaction, which greatly improved the amount of engagement with the reference objects and instruments. In one case, it became clear that before the instruments could be tested, the participant needed to learn that the objects in the sessions could be interacted with. This process took the full six weeks, and insights about that person's responses could only be investigated through additional sessions. Some students began to tire of the reference objects and began to seek out the prototype instrument before the five-minute period had elapsed. As a result, numerical data for these sessions proved to be a poor indicator of mood or engagement without context from the written notes. Smaller variables—such as the position of objects in the room, use of chairs or beanbags, lighting—also changed for each participant. This helped in trying to create an optimum environment for play and engagement, but also limited the potential for the comparison of data between participants in the study. The resulting method, therefore, has more in common with the collaborative approach to experimental music making than it has with methodology for play research. Future sessions will aim to draw even more from this field, to maximize the time participants spend interacting with the prototype instruments.

RESPONSES TO CONGRUENT AND SEMI-INCONGRUENT INTERACTION STYLES

It is too early to determine whether any of the students have a strong preference for either of the two interaction styles supported by the instrument. Due to the methodological changes outlined above, it may be beyond the scope of this pilot study to rigorously investigate this (at least through the collection of quantitative data) without compromising the collaborative approach taken so far. Further testing and refinement of the test-session structure will be needed, but some early observations are outlined as follows.

The participants in general do not seem to have responded negatively to the Si software. If the participants as a group had a strong culturally learned bias towards conventional forms of music making, I would have expected a larger proportion of the participants to have reacted negatively to the semi-incongruent software.

Only one student responded, showing an apparent preference for the C over Si styles. This participant found the lack of direction in early sessions very difficult, responding mainly in later sessions to prompts from the staff and me. It is possible this student's preference was due to the clarity in copying that the C style offers (something which is discussed further below). He is also known to have an aversion to high frequencies, and his preferences might also be down to the differences in high frequency activity offered by the two versions of the instrument—the C software was less active in high registers.

Of the four students who responded consistently to the instruments, three showed signs of interest, enjoyment and focused engagement with both interaction styles. Two showed very positive responses to the Si style, especially when combined with visual feedback. Most notably, one participant has developed a playing style which involves placing the instrument, oriented sensor-down, on the floor, rocking it periodically and sitting back to listen to the responses of the instrument (in this configuration the instrument can activate itself under its own weight). This student has reacted positively to the changing and slightly chaotic responses that the Si software provides. He will often rock the instrument repeatedly, listening to the sounds produced each time, until he finds a sound he likes. On finding a sound, he then leans back to listen to the sound as it evolves over time. It seems reasonable to assume that some form of choice is being made here: sounds are auditioned before one is chosen, which can then be listened to. The Si software version of the instrument was much more conducive to this activity than the C style, because of the greater variety it provided.

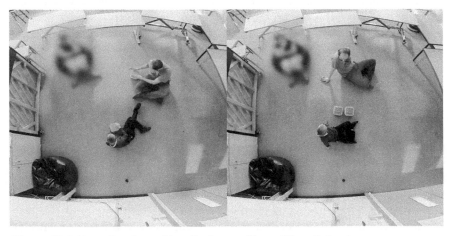

Figure 19.8 Duet Session with a Student at Three Ways School

In the duet sessions (where I would play a second copy of the proto-type), my interactions with students through copying and responding felt very different with each version of the software. I found that with the C software I would try to copy students' actions *and* the sounds, which required greater technical precision. There was also a sense of right and wrong ways to play in this context, which could be useful for skill devel-opment but also provide a barrier to equal collaboration. Conversely, the Si software—by providing changing responses to identical actions—allowed for copying without a right/wrong judgement. I felt that I could copy a par-ticipant as best I can, focusing on listening to the resulting sounds as they emerged and the responses that the sounds provoked. As the Si software results in more complex sounds, it was difficult at times to distinguish what I was producing from the sounds of my duet partner. Copy games were perhaps more difficult as communication in this case felt less direct, but this also felt less invasive for a few of the students involved in the study.

DESIGN SUCCESSES, PRACTICALITIES AND SHORTFALLS

As anticipated, modular parts and easy access to the prototypes' internal components was essential: changes and fixes could be made with little difficulty as the sessions went on. Features gathered from the instruments shown above were effective as anticipated. The soft compressible button generally afforded striking or pressing and was a good primary means of input for the prototype instruments. The cube shape allowed the instru-ments to be rested on floors, chairs and other surfaces (heads, chests and feet!), allowing users who might not be able to hold the prototypes to play comfortably. Hidden layers—provided by three gesture classes and mul-tiple sensitive regions within an apparently singular sensitive surface—accommodated both broad and fine-grained gestural playing styles.

The auditory responses of the prototypes had been limited to simple syn-thesized tones in the sessions. This had excluded one student who seemed disinterested (or perhaps averse) to these sounds. Given the potential to change the sound-output in the sessions, it might have been possible to determine whether other timbres might have achieved better results.

The haptic feedback provided by the embedded speaker was deeply engaging to many of the participants. The gestures in both software types resulting in low frequency/high volume, where this haptic feedback is most intense, were among those that were frequently repeated by the stu-dents. Discovery of vibration-rich responses often coincided with longer or uncharacteristic phases of engagement with the prototypes and positive vocalizations or facial expressions. At this stage in the research, it is too early to tell whether those students that do seek out vibration-rich sounds will settle into a routine of using mostly known gestures and exploring less, or whether the haptic feedback can enhance the attention given to sound exploration. At the very least, the feedback provided by the embed-ded speaker does not seem to have affected participants' preferences for how the instrument is oriented relative to them, as the low-frequency vibrations permeated the prototypes casing and sensor-lid alike.

The prototype used in this phase of tests was an approximate 15 cm cube. Although this is not very large when compared to some conventional instruments, the size relative to some of the younger students may have limited their engagement. The youngest participant in the study, for example, had engaged very little with the instrument presented in the sessions. He did, however, interact more freely with smaller objects; particularly a small wireframe ball with a bell inside, which could easily be grabbed and manipulated by small hands.

Interestingly, although visual feedback provided by LED lights radically improved engagement with the prototypes, it also narrowed the range of actions performed on the instruments. The lights were located at the sensitive regions of the sensor, and attention was drawn to these points as a result. Visual feedback aided participants in finding the active surface of the instrument, but it also limited the desirable orientations of it—students tended to play in such a way that the LEDs could still be seen.

CONCLUSION: FUTURE DEVELOPMENTS

The general session structure will remain the same in future testing sessions, but the method will be developed to accommodate more realistic use of the prototype instruments. Changes will be made that allow small group performance and collaborative play between students and staff to be explored.

As there was no clear preference among participants for the C or Si software, new prototypes will be designed that can provide both types of response. This may give users an opportunity to discover or choose a desired type of response from the instrument. The first prototypes were designed to produce a limited array of sounds, which were controlled by a very limited interface. The next phase of prototypes will explore whether more timbres coupled with multiple sensors, buttons and textured surfaces can be included to provide a greater range of possible interactions with the instruments, without sacrificing the simplicity of the initial prototypes. For example, future prototypes may include active surfaces that can accommodate new materials or parts selected or made by a teacher or parent.

As expected, visual and vibrotactile feedback greatly improved engagement with the prototype instruments. The non-directional nature of the haptic-feedback meant it did not seem to significantly change preferences for how the instrument was oriented. Conversely, it was surprising to see how much the added visual feedback biased students towards maintaining a line-of-sight with the prototype's embedded LED lights. In further iterations of the instruments, it will be necessary to consider how coherent multimodal feedback can be included without this limiting effect.

A significant number of the students preferred or required the involvement of an adult in the sessions. If possible, future sessions will explore how cooperative instrument pairs or modular systems might enable adult-student pairs of musicians to explore sounds.

Although far more time will be needed with the seven participants of this study to determine what features of the prototypes (if any) are effective in facilitating exploratory sonic play, positive responses have been observed. A new iteration of the prototypes will be developed in advance of a second wave of testing with the same seven participants, as will a phase of wider testing in schools and theater productions.

I would like to thank Luke Woodbury and the staff and students at Three Ways School, Bath, for their contribution to the project, without which the research outlined in this chapter would not have been possible.

REFERENCES

American Psychiatric Association. (2017). *What is autism spectrum disorder?* [online] Available at: www.psychiatry.org/patients-families/autism/what-is-autism-spectrum-disorder [Accessed 29 Aug. 2017].

Autism Speaks. (2013). *DSM-5 diagnostic criteria.* [online] Available at: www.autismspeaks.org/what-autism/diagnosis/dsm-5-diagnostic-criteria [Accessed 29 Aug. 2017].

Bela.io. (2016). [online] Available at: http://bela.io/ [Accessed 17 Oct. 2017].

Blatherwick, A. and Cobb, J. E. (2015, June 15). *SenseEgg: A wireless music controller for teaching children with special learning needs.* University of Leeds. [online] Bournemouth University, Fern Barrow, Poole, Dorset, BH12 5BB, UK. Available at: http://eprints.bournemouth.ac.uk/22205/ [Accessed 10 Aug. 2016].

Boucher, J. (2008). *The autistic spectrum: Characteristics, causes and practical issues.* 1st ed. Los Angeles: SAGE Publications Ltd.

Bregman, A. S. and Pinker, S. (1978). Auditory streaming and the building of timbre. *Canadian Journal of Psychology/Revue canadienne de psychologie*, 32(1), p. 19.

Cardew, C. (1971). *Treatise handbook: Including Bun no. 2 [and] Volo solo.* London: Edition Peters.

Charman, T., Baron-Cohen, S., Swettenham, J., Baird, G., Cox, A. and Drew, A. (2000). Testing joint attention, imitation, and play as infancy precursors to language and theory of mind. *Cognitive Development*, 15(4), pp. 481–498.

Cox, C. and Warner, D. (2017). *Audio Culture.* Revised ed: *Readings in modern music.* New York: Continuum International Publishing Group Inc.

Dellacherie, D., Roy, M., Hugueville, L., Peretz, I., Samson, S. (2011). The effect of musical experience on emotional self-reports and psychophysiological responses to dissonance. *Psychophysiology*, 48(3), pp. 337–349.

Ellis, P. and van Leeuwen, L. (2000). *Living sound: Human interaction and children with autism.* Regina, Canada: ISME commission on Music in Special Education, Music Therapy and Music Medicine, pp. 1758–1766. Available at: https://www.researchgate.net/publication/228548295_Living_Sound_human_interaction_and_children_with_autism [Accessed 14 Oct. 2018]

GOV.uk. (2017). *P scales: Attainment targets for pupils with SEN—GOV.UK.* [online] Available at: www.gov.uk/government/publications/p-scales-attainment-targets-for-pupils-with-sen [Accessed 21 Aug. 2017].

Grierson, M. and Kiefer, C. (2013). *NoiseBear: A wireless malleable instrument designed in participation with disabled children*. May 2013. [online] Available at: http://research.gold.ac.uk/8660/ [Accessed 10 Aug. 2016].

Griffiths, E. (2017a). *"Sound symphony" research and development by upfront performance: Experiments film*. [online] Available at: https://vimeo.com/213407929 [Accessed 24 Aug. 2017].

Griffiths, E. (2017b). *'Spinning bowls and milk bottle shoes': Thoughts on 'sound symphony' research and development process*. Upfront Performance Network. [online] Available at: https://upfrontperformancenetwork.wordpress.com/2017/05/18/spinning-bowls-and-milk-bottle-shoes-thoughts-on-sound-symphony-research-and-development-process/ [Accessed 23 Aug. 2017].

Gurevich, M. (2014). Distributed control in a mechatronic musical instrument. *NIME*, pp. 487–490.

Hebb, D. O. (1949). *The organization of behavior: A neuropsychological approach*. New York: John Wiley & Sons.

Hewett, D., Firth, G., Barber, M. and Harrison, T. (2011). *The intensive interaction handbook*. [online] Sage. Available at: https://books.google.co.uk/books?hl=en&lr=&id=f9N0NVkdnb8C&oi=fnd&pg=PP2&dq=intensive+interaction&ots=xlgolFZvmx&sig=j5n1SA9Mhu9tH1iPbo5laYBo8CU [Accessed 30 Aug. 2017].

Hordijk, R. (2015). *The blippoo box: A chaotic electronic music instrument*. [online] Available at: http://hordijk-synths.info/synth/2015/03/25/blippoo.html [Accessed 3 Jan. 2018].

Ihde, D. (2007). *Listening and voice: Phenomenologies of sound*. Albany: SUNY Press.

Jerriel & Fox. (2009). *Interview with Sachiko M.* [online] Available at: http://jameworld.com/uk/articles-58659-interview-with-sachiko-m.html [Accessed 30 Apr. 2017].

Kasari, C., Freeman, S. and Paparella, T. (2006). Joint attention and symbolic play in young children with autism: A randomized controlled intervention study. *Journal of Child Psychology and Psychiatry*, 47(6), pp, 611–620.

Kessler, E. J., Hansen, C. and Shepard, R. N. (1984). Tonal schemata in the perception of music in Bali and in the West. *Music Perception: An Interdisciplinary Journal*, 2(2), pp. 131–165.

Kidd, C., Piantadosi, S. T. and Aslin, R. N. (2010). The goldilocks effect: Infants' preference for stimuli that are neither too predictable nor too surprising. In: *Proceedings of the 32nd annual conference of the cognitive science society*. [online] pp. 2476–2481. Available at: www.bcs.rochester.edu/people/ . . . /ckidd/papers/KiddPiantadosiAslinCogSci2010.pdf [Accessed 1 Nov. 2016].

Klintwall, L., Macari, S., Eikeseth, S. and Chawarska, K. (2014). Interest level in 2-year-olds with autism spectrum disorder predicts rate of verbal, nonverbal, and adaptive skill acquisition. *Autism*, 19(8), pp. 925–933.

Lewis, V. and Boucher, J. (1995). Generativity in the play of young people with autism. *Journal of Autism and Developmental Disorders*, 25(2), pp. 105–121.

Libby, S., Powell, S., Messer, D. and Jordan, R. (1998). Spontaneous play in children with autism: A reappraisal. *Journal of Autism and Developmental Disorders*, 28(6), pp. 487–497.

McCall, R. B. and McGhee, P. E. (1977). The discrepancy hypothesis of attention and affect in infants. In: *The structuring of experience.* [online] Springer, pp. 179–210. Available at: http://link.springer.com/chapter/10.1007/978-1-4615-8786-6_7 [Accessed 1 Nov. 2016].

McLachlan, N., Marco, D., Light, M. and Wilson, S. (2013). Consonance and pitch. *Journal of Experimental Psychology: General,* 142(4), p. 1142.

Meyer, W. (2003). *Toshimaru Nakamura interview.* [online] Available at: www.furious.com/perfect/toshimarunakamura.html [Accessed 3 Jan. 2018].

Musii Systems. (2014). *Musii—multisensory interactive inflatable.* [online] Available at: https://musii.co.uk/ [Accessed 17 Oct. 2017].

Nath, A. and Young, S. (2015). VESBALL: A ball-shaped instrument for music therapy. In: *Proceedings of the international conference on new interfaces for musical expression,* pp. 387–391. Louisiana, USA.

O'Callaghan, C. (2008). Object perception: Vision and audition. *Philosophy Compass,* 3(4), pp. 803–829.

Ockelford, A. (2013). *Music, language and autism: Exceptional strategies for exceptional minds.* Philadelphia: Jessica Kingsley Pub.

Parker, E. (2010). *Whitstable solo. Audio CD.* Available at: www.emanemdisc.com/psi10.html

Phillips-Silver, J. and Trainor, L. J. (2005). Feeling the beat: Movement influences infant rhythm perception. *Science,* 308(5727), pp. 1430–1430.

Plantinga, J. and Trehub, S. E. (2014). Revisiting the innate preference for consonance. *Journal of Experimental Psychology: Human Perception and Performance,* 40(1), p. 40.

Prévost, E. (1995). *No sound is innocent: AMM and the practice of self-invention, meta-musical narratives and other essays.* Harlow, Essex: Copula Matchless Rec.

Prévost, E. (2015). *AMM—historical and theoretical precedents for the workshop.* [online] Available at: www.upprojects.com/media/uploads/workshoppiece-v.3.pdf [Accessed 31 Mar. 2016].

Shaughnessy, N. (2012). *Applying performance: Live art, socially engaged theatre and affective practice.* New York: Palgrave Macmillan.

Skoog Music. (2016). *Skoog in education | Educational sensory toy | Assistive music technology. Skoogmusic.* [online] Available at: http://skoogmusic.com/education/ [Accessed 29 Aug. 2016].

Sound to Music and Yoshihide, O. (2013). *SOUND to music—trailer. Film.* [online] Sound to Music and Yoshihide, O. 2013. Available at: www.youtube.com/watch?v=iMS2sXgL0hQ [Accessed 2 Jan. 2017].

Street, L., Wright, J. and Ellis-Howell, L. (2015). *Under foot.* [online] Available at: www.aboutnowish.com/uFoot.html [Accessed 23 Aug. 2017].

Trump, S. and Bullock, J. (2014). Orphion: A gestural multi-touch instrument for the iPad. In: *NIME.* [online] pp. 159–162. Available at: www.nime.org/proceedings/2014/nime2014_277.pdf [Accessed 23 Apr. 2017].

Ungerer, J. A. and Sigman, M. (1981). Symbolic play and language comprehension in autistic children. *Journal of the American Academy of Child Psychiatry,* 20(2), pp. 318–337.

Vivanti, G., Dissanayake, C., Zierhut, C. and Rogers, S. J. (2013). Brief report: Predictors of outcomes in the Early Start Denver Model delivered in a group setting. *Journal of Autism and Developmental Disorders,* 43(7), pp. 1717–1724.

Walls, B., Manikakis, E. and Gilroy, J. (2003). *Subsonics film*. [online] Sydney, NSW : SBS. 2003. Available at: www.youtube.com/watch?v=Tl8IMc-8-N8

Wittgenstein, L. (2010). *Philosophical investigations*. Oxford: Basil Blackwell Ltd.

World Health Organisation. (2016). *Chapter V Mental and behavioural disorders (F00-F99)*. [online] Available at: http://apps.who.int/classifications/icd10/browse/2016/en#F84.0 [Accessed 29 Aug. 2017].

Wright, J. (2017). *InstabilityExcerpt*. [online] Available at: https://soundcloud.com/joe-wright-music/instabilityexcerpt [Accessed 19 Oct. 2017].

Wright, S. (2016). Available at: www.seymourwright.com/#thesis. [online], www.seymourwright.com/#thesis [Accessed 5 Mar. 2016].

Youth Music and Drake Music. (2015). *Do, review, improve . . . A quality framework for music education*. [online] Available at: http://network.youthmusic.org.uk/resources/do-review-improve-quality-framework-music-education [Accessed 21 Aug. 2017].

Zappi, V. and McPherson, A. (2014a). *Design and use of a hackable digital instrument*. [online] Available at: https://qmro.qmul.ac.uk/xmlui/bitstream/handle/123456789/7204/MCPHERSONDesignandUse2014.pdf?sequence=2 [Accessed 17 July 2017].

Zappi, V. and McPherson, A. (2014b). Dimensionality and appropriation in digital musical instrument design. In: *NIME*. [online] pp. 455–460. Available at: www.nime.org/proceedings/2014/nime2014_409.pdf [Accessed 10 Aug. 2016].

20

Translating Mixed Multichannel Electroacoustic Music With Acoustic Soloist to the Personal Stereophonic Listening Space

A Case Study in Jorge Gregorio García Moncada's *La Historia de Nosotros*

Simon Hall

INTRODUCTION

This chapter outlines the processes undertaken during the production of a commercially released two-channel stereo record of Jorge Gregorio García Moncada´s *La Historia de Nosotros* (in English, *The Story of Us*). As a case study, the production is interesting, as a number of quite specific issues were overcome, technical and practical, as well as musical and aesthetic, that may be pertinent to other practitioner composers and music producers who are working within the electroacoustic domain and are interested in creating recordings for domestic consumption. A strong theme emerged relating to translation: how one may translate eight channels to two; to what degree one should manipulate high dynamic range electroacoustic music to make it work in a record production context; how a live chamber music percussion element should be considered and presented in terms of staging within an electroacoustically driven production; and the fundamentally different, though overlapping, production values that have to be put in place when considering the development of an electroacoustic composition compared to a record.

Issues were considered and resolved using a range of technical and aesthetic principles. These were seated in the literature from the disciplines of both electroacoustic composition and record production, and were applied to create the final commercial recorded artifact.

SPATIALIZATION IN ELECTROACOUSTIC MUSIC AND THE MULTICHANNEL PIECES THAT RESULT

As with most music delivery formats, two-track stereo has historically been the primary designated delivery format for electroacoustically led musical composition that lies within the lineage of the musique concrète tradition. This is directly attributable to the standards across both domestic and professional audio which, since developments beyond mono, have been based and continue to be based around two-channel stereo, whether in the analog formats of vinyl and stereo tape, or digital formats such as compact disc (CD), digital audio tape (DAT) and now, pervasively, the stereo audio soundfile.

Since its first inception in the late 1940s and 1950s, musique concrète has always been devised in a studio context and subsequently presented in a concert hall–type situation. To enhance the audience experience, there has been a strong tendency that goes right back to Pierre Schaeffer (1952, pp. 107–109); Schaeffer et al. (2013, pp. 98–99) to use more than simply one or two speakers when making a presentation of tape music into a concert space. Cage's *Williams Mix* of 1953 made the first aspirations to eight channels of simultaneous playback, and though cumbersome and expensive, various composers through the 1950s to the 1990s developed works with track designations for their master tapes of four channels and upwards.

Since the advent of the Alesis Digital Audio Tape (ADAT) and Digital Tape Recording System (DTRS) modular multitrack formats in 1991 and 1993, respectively, and the subsequent first generation of multiple-output audio interfaces from Digidesign and others (Collins, 2013, pp. 10–11), there has been a strong tendency in acousmatic and other electroacoustically orientated music for fixed media for composers to move beyond two-channel stereo towards multichannel masters for presentation purposes. Modular multitrack formats offered a convenience, and quite sudden flexibility of format, facility and application for the user wishing to disseminate audio beyond two-channel stereo, and many composers and institutions adopted the eight-channel format quickly to facilitate spatialization as an integral part of the composition process, either by point source, multiple stereo images, or by introduction of a level of pre-diffusion in advance of performance. As Otondo's work supports, eight-channel, with full 20Hz–20kHz frequency range (as differentiated from the cinematic 5.1 or 7.1), has now been very much a standard for composition within the genre for some time, and most institutional electroacoustic composition studios and performance systems accommodate this standard (Otondo, 2007).

Although the eight-channel media format has become an accepted standard, the range of speaker configurations that have evolved in parallel for mapping the outputs of those eight channels are less conventionally standardized, though some general protocols have tended to emerge. These include the so-called Circle-styled configurations comprising eight speakers at 45-degree intervals, speaker one either at 0 degrees or −22.5 degrees from the center line, sometimes referred to as "American" or "Double Diamond", and "French" or "Big Stereo", respectively; the "Octophonic Cube", comprising two quadrophonic systems, one raised and one at audience level; and the original "Birmingham ElectroAcoustic Sound Theatre (BEAST) Main 8" that views the speakers in symmetrical pairs designated Main, Wide, Rear and Frontal Distant, as indicated in Figure 20.1 and outlined by Wilson and Harrison (2010) and others (Fielder, 2016; Wyatt, 1999; Zvonar, 2006).

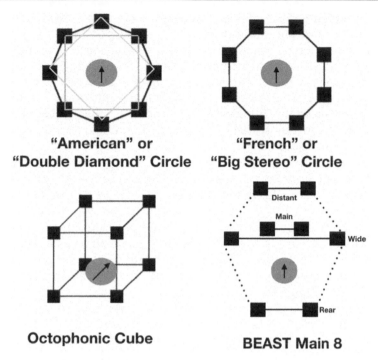

Figure 20.1 Semi-Standard Eight-Channel Track-Speaker Designation Conventions. Arrow indicates center sweet spot and direction of orientation of listener.

JORGE GREGORIO GARCÍA MONCADA'S *LA HISTORIA DE NOSOTROS*

Dealing with an eight-channel piece became the first significant production consideration for *La Historia de Nosotros*. This piece is substantial in concept and scale: an extended four-movement, 81-minute work for eight-channel electroacoustic fixed media tape, for the French loudspeaker configuration. Two movements are purely electroacoustic; two are mixed for solo multipercussionist with electroacoustic octophonic tape.

The work is based on archaeological, historical, literary and mythological sources from a number of South American historical and geographical locations, based on Huitoto history and mythology. The composition incorporates both musical and extra-musical references to its source materials. By incorporating native instrumental material and quotations from contextual literature and anthropological sources, the work alludes to distinctive soundscapes and emblematic cultural symbols. It draws on a wide range of audio sources, including natural sounds, found sounds, instrumental sounds and spoken word narration that collectively make the work almost radiophonic audio theatre in its approach at points.

Timbral design and creation, while determined by the sources, is influenced by composers of spectral music such as Murail, Grisey and Radulescu. A focus in composition is given to conceiving orchestration in terms of timbral gestalt, aiming to deliver a sonorous unity rather than a summation of instrumental and electronic media parts. Special attention is given to developing sound material with similar spectromorphological characteristics across both instrumental and fixed-audio media (Moncada, 2013). Material in the electronic part evolves into new sonorities, and the soundworld created is very processed much of the time, as one might expect with spectralist-influenced electroacoustic music. Much of the material has significant spatial motion, including rotational motion, within the eight-channel sound field.

The piece in performance is, effectively, acousmatic-led chamber music. The multipercussionist is not intended to be amplified, and there is no live processing or live electronics to the acoustic part. As per the French system, the octophonic array surrounds the audience in symmetrical pairs of speakers at 45-degree angles, starting ±22.5 degrees from the center line (see Figure 20.2). Subwoofers are implemented to enhance low frequency content, the odd-numbered channels summed to A and the even-numbered to B. The tape part should be balanced with the acoustic (not amplified) percussionist within the performance space.

As the percussionist performs in only two of the four movements, the performer is at liberty to be seated or to leave the stage for the two acousmatic movements in which they are tacet. It is important to be aware that when playing, the percussion part is wholly embedded into the musical fabric of the piece, and that, in contrast to some other works in the repertoire for instrument and tape, the role of the percussion is more than merely an obbligato to the tape part. Effectively, the piece in performance simultaneously adheres to both the contemporary "classical" chamber music performance practice paradigm and acousmatic music's diffusion practice paradigm.

Using *La Historia de Nosotros* as the centerpiece, a project was undertaken to deliver a performance, recording and practice-as-research collaboration between Bogotá, Colombia and Birmingham, UK, comprising the composer and percussionist from Colombia and engineers and producers from both Bogotá and Birmingham. Personnel involved comprised Jorge Gregorio García Moncada (Composer), Federico Demmer Colmenares (Percussionist), Marcela Zorro (Engineer), Simon Hall (Producer and Engineer), Jonty Harrison (Producer) and Sandro Carrero (Executive Producer).

The recording part of the project was to produce a two-channel stereo Original Master of the work for eventual release on the Sello Disquero Uniandes record label. The team from Colombia traveled to the UK in June 2016, where a performance of the work took place at Birmingham Conservatoire, then developing the record within Conservatoire's recording and production studios.

The record production process followed three distinct stages:

1. Preproduction: The creation of a two-channel stereo tape part stem from the multitrack surround version
2. Production: The studio capture process of the soloist
3. Postproduction: Editing, mixing and mastering

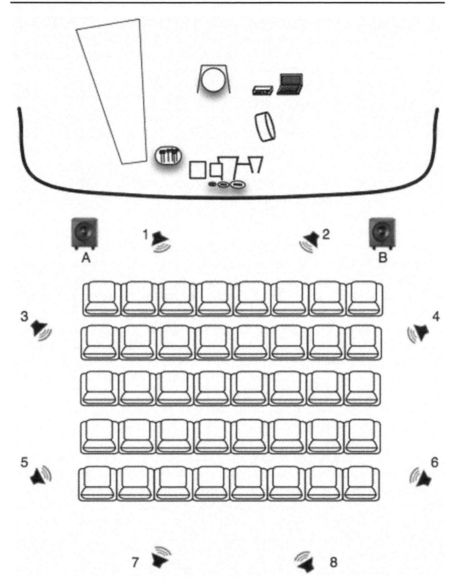

Figure 20.2 Performance Setup for the Work Illustrating Percussionist Setup and Loudspeaker Designation

© Moncada (2013)

AMBISONIC-BASED INTERVENTIONS

As Baalman (2010, p. 209) states, "spatial composition . . . may lose integrity when transferred to another audio technology". As translation had to be undertaken from the French configuration master tape part to two-channel stereo, the team had to consider the best approach to take. The options were to use a two-channel stereo master that had been previously produced by the composer, that he felt never really maintained much sense of the eight-channel spatial elements, or perform a new downmixing process on the eight-channel tape. This could be a straight fold-down mix process of some sort, or an intervention with a form of digital signal processing (DSP). Following experiments with both approaches, the preferred solution employed an intermediary processing stage that applied the principles of Higher Order Ambisonics (HOA).

Ambisonics, with its extended history dating back to Gerzon et al.'s work (1973, 1974), has once again come to the fore as new commercially viable markets have found application for ambisonic audio principles and technologies. Competitively priced virtual reality (VR), immersive gaming and a number of "3D audio" variations of wavefield synthesis have all gained significant traction in both professional and consumer markets (Armstrong, 2016; Ward, 2017).

In the less economically driven world of electroacoustic music, work undertaken by Barrett and Jensenius (2016) applies HOA to facilitate virtual diffusion systems. The *Virtualmonium* is a software prototype codec solution that emulates the traditional loudspeaker orchestras of electroacoustic music such as the BEAST system or Klang Acousmonium within an ambisonic sound field. It goes some way to starting to address the possibilities of being able to diffuse or otherwise position tracks and channels by encoding into a virtual ambisonic sound space to be then subsequently decoded into an installed system of any concert hall or other performance environment. It also paves the way to the potential of undertaking spatialization and diffusion in a VR environment.

The need here, to take the eight-channel speaker designations of the eight channels of the piece and fold down to two channels via an ambisonic-based codec process, could have been addressed by a relatively simple application and adaptation of the *Virtualmonium*. However, the chosen solution was by application of software created by Blue Ripple, a commercial company at the forefront of HOA technology and programming. Blue Ripple (2017) create and supply a range of products for 3D audio, including VR tools and game engines. Their 03A suite of Third Order Ambisonic (TOA) plugins for VST proved a simple but powerful solution that was fit for this purpose.

Implementation of the downmix process utilized Reaper as a digital audio workstation (DAW), due to its ability to host the 16-channel bus width required by TOA, as well as its ability to host the VST plugin format. The "TOA Panner—Eight Channel" was applied, with Azimuth set appropriately for each of the eight-channel track designations to the appropriate virtual speaker angle, Elevation and Gain kept at unity (see Figure 20.3). The 16-channel output was then routed to a parallel 16-channel bus, metered by the TOA Meter, and then decoded in series using the TOA decoder, straight to a "Regular Panner" (as opposed to "Binaural Panner").

The two-channel image was tested utilizing Waves PAZ Analyser to verify phase, and some minor imaging issues with respect to symmetry and rotation were corrected utilizing Waves S1. Reaper's built-in master section was used to confirm mono compatibility.

Figure 20.3 Application of Blue Ripple's Eight-Channel TOA Panner

Azimuth values set to ±22.5°; ±67.5°; ±112.5°; ±157.5°

From an initial qualitative point of view, this fold-down was successful. The biggest potential issue that needed to be avoided was ducking within streams of audio that rotate or otherwise move within the eight-channel sound field. The TOA processing avoided this issue. The composer was happy with this version, and in particular felt better able to hear a more accurate sense of the spatial motion maintained than he had felt compared to two-channel downmix versions that had been produced in the past.

PERCUSSION CAPTURE AND INITIAL POSTPRODUCTION

The studio capture of the live performer across the two movements with percussion parts was quite straightforward in its principles, and in many ways this was quite a conventional engineering and production process as common for any high-end studio recording session. Instruments were arranged across three stations: one for the performance of the shakers, a second for the marimba, and a third for the multipercussion setup of orchestral bass drum, tam-tam, tom-toms, bongos and temple blocks. The relatively dry acoustic of the studio live room was further dried with acoustic screens to further attenuate early reflections.

A close-mic'ed, multi-mic'ed approach was undertaken with all instruments. Shakers were auditioned with stereo pairs of Neumann U87s, Neumann KM184s and DPA4011s (though just U87s were taken forward to mix). Marimba was captured with a pair of custom U87 clones, with an additional Neumann U47FET beneath to capture low-frequency depth and resonance.

Within the multipercussion setup, orchestral bass drum was captured with a Neumann KM184, tam-tam with Neumann TLM103, toms with Sennheiser MD441s (the only dynamic mics employed here), bongos with a stereo AKG422 above and DPA 4011s beneath, and temple blocks with KM184s, as visible in Figures 20.4.a and 20.4.b. A Schoeps Colette pair of overhead microphones with MK2H capsules were captured, but the tracks were discarded at mix.

All microphone recording channels came into a Solid State Logic Duality 24 console. As well as being very transparent with very high amounts of headroom, the desk also facilitated phase reversal for reverse-focused microphones, and additional low cut filtering and functional equalization (EQ) where necessary on the way into the DAW. A relatively standard approach to corrective EQ was taken on capture, including reducing resonant frequencies from toms, low cut filters for the 20Hz–40Hz bottom octave and gentle SSL G-series type enhancement to the sonics of individual mics on specific sources. Analog-to-digital conversion was from a pair of Avid HDio interfaces into the Pro Tools DAW running on a Mac Pro.

Foldback to the performer for synchronization purposes consisted of both a rehearsal mix of the now-stereo tape part via headphones, as well as time position via a video monitor with Pro Tools "Big Counter" (visible in Figure 20.4.b).

The movements were captured in sections, take by take. As synchronization with the tape part was required, the instrumental parts were recorded into Pro Tools playlists, a new playlist for each take across the recording tracks. Some particularly tricky small sections and gestures were captured utilizing a more linear recording methodology to maintain the energy and flow of the performer. These were dropped into the appropriate timecode position as a part of the editing process. The multiple playlists in Pro Tools were then, in postproduction, reviewed and an edited composite derived. Although this was less like the procedure one might expect for a chamber music-style recording, the setup and process could just as easily have been a media session for film or television that had pre-recorded elements to the musical whole to which the performer would have to synchronize.

AESTHETIC AND PHILOSOPHICAL CONSIDERATIONS FOR APPROACHING MIXING: ACOUSMATIC MUSIC COMPOSITION VERSUS RECORD PRODUCTION

The mix stage required significant deliberation in terms of approach. One has ultimately to consider the relationship between the tape part and the recorded soloist and the musical meaning intrinsic within that relationship.

Figures 20.4a and 20.4b Percussion Setup, Microphone Setup and Foldback Via Headphone and Timecode Display

Conceptually, one has to mediate between the inherently acousmatic experience of listening to an audio recording while wishing to balance the perception and portrayal of a live performer performing with an acousmatic tape part.

There are multiple levels of creative agency at work here. An eccentricity with this particular project is that the work passes through a mediatization process twice, with quite separate audio deliverables required at the end of each of these processes: once in the production of the tape part intended for dissemination in the controlled concert hall space; and a second time, and the process that is undertaken here, with the intention of the output being sent into the wild to be consumed in a domestic or personal listening environment. Not unlike producing a vocalist with a backing track, we have two distinct and separate layers of both sonic quality and creative agency that must be conveyed within the mix. As the live performer is visually removed, ultimately all the musical content becomes acousmatic as it moves to the recorded medium. We considered our approach here observing a number of theoreticians in the areas of both acousmatic composition and record production to frame this.

For any piece that is rooted in the electroacoustic domain, Smalley's notions of surrogacy in the context of musical gesture within electroacoustic music cannot be ignored. It is possible to draw on various elements of Smalley's theses, but as an example, within the tape part there are numerous musical gestures that very much fit into Smalley's taxonomy as "second order surrogates", i.e. recorded sound with a traditional instrumental gesture (Smalley, 1997).

Similarly, much of the live percussion part that is performed is very gesturally led in terms of its phrase structure and musical construction, as illustrated by the musical examples in Figures 20.5.a and 20.5.b. As we record them, these performed gestures fundamentally become sound objects in their own right, which of course may also now be seen to be themselves as second order surrogate gestures in Smalley's terms.

So we have a scenario where we must consider the acousmatic element of the work versus the non-acousmatic (i.e. the live performer) in an acoustic performance environment, now transposed to a *portrayal*

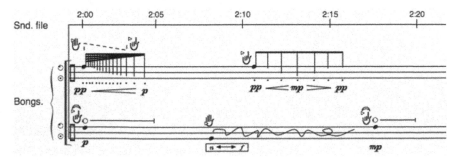

Figure 20.5.a Excerpt From the *La Historia de Nosotros* Score, 2′00″-2′20″

© Moncada (2013)

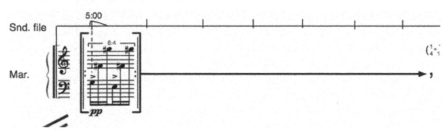

Figure 20.5.b Excerpt From the *La Historia de Nosotros* Score, 5'00"–5'07"

© Moncada (2013)

of the acousmatic element versus non-acousmatic (originally live, now recorded, but technically now also acousmatic) in a studio record production environment.

Eisenberg, in his discussions of phonography and the ramifications for the listener of the recorded artifact, states clearly that the key for the audience is that "the performer is not there" (Eisenberg, 2005, p. 130). This can trace back to R. Murray Schafer's (1994, pp. 90–91) notion of schizophonia, removing a sound source from its creation by the process of recording. More currently, Simon Zagorski-Thomas (2014, pp. 50–51), applying ideas introduced by Frith (1998) and Moore (2012), also notes that:

> when we listen to recorded music, one of [the] complex layers of perceived personae is removed: the reality of the performer's presence . . . becomes less pronounced because we lose the visual aspect—a cartoon-like distortion makes the personae behind the music much less multi-dimensional and that encourages a form of interpretation very different from that of the concert hall.

For this project, the key to the mix approach then has to be the degree to which one should try to maintain (or alternatively deliberately lose) the perception of the agency of the performer, taking into account the representational "cartooning" nature of the recording of the percussionist. The close-mic'ing of the percussion, together with the practicalities facilitating control over timbre via microphone choice, preamplifier choice, EQ and limiting, and with the post-production application of reverb and automation, "creates a sonic image that exaggerates . . . particular features" (Zagorski-Thomas, 2014, p. 55) of the percussion, notably in this case a closeness and enhanced dynamism of the percussion performance, but also in the way the recorded image of the percussion is portrayed to be integrated into the overall mix, in terms of width, perception of space and spatial image, balance and cohesion.

Also drawing on Zagorski-Thomas, one cannot ignore the notion of "staging" here. The intention of the record is very much to be within the first and second categories of Zagorski-Thomas's taxonomy regarding "functional staging". The intention of the production is very much intended for focused listening, while of course also attempting to remain true to the spirit of a chamber music–based live performance atmosphere. But there is, inevitably, and as already alluded to, a need for a level of exaggeration and enhancement to the percussion. Perhaps even, to cite Bourbon and Zagorski-Thomas (2017), to be categorized as a "hyperproduction" of the percussion part to make it sit alongside the soundworld of the tape part.

So in terms of staging and imaging, we have the acousmatic set against the non-acousmatic elements, in an emulated chamber music environment, subsequently transposed and represented in an enhanced manner for a domestic listener. As a recording, this moves some way beyond the discourse of the concert hall, and acknowledges the mediating nature of the recording process. Though as we created the final soundstage and undertook the final mix, we also could not ignore the composer's intention to "create a sonorous unity rather than a summation of the parts" (Moncada, 2013). This had to become the overarching objective.

PRACTICAL CONSIDERATIONS FOR MIXING

The destination for this media, as a commercial recording for distribution, is for domestic individualized consumption in a personal listening environment. As regards the practicalities of this going out into the wild, there has to be a realistic reduction of the dynamic range to take into account the variety of playback systems and contexts that the listener may choose to utilize.

Pro Tools continued to be used as the mixing environment for the project. The edited percussion tracks were balanced as a subgroup within the project. To maintain RMS levels, parallel compression was applied to the subgroup, alongside peak limiting to start to maintain a consistency of level, as was a certain amount of automation of individual elements to allow for internal balance within the percussion part. Artificial acoustic was applied by Sonnox emulations of the Lexicon 480L artificial reverb unit. The two-channel stereo post-ambisonic processed tape part also had parallel compression and peak limiting applied at this stage. The relative levels between the two subgroups, percussion and tape, were then ridden through in Pro Tools, maintaining balance relationships through the duration of the work, and differentiating the two elements as required to maintain the sense of the identities of the respective creative agencies, though one might note that subjectively, some of the most engaging moments of the piece occur when this becomes ambiguous. The final mix was created in-the-box in Pro Tools to generate the two-channel stereo Original Master file.

MASTERING

After the Original Master was produced, the production team took a break and the Colombian contingent returned to their home country. The mixes were then revisited on consumer-level monitors. The composer, while very happy with almost all elements of the production, felt that there was some detail of the high-frequency content of the tape part missing on domestic playback compared to studio playback. The dynamic range of the piece that was still inherent within the Original Master was also deemed too great as it stood.

For a final level of finishing, and the generation of the final Production Master, a mastering process was undertaken by independent engineer Carlos Silva, of C1 Mastering, Bogota. Equalization was applied and dynamic range reduced further by a combination of multiband compression and EQ to readdress the balance between low end and high end.

In terms of the final levels of the work, the Production Master that has now been commercially released has levels equating to -16.5 Loudness Unit Full Scale Integrated (LUFS Integrated), with a Loudness Unit (LU) Range of 15.9. This is compared to the Original Master's -22.7 LUFS Integrated, LU Range of 16.2 (after normalization for comparative purposes); and the summed stereo tape part's—20.8 LUFS Integrated, LU Range of 17.7 (similarly normalized).

Although this is a very extended work across which to make these measurements, subjectively these values are not out of character with the musical content, and they correlate appropriately with the -16 to -20 LUFS target advocated by the Audio Engineering Society (Byers et al., 2015) and being adopted as a loudness normalization value (to a greater or lesser degree at the time of writing) by the major music streaming services.

CONCLUSION

Howlett, in his article, "The Producer as Nexus", claims that at its most generic, the producer's task is to produce a "satisfactory outcome" (Howlett, 2012). To that end, and writing as one of the producers, this has been a successful project from an objective point of view: the album is released on the Sello Disquero Uniandes label (Moncada, 2016) and is commercially available as an album for purchase as well as via streaming services. All parties involved in the project and the processes were satisfied with the final output and deemed the project and collaboration musically worthwhile. The record was also submitted for consideration by The Recording Academy for the 18th Annual Latin Grammy Awards, 2017.

The project provides a tangible potential model for professional standard "instrument and tape" production approaches and values. It also demonstrates viable practice in folding down multitrack pieces to two-channel stereo by application of HOA codec systems, and to some greater degree illustrates an approach to translate some of the contrasting audio requirements of the electroacoustic composer to the needs of the record production process.

The theme running through this whole project has been the translation: from eight-channel tape to two-channel; from concert hall performance to studio performance; from live to acousmatic; from controlled playback environment to uncontrolled. Whether the home listener should be aware of this becomes irrelevant if the record is worthwhile within its own musical terms and is able to be listened to and experienced on that basis. An awareness and understanding, though, of the layers and frameworks that have underpinned the processes may help to enhance the individual listener's experience of the record on playback.

REFERENCES

Armstrong, P. (2016). *Just how big is the virtual reality market and where is it going next?* Available at: www.forbes.com/sites/paularmstrong-tech/2017/04/06/just-how-big-is-the-virtual-reality-market-and-where-is-it-going-next/#40de2214834e [Accessed Sept. 2017].

Baalman, M. (2010). Spatial composition techniques and sound spatialisation technologies. *Organised Sound*, 15(3).

Barrett, N. and Jensenius, A. (2016). The "Virtualmonium": An instrument for classical sound diffusion over a virtual loudspeaker orchestra. In: *NIME 2016 proceedings*. Available at: http://nime2016.org/ [Accessed Sept. 2017].

Blue Ripple Sound. (2017). *3D audio*. Available at: www.blueripplesound. com/3d-audio [Accessed Sept. 2017].

Bourbon, A. and Zagorski-Thomas, S. (2017). The ecological approach to mixing audio: Agency, activity and environment in the process of audio staging. *JARP*, (11). Available at: http://arpjournal.com/the-ecological-approach-to-mixing-audio-agency-activity-and-environment-in-the-process-of-audio-staging/ [Accessed Sept. 2017].

Byers, et al. (2015). Recommendation for loudness of audio streaming and network file playback. In: *AES TD1004.1.15–10*. Available at: www.aes.org/technical/documents/AESTD1004_1_15_10.pdf [Accessed Sept. 2017].

Collins, M. (2013). *Pro tools for music production: Recording, editing and mixing*. London: Focal Press.

Eisenberg, E. (2005). *The recording angel*. New Haven: Yale University Press.

Fielder, J. (2016). *A history of the development of multichannel speaker arrays for the presentation and diffusion of acousmatic music*. Available at: http://jonfielder.weebly.com/ [Accessed Sept. 2017].

Frith, S. (1998). *Performing rites: Evaluating popular music*. Oxford: Oxford University Press.

Gerzon, M. (1973). Periphony: With-height sound reproduction. *Journal of the Audio Engineering Society*, 21.

Gerzon, M. (1974). Surround sound psychoacoustics. *Wireless World*, 80. Reproduction Available at: www.audiosignal.co.uk/Resources/Surround_sound_psychoacoustics_A4.pdf [Accessed Sept. 2017].

Howlett, M. (2012). The record producer as nexus. *JARP2012*. Available at: http://arpjournal.com/the-record-producer-as-nexus/ [Accessed Sept. 2017].

Moncada, J. (2013). *Ukhu Pacha & La Historia de Nosotros electroacoustic music composition portfolio*. University of Birmingham. Available at: http://etheses.bham.ac.uk/4867/ [Accessed Sept. 2017].

Moncada, J. (2016). *La Historia de Nostros*. Bogotá: Sello Disquero Uniandes. Available at: www.amazon.com/Historia-Nosotros-Gregorio-Garc%C3%ADa-Moncada/dp/B01M8F8NI7 [Accessed Sept. 2017].

Moore, A. (2012). *Song means: Analysing and interpreting recorded popular song*. Farnham: Ashgate.

Otondo, F. (2007). Recent spatialisation trends in electroacoustic music. In: *EMS07 proceedings*. Available at: www.ems-network.org/spip.php?article272 [Accessed Sept. 2017].

Schaeffer, P. (1952). *À la Recherche d'une Musique Concrète*. Paris: Seuil.

Schaeffer, P., North, C. and Dack, J. (trans.). (2013). *In search of a concrete music.* Los Angeles: University of California Press.

Schafer, R. M. (1994). *Our sonic environment and the soundscape: The tuning of the world.* Rochester: Destiny.

Smalley, D. (1997). Spectromorphology: Explaining sound shapes. *Organised Sound,* 2(2).

Ward, P. (2017). The rise of 3D. In: *Pro sound news Europe.* Audio Available at: www.psneurope.com/the-rise-of-3d-audio/ [Accessed Sept. 2017].

Wilson, S. and Harrison, J. (2010). Rethinking the BEAST: Recent developments in multichannel composition at Birmingham electroacoustic sound theatre. *Organised Sound,* 15(3).

Wyatt, S. (1999). Investigative studies on sound diffusion and projection. In: *eContact 2.4.* Available at: http://econtact.ca/2_4/Investigative.htm [Accessed Sept. 2017].

Zagorski-Thomas, S. (2014). *The musicology of record production.* Cambridge: Cambridge University Press.

Zvonar, R. (2006). A history of spatial music. In: *eContact! 7.4.* Available at: http://econtact.ca/7_4/zvonar_spatialmusic.html [Accessed Sept. 2017].

21

Score Scroll

Replacing Page-Based Notation With a Technology-Enhanced Solution in Composition and Performance

Bartosz Szafranski

INTRODUCTION

In the author's practice-led research into timbre transformation and the devices required to construct a coherent musical form with a very slow distribution of structural sound events, the aesthetic of the music conflicts with the physical limitations of human ensemble performance. The music is structured based on precise timings, yet the material does not provide regular cues for player synchronization; what is more, very detailed and evolving instructions regarding articulation demand full attention, creating problems for the conductor-performer communication. This music, which focuses on slow transformation of parameters, requires an alternative approach to presenting the score and facilitating ensemble performance.

The problem is addressed by producing animated scrolling scores as video files, to be projected during performances. This simple and elegant solution, while fitting very well with the nature of the material and avoiding conducting issues, presents multiple challenges regarding the technology, logistical execution and reliability of performance. The score is required to move horizontally across a static play-line in a steady, smooth and continuous manner, for the complete duration of a composition, between 12 and 15 minutes. Therefore, it must be professionally edited and exported in a format that is high-definition and compatible with these requirements and with video editing software. Professional score editors have limitations and default behaviors that need to be overcome to make this possible; for example, bars need to be automatically resized to accommodate notes and the process of exporting images needs to be determined. Moreover, the video score must be presented to the players in real time, and in a size ensuring full legibility, without jeopardizing the immersive quality of the concert. In this respect, the composer's concern was that a large display, for example, a flat-screen TV measuring more than 40 inches, would obscure the visual contact between the performers and the audience to the point of breaking the psychological relationship which lies at the core of live performance. This is comparable to a basic issue of musicians performing in a recording studio, removed from an audience, as discussed by Frederickson (1989):

> For the musicians, the microphone becomes the audience, and an unforgiving one at that, for the microphone does not respond, it does not register the personal presence, expression, and communication of the performer.

This approach also raises questions connected to the interpretative value of human performer input in the context of automated presentation.

It is important to note that the idea of video score presentation in live performance is not new, but it has usually been combined with non-standard graphic notation, whose effect on the dynamic of execution is very different. Vickery (2012) identifies 2007 as the year software solutions became capable of the satisfactory manipulation of notated music, leading to the emergence of the 'screen score'; however, this chapter traces the development back to the 1950s and the 'mobile score' experiments of Earle Brown, John Cage and Morton Feldman. This shows a line of development in the area of graphical notation, which Vickery amplifies in the definition of the 'scrolling score'. Just as the description of this concept identifies the key elements relevant to the author's project – in other words, a continuous sideways motion with events striking a playhead – it also proceeds to observe that the technique is successfully applied to graphical notation in particular, because traditional notation does not maintain proportionality in the duration of notes and events.

Vickery's role in the Decibel ScorePlayer application project makes it convenient to find further evidence of the strong connection between the video score and graphical notation. Examples of application are focused exclusively on graphical notation both in the conference paper introducing the application (Hope et al., 2015) and on the relevant web page on Apple's App Store (App Store, 2018), where three screenshots are provided for illustration. Similarly, other papers on the topic rely on a wide variety of graphical approaches to notation, for example, Kim-Boyle (2014), Hope (2017), as does the most comprehensive online resource on animated scores, Animatednotation.com (Smith, 2018), which offers an impressively broad overview of composers utilizing video for score presentation.

By contrast, the scores investigated in this chapter are predominantly presented in standard practice traditional notation, with consistent barring, even if they contain symbols and elements associated specifically with 20th-century practice.

The two case studies discussed in detail in the following sections serve to provide the context and justification for the necessity of introducing this idea into the composer-performer relationship. The reasoning behind it is explained from the composer's perspective and based on the players' feedback, without which it would be impossible to appreciate the full impact of this performance practice. The first composition, *Eight*, was workshopped, rehearsed and performed by Konvalia Quartet with Alina Hiltunen on first violin, Agata Kubiak on second violin, Marietta Szalóki on viola and Sam Creer as a stand-in cellist; the composer played the piano. *Six Spiders* is a duet which features the composer in the role of the guitarist, and Agata Kubiak, who commissioned the inclusion of a unique dual part of voice and violin for herself to perform and who became closely involved in the creative process and experiments with score presentation. These musicians' contributions have been of great importance and have undoubtedly enhanced the result of the investigation.

CASE STUDY 1: EIGHT

The first composition of the project is entitled *Eight*, which refers simply to the arithmetical value—a structurally relevant number. While a low-pitched pre-recorded electronic drone is engaged in an extremely slow glissando of a whole step from G to A, the piano sounds non-triadic chords at large distances, and the string quartet creates a polyphonic texture which, in its slowness, mimics the idea of time-stretching in digital audio processing. The music can be described as slow and stretched out, quiet and fragile, tense and sustained; it is moderately dissonant, with non-directional harmony projected as cyclical transformations of two starting-point chords. There are no programmatic influences or references, although the extramusical meaning of inability to externalize emotion through music is suggested by the performance direction "Slow and yearning for lyricism".

The processes that led to *Eight*'s creation laid the foundations for the methodology of composition guiding the practical work behind the entire 90-minute portfolio for this research project, simultaneously exposing the major challenges of score presentation and performance that have led to the notational investigation described in this chapter. Apart from the general aesthetic of the stretched-out sound-world, which by itself need not enforce the assumption of technical problems for 21st-century performers, it was the approach to balancing, in visual presentation, the intended effects of timbre, line and harmony that made traditional paper score presentation impractical for live performance.

MOVING SLOWLY

The composition has a slow perceived tempo of unfolding. This is achieved by means of a careful placement of stronger musical gestures at timed intervals averaging eight seconds, but ranging from 0.5 seconds to 24 seconds. The backbone of this structural conceit is laid bare by the sparsely distributed chords in the piano part, although they do not project the form in full; there are a number of melodic steps in the string parts which do not coincide with the chords, but have sufficient textural strength to be formally relevant.

Stretched between these remote tentpoles are the quartet's individual lines, finding a sense of counterpoint in predominantly scattered, overlapping phrasing. Note durations, dynamic envelopes and timbre

Figure 21.1 Bars 1–8 of *Eight* With Hit-Point Markers Indicating Structural Events

transformations are implemented with a strong sense of independence between the parts, in order to create the effect of continuous morphing of colors in a balanced overall texture. The downside of this approach is that shared rhythmic cues are rare, making temporal synchronization within the unaided ensemble impractical. Forcing the ensemble to seek out cue points within this texture would carry the risk of shifting the performers' attention away from the key expressive aims, especially the implementation of controlled gradual changes of timbre, and towards alleviating the undesired pressure stemming from striving to stay in time. The potentially simple solution of employing a conductor is made less tempting by the exceptionally slow tempo and complete avoidance of pulse, as well as the fact that conducting might create a visual distraction by introducing onto the stage aspects of movement too vigorous to correspond to the physicality of the players' actions or, more importantly, to the resultant sound.

SLIDING SLOWLY

Excessively slow *glissandi* covering small pitch intervals are a feature of the composition. More than a modernist cliché, they are a key contributing factor in the engineering of the taut but fragile auditory experience; additionally, they become a time-stretched version of portamento detail in lyrical string performance.

This device creates problems at both individual and ensemble levels. An individual string player is likely to find it challenging to anticipate and maintain the correct speed of a stretched-out glissando, especially if a small intervallic bend forces them to zoom in on microtones; and it is important for the integrity of the music to ensure this speed is as close as possible to consistency. When it comes to performing these events in an ensemble, the prolonged duration of a player's glissando makes it challenging for the rest of the group to remain aware of the exact rhythmic placement of events in their own parts while the slide is continuing.

TRANSFORMING SLOWLY

Whether a long note remains static in pitch, undertakes a slow glissando or reaches a point of standard fingered note change, it is most likely also subjected to a slow gradual change of articulation parameters, resulting in timbre transformation. Predominantly, pitch shifting and timbre transformations occur simultaneously or in an overlapping fashion.

As in the case of *glissandi*, a major part of the challenge to the performer is the expectation that the full duration of a timbre transformation will be executed faithfully to the notated score, requiring a special degree of control. The types of gradual timbre transformations required of the string quartet in *Eight* are listed below; they occur in a wide array of combinations.

1. Bow position, for example, *sul tasto* to *sul ponticello*
2. Vibrato intensity, for example, *non vibrato* to *molto vibrato*
3. Dynamic level

In addition, the standard articulation on-off effects are in use: unmeasured tremolo, half-step trill, natural and artificial harmonics. The key impact of this on the notated presentation of the score is that a lot of information is being relayed to the performer, occasionally with considerable levels of independence between different aspects of articulation. Not only is the performer tasked with keeping track of the start and end points of the different layers of timbre control, but also they are expected to execute the gradual transitions at the right speed, and combine these with ensemble timing and pitch control, as discussed in the previous two sections.

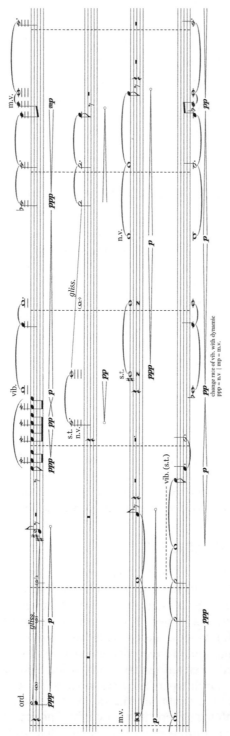

Figure 21.2 Bars 23–27 of *Eight* With Slow *Glissandi* in Violins 1 and 2

Figure 21.3 Bars 8–14 of *Eight*, Violin 2 and Viola Only—Timbre Transformations

From the composer's perspective, the importance of an accurate execution of these parameters results from two fundamental aspects of the composition process: (1) the structure is based on precisely calculated timings, in order to adhere to the average of eight seconds between strong gestures and (2) the composition uses a points-based grading system controlling the timbre complexity, which is projected as another formal device onto the timed structure (harmony also plays a significant role, which is beyond the scope of this chapter).

At the same time, it is of importance to note that the aim is not surgical precision at the cost of interpretation and expression. The structural calculations and the timbre grading system are actively involved in creating the general atmosphere of fragility and introspection, which should be emphatically projected by the performers; therefore, pursuing the idea of animated scores becomes a quest for preserving expressive freedom within a performance that is faithful to the composition process. It is a technical solution to the question of offering the performers the maximum level of comfort, so that expressive involvement of the players is unhindered by the challenges of timing and ensemble synchronization.

CASE STUDY 2: SIX SPIDERS

Several primary features of the second composition, *Six Spiders*, were developed as solutions to the unique problems presented by the initial commission, by Agata Kubiak, to include a prominent dual part of violin and voice by one performer. In order to avoid logistical issues with rehearsals and performances, the ensemble was kept very small, adding only an electric guitar and pre-recorded electronics. In this context, much emphasis was placed on ensuring that compositional choices regarding the violin/voice part made this part fully responsible for controlling the primary parameters of time, timbre and harmony. As the effect would have to be in agreement with the general aesthetic of the project, with slow development and interest in timbre, the burden on the dual soloist part created a set of challenges which put to test the methods created in the initial stages of composing *Eight*. The individual's near-complete responsibility for maintaining musical interest on the one hand, while tackling the dual performance challenge on the other, affected the composition process to the point where the ambition of reproducing the methodology that enabled the detailed pre-planning of the structure of *Eight* was not implemented. The problem of building a correspondingly slow musical experience with very limited instrumental means became a priority and a driving force behind an increased flexibility concerning the application of the existing methods. Placing this music in the medium of a scrolling video score would prove transformative in the composer's endeavors to address the challenges; more importantly, it would change his understanding of the way the different component parts should correspond to project the intended sound-world. The work is in five sections and, for reasons explained below, the lyrical content is very limited. The text is reprinted here, due to its importance for considerations of timbre and notated representation.

Section 1:

Interact
with
human beings.

Section 2:

Get over
it.
We've been
through this.

Section 3:

Have faith
in you.
You will never
let you down.

Section 4:

Achieve!
Achoo!
Bless you.

Section 5:

No more fear.

MOVING SLOWLY

Settling on a single average duration of a sound event at the pre-composition stage was one aspect of the formal method that proved unsatisfactory early in the process. In order to maintain musical interest over the course of the composition, the decision was made to explore varying textural relationships between the main instruments. With this in mind, the time control was revised to include two types of average time periods.

In movements 2 and 4, this reflects the approach of *Eight*, as the average distance between prominent chord attacks in the violin part was measured, in order to clearly declare the musical spaces enclosed between them as the moments of interrogating the timbre transformations. In movements 1, 3 and 5, the pre-compositional average time distance between events was replaced by controlling the average duration of a single-breath vocal phrase; the phrases are separated by moments of repose for the voice, and these are not included in the calculation.

Compared to the ensemble of *Eight*, the unusual idea of assigning independent vocal and instrumental parts to a single performer changed the reasons for the material, posing challenges of presentation, but it did not change the nature of the challenges. The work traces the expansion from an average of 12 seconds in the first movement to an average of 13 seconds in the final movement. Again, stronger musical gestures provide tentpoles, between which gradual timbre transformations are stretched. As the electric guitar part is harmonically fluid and supportive in nature, the voice and violin performer is not concerned about ensemble synchronization. Instead, the

Figure 21.4 Bars 8–15 of *Six Spiders* With Time Indications

context of slow development presents a challenge to maintaining the correct timings while a single player controls two independent instrumental lines.

The nature of the challenge resulting from the slow development is similar to *Eight* in one respect—ensuring that the timings of the events are performed correctly, despite the large distances between them—and very different in another. Ensemble synchronization is not a concern, especially as the guitar part is designed to follow the central voice/violin part, but the ability of a single performer to sustain the stretched-out events in two independent parts, while looking after intonation, is emphatically put to the test. In addition, if the performer is required to engage in page turning, the level of pressure may reach a level detrimental to the quality of the performance.

Taking these issues into consideration, an intention was formed to ensure, first, that the video presentation should support the player in maintaining the structural integrity of the piece—by nearly guaranteeing the various set timings of events would be executed correctly with the correct timbre transformations achieved. Second, it should make it convenient for the voice and violin dual part player to read from the two staves, thus removing a potential source of anxiety and helping them focus on expression and execution of articulations.

Furthermore, the player is asked to explore several textural relationships, varying the way in which attention must be divided between the two staves—it is not possible to maintain a consistent approach to reading the music from the dual part throughout *Six Spiders*. The global structure of the piece is inspired by Béla Bartók's symmetrical approach (Locke, 1987), as there is a movement three which is central in both formal position and importance; this is surrounded by a two and a four which are related to each other; finally, a one and a five related to each other on the surrounding outer layer, and also related to the 'kernel' movement. These movements and how they connect with each other are characterized by the type of textural relationship between the violin and the voice; whether the voice or the violin part is subjected to the control of the average duration of events; and by soft or hard quality of note attacks in the violin:

Section 1 (2m. 52s.): soft: unison
Section 2 (2m. 36s.): hard: melody and accompaniment
Section 3 (3m. 44s.): soft: unison to counterpoint
Section 4 (2m. 56s.): hard: melody and accompaniment
Section 5 (2m. 52s.): soft: unison expanding

With a video score presentation in front of the players, all the timings for these changes could be rendered very accurately.

SINGING SLOWLY

The sense of slowness is then enforced by placing, usually, two to four syllables under a phrase of this length, which is the reason behind the limited lyrical content. As clarity of words is prioritized over phrase complexity, there is a maximum of only two notes to a syllable; therefore, this links to harmony in the number of pitches per vocal phrase, which is between

Figure 21.5 Bars 211–218 of *Six Spiders* With a Large Amount Information for the Singer-Violinist

two and five. The human voice within an ensemble inevitably draws more attention to itself than the instrumental colors, creating an opportunity to use the economy of lyrical content to support the main structural characteristics produced by the control of the placement of notes in time. The decision was made to select this idiomatic parameter, the lyrics, as a key feature of *Six Spiders*.

This adds another stream of information being sent to the dual-part performer alongside pitch, timing, dynamics, articulation and texture—this is a lot to read from the printed pages, especially as much has to be stored in memory from looking ahead of the play point in the score. Therefore, the risk is that a performance might become an inaccurate representation of the artistic intentions—an approximation rather than an execution of the parameters.

SLIDING SLOWLY

Similarly to *Eight*, lengthy *glissandi* covering very small intervallic distances are a feature and a potential problem during performance. It is challenging to stay in control of the progress of a long slide at this slow tempo,

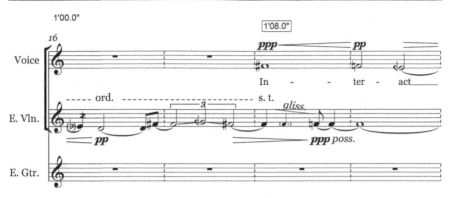

Figure 21.6 Glissandi in Bars 16–19 of *Six Spiders*

in order to ensure the correct durations; however, in *Six Spiders* a violin glissando potentially coincides with an independent vocal phrase executed by the same performed, making it considerably more valuable to introduce a visual aid capable of supporting both parts simultaneously and securing their synchronization.

TRANSFORMING SLOWLY

Building on the methods developed for *Eight*, slow changes of articulation parameters are implemented to achieve expressive timbre transformations. The method of grading timbre complexity remained an efficient way of making compositional choices, but application required more flexibility than in the case of *Eight*, primarily due to the efforts to maintain a meaningful relationship between the violin and the voice and, once again, the size of the ensemble. Additionally, the existence of the words meant that the more complex articulations had to be used with caution, lest the combined effect become grotesque. In the vocal part, very complex articulations were avoided entirely, due to the potential of conflict with the emotional color of the material.

The nature of the challenge in *Six Spiders* is different primarily because the voice/violin player is in a very exposed performance context, having to cope with large amounts of changing information without the comforting support of a larger ensemble. The material is tense and fragile, requiring a highly developed level of control from the player, making meaningful the effort to develop an approach to score presentation that relieves some of the pressure of keeping track of all the information.

DEVELOPING THE SOLUTION

Most of the problems listed above were clearly identifiable at the composition stage, but the consequences and potential solutions became considerably more defined during the initial discussions with prospective performers. In particular, *Six Spiders*, being commissioned by violinist, vocalist and researcher Agata Kubiak, was scrutinized in depth, due to the unusual nature of the instrumentation. The other key influence was internal—resulting from the composer's urge to present the music visually in such a way as to make it reflect the auditory experience—and the drive was to design a mode of delivery of the score that had the relevant physically to it. The fragile, tense atmosphere of the music, emphasized by the protracted timbre transformations, with the slow distribution of strong gestures, creates a sound-world in which page-turning breaks the immersion and interpretive efficiency of the players. In the case of *Eight*, the combination of voice-leading and texture characteristics means that convenient page-turns cannot be located, so relying on traditional paper-based part scores would require modifications of the musical material, which was not an acceptable solution in this project. Additionally, the activity would take away some of the ceremonial quality of performance experienced by the audience, as the page-turn gestures are a highly exposed physical element not directly involved in producing sound and not designed to supplement the key composition devices—page-turning is purely mechanical.

The solution has been to reproduce the scores as animated side-scrolling graphics, encoded as video files to be broadcast to the players in rehearsal and performance. This idea was inspired by the recent, yet already well-developed, movement to publish various types of animated scores with synchronized music online on the YouTube platform—the most prominent channel devoted to this is Score Follower (Scorefollower.com, 2018). With only basic synchronization features, the existing examples of traditionally notated music in video format are primarily designed for score study and music appreciation activities, as well as promotion of new music and emerging composers, so the aim to present this type of notation medium to players as a full-scale replacement for hard-copy parts has created a new, specialized area of investigation within the associated composition project.

Compared to video scores for study purposes, employing the medium as part-score replacement in performance imposes unique requirements:

- The score has to scroll right to left smoothly and at a constant speed, as any change in scrolling pace, if there is no tempo change precisely indicated, would risk breaking the focus established between the player and the display. Issues of timing and synchronization would be likely.
- Unless the music is intended to stop for a time, there should not be any page-turn effect, as it is visually too startling and confusing in a video file. The effect must be that of the score as a single extremely wide image gliding steadily across the screen.
- There must be a playhead (play line) showing the current playing position to all performers, in order to maintain synchronization. A degree of transparency is required to ensure it does not obscure any notated information.
- The points above will only ensure a true representation of the music if the note spacing in the score editor is consistent and there is a constant bar width in place.

- The video must be in high definition (1080 p), and a high frame rate (30 fps) should guarantee a smooth reading. The clarity of presentation should be maximized to satisfy player eyesight requirements and to compensate for lighting problems.
- Any screens used for playback of the scores would have to be sufficiently large to make the music very comfortable to read, especially considering the amount of additional information about changes in articulation and the unique dual part in *Six Spiders*.

A key early decision was that all players would read from the full score. While the primary reason was ensemble synchronization—being able to glance at other musicians' parts to gain a better understanding of the music—it also meant that only one video file per composition would be needed and that multiple players would be able to read from a single screen. The additional workload connected to using part scores would be a great challenge, particularly when score revisions are considered, and the logistics of setting up multiple video displays would be a considerable complication. What is more, it would contradict the key achievement that the musicians were most enthusiastic about—the ability to see what the others are doing.

STAGE 1: SCORE EDITING

The score editor of choice for this project was *Sibelius*. No major issues were encountered during the process of engraving, but preparing it for video editing proved a significant challenge. The first step was to consider note spacing and automatic layout options governing staff spacing and bar width—the fundamental topic of rhythmic spacing in music editing (Smith, 1973). The first two are basic modifiable options under house style and only took a small amount of trial and error to adjust; the bar width on the other hand, while controllable, does not have an option to set a fixed width. Sibelius would automatically choose a default width based on the contents of the bar, and changing every barline by hand would be inconsistent. Therefore, a workaround solution was needed, and the only one to prove effective was to create an unused empty staff, fill its bars with a string of quaver notes for the entire duration of the score, then use the Layout option *Focus on Staves* to force Sibelius to only display the staves needed in the score. The rhythmic value of the notes on the empty staff should depend on the shortest value found in the score. The effect of this, a fundamental requirement for a scrolling score, corresponds to what Smith describes as an 'extreme case' of 'strict rhythmic spacing', which highlights the contrast of media.

The second step became the most time-consuming element of the whole process—exporting image files of the scores for video processing. The aim was to achieve a single image file of an entire score, in order to set the parameters for the scrolling motion only once and avoid trying to join separate images while maintaining a consistent speed of motion. It proved impossible to achieve this aim on the computers available to

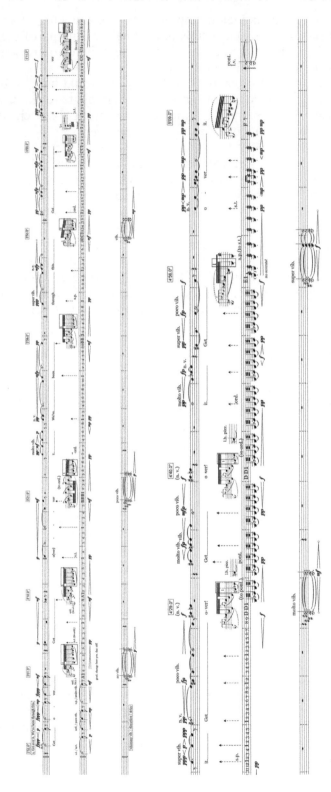

Figure 21.7 Section 2 of *Six Spiders* as a 'Sausage' Graphic Exported From Sibelius

the author, although it is an option that may be practical on a high-specification graphics workstation. Sibelius gives the option called *Select Graphic*, and the selection may be exported in a small range of image formats; viewing the score in Panorama mode makes it possible to zoom out until the whole score is visible, which can be selected and exported, with options for size and quality. While to achieve a quality image it is customary to aim for the size 300 dots per inch, it proved necessary to decrease it gradually until the employed computer's GPU was able to deal with the size of the resultant files. Through lengthy trial and error, respecting the limitations of the available hardware, a reliable process crystallized which involved exporting even sections of just under three minutes of score length.

STAGE 2: VIDEO EDITING

Video editing took place in *Lightworks* and, once the trial and error to find the right image size was resolved, it was a process satisfying in its efficiency. By using a keyframe effect, it was possible to achieve the smooth scrolling motion, as well as an opacity-controlled playhead, and most of the work was focused on setting the right speed of motion. Finally, it was necessary to ensure the score sections exported from Sibelius had equal number of leading and trailing bars (evident in Figure 21.7) in order to create cross-fade effects when joining the separate video-files into a single full-score experience. The pre-recorded electronic drones were embedded in the video files to provide reference for setting the scroll speed and to make it convenient for the musicians to rehearse with the playback.

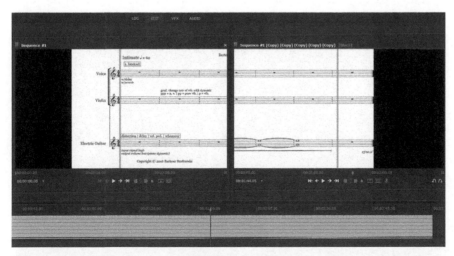

Figure 21.8 Editing and Calibrating Sections of *Six Spiders* as Video in Lightworks

STAGE 3: PERFORMANCE

Even though the approach was new to all parties involved, presenting the completed scrolling scores for both *Eight* and *Six Spiders* to the musicians during rehearsals was very rewarding. Having set up a duplicate HDMI display feed from a laptop PC into a large TV screen, it was possible to complete the first read-through attempts with a minimal amount of preliminary discussion, which primarily focused on details of articulation and expression.

In the case of *Eight*, the key achievement was ensemble synchronization, which the scrolling score made seamless and automatic. There was no need to compare the parts and agree cues, which removed considerable pressure, making it possible to focus more closely on details of timbre. Agata Kubiak, the second violin, summed up this improvement in rehearsal logistics when asked about the usefulness of this approach beyond the compositions currently at hand:

> I think this is going to save so much time, because if you looked at what we do when we work on something in the first few rehearsals, we spend half of the rehearsal flicking between the score and our parts. Because we're making little cues and writing little notes on who's doing what . . ., so just seeing everything laid out as a score as you play it, it will save an enormous amount of time. . . . It gives you that additional help with seeing what everyone else is doing, and you need to know that anyway.

Any potential issues with timing and executing various articulations over extended periods, as well as the problematic *glissandi*, had thus been solved before the first rehearsal. The chords in the piano part, frequently separated by considerable amounts of time, would approach the playhead

Figure 21.9 Rehearsing *Eight* with Konvalia Quartet (Piano out of Frame)

Figure 21.10 Six Spiders Performed in St Mary's Church, Ealing, London, December 7, 2017

gradually with no need for the performer counting the beats, and, on second attempt, the ensemble timing was executed with nearly surgical precision.

Agata was in a very different situation when it came to rehearsing *Six Spiders* with the scrolling score. She had already performed a well-formed draft from a printed part score and she found herself re-adjusting her approach, almost re-learning it, which was evidence of just how different the two media are, despite the fact that both contain traditional notation. Nonetheless, the unusual voice and violin role became easier to play, as Agata felt it allowed her to "look in between the two parts", rather than shifting her focus from one to the other. She was also optimistic about the expressive interpretation of the music presented in this fashion:

> I actually found it easier—I would love this as a practice device if not for everything. It's like having a visual silent metronome . . . fluidly, without the rigid beats.

The fact that the composer participated as a musician allowed him to maintain a very good level of understanding with the other musicians throughout the rehearsal process, and he also benefitted greatly from this medium of score presentation as a performer. The ensemble were able to successfully premiere *Eight* after only two rehearsals of 45–60 minutes.

CONCLUSION

The premiere performance of *Six Spiders* and *Eight* took place on December 7, 2017, at St Mary's Church in Ealing, London, and of the experience was reassuring concerning the effectiveness of these scrolling scores. While the pianist comfortably read the music from a 15-inch laptop display, Agata and the quartet had a 42-inch TV placed in front of them, and the setup worked successfully, even though it was logistically challenging to implement. While there remain interesting issues for further investigation – for example, working with a larger ensemble, the potential role of a conductor, audience engagement – it can be noted with satisfaction that, in a first performance situation, the issues previously identified as the reasons behind the scrolling score approach had all been solved and the music produced was very close to what the composer had intended.

It is important to note that this solution was developed specifically for the stretched-out textural music exemplified by *Eight* and *Six Spiders*, and it is possible that there are approaches to composition for which it would be a failure. At the same time, the musicians' feedback leads to the belief that the scrolling score has potential applications in rehearsal and performance strategies considerably broader than the author's own compositional output.

The key achievement of this project has been to test and confirm the potential of the scrolling score to be used for reading traditional musical notation in rehearsal and live performance, especially that involving an ensemble. The question of traditional notation is important, because the well-explored use of video for graphical notation limits the application of the technique to music associated with radical modernism and aleatory, leaving unexplored a great number of combinations of this mode of score delivery with other aesthetics. In the context of a rehearsal, the approach creates a new experience as regards the simplified logistics and increased speed of practice, as well as ensemble dynamic, where the need for preparatory discussion and cue planning is considerably smaller. Additionally, with neither page turning nor memorization required, and with the greatly simplified timing of events, live performance is also strongly affected.

There remains much to improve from the technological standpoint. The process of preparing these scrolling scores was very cumbersome and suffered from recurring issues of processing power and inflexible settings individual to each image processed. Whenever revision was needed, the entire process might have to be restaged. Streamlining by means of bespoke software is a future development without which a wider application of the technique might prove impractical; for larger ensembles, this would require the ability to handle individual parts dynamically. At the same time, improving the technology behind the process is likely to create further opportunities for new developments in the efficiency of composer-performer communication.

REFERENCES

App Store. (2018). *Decibel scoreplayer on the app store.* [online] Available at: https://itunes.apple.com/us/app/decibel-scoreplayer/id622591851?mt=8 [Accessed 10 May 2018].

Frederickson, J. (1989). Technology and music performance in the age of mechanical reproduction. *International Review of the Aesthetics and Sociology of Music*, 20(2), p. 193.

Hope, C. (2017). Electronic scores for music: The possibilities of animated notation. *Computer Music Journal*, 41(3), pp. 21–35.

Hope, C. and Vickery, L. (2011). Visualising the score: Screening scores in real-time performance. In *Diegetic life forms II conference proceedings*.

Hope, C., Wyatt, A. and Vickery, L. (2015). "The decibel scoreplayer: A digital tool for reading graphic notation." In *Proceedings of the international conference on technologies for music notation and representation*.

Kim-Boyle, D. (2014). Visual design of real-time screen scores. *Organised Sound*, 19(3), pp. 286–294.

Locke, D. (1987). Numerical aspects of Bartok's string quartets. *The Musical Times*, 128(1732), p. 322.

Scorefollower.com. (2018). *Score follower—new music resource.* [online] Available at: https://scorefollower.com/ [Accessed 10 May 2018].

Smith, L. (1973). Editing and printing music by computer. *Journal of Music Theory*, 17(2), p. 292.

Smith, R. (2018). *animatednotation.com.* [online] Animatednotation.com. Available at: http://animatednotation.com/index.html [Accessed 10 May 2018].

Vickery, L. (2012). The evolution of notational innovations from the mobile score to the screen score. *Organised Sound*, 17(2), pp. 128–136.

SCORES

Szafranski, B. (2016). *Eight. For string quartet, piano, and electronics.* Score available on request.

Szafranski, B. (2017). *Six spiders. For singer-violinist, electric guitar, and electronics.* Score available on request.

RECORDINGS

Kubiak, A. and Szafranski, B. (2018a). *Eight live premiere December 2017 backup video.* [Online Video] 7 Dec. 2017. Available at: https://youtu.be/zgf-ZLppJRRo. [Accessed 15 January 2018].

Kubiak, A. and Szafranski, B. (2018b). *Six spiders live premiere December 2017 backup video.* [Online Video] 7 Dec. 2017. Available at: https://youtu.be/EvqfHh6hCUg. [Accessed 15 January 2018].

22

Everything Is Musical

Creating New Instruments for Musical Expression and Interaction With Accessible Open-Source Technology— The Laser Room and Other Devices

Alayna Hughes and Pierluigi Barberis Figueroa

INTRODUCTION

Musicians and artists are integrating the use of new controllers and instruments into their performance. In recent years, mainstream artists such as Bjork, Imogen Heap, Nona Hendryx and Laurie Anderson have begun to use new instruments and wireless controllers in order to create music.

The live performance of an artist is the opportunity to connect with their audience as well as give an exceptional and innovative show. While many mainstream artists have developed their performances into impressive shows featuring state of the art visuals, some artists have opted to interact with their music by adding new controllers and instruments.

MAINSTREAM USE OF NEW INSTRUMENTS AND TECHNOLOGY

Bjork

Bjork is known as an artist who has integrated interesting technology and new instruments into her live shows and given the audience new ways of experiencing her music through apps and virtual reality.

During her 2011 Biophilia tour, Bjork performed with several new instruments. The Gravity Harp, designed by Andy Cavatorta, was a 20-foot high set of pendulums that swung back and forth to pluck strings with motorized assistance to keep in correct tempo (Marantz, 2019).

Another instrument used during this tour was a hacked Tesla coil, which received a signal to activate and play a musical note when a synth was played. Bjork has also performed with the Reactable (Technology, 2019), an instrument developed by the Music Technology Group at Universitat Pompeu Fabra that uses blocks to control sounds and patterns through an interactive tabletop screen.

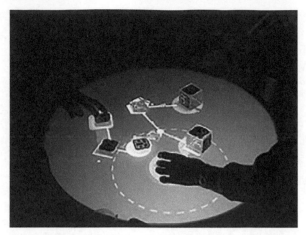

Figure 22.1 Reactable Instrument

Source: Reactable Systems SL

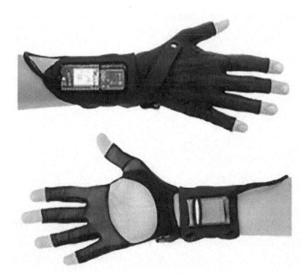

Figure 22.2 MiMu Gloves

Source: Adrian Lausch/MI·MU Gloves Limited

Imogen Heap

Imogen Heap is known as a successful musician and has been involved in developing wearable controllers known as MiMu gloves. These gloves are used in live performance in order to manipulate vocal performance (or an assignable input) with the aid of sensors embedded into the gloves.

A wearable device such as this is beneficial for vocalists who need to have a means of controlling effects within their set. A solution like this

Figure 22.3 Nona Hendryx Performing With Prototype of "The Music Suit"
Source: Berklee College of Music

allows the musician to control another instrument or to use the gloves to give the audience the visual that the performer can control their sounds by performing hand gestures.

Nona Hendryx

Nona Hendryx is a performer who gained recognition through the group LaBelle and later continued as a solo act and who combines funk, soul and electronic genres. She has incorporated technology into her performances over the past few years by working with technologists. She uses several devices in live performance, such as an audio tutu, wireless glove and a wireless bodysuit that acts as a MIDI controller (*Technology—Nona Hendryx*, 2019).

"MAKER" TECHNOLOGY AND THE POSSIBILITIES OF NEW INSTRUMENT CREATION

The emergence of open-source and easily accessible microcontrollers and sensors has allowed hackers and musicians the ability to create their own instruments and controllers.

Through the use of Arduino and Raspberry Pi, makers have been able to create their own instruments and controllers for use in their music. The cost of building a custom device is affordable and is a feasible option for hobbyists and musicians looking to innovate their act or create a new product. By using small microcontrollers and sensors, any object can be made into a musical instrument.

For example, an Arduino Uno or Adafruit Flora can be used in conjunction with a small electret microphone and attached to any physical

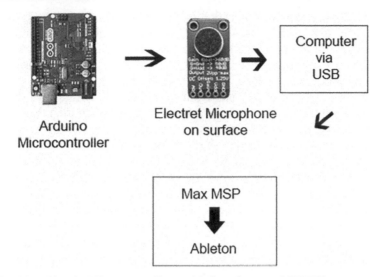

Figure 22.4 Simple Microcontroller and Microphone to MIDI Flow

object. This acts like a contact microphone and can be connected to Max MSP or Ableton Live in order to turn the sound input from the microphone into MIDI data, thus turning the object into an instrument. If the device is attached to a table, for instance, a user can tap on the table in varying places with various velocities. The device will then read those taps as an analog level and the user can read this information in MAX MSP and convert it to MIDI information playable by a software instrument or synth.

CASES IN UNCONVENTIONAL MUSICAL INSTRUMENTS—THE LASER ROOM, MUSIC SUIT, ROBATON ROBOTIC INSTRUMENT

Case 1: The Laser Room—An Interactive Musical Space

The Laser Room was a musical room concept by Curiosibot (Barberis and Hughes). The room was designed as an interactive area in which multiple users could move around a space and create music and visuals without the use of a controller.

Inside the room on one side was ten ultrasonic distance sensors wired to pairing laser sensors. Each laser is pointed towards a photocell on the opposite wall, which acts as an 'absorbent' sensor keeping the laser inactive until it is interrupted by a physical object. The lasers shoot across the room and give the user the appearance of a "Mission Impossible" type scenario. In fact, this is to show lines of interaction, and each laser represents a different sound or set of sounds. The software used for the project is Max MSP; Arduino runs the programming of the sensors and Ableton Live reproduces the sounds.

The idea of this project was to create an interactive experience through creation of sound. The project was first exhibited at Maker Faire Rome in

October (2016). In the first iteration of the installation, a higher-powered laser usually used for DJ performance was mounted at the front of the room on top of the frame. The laser was then run through a DMX box and connected to the computer running MAX and Ableton programming of the installation. This laser was used to provide a visual experience for the project. In the first iteration, data from the distance sensors on the walls was sent as MIDI data from Max to the DMX program, which then varied the projected image based on the note received.

During the first presentation of this version (which was closely followed by a second presentation at Maker Faire Bilbao), observations were made regarding the way that people interacted with the project, how this differed by age and how to improve the project for the next presentation.

The first observations concerned the age of the users and how they interacted with the room. The project was visited by a wide age range—from babies to adults. The children who were of pre-school to kindergarten age tended to want to interact with the project like a game and did not show interest in the cause and effect of the sensors and the sounds or visuals.

Children who were of primary school age tended to be the most intuitive users of the room. These children would walk into the space and interact freely and with enthusiasm to create sounds and enjoyed the visual projections. This age range needed little to no explanation and did not care so much about the technical premise of the project; they merely recognized that they were able to create music by moving around the room and touching laser lights.

Interestingly enough, adults did not prove to be the most intuitive users of the installation. For the most part, they entered the room and simply stared at the projection of the laser against the wall and did not move around the space. The adults required more explanation of the ideal use of the instrument and how it functioned.

During the second iteration of the Laser Room, a few changes were made in order to make the visuals more interactive to the user and to

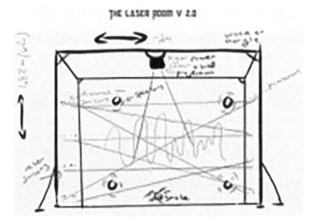

Figure 22.5 Laser Room Design Drawing

allow the users to hear the changes in sounds more directly. To allow for the users to see an immediate and direct relation to their actions and the visuals, a feature was added using a Kinect camera that translated the outline of the body into a laser projection. In order for the users to hear the sounds clearly and also to abide by the regulations of the Sonar+D Barcelona Festival where this was shown, headphones were added to replace speakers.

By making these few changes, the users, who were nearly all adults (only two children visited), were able to understand the project with explanation and to directly see and hear the result of their interaction with the room.

Figure 22.6 Laser Room in Use at Sonar+D Barcelona 2017

Case 2: The Music Suit—Building a Wearable Controller

Another unorthodox controller from Curiosibot was born from the collaboration with the musician Nona Hendryx. The concept from the artist's side was to create a music suit that she could use wirelessly in conjunction with her other controllers in a live performance.

To approach this project, the best method of using conductive materials was researched by testing cotton fabrics and spandex with Bare Conductive paint, conductive sewing thread and flexible conductive fabric.

The artist wanted to use a spandex full body suit, so spandex material was used to make a prototype.

The largest issue working with the conductive steel threads and stretch material proved to be the continuity of the connection. This was partially remedied by stretching the material while sewing to replicate the stretch from the body.

A sleeve prototype was constructed and demonstrated at a Music Hack Day event utilizing three fabric pads, an accelerometer and an Adafruit Flora with good results; however, Bluetooth connectivity would need to be added for the wireless feature. The Flora was chosen to be the main board for the suit due to its flat shape and wearable friendly design. At the time, capacitive touch style boards were not yet available, so the Flora was used and wired with resistors to facilitate capacitive touch with the fabric.

For the final design, three Adafruit Floras were used with one assigned to three patches. Sewable LEDS were added to the legs and arms for an aesthetic purpose and were also wired to a small microphone in the rear so that they would react to sound levels. On the left arm, eight patches were added to act as a piano.

For this, an Arduino FIO was used, as it was the only small board that provided the appropriate number of pins needed, and wired in the same manner as the Floras. In addition, a small Adafruit accelerometer was added to one of the sleeves to act as a variable send for effects.

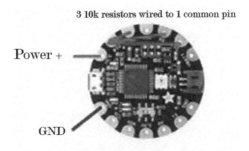

Figure 22.7 Capacitive Touch Wiring Configuration

Once the suit was sewn and wired, Bluetooth SMIRF chips were added to the boards using the capacitive touch configuration. The boards were then read via serial ports in MAX MSP and the information was scaled into MIDI information. The data was then converted into reading as a controller into Ableton Live so that it could be MIDI mapped in the program.

To increase ease of use for Nona, the MAX patch was converted into a Max for Live patch so that she could select the routing of the tracks and effects as she wished. The outcomes of the suit were mixed, as the biggest issues of creating this type of controller were dealing with multiple Bluetooth ports and with material that stretches, causing the conductive thread to break. Further iterations of this type of device would benefit from newer boards specific for capacitive touch, use of only one Bluetooth device, as well as a change in design of the actual suit.

Case 3: RoBaton and Curiosibot—Robots as Musicians

A case of using machinery or robotic components as musical instruments or to act as the musician is the 2014 project of Pierluigi Barberis and Alan Tishk known as Robaton. The project was conceived as an interactive tool for conductors, as well as a tool to create bridges for other professionals such as electronic engineers to develop their creative side.

The idea of a musician interacting with a machine in real time was to open new opportunities of performances for musicians and to create a new style of performance for the audience to see. The concept of working with robotic parts is often to showcase that anything can be transformed or adapted into a musical instrument.

The instrument utilized two octaves of a set of a symphonic instrument called crotales, which allowed for a design that would be a strike up method. Solenoids of 12 and 24 volts were used and mounted below the crotales along with analog LEDS to present a visual aspect as well as a practical confirmation of which note was being struck.

The operation of the instrument relied on an Arduino Due as the motherboard and MAX MSP as the programming software. In addition to the solenoids acting as the 'fingers' or mallets of the instrument, a robot arm by Lynxmotion was used to act as a head that followed the action of the notes and, with the addition of a 3D printed face, gave a 'human' aspect to the instrument.

The instrument runs from MIDI information sent from Ableton Live and transmitted through Max, which was then sent as voltage to the solenoids. Each solenoid was assigned to a note; therefore, as MIDI information was written from Ableton, the correlating crotale was struck. The instrument can be activated by sending MIDI from a program; in the Robaton project,

Figure 22.8 Curiosibot/RoBaton in Performance at Sonar+D 2014

it was activated and followed the tempo set by the conductor, who was placed in front of a Kinect camera (2019).

The Arduino-based robotic instrument project also became accessible and open source to anyone who wanted to replicate or improve upon the device for themselves (Barberis, 2019).

CONCLUSION

Open-source microcontrollers and inexpensive parts allow for inventors and musicians to create their own unique instrument, and processes like 3D printing allow for inexpensive prototyping and manufacturing. Musical devices and instruments no longer need to be conventional or restricted only to what can typically be found in a band or orchestra—anything can be musical.

REFERENCES

Barberis, Pierluigi. (2019). *Curiosibot. Instructables.com*. Available at: www. instructables.com/id/Curiosibot-1/ [Accessed 12 Jan. 2019].

Marantz, A., Chotiner, I., Sanneh, K., Cassidy, J., Wiley, C. and Halpern, J. (2019). Inventing Björk's gravity harp. *The New Yorker*. Available at: www. newyorker.com/culture/culture-desk/inventing-bjrks-gravity-harp [Accessed 12 Jan. 2019].

Robaton (Curiosibot) | Berklee College of Music Archives *Robaton (Curiosibot) | Berklee College of Music Archives*. (2019). *Archives.berklee.edu.* Available at: https://archives.berklee.edu/robaton-curiosibot/24 [Accessed 12 Jan. 2019].

Technology, R. (2019). *Reactable-music knowledge technology.* Available at: http://reactable.com/ [Accessed 12 Jan. 2019].

Technology—Nona Hendryx (2019). *Nona Hendryx.* Available at: https://nona-hendryx.squarespace.com/technology [Accessed 12 Jan. 2019].

23

The Impact of a Prototype Acoustically Transparent Headphone System on the Recording Studio Performances of Professional Trumpet Players

Andy Cooper and Neil Martin

INTRODUCTION

> Headphones often become the locus of a musician's discomfort in the studio, and the first instance of dissatisfaction concerns the sound of their instrument as mediated by technology and technician, compared to the sound of their instrument in the room.
>
> (Williams, 2009)

In commercial recording studios that focus primarily on recording instrumental and vocal performances, it is widely accepted that the foldback mixes created in the studio are a compromise necessitated by the need to isolate cue-mixes and click tracks from the recording microphones through the usage of closed-back headphones. Many performers will also shun the provision of the recording microphone feed as part of their monitor mix in favor of removing one ear cup of the headphones (or wearing single-sided models) to allow the cue-mix to be presented to one ear and to monitor their own instrumental performance with the other. In a recent study, 12 international conductors were surveyed with regard to headphone monitoring within a studio recording situation, and ten of these indicated that they did not use both of the headphone ear cups while performing. In Soudoplatoff and Prass (2017, p. 6), the authors state that

> Eight out of twelve participants reported that they always wore only one ear cup of the headphones when conducting. One participant specified that he sometimes wore only one ear cup; another that he covered only half of each ear. Five conductors explained that wearing only one ear cup allowed them to hear the direct sound of the orchestra. Two of them mentioned the importance of connecting with the orchestra through the acoustic sound and one specified that he wanted to enjoy the 'purity of timbres.'

The team's research thus far has concentrated on trumpet performance, this being the primary instrument of the second author. The trumpet provides an interesting starting point for the research given the instrument's highly directional sound propagation characteristics and the importance of the tone quality produced by the musician. It is also common practice for trumpet players to resort to the single-sided headphone monitoring technique in recording situations. The timbral qualities of the instrument are well known and have been the focus of extensive studies from Risset (1966), Moorer and Grey (1978), Morrill (1977), Meyer (2009) and others.

BACKGROUND

The research conducted for this project relates to both the timbral qualities of the trumpet, in an attempt to define the physical elements which may define a good trumpet tone quality, and the potential performance benefits of pseudo-binaural real-time headphone monitoring.

Trumpet Tone Quality

The spectral centroid of the harmonic series produced by the trumpet increases with intensity (Beauchamp, 2007); in other words, the position of the stronger harmonics present move upwards in the series in a direct relationship to the amount of vibrational energy created in the bore of the instrument by the musician.

The production of high-quality trumpet tone is achieved by a combination of the correct vocal tract position, the lip-reed mechanism and the player's breath control (Fox, 1982). Kohut (1985, p. 163) suggests that breath control is the most important element of the three:

> Breath directly affects intonation, articulation and diction, vibrato, dynamic level and intensity of the tone as well as phrasing, accents, and other aspects of musical expression . . . correct breathing, therefore, is an essential requisite to good performance, since it affects practically every aspect of tone production and musical expression.

The playing of the instrument also involves sensory feedback loops, as illustrated in Figures 23.1 and 23.2. However, the body of existing research into the importance of auditory feedback in musical performance has largely concentrated on auditory deprivation with studies by Repp (1999), Finney (1997) and Kristensen (20014) being notable in this area. For the purposes of this particular research investigation, we are of course examining the effects of compromised auditory feedback, rather than complete removal, on the resultant performance. It is interesting to note Kristensen's suggestion that the auditory feedback mechanism is the only feedback loop directly associated with timbre production.

Binaural Monitoring Systems

While the idea of mounting microphones on headphones is not new, it has traditionally been implemented for purposes other than studio performance

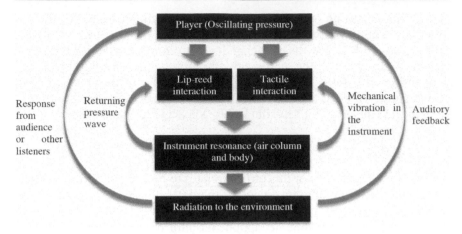

Figure 23.1 Trombone Feedback Model

Source: Kristensen 2014, p. 45

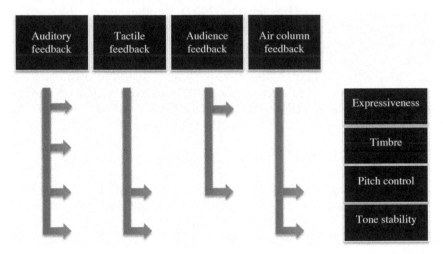

Figure 23.2 Suggested Relationship Between Feedback Loops and Sound Production in Wind Instruments

Source: Kristensen 2014, p. 46

monitoring. Commercially, the most notable existing system, specifically designed for the binaural recording enthusiast, is the long-discontinued JVC HM-200E model. This closed-back headphone system, released towards the end of the 1970s, featured microphones mounted towards the front of the ear cups within a molding imitating the function of the pinna. The attached cable terminated in two TRS jack plugs, one carrying the output from the two microphones to attach to a stereo recorder, and the other to return an amplified monitoring signal to the headphone drivers (Usami and Kato, 1978).

Various noise-cancelling headphones have also had microphones mounted on the ear cups to allow the electrical cancellation of external sounds via polarity inversion and also to allow the wearer to switch to the external signal to hear their local environment without removing the headphones. The Sensaphonics (Sensaphonics, 2019) 3D range of in-ear monitors also employ a binaural microphone system, allowing the sound from the stage to be mixed with the monitor mix to provide performers with a more natural sonic environment and to aid connection with the audience.

METHOD

The methodology followed was first to ascertain whether there were any subjective performance improvements with a trumpet monitoring signal derived from close to the player's ear, in comparison to that picked up by a recording microphone. Following on from these initial casual tests, several iterations of prototype binaural headphone monitoring systems were experimented with, the aim being to design a system that presented the wearer with a listening experience that approached that of the playing experience without headphones. Once the prototype designs were able to recreate this experience reasonably accurately, professional performers were brought into a studio recording session to allow comparative trumpet performances to be recorded using both conventional studio headphone monitoring arrangements and the prototype pseudo-binaural system. The aim of this testing was both to gather qualitative feedback from a selected sample of expert-performer subjects, and also to analyze qualitatively the resultant audio recordings in terms of pitch, timing and timbre to investigate any differences in the performance.

Preliminary Design Justification

The first preliminary investigation the team conducted was to determine whether a microphone set up close to the ear position and folded back into the cue-mix would provide a more satisfactory solution for monitoring with both headphone ear cups than that of the recording microphone signal being folded back. This arrangement provided a more natural representation of the trumpet when routed back into the performer's headphones, which was much preferred to that of the recording microphone signal by the second author, and also created recordings that were subjectively favored over those provided by the other technique. Most noticeably, these recordings demonstrated a greater control of breath energy and therefore better tone qualities than the recordings acquired using the conventional approach. This discovery led to the idea of developing a more practical solution for providing the signal at the ear position to the performer by employing head-mounted microphones. The two spectrograms provided here as Figures 23.3 and 23.4 demonstrate the radically different spectra propagated by a single trumpet note at a reasonably conventional recording position 15 cm to the front of the bell and a position alongside the performer's ear. Of note here is the presence of less inharmonic content at the

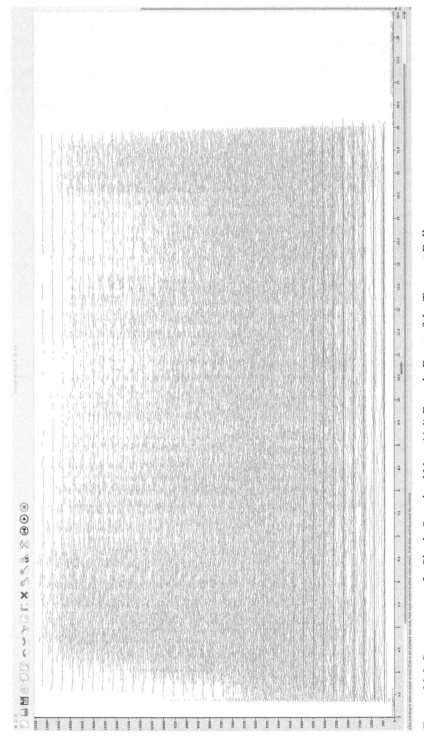

Figure 23.3 Spectrogram of a Single Sustained Note (A4) From in Front of the Trumpet Bell

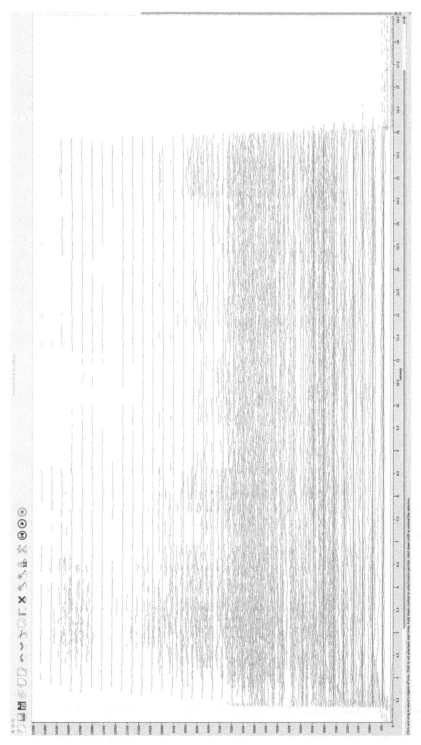

Figure 23.4 Spectrogram of a Single Sustained Note (A4) From the Ear Position of the Player

ear position in relation to the bell, combining with the amplitude reduction of the higher frequency components.

Prototype Development

In order to ascertain a suitable microphone orientation for headphone mounting, a pair of Audio Technica AT803 omnidirectional lavalier microphones were mounted onto a pair of AKG K44 semi-open circumaural headphones in five different configurations. A small-format audio mixer was used to provide phantom power and amplification to the microphones; this also allowed cue-mix signals to be mixed with the signals from the microphones. Each orientation was tested during instrumental performance and also in carrying out normal conversation in various rooms, in order to ascertain the perceived 'naturalness' of each.

It was agreed that Position 2 provides the most natural results based on these tests, despite the relative level of attenuation of signals from the rear being greater than that of normal hearing and thus adding an element of directionality into the perceived sound field. As an extension of these tests, directional (cardioid) microphones were experimented with during the development of Prototype 1, but these failed to give a particularly

Figure 23.5 Position 1: Central on Outside of Ear Cup and Forward Facing. This orientation created a spatial image that was perceived as being rather unnatural as it created the disconcerting perception of an artificial widening of the wearer's head.

Figure 23.6 Position 2: Inside Pad at Front of Ear Cup, Forward Facing. Placing the microphones in close proximity to the side of the wearer's head created a more accurate and natural-sounding spatial image. The acoustic shadowing of the ear cups provided noticeable attenuation of sound incident from behind.

Figure 23.7 Position 3: Above the Ear Cup Centrally, Forward Facing. With the microphones in close proximity to the head (mounted on the side of the headband) a reasonably natural, although slightly wide in the case of the lavalier microphones due to mounting restrictions, spatial image was attained. The room ambience is slightly more exaggerated in comparison to Position 2 and there is no attenuation of sound incident from behind the head.

Figure 23.8 Position 4: In Front of the Ear Cup (Upper-Side) Facing 45 Degrees Downwards. This mounting orientation was found to be very similar to that of Position 2 but the perceived naturalness of the environment was felt to be overly influenced by the nature of the flooring surface.

Figure 23.9 Position 5: XY Coincident Arrangement Above Head. This arrangement created an experience of an unnatural room ambience, the partial HRTF (binaural) function of the other configurations is completely negated with this orientation.

natural sounding environment; we suggest this may be due to the off-axis colorations associated with directional pick up.

Prototype 2 was developed using small electret microphones and AKG K44 headphones. A control box was designed to be powered by a 9V cell battery which provided 'plug-in' phantom power and amplification for the

microphones and the ability to mix the microphone signal with that of a
stereo cue-mix. Prototype 2 proved effective in a range of tests but suffered from poor signal to noise ratios, a lower than ideal maximum SPL and
non-uniform frequency response characteristics. This was mostly due to the
compromised quality of the range of small electret microphone transducers
that are available to be sourced directly from electrical component suppliers.

For Prototype 3, Soundman OKMII microphones were employed, again
in conjunction with the inexpensive AKG K44 headphones, and now in
combination with an improved control box design (lower noise, increased
dynamic range, greater power efficiency). However, the OKMII microphones were difficult to mount directly to the front of the ear cups, so a
compromise was made by placing these in Position 3. Due to the flat shape
of these transducers, it was possible to mount them closer to the side of
the head than it was with the AT803s used in the positioning tests and, as
a result, a slightly more natural sound field was realized.

Experiment 1

A professional trumpet player was brought in to perform on a studio session. The overdubbed melody was toward the top of the musical range of
the trumpet between the notes of B6 (concert) and E6 and featured a fast
staccato rhythm on the note of B. The takes were performed to a mixed
stereo backing track and a metronome click, using two differing monitoring techniques:

1. using foldback from the recording microphone with the headphones
   on both ears
2. using the Prototype 2 system with the headphones on both ears

Each performance was recorded with an AKG C414XLS microphone via
an Audient ASP008 pre-amplifier and Avid HD192 A/D Converter to Avid
Pro Tools at 24bit/44.1kHz resolution.

Experiment 2

Two further professional trumpet players were brought in to perform on a
studio session. Initially a three-part arrangement was performed to a click
track, using conventional headphone monitoring techniques and recorded
as a section via a stereo microphone (XY) array. The musical arrangement
was deliberately designed with close voicings in order to create a challenging backing track, where it would be difficult for the performers to
identify their own overdub melodies when playing. Each performer then
overdubbed a melody line to a cue-mix containing the click track and the
initial stereo recording. A single take was performed with each of three
monitoring approaches:

1. wearing both sides of the headphones with the signal from the recording microphone returned via an auxiliary send from the recording
   console

2. wearing one side of the headphones with just the cue-mix provided
3. using the Prototype 3 system with the headphones on both ears

After the session, the players were asked for their personal appraisal of each variation, and the recorded files were kept for further analysis. Each performance was recorded with AKG C414XLS microphones via an SSL Duality console and Antelope Orion A/D Converter to Avid Pro Tools at 24bit/44.1kHz resolution.

RESULTS

Experiment 1

Analysis of the recordings did not determine any notable differences in pitch accuracy between the two takes; however, the note timing of the staccato passage was more accurate in the performance using binaural monitoring, and this take also exhibited greater transient definition at the start of these notes. The sustained B (concert) notes at the end of the staccato phrase in each recording were measured in peak and maximum RMS level, and the harmonic balance quantified by an FFT process at 350 ms after the start of the onset (steady state). Although, subjectively, the take performed with the Prototype 2 monitoring system exhibits a more pleasing tone, analysis of the spectral content alone does not provide any specific clues as to why this may be the case.

The magnitude of the difference in playing intensity between the two takes, for this particular note a difference of 3 dB, makes direct comparison of the resultant spectra rather meaningless here, although it is interesting to note the much-reduced relative level of the 2nd harmonic produced on this particular note in the binaural performance.

Experiment 2

A representative phrase was selected from each of the alternative melodies. The phrase from Performer 1 ranged from F4 to Eb5, and the phrase from Performer 2, Eb4 to C5. The analysis of these recorded phrases revealed very little deviation in terms of pitch and timing from either performer, although interestingly, the performance from each performer using the auxiliary send signal from the recording microphone for monitoring tended to be very slightly flatter, on average, in comparison to the

Table 23.1 Sustained B6 at 350 ms

Monitoring Type	Peak Level dBfs	RMS Level dBfs	Level of Harmonic (n) dBfs									
			1st	2nd	3rd	4th	5th	6th	7th	8th	9th	10th
Single Side	−10	−14	−19	−19	−22	−33	−30.5	−34	−37	−37	−44	−46
Binaural	−7	−11	−15	−27	−24.5	−33	−28.5	−31	−39	−40.5	−35	−39

other takes; but still with less deviation than the accepted just noticeable difference (JND) threshold of +/−10 cents. However, there are noticeable variations in the subjective quality of each take, which relate to the 'quality of tone' produced and the impression of 'confidence' in the performance of the musicians. Of particular interest in the results of this experiment are the quantifiable differences in the frequency spectra produced when using the differing techniques. It is expected, and well documented, that as the dynamic level increases, the higher partials of trumpet tones will be raised in level relative to the other partials, but in the analysis of these performances it may be seen that the harmonic balance is altered in other ways, and often without a notable difference in the relative intensity.

The final note of each phrase shown was measured in peak and maximum RMS level, and the harmonic balance quantified by an FFT process at 350 ms after the start of the onset (steady state). Both players were remarkably consistent in terms of the peak RMS level measured, with Performer 1 at −22.5 dB RMS for each take and Performer 2 at −19 dB RMS for all but the auxiliary send example, which was 1 dB higher at −18 dB RMS. Peak levels had a slightly greater variance, with Performer 1 having a variance of 2.5 dB between the highest and lowest measurement, and Performer 2, 2.3 dB. The binaural examples saw both players with the fundamental at the lowest measured level, although for Performer 2 this happened to be the example with the highest overall peak level, and the greatest difference between the level of the fundamental and the 2nd harmonic (+11.5 dB and +9 dB, respectively). From the 10th harmonic upwards, both of the binaural examples possessed the least energy of the set.

Table 23.2 Performer 1—Final Note of Analyzed Phrase (A4) at 350 ms

Monitoring Type	Peak Level dBfs	RMS Level dBfs	Level of Harmonic (n) dBfs									
			1st	2nd	3rd	4th	5th	6th	7th	8th	9th	10th
Aux Send	−13.3	−22.5	−38	−32	−28	−30	−34.5	−34.5	−34	−39.5	−42.5	−49.5
Single Side	−11.4	−22.5	−40	−32.5	−31	−25	−32	−31	−30.5	−41	−41.5	−45.5
Binaural	−13.9	−22.5	−41.5	−30	−31.5	−30	−34	−38.5	−53	−49	−45.5	−58

Table 23.3 Performer 2—Final Note of Analyzed Phrase (A#4) at 350 ms

Monitoring Type	Peak Level dBfs	RMS Level dBfs	Level of Harmonic (n) dBfs									
			1st	2nd	3rd	4th	5th	6th	7th	8th	9th	10th
Aux Send	−12.1	−18	−33	−26.5	−25	−26	−34.5	−32.5	−36.5	−43	−51	−53
Single Side	−13.1	−19	−33	−28	−27	−32	−34	−36	−42.5	−53	−49.5	−56
Binaural	−11.4	−19	−33.5	−24.5	−26	−25	−30	−29	−39.5	−46	−51	−59.5

Figure 23.10 Experiment 2: Performer 1, Aux

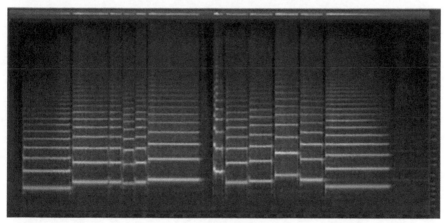

Figure 23.11 Experiment 2: Performer 1, Single

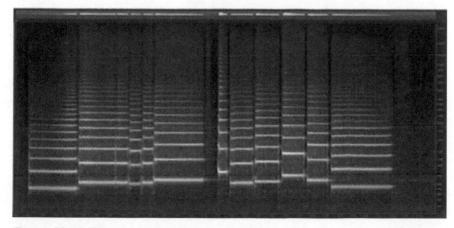

Figure 23.12 Experiment 2: Performer 1, Binaural

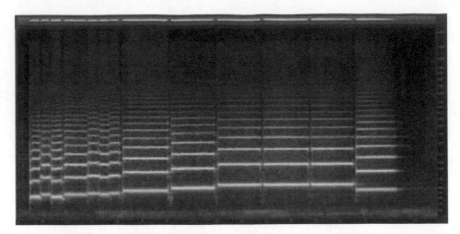

Figure 23.13 Experiment 2: Performer 2, Aux

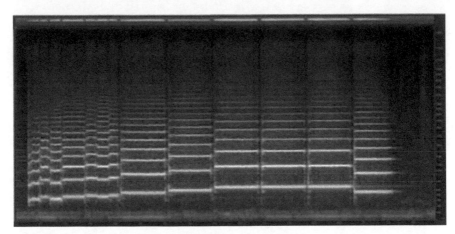

Figure 23.14 Experiment 2: Performer 2, Single

Figure 23.15 Experiment 2: Performer 2, Binaural

It could be suggested that the 'rounder' tone produced by both players using the binaural monitoring system could be due to the relatively higher energy of the 2nd harmonic and the reduced energy in the very high harmonics present in both these examples, but further research and analysis would need to be carried out to prove a definite correlation between these phenomena and what is considered to be good trumpet 'tone quality' in this particular note range.

COMMENTS

Experiment 1: Performer's Comments

"This is better than when you have a really good sound engineer who is sympathetic to the needs of the performer", Professor Mike Lovatt, Hon ARAM, lead trumpet of The John Wilson Orchestra and The BBC Big Band.

Experiment 2: Performers' Comments

Performer 1: "During the test recordings it was good to have the extra control. The chance to hear oneself so clearly meant that I was able to concentrate more on the music in front of me. I suspect it would be most useful in sessions with multiple musicians where quite often one's individual sound is lost".

Performer 2: "I found it easier to play in tune with this new kit. And it was easier to get the notes to speak, especially up high. I didn't have to blow so hard".

Comments From Angel Studios and Abbey Road Studios

The recordings from Experiment 1 were taken up to Abbey Road Studios, where one of the senior recording engineers was asked to listen to them and identify any noticeable differences in quality of performance. The tracks were labelled in such a way that they could be identified clearly after the blind test, but so that the filenames would not indicate which recording was which. The engineer happened to listen to the recording made with traditional monitoring first. When he listened to the second track, it took him only a few seconds to identify it as the superior performance.

The Prototype 3 system was taken to Angel Studios and used by the lead trumpet player of the BBC Big Band for a commercial overdub session. The recording engineer set up a separate mix for the headphones: some of the sound from the saxophones section—which was screened and to the right of the trumpets and trombones section—was added to the right side of the foldback mix.

It was an intense three-hour session, involving a significant amount of high-note playing for the lead trumpet, with emphasis on a repeated D6 (concert). In an informal interview immediately following the session, it became clear that the musician felt consistently more comfortable and in control compared to using more usual foldback setups. Other members of the brass sections—trumpet players and trombonists—tested the headphones briefly during the session; each one felt it to be superior to more usual setups.

DISCUSSION

The results detailed above seem to suggest that the use of the acoustically transparent headphone (ATH) system does indeed have a notable impact on both the quality of tone production and the confidence of the performer when using headphone monitoring. The fact that the control unit of this system provides the performer with the ability to balance the mix between the cue-mix and their own performance should not be overlooked, as this may well be an important factor in their preference of this monitoring strategy, and this is directly referred to in the comments of Experiment 2: Performer 1. However, in the case of the single-sided headphone monitoring performances detailed in Experiment 2, the performers also had control, via the headphone amplifier in the studio, to alter the relative level of the cue-mix in the headphones, as there was no microphone signal being sent to the headphones.

In terms of the harmonic analysis of the performances, the different registers in use between the phrases analyzed in both experiments renders a direct comparison between both in terms of the relationship between the timbre and the perceived tone quality impossible. However, the commonality in the results of Experiment 2 with regards to the effect of the differing monitoring approaches relating to the timbres produced presents some interesting questions for further research. The results of the binaural approach for both performers effectively exhibit a similar upward shift in the spectral centroid, but this is without the increase of intensity that is expected to accompany the timbral change. It could be suggested then that the timbral change has been a result of a change in the interaction between the three methods of note production: vocal tract shaping, the lip-reed mechanism and breath control, influenced by the difference in auditory feedback quality.

This assumption is also supported by the universal comments from the performers regarding their control over their individual performances being enhanced through the use of the ATH system. The suggestion here is therefore that improving the quality of the auditory feedback assists the trumpet-playing musician in their perception of tone production and allows them to modify the three interacting elements of this production with greater confidence and to superior effect.

CONCLUSION

The research presented here should be considered as a preliminary, rather than definitive, investigation into the effects of the quality of headphone monitoring on musical performance. The number of subjects involved in the experiments here is too low to state any of the findings as being indicative towards the tendencies of all trumpet players. However, the outcomes can be taken as positive encouragement that this line of research is worth pursuing. There appear to be correlations between performance monitoring accuracy and the level of control possible for the musician to realize in their performance, and therefore it is likely that the use of the ATH system, or a development thereof, will be able to generate better studio performances from wind instrument players and perhaps singers in comparison to the current monitoring techniques generally employed.

Informal testing has been undertaken with a variety of other performers, including instrumentalists, singers and conductors, and the comments have all been positive and encouraging.

It is intended to undertake further testing with a set of four pairs of the prototype headphones. The authors hope to establish that the spatial information provided by the ATH system will make it easier for performers to differentiate their own sound from the sounds of the other performers, initially in a standard trumpet or trombone section of four performers. We hope that the performers will find it easier to play in tune and to control their tones and volumes relative to each other. We also hope that the performers will find it beneficial that they have individual control of both their own volume in the headphones and that of the foldback.

REFERENCES

Beauchamp, J. (2007). *Analysis and synthesis of musical instrument sounds in analysis, synthesis, and perception of musical sounds: The sound of music* (Ed. Beauchamp, J). New York: Springer Science & Business Media.

Dodge, C. and Jerse, T. (1985). *Computer music: Synthesis, composition and performance.* New York: Schirmer Macmillan.

Finney, S. A. (1997). Auditory feedback and musical keyboard performance. *Music Perception: An Interdisciplinary Journal*, 15(2).

Fox, F. (1982). *Essentials of brass playing: An explicit, logical approach to important basic factors that contribute to superior brass instrument performance.* Harlow, Van Nuys: Alfred Music Publishing.

Kohut, D. L. (1985). *Musical performance: Learning theory and pedagogy.* Englewood Cliffs, NJ: Prentice-Hall.

Kristensen, E. (2014). *An acoustical study of trombone performance, with special attention to auditory feedback deprivation.* Norwegian University of Science and Technology, Department of Physics. Available at: https://daim.idi.ntnu.no/masteroppgaver/008/8684/masteroppgave.pdf [Accessed 19 July 2017].

Meyer, J. (2009). *Acoustics and the performance of music: Manual for acousticians, audio engineers, musicians, architects and musical instrument makers.* New York: Springer Science & Business Media.

Moorer, J. A. and Grey, J. (1978). Lexicon of analyzed tones (Part 3: The Trumpet). *Computer Music Journal*, 2(2). Cambridge, MA, MIT Press.

Morrill, D. (1977). Trumpet algorithms for computer composition. *Computer Music Journal*, 1(1). Cambridge, MA, MIT Press.

Pavel, A (1983). *High fidelity stereophonic reproduction system*, US4412106A.

Repp, B. H. (1999). Effects of auditory feedback deprivation on expressive piano performance. *Music Perception: An Interdisciplinary Journal*, 16(4). Oakland CA, University of California Press.

Risset, J. (1966). *Computer study of trumpet tones*. Murray Hill, NJ: Bell Telephone Laboratories.

Sensaphonics. (2019). *In ear monitoring solutions*. [Online] Available at: www.sensaphonics.com/3d [Accessed 21 Jan. 2019].

Soudoplatoff, D. and Pras, A. (2017). *Augmented reality to improve orchestra conductor's headphone monitoring*. AES Convention paper 9720 142nd AES Convention, 20–23 May, Berlin, Germany.

Usami, N. and Kato, T. (1978). *Headphone unit incorporating microphones for binaural recording*, US4088849.

Williams, A. (2009). I'm not hearing what you're hearing: The conflict and connection of headphone mixes and multiple audioscapes. *Journal on the Art of Record Production* (4).

24

Evaluating Analog Reconstruction Performance of Transient Digital Audio Workstation Signals at High- and Standard-Resolution Sample Frequencies

Rob Toulson

INTRODUCTION

High-resolution (or hi-res) audio can be defined as digital audio data that has greater amplitude resolution than 16-bit or greater time-axis resolution than 44.1 kHz (Rumsey, 2007). As the compact disc (CD) delivery format itself delivers 16-bit and 44.1 kHz accuracy, hi-res can sometimes be described simply as "greater than CD" resolution. In modern music production projects, 24-bit recording and reproduction is standard, but considerations among 44.1 kHz, 96 kHz and 192 kHz sampling rates are still made by many professionals.

The performance of digital audio sampling continues to generate debate among experts, even given the thoroughly documented theories developed first by Nyquist (Nyquist, 1924) and Shannon (1949). Rumsey (2004, pp. 34–35) reflected on this debate, reporting that strong arguments remain for sampling at no higher than a rate at double the bandwidth of human hearing, as specified by Nyquist's and Shannon's sampling theories. Conversely, Rumsey also notes that many professional studio engineers and music producers are in favor of higher sample rates in practice.

Psychoacoustics research by Moylan (1987) indicated that listeners can hear the difference between audio sampled at higher resolutions than the Nyquist frequency for the threshold of human hearing (40 kHz), particularly with the onset of transients. Higher sampled audio is described as sounding warmer, sweeter and fuller. More specifically, Moylan experimented and deduced that humans can hear the influence of a 45 kHz frequency superimposed on a 15 kHz fundamental, even though humans cannot hear the 45 kHz frequency when it is emitted alone. Moylan concluded that humans can hear above the standard 20 kHz range in more complex sounds, though not specifically on single sinusoids. A significant challenge to Moylan's results is that the rationale for subjects hearing the additional ultrasonic components is owing to intermodulation distortion caused by non-perfect playback electronics and hence additional distortion components present within the audible (i.e. sub-20 kHz) range, as discussed by Colletti (2013) and Lavry (2012).

385

A number of listening tests have been conducted to help decide if high-resolution audio is identified and perceived as an improvement by listeners. Jackson et al. (2014) evaluated the audibility of digital audio filters in order to identify if every aspect of audio signals can be conveyed using only frequencies below the Nyquist limit of a standard CD recording (i.e. 22.05 kHz). Their conclusion was that audible signals do exist that cannot be encoded transparently by a standard CD. Reiss (2016) gives an overview and meta-analysis of published psychoacoustic testing for evaluating audio signals sampled at higher than the CD standard of 44.1 kHz. Reiss' study observed the findings of 12,000 different individual listening tests and draws the conclusion that high-resolution audio has a small but important advantage in its quality of reproduction over standard-resolution audio content.

Stuart (2015) makes a number of important observations and claims that are under-investigated, arguing that "the sounds that are important to us" are not represented by the frameworks described by Nyquist and Shannon. Stuart emphasizes that sound is not inherently band-limited, and hence some audio signals extend beyond the documented human threshold of hearing. Equally, Stuart explains that music and audio sounds do not have an infinite nature, a predictable occurrence or repetition, which are all factors that the conventional signal reconstruction theories rely on.

In this paper, we look specifically at analyses of Shannon's reconstruction theories for common audio signals that do not adhere to the ideal framework—signals which are neither infinitely repeating nor predictable in occurrence. The evaluation is intended to explore how the ideal reconstruction theory performs when considering data that more closely represents produced music than band-limited infinite sine waves.

AUDIO SIGNAL RECONSTRUCTION AND MODERN MUSIC PRODUCTION

The Whittaker-Shannon Interpolation Formula (WSIF)—sometimes called the Ideal Interpolation Formula or the Sinc Interpolation Formula—states that a band-limited continuous-time signal x(t) of bandwidth B Hertz can be discretely sampled and uniquely recovered by Equation 1, providing that the sampling rate Fs > 2B (Shannon, 1949). The WSIF is expressed as

$$x(t) = \sum_{n=-\infty}^{\infty} x(nt) \frac{\sin\left[\left(\frac{\pi}{T}\right)(t-nT)\right]}{\left(\frac{\pi}{T}\right)(t-nT)} \tag{1}$$

where the continuous-time signal *x(t)* is sampled at discrete *nT* intervals. *T* is the sampling period given by T = 1/Fs.

The WSIF assumes an infinite number of samples and an infinitely repeating time signal, as well as a history of infinite data (as shown in Equation 1, data for all values of n between -∞ and +∞). Hence, the WSIF is "non-causal and physically non-realisable" (Proakis and Manolakis, 1992, p. 425).

Audio signals in many applications are neither predictable nor guaranteed to be repeated, and hence do not deliver an infinite history of data. Real audio signals therefore do not guarantee to obey the concepts of ideal signal reconstruction at all sample frequencies. Despite this limitation, reconstruction techniques are rarely evaluated for transient signals at sample frequencies nearing the Nyquist limit. Given the continuing debates regarding hi-res audio, it is therefore important to model and evaluate the reality of signal reconstruction with respect to transient and non-sinusoidal signals, in order to acquire a better representation of how signal reconstruction performs in practical audio systems. If it has been agreed that the whole bandwidth of 20–20,000 Hz should be uniformly reconstructable for high-fidelity audio, then the reconstruction of a signal containing 20 kHz content should be equally as successful as that of a 100 Hz signal, irrespective of whether listeners can generally perceive a difference or not. The arguments for standard-res audio sampling being sufficient are based substantially on Shannon's theories and the ability for the WSIF to perfectly reproduce sampled analog signals.

In a practical context, and with reference to music production, modern digital audio workstations (DAWs) and associated processing tools (plugins) have the ability to create and manipulate signals with no regard for standard sampling and reconstruction criterion. For example, a transient percussion signal might be sampled at 44.1 kHz through a 22 kHz band-limiting (anti-aliasing) filter, yet, once in the digital domain, it might be processed with digital dynamic range or wave-shaping tools that attempt to extend the theoretical bandwidth beyond the previously enforced Nyquist limit. A second example is a software synthesizer tool that attempts to create a square wave output, which would theoretically incorporate unlimited odd harmonics. The audio signal presented to the reconstruction filter within the digital-to-analog convertor (DAC) is digitally manipulated post-anti-aliasing, so the performance of reconstruction could become unpredictable. In this scenario, the reconstruction filter may simply not be capable of realizing the effect of non-linear processing tools that have been introduced, or the reconstruction of such processed signals might introduce artifacts that could become audible. Additionally, in music production (particularly scenarios utilizing multitrack synchronized audio), temporal accuracy is of significant importance. Indeed, temporal errors in reconstruction may be more audible and subjectively detrimental than artifacts and distortions identified and measured in the frequency domain.

Conventional testing and analysis of audio systems and signal processes is usually conducted with a continuous 1 kHz test frequency. However, in order to evaluate the reconstruction of DAW processed audio signals and potential benefits of utilizing high-resolution sample rates, the WSIF's performance for all audio signal types should be considered in more detail than has previously been conducted in prior published research and analyses. Of particular interest for modeling and analysis are transient waveforms and those with fundamental frequencies approaching the Nyquist limit.

EVALUATING THE RECONSTRUCTION OF TRANSIENT AUDIO SIGNALS

In the following modeling experiment, an 8 kHz sine wave with a transient decay is chosen for evaluation, as shown in Figure 24.1.

The 8 kHz sine wave is chosen because it is a suitably high-frequency signal that comes sufficiently close to the standard (44.1 kHz) Nyquist sampling limit, while still being audible to healthy listeners. The 8 kHz frequency is also a significant frequency in music production, and manipulation of this frequency with equalization can make a substantial difference to the audible attributes of produced music; indeed, many music producers recommend manipulating this frequency to reduce sibilance in a singer's voice or to add clarity, presence and sparkle to recorded instruments (Izhaki, 2008, pp. 212–213; Owsinski, 2013, p. 27). The test signal was created with a 1 MHz sample frequency in order to—as close as possible—mimic an analog signal that can be sampled and reproduced at a number of reconstruction frequencies.

Figure 24.2 shows the success of the WSIF reconstruction for the finite decaying 8 kHz sinusoid signal sampled at Fs values of 44.1 kHz, 96 kHz and 192 kHz. It can be seen that at 192 kHz, the reconstructed signal is superimposed almost exactly over the original continuous-time signal and the sampled data. With 44.1 kHz sampling and reconstruction, discrepancies between the original signal and the reconstructed signal can be seen.

Most notably, the 44.1 kHz reconstruction is unable to reproduce the rapid attack profile of the transient signal. Its temporal profile has been altered, in that the attack onset is at a reduced gradient and the peak value occurs later than that of the original signal. This temporal error could potentially impact on the accuracy and synchronicity of audio playback, particularly in a multichannel setup, and alter subjective musical attributes such as perceived "tightness" and "crispness".

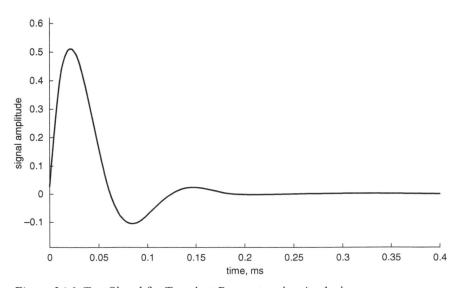

Figure 24.1 Test Signal for Transient Reconstruction Analysis

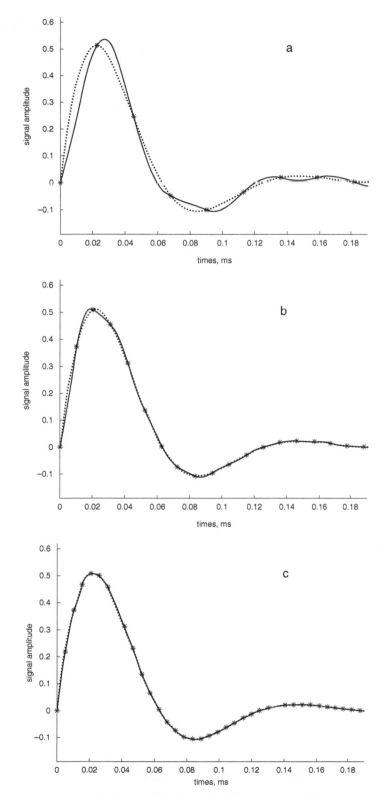

Figure 24.2 Reconstruction Profiles for an 8 kHz Transient Sinusoid Sampled at (a) 44.1 kHz, (b) 96 kHz and (c) 192 kHz. Reconstruction signal shown as solid line with analog waveform as dash-dot line. Sample points shown as asterisks.

Figure 24.3 shows the Fast Fourier Transform (FFT) frequency spectra of the 1 MHz sampled test signal, as well as the frequency profiles of the reconstruction data shown in Figure 24.2. Each spectra in Figure 24.3 shows the broad fundamental 8 kHz peak of the test signal. Given the transient nature of the signal and the small number of oscillations before the

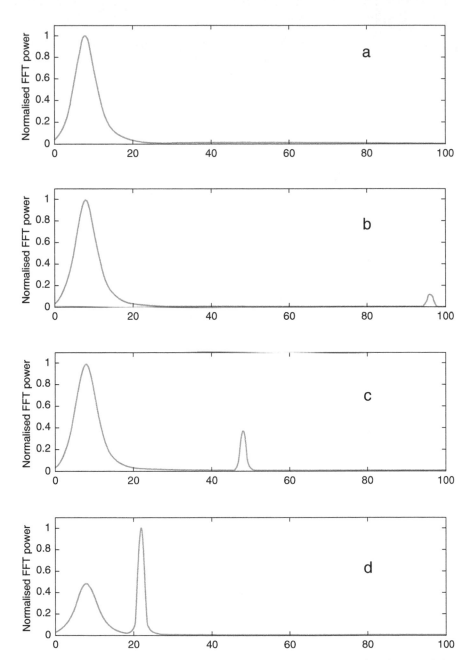

Figure 24.3 Frequency Spectra of the 1 MHz Sampled Test Signal (a) and the Reconstruction Models Calculated for (b) 192 kHz, (c) 96 kHz and (d) 44.1 kHz Sample Frequencies

test signal decays to zero, a broadband peak is expected from the FFT calculation, and it can be seen that the 8 kHz peak occupies all of the audible
range up to and above 20 kHz.

It can be seen that the WSIF reconstruction for discrete samples introduces an alias peak at the Nyquist frequency for each of the three sample
frequency models. These aliases are at high (inaudible) frequencies for
192 kHz and 96 kHz models, but the 44.1 kHz alias is seen to be significantly powerful and close to the human threshold of hearing. While
these artifacts are anticipated to be inaudible to human listeners, there is a
possibility that intermodulation distortion components, within the audible
range, could be introduced in an onward processing system as a result of
errors in the transient reconstruction.

To evaluate the WSIF as the sampling ratio approaches that of the
Nyquist limit, i.e. at Fs/B = 2.2, a 20 kHz transient test signal is also considered with a sample frequency of 44.1 kHz (shown in Figure 24.4). It can
be seen that the reconstructed signal is substantially different from that of
the original time-domain profile, meaning that accurate reconstruction is
not achieved for this transient signal with a fundamental frequency close
to the Nyquist limit.

It is therefore shown that the WSIF, i.e. the transfer function of an ideal
reconstruction filter, does not uniformly reconstruct signals for the entire
Nyquist bandwidth, if the signal is transient and made up of a finite number of data samples. The analysis indicates that, when considering transient
signals, higher sample rates above the Nyquist criterion (i.e. at > 2*B) are

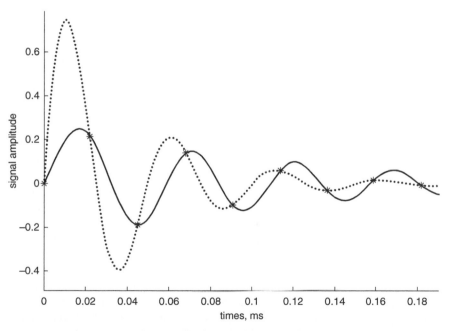

Figure 24.4 Reconstruction Profile for a 20 kHz Transient Sine Wave Sampled at
44.1 kHz. Reconstruction shown as solid line with analog waveform as dash-dot
line. Sample points shown as asterisk.

required for accurate reconstruction of the entire audible (20–20,000 Hz) frequency range.

HARDWARE VERIFICATION

To verify the accuracy of the WSIF and transient signal reconstruction in practical digital-to-analog audio conversion systems, a hardware verification exercise was conducted. Digital signals were generated at the designated Fs rates in Matlab and rendered as Microsoft Wave (.wav) audio files. These test wave files were loaded into Fidelia audio playback software and output from a number of different audio interfaces, including TC Electronic Studio Konnekt 48, Focusrite Saffire 6 and Cambridge Audio DacMagic Plus devices. Analog signal profiles were captured with a GW Instek digital oscilloscope utilizing a 5 MHz sample rate.

For each hardware interface, the transient test signal shown in Figure 24.1 was captured after analog reconstruction for 44.1 kHz, 96 kHz and 192 kHz sample frequencies. Time and frequency domain results for signal reconstruction with the TC Electronic Studio Konnekt, Cambridge DacMagic and Focusrite Saffire are shown in Figures 24.5, 24.6 and 24.7, respectively.

Looking first at the temporal performance of the TC Electronic DAC (Figure 24.5), it is seen that the 192 kHz reconstruction most closely resembles the profile of the test signal shown in Figure 24.1. The 96 kHz reconstruction is very similar, whereas the 44.1 kHz reconstruction shows a much reduced attack gradient and hence a greater period between the onset of the transient and the peak value, which corroborates properties of the 44.1 kHz model result shown in Figure 24.2a. The frequency domain data in Figure 24.5 shows no significant artifacts for the 192 kHz and 96 kHz reconstructions, though the 44.1 kHz reconstruction does show evidence of the 22 kHz artifact that is seen in the model results shown in Figure 24.3d.

The Cambridge DacMagic reconstruction profiles (Figure 24.6) similarly show a most accurate time-domain reconstruction for the 192 kHz sample rate. Again, the 44.1 kHz reconstruction has a significantly reduced attack gradient and hence a greater period between the onset of the transient and the peak value. The 96 kHz and 44.1 kHz signals show clear sinusoid artifacts on the waveform, which are evident in the frequency domain plots and matching those shown in the model results of Figures 24.3c and 24.3d. The artifact frequency peak of the 44.1 kHz reconstruction is seen to bleed significantly into the audible (sub-20 kHz) range.

The Focusrite Saffire (Figure 24.7) also shows significantly better reconstruction performance at 192 kHz, though the 96 kHz reconstruction also closely resembles the ideal signal profile shown in Figure 24.1. The Focusrite DAC is USB powered and unsurprisingly displays a lower signal-to-noise ratio, which can be seen on both the time-domain signal and as spurious data on the frequency domain plot. As with all the DACs tested, the 44.1 kHz reconstruction has a significantly reduced attack gradient and hence a greater period between the onset of the transient and

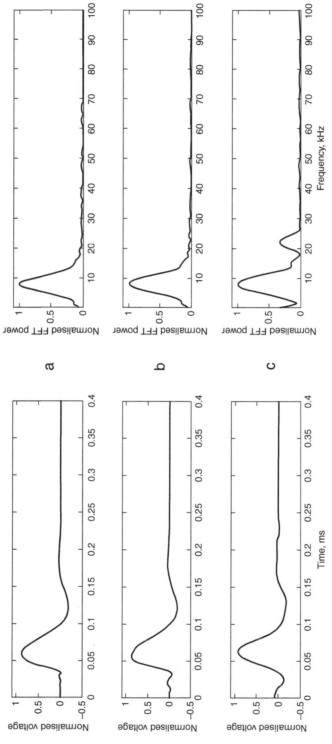

Figure 24.5 TC Electronic Studio Konnekt Reconstruction of Transient Waveform Data at Fs = 192 kHz (a), 96 kHz (b) and 44.1 kHz (c)

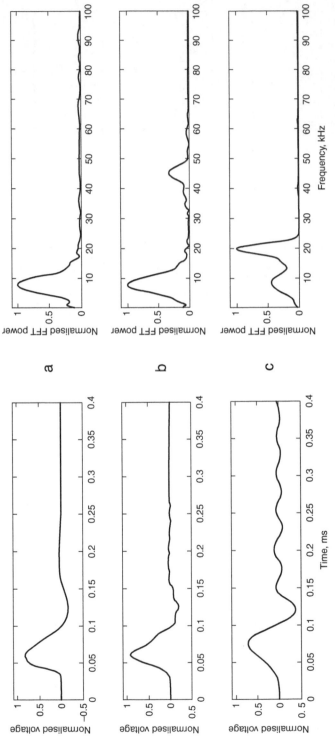

Figure 24.6 Cambridge DacMagic Plus Reconstruction of Transient Waveform Data at Fs = 192 kHz (a), 96 kHz (b) and 44.1 kHz (c)

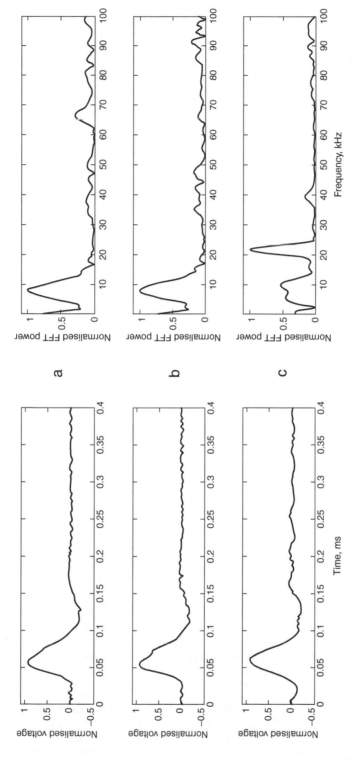

Figure 24.7 Focusrite Saffire 6 Reconstruction of Transient Waveform Data at Fs = 192 kHz (a), 96 kHz (b) and 44.1 kHz (c)

the peak value. The frequency domain data for the 44.1 kHz reconstruction again shows the alias component, as seen in the model shown in Figure 24.3d. Here it is also seen that the 8 kHz test frequency is significantly altered and corrupted in the reconstruction.

The purpose of this analysis is not specifically to evaluate system performance between devices, but moreover to verify the WSIF model results and identify if higher sample frequencies yield performance benefits in terms of reconstruction accuracy. In all cases the 192 kHz reconstruction shows greater temporal accuracy and less introduced artifact in the frequency domain.

DISCUSSION, CONCLUSIONS AND FUTURE WORK

It is shown by the presented analyses that, in practice, the WSIF becomes compromised for high-frequency signals which are transient and non-infinite in repetition—i.e. signals that could be described as authentically produced audio or music signals, or components thereof. In particular, transient signals, such as those associated with percussion instruments, can therefore expect to be more authentically reproduced when stored and replayed digitally at higher sample frequencies. It is seen in both the mathematical models and the hardware testing that the transient signals reproduced at lower sample frequencies have more shallow gradients of attack with delayed positioning of the transient peak, as well as having potentially audible artifacts introduced in the frequency domain. This is generally owing to the fact that in order to perform accurately, the WSIF requires a pre-filled history of data that perfectly reflects the signal to be reconstruction, which is not likely to be the case for real audio and music signals that are reconstructed through a non-ideal (i.e. non-brickwall) analog filter. In the hardware DAC test results, differences were seen among the types of artifacts and errors of reconstruction at 44.1 kHz, most likely owing to subtly different reconstruction filter designs being implemented by different manufacturers. All hardware DACs, however, reproduced the test signal very well at 192 kHz.

In practice, transient audio signals are generally unpredictable and encountered after a brief period of silence, especially those relating to percussion signals such as kick and snare drums, so temporal accuracy is of importance when considering music production projects and particularly multitrack audio that relies on precise synchronization of audio data. This investigation shows that the time-domain accuracy of reconstruction is reduced for transient signals that have frequency components approaching the Nyquist limit, so raising the Nyquist limit (by implementing a higher sample frequency) can be beneficial in practical scenarios.

As Stuart (2015) emphasizes, audio signals are not inherently band-limited and, even if a band-limiting (anti-aliasing) filter is used at the point of digital conversion, once in the digital domain, today's music producer has many tools to manipulate the waveform samples regardless of ideal sampling theories. For example, compression and distortion tools can easily turn pure sine waves into hard-clipped waveforms, transient envelope tools can sharpen the attack profiles of percussion transients, and synthesizer tools can generate harsh square wave signals with full-scale slew between individual samples. If the authenticity of the reconstruction is desired to be most accurate, then the use of higher sample frequencies, as shown by this research, can be expected to give a performance advantage.

The analysis presented here therefore, in many ways, aligns with the perception of critical listeners in the music production industries who claim that transients and temporal accuracy are audibly improved with high-resolution audio utilizing sample frequencies at 96 kHz (4.8 times the audible frequency bandwidth) and 192 kHz (9.6 times the audible bandwidth).

It is common to quantify the performance of audio systems by electronic measurements, i.e. rather than through subjective listening. For example, total harmonic distortion is regularly quantified by electronic signal analysis with a test signal at 1 kHz. Such quantified measurements are used to give an indication of the performance or "quality" of the audio device, and an assumption is made that the performance parameters correlate with the subjective listening experience. With this regard, it is therefore suggested that quantified performance metrics could be developed for measuring the capability of DACs to accurately reproduce digital waveforms at standard- and high-resolution sample frequencies—particularly transient test signals as opposed to continuous sinusoidal signals. While it is necessary to evaluate performance at the standard 1 kHz test measurement, it is suggested that all audio equipment should also be graded on performance for signals close to the Nyquist limit, and particularly with respect to the reproduction and signal processing of transient signals.

The results gathered show that the WSIF model and the actual DAC reconstruction of an 8 kHz transient signal is more accurate when using a 192 kHz sample frequency as opposed to a 44.1 kHz sample frequency. It is not proven, however, whether healthy listeners are able to repeatedly identify this performance improvement in listening tests, though analysis has shown that temporal errors and spectral artifacts within the human hearing range are potentially generated. It could therefore be hypothesized that, for some test material, a difference could feasibly be identified by listeners. This hypothesis will form the basis for future testing.

In general, there are still two fundamental unanswered questions with regards to the human perception of high-resolution audio:

1. Can ultrasonic (i.e. >20 kHz) audio components and qualities be perceived when using high-resolution audio sample rates?
2. Can high-resolution sample rates noticeably improve the reproduction quality of 20 kHz band-limited audio?

These two questions will be considered in future experimentation related to the research results presented here and with respect to common processes in music production. In particular, the design of listening test experiments is critical in achieving conclusive results. When considering listening tests with produced music material, it is important to ensure that the entire music production signal chain extends to suitably high ratings, for example, using ultrasonic microphones, suitably high sample rates, anti-aliasing filters and low-distortion hardware to avoid perceivable sub-20 kHz intermodulation artifacts. These testing methods will be evaluated and perfected in future studies in order to confirm whether the theoretical and measured advantages of high-resolution audio shown in this paper can be perceived by listeners in both laboratory and social listening scenarios.

REFERENCES

Colletti, J. (2013). The science of sample rates (When higher is better—and when it isn't). *Trust Me I'm a Scientist* (website). Available at: www.trust meimascientist.com/2013/02/04/the-science-of-sample-rates-when-higher-is-better-and-when-it-isnt/.

Izhaki, R. (2008). *Mixing audio*. 2nd ed. Oxford: Taylor & Francis, pp. 212–213.

Jackson, H. M., Capp, M. D. and Stuart, J. R. (2014). The audibility of typical digital audio filters in a high-fidelity playback system. In: *Proceedings of the 137th audio engineering society convention*, Los Angeles.

Lavry, D. (2012). The optimal sample rate for quality audio. *Lavry Engineering Inc.* (website). Available at: www.lavryengineering.com/pdfs/lavry-white-paper-the_optimal_sample_rate_for_quality_audio.pdf.

Moylan, W. (1987). A systematic method for the aural analysis of sound in audio reproduction/reinforcement, communications and musical contexts. In: *Proceedings of the 83rd convention of the audio engineering society*. New York.

Nyquist, H. (1924). Certain factors affecting telegraph speed. *Bell Systems Technical Journal*, 3, p. 324.

Owsinski, B. (2013). *The mixing engineer's handbook*. 3rd ed. Boston: Cengage Learning, p. 27.

Proakis, J. G. and Manolakis, D. G. (1992). *Digital signal processing: Principles, algorithms and applications*. 2nd ed. New York: Palgrave Macmillan, p. 425.

Reiss, J. D. (2016). A meta-analysis of high resolution audio perceptual evaluation. *Journal of the Audio Engineering Society*, 64(6), pp. 364–379.

Rumsey, F. (2004). *Desktop audio technology: Digital audio and MIDI principles*. Oxford: Focal Press, pp. 34–36.

Rumsey, F. (2007). High resolution audio. *Journal of the Audio Engineering Society*, 55(12), pp. 1161–1167.

Shannon, C. E. (1949). Communication in the presence of noise. *Proceedings of the Institute of Radio Engineers*, 37, pp. 10–21.

Stuart, B. (2015). High resolution audio: A perspective. *Journal of the Audio Engineering Society*, 63(10), pp. 831–832.

25

Acoustic Transmission of Metadata in Audio Files Using Sonic Quick Response Codes (SQRC)

Mark Sheppard, Rob Toulson and Jörg Fachner

INTRODUCTION

Sonic Quick Response Code (SQRC) algorithms offer a methodology for introducing inaudible metadata within a high-definition 96 kHz sampled audio file (Sheppard et al., 2016). This metadata insertion concept is analogous with visual Quick Response (QR) codes, which display binary image data representing an internet web-link (18004:2000, 2019). QR codes are two-dimensional matrix barcodes that are read by smartphone and tablet-based applications, along with dedicated QR reading devices. The encoded information contained within the QR code can consist of any alphanumeric combination and represent a variety of information such as website addresses, email links and catalogue information. Visual QR codes can be read by any digital camera system that has sufficient resolution to capture the image. The proposed SQRC holds organized acoustic energy in the 30–35 kHz bandwidth range in order to perform a similar function. This audio-embedded metadata can be transmitted and decoded efficiently using File Transfer Protocol (FTP) or via acoustic transmission over distance using a 48 kHz rated loudspeaker, and decoded after capture with a high-resolution (40 kHz bandwidth) microphone. In the research presented here, other loudspeaker/microphone combinations are also trailed, and their SQRC decoding efficiency is quantified.

The proposed benefit of embedding an SQRC within 96 kHz audio and music files is that any receiver with sufficient bandwidth and decode software installed can immediately find metadata on the audio being played, without the need for complex audio fingerprinting algorithms, such as those used by Shazam (Wang, 2006), which rely on the network transmission of audio data and large databases of catalogue fingerprints to identify an audio source.

Psychoacoustic watermarking is another current method for embedding metadata within an audio waveform, with the caveat that the data is imperceptible to the human auditory system (Cvejic and Seppanen, 2001). However, psychoacoustic watermarks, to date, have been applied to frequencies within the limits of the human hearing range (20–20,000 Hz), bringing the potential to add distortions and audible artifacts to the carrier audio signal.

These sub-20 kHz watermarking techniques include those described by Bender et al. (1996) and Sinha et al. (2014), and are particularly effective despite having limited space for metadata insertion without compromising the integrity of the 44.1 kHz or 48 kHz sampled carrier audio. Adding metadata to frequency ranges above 20 kHz negates this space limitation issue, but it also has the potential to introduce intermodulation distortion artifacts into the sub-20 kHz audio range (Toulson et al., 2013).

Sampling the carrier audio at more than twice the rate of the commercial compact disc (CD) standard (i.e. sampling at 96 kHz) theoretically gives greater resolution and accessibility of higher frequency ranges as defined by the Nyquist criterion, which states that the upper limit of the sampling rate is twice the highest frequency within the signal. Thus, a sampling rate of 96 kHz allows a maximum metadata frequency of 48 kHz to be embedded. Currently, the SQRC encoding strategy is deployed in a stenographic manner of hiding data in plain sight (Brann, 1981), but it can easily be expanded by using a multitude of encryption strategies, such as embedding a spread spectrum watermark (Wojtuń, 2011) or using a more linear pulsed approach (Lopes et al., 2015). Embedding high-frequency sounds at a greater bit depth to increase signal clarity could in turn expand these strategies.

Digital music presented as 96 kHz pulse code modulation (PCM) audio is envisaged to become the future standard audio format for both industry professionals and the consumer (Albano, 2017). As a front-runner in this development, the Apple Mastered for iTunes program has already implemented the 96 kHz standard for professional delivery of files to the iTunes Music Store (Katz, 2013). Additionally, many online music stores specializing in high-resolution audio also exist for delivering 96 kHz music to consumers, for example, HD Tracks (www.hdtracks.com), Qobuz (www.qobuz.com) and Pro Studio Masters (www.prostudiomasters.com). One unique application of SQRC is in embedding International Standard Recording Code (ISRC) data within an audio waveform, so that broadcast reporting and music cataloguing processes can be more easily automated, which is of significant value to the music industry, as discussed previously by Toulson et al. (2014). This chapter gives a description of the SQRC method, presents the results of audio encode/decode experiments conducted to date, and proposes a number of improvements that will be evaluated in future experiments.

ENCODING SONIC QUICK RESPONSE CODES (SQRCS)

The SQRC Algorithm

Previously, perceptual audio investigations have been carried out on pulsed frequencies up to 22 kHz (Lopes et al., 2015), but it is suggested in this research that it is potentially possible to utilize the inaudible frequency range of 22–48 kHz in 96 kHz sampled recordings for pulsed metadata insertion (Figure 25.1). This can be achieved by inserting a series

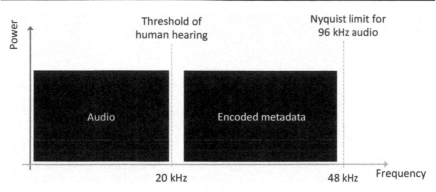

Figure 25.1 A Schematic Showing the Frequency Range Available for Insertion of High-Frequency Metadata When Using 96 kHz Sampled Audio

Figure 25.2 Spectrogram of 30–32.25 kHz Frequencies Representing Alphanumeric Characters. The alphanumeric sequence of characters encoded in this figure are "abc defghijklmnopqrstuvwxyz\1234567890Δ./:-+$*%" (note Δ = Space).

of sequentially encoded alphanumeric characters as discrete signal data, using Matlab R2015a.

As shown previously by Sheppard et al. (2016), a 96 kHz 24-bit source audio wave (WAV) file is encoded with the SQRC frequencies that refer to the 26 characters of the English alphabet. Alphanumeric characters are encoded as 100 ms sinusoid bursts with a unique inaudible frequency representing each character. Characters are encoded upwards from 30 kHz at 50 Hz intervals; for example, a 100 ms burst at 30,000 Hz represents the character 'A' and 30,050 Hz therefore represents the character 'B'. A frequency of 31,250 Hz subsequently represents the character 'Z', with numerics and symbols at higher intervals up to 32,250 kHz. A spectrogram of all the alphanumeric signals being played sequentially through an Adam A7X loudspeaker and recorded with a Earthworks SR40 microphone are shown in Figure 25.2.

Characters can be combined sequentially to form a website URL, ISRC data or descriptive text, as shown in Figure 25.3, which contains the sequence "www.anglia.ac.uk/code".

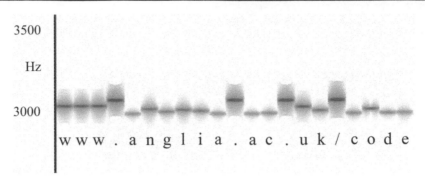

Figure 25.3 Spectrogram of Website URL Encoded as SQRC Data

ACOUSTIC TRANSMISSION RESULTS USING HIGH-RESOLUTION AUDIO DEVICES

Three sets of high-frequency encoded SQRC were transmitted acoustically over a range of distances. The SQRC were transmitted at a calibrated 85 dB LUFS from the primary source monitor (ADAM A7X loudspeaker), and samples were recorded at each distance interval using an Earthworks SR40 high-definition microphone, and decoded using a Matlab SQRC algorithm. Background noise levels throughout the investigation were in the range of 30–35 dB (A-weighting). The three samples acoustically transmitted over a distance of 0.1–5 meters in this experiment were:

- 26 sequential characters of the English alphabet (A to Z)
- URL: www.bbc.co.uk
- ISRC: gbpaj1500001

ACOUSTIC TRANSMISSION OF SQRC WITH STANDARD-RESOLUTION DEVICES

The viability of using lower resolution (22 kHz rated) equipment for audio transmission and reception, with effective coding and decoding of the SQRC metadata, was investigated. For capture, Samson C01 and iPhone 5S standard-resolution microphones were trialed against the high-resolution Earthworks SR40. For transmission, generic 6 cm and 20 cm loudspeakers (Sony model 1-826-115-11 and Peavey Blazer 10W, respectively) were compared against the ADAM A7X 48 kHz monitor. The acoustic transmission testing was carried out over a distance of 5m, with different microphone/monitor speaker combinations (Figure 25.4).

The SQRC used in the following acoustic transmission experiments consisted of a website URL, an ISRC and a 26-character English alphabet set. These SQRC were transmitted at a loudness of 85 dB LUFS from the primary source speaker, with background noise levels in the range of 30–35 dB (A-weighting). In each transmission experiment, the microphone under test captured the monitor output as a 24-bit 96 kHz WAV audio file, which

Table 25.1 Three Different Audio Samples Are Acoustically Transmitted Over Distance and Include a 26-Character English Alphabet (A-Z), an ISRC and a URL

Distance (m)	Decode Efficiency (%)
0.1	100
1	100
2	100
3	100
4	100
5	100

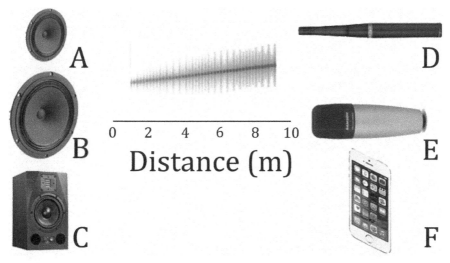

Figure 25.4 Acoustic Transmission of SQRC From Various Speaker Sources Over 5m. Speaker/ Microphone combinations: A = 6cm speaker, B = 20cm speaker, C = ADAM A7X monitor, D = Earthworks SR40, E = Samson C01, F = iPhone 5s.

was then translated using a Matlab-derived SQRC decode algorithm. In the following experiments, decode efficacy is scored as a percentage of correctly translated characters.

The decode efficacy of acoustically transmitted SQRC over a distance of 5m is detailed in Figure 25.5. In this scenario, a 6cm (Sony model 1-826-115-11) speaker is used as the transmission device, and three recording microphones are compared: a high-definition Earthworks SR40 microphone against two standard-definition microphones (Samson C01 and the iPhone 5S). The decode efficiency at each distance is calculated from the number of correctly translated SQRC and displayed as a percentage. Note that the closest measurement to the transmission speaker cone is taken at a distance of 0.1 meters. The high-definition Earthworks SR40 microphone has the highest decode efficiency, though in this scenario is only 100% effective at a distance of 0.1 meters from the transmission speaker. Interestingly, the standard-definition Samson C01 is able to partially decode

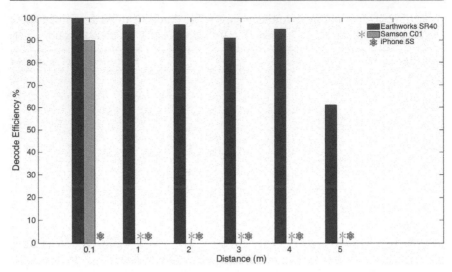

Figure 25.5 Decode efficiency of SQRC after acoustic transmission over 5m using a 6cm speaker with Earthworks SR40/Samson C01/iPhone 5S microphones.

* Denotes zero decode.

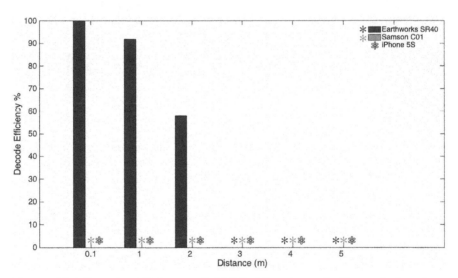

Figure 25.6 Decode Efficiency of SQRC After Acoustic Transmission Over 5m Using a 20cm Speaker With Earthworks SR40/Samson C01/iPhone 5S Microphones

* Denotes zero decode.

SQRC at this distance. The iPhone 5S microphone is unable to translate any SQRC.

Figure 25.6 shows the acoustic transmission decode results from the scenario where a 20cm (Peavey Blazer 10W) speaker is used as the transmission device, and the same three recording microphones are compared

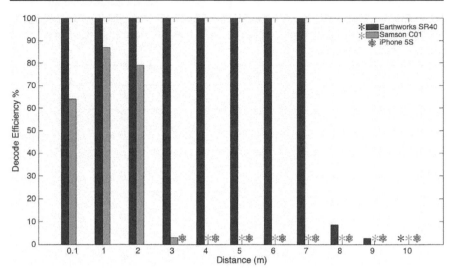

Figure 25.7 Decode Efficiency of SQRC After Acoustic Transmission Over 5m Using a ADAM A7X Monitor With Earthworks SR40/Samson C01/iPhone 5S Microphones

* Denotes zero decode.

(Earthworks SR40, Samson C01 and the iPhone 5S). The decode efficiency at each distance is significantly reduced with the Earthworks SR40 and the Samson C01. The iPhone 5S still displays zero decode efficacy at all the sampled distances.

Figure 25.7 shows the effect of increasing decode efficacy when two high-definition components are combined, i.e. when the ADAM A7X monitor is utilized for acoustic transmission of SQRC and the Earthworks SR40 microphone is used to record them. Decode efficiency is also increased in the Samson C01 recording scenario, though the iPhone 5S microphone still shows no improvement in decode efficacy at any distance even when the transmission monitor is high definition.

FULL-BAND ULTRASONIC WHITE NOISE DEGRADATION ON SQRC EFFICACY

A variable power 16–48kHz white noise degradation experiment was carried out on an SQRC recorded by an Earthworks SR40 microphone at a distance of 1m from the transmission loudspeaker (ADAM A7X monitor). The source audio was sampled at 96 kHz and represented a 26-character A-Z English alphabet. The 16–48 kHz white noise was algorithmically added to the recorded SQRC at variable power magnitudes, expressed in dB, and the decode efficiency calculated as a percentage (Figure 25.6). For both the acoustic transmission and FTP wave file analysis, 16–48 kHz white noise was combined algorithmically using a Matlab script.

Table 25.2 Decode Efficiency of SQRC After Exposure to Varying dB Levels of 16–48 kHz White Noise. Both on SQRC embedded audio transmitted over FTP and acoustic transmission

Noise Level (dB)	FTP Efficacy (%)	Acoustic Efficacy (%)
0	100	100
−10	0	0
−20	0	0
−30	0	0
−40	0	0
−50	0	100
−60	0	100
−70	0	100
−80	0	100
−90	0	100
−100	0	100
−110	0	100
−120	0	100
−130	100	100

In order to do a direct comparison of exposure to high-frequency white noise on acoustically and FTP transmitted metadata, a Matlab-generated white noise algorithm was embedded into both audio file types. The threshold limit at which the white noise had an impact was elucidated, and it was found that relatively low dB white noise insertions reduced decode efficiency markedly.

DISCUSSION AND CONCLUSION

SQRC data can be both effectively encoded and decoded over FTP exchange and acoustic transmission, using high-resolution transmission and recording apparatus. This can be used in conjunction with lower resolution equipment to attain a less optimal level of decode. The decode process is robust and resilient in noisy environments, as the majority of environmental noise is below the threshold of the SQRC. Thus, with this rationale, any receiver that has sufficient bandwidth and decode software installed can immediately find SQRC-derived metadata embedded in the audio being transmitted, without requiring fingerprint analysis or an active Internet connection. The research shows that the SQRC method is a viable acoustic protocol that could feasibly be utilized in a variety of practical applications.

The SQRC are very susceptible to white noise attack in the same transmission frequency range, with even low levels of white noise disrupting the decoding process. SQRC decode efficiency over FTP is less robust from 16–48 kHz white noise attack than SQRC transmitted acoustically. This may be explained by the fact that acoustically transmitted sound has additional energy from reflected surfaces, which in turn allows it to be less affected by direct summation via a digital algorithm.

The high-definition Earthworks SR40 microphone was able to achieve greater decode efficacy over 5m across all three speaker types (i.e. ADAM A7X monitor, 6cm Sony model 1-826-115-11 speaker and the 10 W Peavey Blazer 20cm speaker) than the Samson C01 and iPhone 5S microphones. Translation efficacy was at its most optimal with the ADAM A7X speaker. Conversely, the Samson C01 was only able to produce partial decode of SQRC metadata even when the high-definition ADAM A7X monitor was utilized as the transmitter source. Though decode is sub-optimal with standard speakers at present, there is potential to develop the decode algorithm further to increase translation efficacy and facilitate the use of standard-definition speakers for SQRC transmission.

The SQRC decode efficiency in the iPhone was shown to be poor, due to the maximum frequency range of the microphone to be 22–24kHz (Aguilera et al., 2013). Reducing the upper frequency limit of the SQRC to 22–24kHz could optimize the iPhone for use as an SQRC receiver. Conversely, if the manufacturer were to extend the microphone frequency range, then the iPhone could be used to receive SQRC metadata.

FUTURE WORK

Investigations into the performance of SQRC in an acoustic transmission and reception scenario will be continued, and performance in increasingly high-frequency noisy environments will be evaluated to define the resilience and robustness of the proposed algorithm as a method of embedding metadata into suitably high-definition digital media. This chapter does not concern itself with high-resolution audio perception (Reiss, 2016), but will investigate this aspect with EEG analysis of SQRC. This will be conducted to verify that no subconscious perception of SQRC is encountered, building on the past work by Oohashi et al. (2000, 2002).

REFERENCES

18004:2000, I. (2019). ISO/IEC 18004:2000. [online] ISO. Available at: https://www.iso.org/standard/30789.html [Accessed 15 Feb. 2019].

Aguilera, T., Peredes, J. A., Alvarez, F. J., Suarez, J. I. and Hernandez, A. (2013). Acoustic local positioning system using an iOS device. In: *International conference on indoor positioning and indoor navigation*: (*s.n.*).

Albano, J. (2017). *How high is high enough for hi-resolution audio?* Available at: https://ask.audio/articles/how-high-is-highenough-for-hiresolution-audio [Accessed Oct. 2017].

Bender, W., Morimoto, N. and Lu, A. (1996). Techniques for data hiding. *IBM Systems Journal*, 35, pp. 313–336.

Brann, Noel L. (1981). *The Abbot Trithemius (1462–1516): The renaissance of monastic humanism*. Vol. 24. Brill: (*s.n.*).

Cvejic, N. and Seppanen, T. (2001). Improving Audio Watermarking Scheme Using Psychoacoustic Watermark Filtering. *Information Processing Laboratory*, pp. 1–6.

Katz, B. (2013). *iTunes music: Mastering high resolution audio delivery: Produce great sounding music with mastered for iTunes*. Focal Press: (*s.n.*).

Lopes, S. I., Vieira, J. M. N. and Albuquerque, D. F. (2015). Analysis of the perceptual impact of high frequency audio pulses in smartphone-based positioning systems. Industrial Technology (ICIT), In: *2015 IEEE international conference on*, pp. 3398–3403.

Oohashi, T., Nishina, E. and Honda, M. (2002). Multidisciplinary study on the hypersonic effect. *International Congress Series*, 1226, pp. 27–42.

Oohashi, T., Nishina, E., Honda, M., Yonekura, Y., Fuwamoto, Y., Kawai, N. and Shibasaki, H. (2000). Inaudible high-frequency sounds affect brain activity: Hypersonic effect. *Journal of Neurophysiology*, 83(6), pp. 3548–3558.

Reiss, J. D. (2016). A meta-analysis of high resolution audio perceptual evaluation. *Journal of the Audio Engineering Society*, 64(6), pp. 364–379.

Sheppard, M., Toulson, R. and Lopez, M. (2016). Sonic Quick Response Codes (SQRC) for embedding inaudible metadata in sound files. *Audio Engineering Society Convention*, 141. Audio Engineering Society, pp. 1–7.

Sinha, M., Rai, R. K. and Kumar, P. G. (2014). Study of different digital watermarking schemes and its applications. *International Journal of Scientific Progress and Rese*arch, 3(2), pp. 6–15.

Toulson, R., Campbell, W. and Paterson, J. (2013). Evaluating harmonic and intermodulation distortion of mixed signals processed with dynamic range compression. *KES Transactions on Innovation in Music*, 1(1), pp. 224–246.

Toulson, R., Grint, B. and Staff, R. (2014). Embedding ISRC identifiers in broadcast wave audio files. In: *Innovation in music 2013*: (*s.n.*).

Wang, A. (2006). The Shazam music recognition service. *Communications of the ACM*, 49(8), pp. 44–48.

Wojtuń, J., Piotrowski, Z. and Gajewski, P. (2011). Implementation of the DSSS method in watermarking digital audio objects. Annales UMCS, *Informatica*, 11(3), pp. 57–70.

Part Four

Business

26

Can Music Samples Be Cleared More Easily?

Development of an Automated Process to Clear Music Samples for Legal Creative Reuse

Stephen Partridge

INTRODUCTION

This research is focused on issues pertaining to development of an automated process to clear music samples for legal creative reuse. The contexts that will be considered concern the law, music products and audiences. Discussion under these themes will be infused with qualitative insights gathered via interviews with key stakeholders, namely industry-based lawyers, publishers, label owners and artists. The applied focus for this paper concerns development of the online music samples repository Mashupaudio.com, the intention for which is to provide a better solution to the problem of music sample clearances.

Mashupaudio.com has evolved to become a research and development project, but was initiated as an enterprise project based on a simple idea that occurred to me early in 2009. There is an enormous and largely untouched frozen store of audio files that combine with master recordings when songs are mixed for commercial release. These are commonly known as *stems*, and referred to as such herein. Stems entail recordings of various vocal parts, keyboards, guitars, bass lines, drums, strings, brass, percussion, special effects/ambient noise etc. Stems are often reused creatively when they appear as samples within newly created works.

Before samples can legitimately be incorporated into recordings that are intended for commercial release, they have to be cleared. Sample clearance is a process by which those who own the rights to the sampled work agree to its creative reuse within a new work, typically via a negotiation that will entail a fee and/or acknowledgement via a song credit. Sample clearance is the responsibility of a producer of sample-based music, but it can be fraught with difficulties for those who have limited insight into the requisite systems, processes and related legal ramifications linked to sampling. To clear a sample in advance, prior to publishing a work that contains samples, can be a time-consuming and expensive process. Failure to clear a sample in advance can be even more expensive, to the extent that it is possible for a producer of sample-based music to receive zero royalties and then face additional charges that can entail significant financial losses.

The initial development of Mashupaudio.com is perhaps best described as an enhanced iTunes for producers of sample-based music, an application that would allow producers to sift through stems that they may wish to incorporate into their own productions, and then download these for creative reuse. This project has consisted of two distinct phases, the first of which was to create and subsequently test a fully functional website that would achieve this goalThis research now concerns itself with the next phase of development, which explores how sample clearance might become part of a fully automated process that would duly reward rights' holders.

Ownership of a track that contains samples can be a highly contentious and complex area. The owner of a track (or, as is often the case, the multiple owners) can expect to generate revenue via the track being played on the radio, in TV or in public. The royalty payments that remunerate rights' owners in this way, known as neighboring rights, reward both the owners of master recordings (typically record labels) and the performing artists/ producers who played on and created the track. Development of a system that would divide royalties between multiple co-owners of a track that contains samples, acknowledging and rewarding the owners of the original track that was sampled as well as the new track that contains the sample, is therefore very difficult to conceptualize and design.

Processes via which a prospective sample might be accessed add even more complexity for would-be producers. At the moment, producers of sample-based music struggle to access stems, let alone assess whether they might work in a mix. Music and entertainment lawyer and consultant Ann Harrison identifies this as an essential part of the production process in this context: "you're going to need a recording of what the sample work is going to sound like in your version of it, even if it's only a demo" (Harrison, 2011, p. 343). While Harrison reflects on convenience for producers of sample-based music, rights' holders would also need to benefit before agreeing to provide access to their stems.

Even where tracks contain samples that have been legitimately cleared, those whose creativity is reused often fail to benefit financially. Despite James Brown's 'Funky Drummer' being one of the most sample tracks of all time, drummer Clyde Stubblefield received only modest remuneration, and as Scholz observes, "virtually no royalties from these mega-hits have found their way into the hands of the musicians who actually created the music" (Scholz, 2015).

Technological changes in the digital age have fundamentally altered how audiences expect to experience media. As early as 2005, music industry executive, innovator and educationalist David Kusek considered the likely impact of the increased availability of computer-based music studios (or audio sequencers) upon the monetization of musical creativity in the context of producers of sample-based music:

There is no doubt that ubiquitous music production software that brings with it the ease of recording, mixing and editing, combined with the ability to effortlessly distribute music online, is playing a role in propelling the music industry into the twenty-first century. . . . Once again, the pie gets better—more options and bigger markets.

(Kusek, 2005, p. 145)

Yet the pie that Kusek refers to seldom yields a palatable slice for rights' holders in the context of revenues that might feasibly be generated via music samples.

METHODOLOGY

The methodology adopted to negotiate these challenges started with a review of literature that conventionally underpins my PhD thesis (scheduled for completion in 2019). This informed questions that were posed to a total of 20 interviewees: five music lawyers, five publishers, five music business executives and five artists/producers. Gathering of qualitative data helped to update and further inform my assessment of key issues and debates, as well as how the solution might be modeled. For example, the need to be sympathetic towards the prevailing view that writers/creators should retain control over that which might be sampled informed the need for the system to be 'opt-in'. While I acknowledge that such a system limits control once a writer/creator has opted to allow a piece to be sampled, at least the writer/creator is empowered to decide which of their creations will be made available for creative reuse. The writer/creator of the original work is empowered to choose in relation to any of their works, therefore, whether they are prepared to offset any reputational risk (such as were the sample to be used inappropriately, or in a work that they would prefer not to be associated with) with the prospect of delivering additional income via royalty payments.

The aspect of each interview that linked most directly to the evolution of an actual solution, however, was necessarily quantitative in nature (given the intention to develop a process that would be fully automated, and as such devoid of any subjective or human input). The final section of each interview required each industry-based expert to negotiate an app-based process that I devised to help generate algorithms. This required respondents to determine how a subset of key factors should be prioritized, including:

- Whether the artist being sampled is established/well known
- Whether the sample used is from a vocal as opposed to an instrument part
- Whether the sample used is taken from the verse of a song or from its chorus

Data gathered via these questions has been aggregated and considered in conjunction with more easily measured factors such as the duration of a sample, how often it is repeated and the proportion of the new work that is made up of samples.

The aim is for the data gathered to inform a process that can quantify the fiscal value of a music sample. The problem is sufficiently complex to ensure that the solution will inevitably be imperfect in some instances, but the aim is to create a model that works and could feasibly benefit all of the stakeholders concerned.

My research has become organized under four interrelated themes: the law, audiences, music products and the business. I will consider the first three of these within this chapter, the fourth being a work-in-progress at this stage.

THE LAW

In relation to the law, I have sought to explore:

- The extent to which the law serves the best interests of audiences, as well as other stakeholders linked to the rights' holders, including artists, record labels, lawyers and publishers
- The prioritization of moral rights as the most significant factor that defines outcomes of cases that concern infringement and clearance of music samples
- Models and solutions that might enable more efficient application of the law

In the late 1980s, audiences found nascent forms of sample-based music compelling, with prominent artists sampling in a relatively uninhibited manner while generating significant sales. A much more litigious period followed that stymied creativity in this form, and creative constraints pertaining to this form have been manifest since. Kembrew McLeod and Peter DiCola explore the many complexities concerning how samples are licensed in their 2011 book *Creative License: The Law and Culture of Digital Sampling* (McLeod and Dicola, 2011). Within their case studies, they establish that had the releases of ground-breaking sample-based albums *Three Feet High and Rising* (De La Soul, 1989), *Paul's Boutique* (The Beastie Boys, 1989) or *It Takes a Nation of Millions to Hold Us Back* (Public Enemy, 1988) been attempted in the current era, they would have achieved *negative* sales (McLeod and Dicola, 2011, p. 208), in that for each unit sold a financial loss would have resulted.

Meanwhile, remix artist Dangermouse only achieved *non*-sales with 2004 bootleg *The Grey Album*, which mixed samples from the album *The Beatles* (The Beatles, 1968) with samples from an a cappella version of Jay-Z's *The Black Album* (Jay-Z, 2003), despite hypothetical sales (downloads, at least) exceeding 1 million (McLeod and Dicola, 2011, p. 180). This would suggest that the law does not adequately serve the public interest, given that these seminal albums remain both commonly revered and critically acclaimed, but production of their like is no longer feasible from a legal perspective. Of DJ Dangermouse's *Grey Album*, Jeremy Silver asserts that a commercial release was prohibited by a "combination of label resistance and artists refusals", yet if a commercial entity were able to "create something that put its focus on discovering and surfacing this kind of underground, eclectic content, everyone would love it and no one would sue" (Silver, 2013, p. 85). Silver perceives commercial potential in relation to such sonic mash-ups, but recognizes with informed clarity the nature of institutional resistance to such forms.

As soon as sample-based music appeared significant and was no longer perceived as being a mere fad, the crisis in sampling culture was established. Writing at this time, Simon Frith notes in relation to music sampling how "people are getting pleasure from music-makers who are getting nothing back in return" (Frith, 1988, p. 62). A greater degree of interest and professional sophistication ensued from 1989 onwards, however, from the perspective of established industry-based stakeholders. Record labels, the producers and especially owners of music catalogs that have been heavily sampled are now considerably more vigilant. When realization within the music industry concerning the fiscal value of sampling struck, increasingly forceful application of the law led to perceived negative implications for pioneering producers of sample-based music. As Joanna Demers notes, "expensive litigation has fundamentally changed Public Enemy's sound by making the group unwilling to sample music anymore" (Demers, 2006, p. 10). While it is impossible to speculate on the quality and quantity of sample-based music that might have been manifest had the process of sample clearance been more sympathetic towards creative reuse, it seems reasonable to conclude that audiences have been deprived of optimum development of a form that generated significant sales during a less litigious era. While artists' property was duly being protected via application of the law, this was perhaps a period when the possible promotional benefits of being sampled (such as to audiences who would otherwise not have experienced the original sampled artist) had not been recognized or appreciated.

Legal reuse of any of samples is not possible until an agreement has been reached with the rights' holder(s). Such agreements are negotiated by lawyers with a keen eye on remuneration for the sampled artist, as noted by intellectual property lawyer Ben Challis, who observes: "No-one wants to see their track used in another sound recording and get nothing back for it" (Challis, 2016). While not the sole facet to be considered within this context, monetization of intellectual property that has been creatively reused appears to be the foremost priority for rights' owners, their record labels, publishers and their legal representatives.

The status quo continues to serve stakeholders that resist change. This may be understandable, given prevailing uncertainty over recent years that have seen fiscal devaluation of physical sales of recorded music. But to stymie creativity in the context of popular genres might also suggest that new potential revenue streams are not being explored or exploited due to this conservative milieu. Challis questions the motivation that lies behind this resistance to change: "I know there are teams of musicologists who seem to spend their whole life listening to new songs going 'oh that might be Quincy Jones, that might be a funk band that I could represent if I can work it out'. That's an industry in itself growing up" (Challis, 2016). Publisher John Truelove succinctly echoes this sentiment, observing that "being a lawyer for Kanye West must be great business" (Truelove, 2016). Jeremy Silver also observes a current system that is particularly favorable towards (and lucrative for) the legal profession, noting that "domination is all about details and labels' lawyers love the minutiae" (Silver, 2013, p. 41). Silver implies that drawn-out negotiations and legal cases afford opportunities for lawyers to charge significant fees, summarizing that "those that do benefit of course are the legal fraternity themselves" (Silver, 2013, p. 157).

Despite this perceived cynicism, however, music business executives acknowledge the need for a system that continues to protect rights' holders and their works from unchecked exploitation at the hands of those who wish to sample. Record label executive and music publisher Andy Heath notes" "copyright exists in order to free the creator to benefit from their work. That is an edifice that I will defend to my dying day. I believe it is a good thing, I believe it is a democratizing thing" (Heath, 2016). Heath does note the difficulties that current systems and processes present to those who wish to produce sample-based music, however, acknowledging that the current system does not serve producers of sample-based music well:

> whilst it would be an inconvenience and a disadvantage in certain circumstances for some music makers, tough—that's just how it is. And I am sympathetic with modern music makers because thirty years ago you wouldn't be clearing a ton of stuff you have to clear now.
>
> (Heath, 2016)

This musical form is a low priority for this particular publisher/label owner, who asserts that the current system runs as he'd like it to and should therefore remain as is.

Interviewee C suggests that, in a perfect world, sample clearances should not be problematic:

> If the composer reaches out early enough and their publisher does their job, they should be interacting with the fellow publishers who control the other compositions and work an agreeable rate that is suitable for the new work that is being created.
>
> (Interviewee C, 2016)

But the prevailing view, as expressed by those in the legal profession, is that administration of sample clearances is not nearly as straightforward in practice, as Challis observes:

> the actual systems are so complex and by their very nature they will therefore restrict access or push it underground. . . . And then of course you have got this awful fear, the person who has done the sampling,

that he's thinking I've finally got a hit and now I am going to lose everything from my hit.

(Challis, 2016)

Challis continues, from the perspective of the producer of sample-based music that "the person is asking 'if I can't get these licenses quickly and affordably I just won't do it. Or I'll do it without permission'. Well that can't be sensible for either side. No one's getting paid, and everyone's getting sued" (Challis, 2016).

Music lawyer Mulika Sannie's conclusion suggests that a system designed to make it easier to create sample-based music would need to be demonstrably beneficial to the power base of stakeholders: "Yeah you could definitely say that the law serves those particular people who have the money, the clout, the gravitas, the negotiation power, the bargaining position to actually exploit those particular works" (Sannie, 2016). Sannie recognizes not just a poor system, but also an imbalance of power between those who wish to sample and those who claim to represent the interests of rights' holders.

To provide context to this discussion and that which follows, Table 21.1 shows a summary of sales data comparing three best-selling hip-hop albums in the US in 1990 to sales volume in 2015 for comparably prominent hip-hop artists.

While the historical context is notably different in some ways, especially given the emergence of download sales in the interim period, the difference in sales volume between the two eras is considerable with figures less than halved today compared with the late 1980s. The year 2014

Table 26.1 Sales Then and Now: Comparing Best-Selling Hip-Hop Album US Sales Figures Between 1988–9 and 2014–5

Artist Album Title (Year of Release)	Sales Indicator
De La Soul *Three Feet High and Rising* (1989)	US Platinum (over 1 million sales)
Public Enemy *It Takes a Nation of Millions to Hold Us Back* (1988)	US Platinum (over 1 million sales)
Beastie Boys *Paul's Boutique* (1989)	US 2 x Platinum (over 2 million sales)
J. Cole *2014 Forest Hills Drive* (2014)	577,000 sales
Iggy Azalea *The New Classic* (2014)	485,000 sales
Nicki Minaj *The Pinkprint* (2014)	300,000 sales

(Sources: Tsort, 2015; Kyles, 2014; Where's Eric, 2015)

appeared to be when downloads exceeded CD sales for albums sold in the US (Caulfield, 2014), and while this is noteworthy it is worth adding that the data compiled (sourced from Billboard statistics) includes sales both via CD purchase and download. More recently, however, the perceived threat prevalent within the post-Napster download era has eroded due to growing acknowledgement of the opportunities that have been afforded by subsequent music streaming and app-based initiatives. "The Official Charts Company says that the growth of streaming consumption over the last year has gone from 600 million a week in January 2016, to 1.2 billion a week in June" (Need, 2017, p. 24).

The key technological difference between Spotify and download sites is a consumer choice that has moved away from a desire to own to instead merely have access. Streaming audio is technologically viable now in ways that would not have been considered when the phenomenon of music file-sharing was first manifest, and services such as Spotify and Deezer typically generate revenue via two concurrent models: (1) listeners can either use a free service, whereby revenues are generated via advertising or (2) can pay monthly as subscribers in order to skip the adverts. While online innovation presents new opportunities, however, new threats also lurk in cyberspace:

> The value gap describes the growing mismatch between the value that user upload services, such as YouTube, extract from music and the revenue returned to the music community—those who are creating and investing in music. The value gap is the biggest threat to the future sustainability of the music industry. . . . Estimated revenue per user, Spotify ($US20) versus YouTube ($US1).
>
> (Anon. IFPI, 2017, p. 24)

The value gap that has become a prevalent cause for lobbying within the music industry. Google has recently made changes to its search engine in response to criticism that it enables people to find sites where they can download entertainment illegally. A BBC News article cited the following statistics in its analysis of the problem:

- The BPI made 43.3 million requests for Google to remove search results in 2013 (the US equivalent group, the RIAA, made 31.6 million).
- Google removed 222 million results from search because of copyright infringement.
- Google's Content ID system, which detects copyrighted material, scans 400 years-worth of video every day.
- 300 million videos have been "claimed" by rights' holders, meaning they can place advertising on them.

(Lee, 2014)

While the scale of the activity appears significant as regards the large numbers involved, the BBC article suggests that trade body the British Phonographic Institute (BPI), which represents the rights of artists in the UK, has for many years been seeking a more robust mechanism from Google to address piracy. For its part, Google's own stated aim is to do more to address this complex problem, as described under the heading "Google's Anti-Piracy Principles" within the "How Google Fights Piracy" report: "The best way to battle piracy is with better, more convenient, legitimate alternatives to piracy" (Google, 2014). This summarizes their current approach, albeit with a financial caveat whereby for legitimate sites to be promoted to the top of the search results page, they are effectively required to pay for the advertising space.

Google states that:

> Copyright owners can use a system called Content ID to easily iden-
> tify and manage their content on YouTube. Videos uploaded to You-
> Tube are scanned against a database of files that have been submitted
> to us by content owners. Copyright owners get to decide what happens
> when content in a video on YouTube matches a work they own. When
> this happens, the video gets a Content ID claim.
>
> (Google, 2017)

Challis speculates over the motivation, or otherwise, behind Google's commitment towards anti-piracy measures, observing:

> you can do content ID with YouTube. It always surprises me as one of
> the world's leading technology companies how appallingly poor their
> technology solutions are. It's true, you know, that they have chosen
> to make them poor. But they're a business and they've grown hugely
> out of it.
>
> (Challis, 2016)

Beyond implying that if the Content ID system were slicker, then Google would make less money, Challis considers the dynamic power relationships that exist between Google and European governments:

> the EU look at America and think that's where Google's money goes
> to eventually and we'd like to come here really so maybe we should
> be nice to Google after all, because then we'll be part of a new wave
> of technology. It's a political process as well at the top, at the higher
> level, it's a political process.
>
> (Challis, 2016)

While not categorized as piracy, litigious responses towards instances where music sampling has generated revenue have often been aggressive. If Google's content ID systems were to evolve sufficiently in order that they could identify and challenge the presence of music samples, then the publishing of sample-based music could certainly become much more difficult. The power relationships here focus on defending rights, rather than seeking to exploit their potential for the promotional or fiscal benefit of rights' owners. Such power relationships, therefore, do little to serve the interests of rights' owners over the long term.

Stream-ripping is a contemporary phenomenon that could be applicable to the context of Mashup, and as such needs to be considered as a potential security threat. Stream-ripping "enables people to 'rip' their favorite music via streaming platforms and keep the files on their computer" (Anon, 2017). Any rights' holder would be cautious when faced with the prospect of releasing their back catalog of instrument parts from archives of sound recordings given that a website or app that would preview these for prospective purchasers would, at that point, open itself up to the threat of piracy. If the 'try before you buy' process involves listening to the music that a user might want to download for creative reuse, then there would inevitably be a threat that a previewing listener could stream-rip the preview and reuse it without paying. The development team for Mashup undertook a crude listening test whereby the resolution at which a prospective downloader might preview a sample was carefully considered. Our assessment focused on a user's need to preview audio at a resolution that would adequately represent the download but at a resolution that would be lower than that which they could credibly incorporate into a mix for commercial release, and we settled on an MP3 file resolution of 256 mega bits per second. There remains a risk that a low-resolution file could also be pirated in the form of file sharing, and indeed this has been a preferred aesthetic within some forms of sample-based music, but the robust audit trail within the site can easily identify which samples have been downloaded by whom and when in order that such instances of piracy are track-able via Mashup.

So the context for creative reuse of samples remains challenging, given that in most cases a composer of sample-based music can neither sample easily nor legitimately publish works that contain samples without having first undertaken complex (and commonly expensive) negotiations. The precise nature of such negotiations, and the broader difficulties faced by producers of sample-based music, is summarized by Ben Challis:

> to get individual samples done, particularly on a low budget and without education . . . in the musical sense, even where to go . . . the horror dawns on you that it's not just one place you can go to, you've got to go to three publishers, two record labels, a manager and an ex-manager, and then multiply that into a global market where you have twenty-seven collection societies to deal with—that's before you leave Europe! Which drives people just to do it and take the risk. The problem is it's not a risk if you are not successful, but if you are successful be prepared to lose a hundred per cent of your royalties and perhaps pay more on top.
>
> (Challis, 2016)

In the midst of such uncertainty, there would need to be compelling reasons for rights' owners and their representatives to elect to support an initiative to make production and distribution of sample-based music easier. But even beyond the context of sample-based music, many frustrations are felt in relation to the flow of royalties. Music lawyer Nigel Davies questions the relevance and suitability of collection societies that have failed to coordinate their global efforts despite the technological opportunities afforded by increasingly sophisticated internet services:

> I think the concept of collecting societies is out-dated, outmoded and ridiculous. . . . I do a lot of work with new artists and when you try and explain to them even if they have a hit that they might not see a penny for a year and a half, because that income is generated in (let's say) America and it's collected by collecting societies and they account half-yearly to another collection society and they account half-yearly to your publisher who may account half-yearly to you, you know that's ludicrous—it's in conceivable that there shouldn't be a system in place where all digital income is collected and accounted for on a far more regular basis. . . . I think it's about time the system is completely overhauled. Payments paid more quickly, people can put food on the table for their kids.
>
> (Davies, 2017)

Interviewee C perceives the future role of collection societies with greater optimism, however, opining:

> as we work towards global repertoire databases and more transparency it should be easier to identify who actually is the publisher and the original composers of works, which has been lacking in the streaming era. I do feel that this is being corrected at the moment.
>
> (Interviewee C, 2016)

While collection societies appear to have coordinated their efforts across international boundaries more effectively of late, as Interviewee C suggests, greater efficiency in the flow of royalties remains a serious problem for artists. The Global Repertoire Database (GRD) initiative, whereby "an alternative and more comprehensive database was the goal of a group of music industry entities" (Milosic, 2015), was an attempt to address shortcomings of this nature. Andy Edwards of the Music Managers Forum assesses the dynamic power relationships that were prevalent throughout this fraught initiative, suggesting that the endeavor should have been

> about solving problems and setting common standards. It is not about power and control, which obsesses too many stakeholders and ultimately constrains everyone. Establish common standards through a GRD and creativity and wealth will scale new heights for the benefit of everyone.
>
> (Edwards, 2016)

Edwards expresses disappointment over an opportunity missed through the failure of the GRD, while Milosic assesses probable reasons for its failure, suggesting that "collection societies feared losing revenue from operational costs under a more efficient GRD system. Another reason could be a dispute over control of the global database" (Milosic, 2015).

Artists appear similarly frustrated by structures and processes that seem inefficient, especially where the flow of royalties is concerned, as expressed by 1997 Mercury prize winner Kirk Thompson (aka DJ Krust) of Bristol-based group Roni Size/ Reprazent:

> I think that the people who collect all the money, they're the ones who benefit because they're collecting stuff from twenty five or whatever different sources and then they're holding it for six to twelve months getting the interest on it and then maybe they'll give you it if you ask for it, but if you don't ask for it they wont give it to you and they share it out at the end of the year between themselves. It's a weird system.
>
> (Thompson, 2017)

These challenges are compounded further in the context of music sampling by recent cases concerning tracks that don't even contain samples, but are alleged to have been so significantly influenced by previous works that copyright claims have been supported in the courts. Music Lawyer Florian Koempel suggests problems of subjectivity in relation to two recent controversial outcomes of music-related cases:

> you never know the outcome . . . the whole Marvin Gaye/Farrell stuff. I think it's the wrong decision. They didn't take the substance from a copyright point of view. On the other hand I think that Led Zeppelin nicked 'Taurus' from Spirit and the judges went completely against what I thought was the right thing . . . they are indeed blurred lines.
>
> (Koempel, 2016)

Sannie's suggestion that the subconscious should be considered further compounds the complexity of this contentious legal territory, pondering

> there is a lot of information that a human being processes and we don't always know when it has come into our head, how it has come into our head or where it came from. So sometimes we think "yes I just thought that, and that is brand-new", and it might not necessarily be because you've heard it at some point in your life and it has just sat there in the back of the head.
>
> (Sannie, 2016)

One might argue that if an individual can't be sure of the extent to which they might have been influenced by what they may have heard at some point in their past, how could a law court possibly preside over such cases with any degree of confidence or consistency? Can any of us ever know for sure whether we have created something original? Truelove notes, "of course before the electronic era there was the Rolling Stones cadging off Muddy Waters et cetera" (Truelove, 2016), yet the Rolling Stones' legal representatives are reputedly some of the most aggressively litigious in the music business. The irony appears to be as rich as the problem is deeply complex.

So while there is growing optimism in recognition of the potential of streaming services to generate significant revenue for labels and rights' owners, recent history has been turbulent for the music industry in the digital age. The long-term profitability of streaming services such as Spotify remain questionable. While revenues in 2016 reached €2 Billion, the service "is yet to prove that it can deliver profits: (Titcomb, 2016). "Spotify has operated at a loss for the past decade, reporting a $195.5m operating loss in 2015, though some are predicting that will soon change" (Ellis-Petersen, 2017).

Meanwhile, however, growing revenues for rights' owners seem to have heralded renewed openness and acceptance of such new app-based music initiatives. Michael Nash, executive vice president of digital strategy at Universal Records, suggests "the only reason we saw growth in the past two years, after some fifteen years of substantial decline, is that music has been one of the fastest adapting sectors in the digital world" (Ellis-Petersen, 2017). Nash cites the need to continue to adapt, and the avoidance of complacency, as key to the sustainability of this now rapidly evolving sector. Ole Obermann, Chief Digital Officer and EVP Business Development, Warner Music, enthuses in response to opportunities presented by developers of music-based apps: "As long as these new use cases don't put our core business at risk, we can be pretty aggressive and we're willing to experiment". Obermann notes previous "problems with unlicensed content" but see in "the success of apps such as Musical.ly . . . the demand for innovation around user experience that is built around UGC. This is an area that has tremendous unlocked value" (Anon. IFPI, 2017, p. 20). On this premise, the development of a music app that might present a solution to the complex problems associated with sample clearances was pursued with a degree of optimism.

MUSIC PRODUCTS

Within this chapter I will seek to contextualize key points in the development of a music app designed to automate the process of clearing music samples for legal creative reuse, focused initially on Mashupaudio.com. In relation to this, I have sought to explore:

- The extent to which sampling is actually a (relatively) new phenomenon from the perspective of composers, or contrarily the extent to which composers have always been influenced by, and built upon, the creative endeavors of those who have preceded them
- Ways in which technology has influenced how composers undertake their work
- The extent to which digital sampling can be considered valid as the craft basis for an art form

Ownership of a piece of music that has been produced collaboratively has become an increasingly complex issue as the music business has evolved, and far more difficult to define or decide since the inception of multi-track recording. Temporal relationships among different instrument parts that comprise a track have become inexorably more fractured when compared to the days when the group, or house band, would rehearse the backing track before capturing a live recording. The ability to easily sample music via the use of digital tools has exacerbated this from a legal perspective, a development that label owner, publisher and former session musician Andy Heath considers:

> I remember being in recording studios going "you remember that string line that The Hollies had on that single that they had out last week, why don't we just copy that string line but just play it in reverse or just kind of move it around." We did that all the time, and no one thought twice about that. No one thought that was anything other than completely legitimate. And it wasn't until sequencers and sampling machines came in, whereby you literally took someone else's music and placed it in your own, and frankly the whole thing has got out of hand legally.
>
> (Heath, 2016)

There is a suggestion implied by Heath concerning craft and expertise, whereby those who have taken the time and benefited from the opportunity to learn an instrument might be allowed to get away with a greater degree of appropriation than would be the case for a less-trained sample-based music producer. As Heath summarizes, "If someone's creative with a sample I'm much more comfortable with it" (Heath, 2016). Publisher and sample musician John Truelove echoes this, condemning what he sees as the sort of uncreative sampling that should be challenged or legislated against: "I think it's impossible because some people's idea of a remix is to take *Just Like A Prayer* and just stick a hi-hat on top of it" (Truelove, 2016).

While some contemporary groups reach agreement to share songwriting rights early in their careers, there remain many instances of contention between musicians who cannot agree on proportions of co-ownership. And since the development of the role of producer as a creative entity within the production process, this additional contributor may also claim a proportionate share of rights' ownership. As Frith observes "Since the development of multitrack recording few records represent a single musical event . . . and the person who 'fixes' the sound of the final product these days is, thus, the remix engineer (and even this 'finished' sound can later be remixed)" (Frith, 1988, p. 67). It is very difficult, therefore, when considering a producer's role in the context of Mashupaudio.com, to explore a justifiable percentage payment that might be awarded to a producer of sample-based music that may comprise multiple stems from a variety of sources.

This is something that was tested via the process of interviewing the various stakeholders: label owners, lawyers, musicians and publishers. As a starting point I proposed to interviewees that royalty payments might be allocated on the basis that the producer of music that contains samples (Mashupaudio.com customer, or user) will be entitled to a minimum of 10%, with pro rata division of income among the owners of the samples used. Responses to this proposal were varied, but largely supportive towards the creative endeavors of producers of sample-based music. Music lawyer Florian Koempel's initial response was to restate the manner in which such negotiations currently proceed, asserting that "the percentages will be dealt with in a negotiation" (Koempel, 2016). Another music lawyer Ben Challis elaborated on this, noting that "royalty accounting, particularly for sound recordings, is a dark art. I always think everyone in the creative chain should get paid. Unfortunately at the moment it's very hard to work out *what* they should get paid" (Challis, 2016). Challis highlights the complexity of such process, suggesting that a different approach might be beneficial given the problems currently manifest when it comes to ensuring that the right people are duly compensated.

Another interviewee goes on to express candid skepticism in relation to practices within their own profession: "I have to deal with record labels and they speak in the name of the artist and you think, your artist would never do that. They don't even know what you're talking about. You're just making it up as you go along" (Interviewee B, 2016). This would suggest that individual law professionals are often inclined to self-serve ahead of serving the interests of the artists whom they are supposed to represent. Interviewee A implies similar concerns" "I shouldn't say this, I doubt very much whether some people would like that because it removes the bargaining or the negotiation process, and negotiation processes obviously make people money" (Interviewee A, 2016).

But despite broad recognition of the problem and the cultures and practices that are currently prevalent in this field, there appears to be no obvious solution or consensus on how to address this. When I suggested to label owner and publisher Andy Heath that within Mashupaudio.com producers of music that contains samples might be entitled a minimum of 10%, he replied "I think that's a terrible idea" (Heath, 2016). Interviewee C suggested to the contrary, that "in principle I think that's a great idea, if people know that if you sample this you have to give away fifty per cent of your copyright, or five per cent or ten per cent. In principle I think that's fantastic if it could work" (Interviewee C, 2016). With expert opinion so divided, this suggests that a solution that might be considered in any way perfect does not exist. Interviewee C notes a specific example that highlights the complexity of the challenge, noting that "drums in particular, given how audio loops are so regularly used by everyone, it's hard to justify taking much of a percentage" (Interviewee C, 2016). And Challis concludes more broadly that "the system is far from perfect. In fact administration of sampling is as far from perfect as you could probably imagine in the music industry" (Challis, 2016).

When rights' owners benefit financially from their work being sampled, they might be more inclined towards condoning creative reuse. Camille Yarbrough's 'Take Yo' Praise' (Yarbrough, 1975) was sampled in Norman

Cook's (aka Fatboy Slim) 1999 hit 'Praise You', in which he sampled a short section of Yarbrough's vocal. Katz notes that

> Yarbrough received (and still receives) a considerable amount of money from the song not only from album sales, but from licensing fees paid by the many film and television producers who have used the song. She does not downplay the significance of this windfall, which she has described as 'a gift'. She later joked, 'I have a platinum card, so now I praise Fatboy Slim!'
>
> (Katz, 2004, p. 150)

In this case Cook deliberately acknowledges Yarborough both with a songwriting credit and a 60% cut of royalties from the song, suggesting that fair treatment and financial reward for songwriting rights' holders helps to engender support for the creative endeavors of producers of sample-based music. As echoed by the Performing Rights Society in their members' monthly publication in 2017, "The use of samples in new songs has increased the popularity of the original titles. That brings other licensing opportunities" (Need, 2017, p. 26).

Suzanne Vega's track 'Tom's Diner' (DNA, 1990) was sampled for a remix that had a considerable impact upon her popularity as an artist and sales of earlier albums that were propelled via this single release. There have been many similar examples throughout the digital media era in particular, from Eminem sampling Labi Siffre's 'I Got The . . .' on his track 'My Name Is' to Kanye West sampling Ray Charles' 'I Got A Woman' on his track 'Gold Digger' (Capital Xtra, 2014). The fundamental problem for the remix market, however, is that it is not free to function to its potential. The custom and practice surrounding the law, in conjunction with the law itself, similarly restricts income generation both in the UK and in the USA for rights' holders and in turn producers of sample-based music.

By 2017, the applied component of this thesis (Mashupaudio.com) had been developed in beta form by way of an attempt to present a solution to the problem of music sample clearances, albeit with limited scope at that stage. For example, rights' owners were able to upload music content for creative reuse and could be remunerated to some extent, but would not automatically receive royalty payments pertaining to any sample/stem that would be contained within a future release (such as by a remix artist/producer of sample-based music). Having produced a track, a Mashup customer (user) could seek to exploit their composition via a commercial release, but would first need to negotiate a deal directly with rights' owners in order to legitimize this.

There was also no method by which rights' holders for the original/sampled work being credited on the new composition. Interviewees deemed this a crucial area of development, with an array of comments made across the stakeholder groups interviewed. Challis reasons the need to acknowledge original works, asserting that "if you take something and you don't credit someone or pay them a fair share then you're stealing" (Challis, 2016). Another music lawyer, Florian Koempel, concurs, suggesting that "when you declare the recordings that have been sampled you have to acknowledge the original composer" (Koempel, 2016). Andy Heath reinforces a recurring emphasis on the need for acknowledgement as a prerequisite, stating "I think it's always correct to acknowledge the samples" (Heath, 2016). Clear acknowledgement of those who created the original works that are to be sampled would, therefore, have to be incorporated into the next iteration of the platform.

The subsequent phase of development would seek to address these challenges, comprising two additional functions:

1. A refined upload function, to serve both rights' holders and producers of sample-based music
2. A refined payment function, whereby royalties are distributed to rights' owners based on declared ownership percentages

While focusing on these developments, it remained crucial to consider reasons why rights' owners might feel compelled to engage, given that such engagement could feasibly impact their original works negatively. Rights owners who would opt to upload their music for others to download for creative reuse, for example, choose to do so on the understanding that at this point they relinquish control of how the content concerned will be repurposed. While moral rights might be supported to some extent via technological intervention, such as through the use of blockchain technology, such interventions would prove difficult to automate.

A blockchain is a digital ledger in which transactions made in bitcoin or another cryptocurrency are recorded chronologically and publicly (Iansiti and Lakhami, 2017). These distributed databases are used to maintain continuously growing lists of records, called 'blocks'. Each block contains a timestamp and a link to a previous block. A blockchain is typically managed by a peer-to-peer network, collectively adhering to a protocol for validating new blocks. By design, blockchains are inherently resistant to modification of the data, so deployment of such technology might permanently record a rights' holder's preference when it comes to the appropriateness of subsequent usage. So a rights' owner might choose to preclude incorporation of a sample into a new track that represents a specified ideology that would be at odds with the rights' holder's own beliefs. This would merely serve as a record of what's permissible, however, rather than being a technology that could enforce the rights' holder's will.

More broadly, questions concerning the validity of sample-based music as an art form elicited a range of different but largely supportive responses from interviewees. Interviewee C enthuses "it's certainly an art form . . . some of the early hip-hop stuff with the James Brown samples, it was incredibly inventive of its time" (Interviewee C, 2016). Heath doesn't absolutely concur, noting "it's a subjective issue. But I think that some music makers can use sampling extremely creatively. I think a lot of it though . . . is lazy" (Heath, 2016). Truelove shares both perspectives, suggesting that this form can be valid as art but often isn't:

There's an awful lot of digital sampling that is effectively just boot-
legging. There is a lot of it that is not creative . . . but then there's a
lot more where people are being incredibly creative and imaginative
and pushing the boundaries and pushing the envelope. So overall yeah
I think the answer is considerably, but with qualification.

(Truelove, 2016)

Noteworthy here is an underlying skepticism, even from a publisher whose
early career as an artist was based upon the commercial success of sample-
based records. Koempel echoes this dichotomy, stating on the one hand,
"the whole area of using other people's work from a legal perspective, it's
straightforward taking", before echoing Truelove's balanced assertion that
"the interesting thing about samples . . . is it depends on a case-by-case
study" (Koempel, 2016).

Such is the contention that one lawyer, within one interview topic, presents differing perspectives that sum up the ambiguity and inconsistency in
how the law is applied and legal decisions are reached.

Besides artistic curiosity, the likely motivation for rights' owners to
engage with a platform such as mashupaudio.com would be both to seek
promotion of their original works by association with/inclusion within
a new composition, and to receive royalty payments for their own creation. Revision of the upload function will optimize the need to capture
the data necessary to inform subsequent royalty payments to rights' owners. Having produced a track, a Mashup customer (user) may then seek to
exploit their composition via a commercial release. The upload function
will require the user to specify the extent to which the track contains their
own work, as well as elements that are owned by other rights' holders. In
addition, any samples used will automatically lead to the rights' holders
for the original/sampled work being acknowledged via credits on the new
composition.

At the point of upload, when rights' owners elect to volunteer their
stems and/or master recordings for creative reuse, they would be required
to state the extent to which they own (either solely or partially) the content concerned. This is not a particularly complex or problematic process,
other than when it is neglected, such as due to writers' procrastination or
neglect. The CD Baby Royalty Split Sheet (Appendix 1) performs a comparable function, and development of something similar but duly revised
for this platform would suffice as regards a process for data capture from
rights' owners.

McLeod and Dicola highlight the profile or commercial success of the
artist as a key component of any attempt to automate sample-clearance
processes, listing:

- Recongnizability of the portion sampled
- Whether the sampled musician had a major label or distributor
- Popularity of the sampled recording or composition
- Level of the sampled musicians commercial success or fame

(McLeod/Dicola, 2011, p. 154)

Mcleod and Dicola imply that more prominent works would almost inevitably carry greater fiscal value than less prominent works. To optimize the process of automation, therefore, linkage between the Mashup platform and a live or regularly updated index that records sales of published master recordings would appear critical.

In order to automate processes pertaining to these facets defined by McLeod and Dicola, the process will specifically incorporate the value of sales (physical, download and via streaming) that the track concerned has generated. Ideally this will be fully automated, such as via a link to another index or data stream. Artist Dan Stein (aka DJ Fresh) noted during our interview (Stein and Fresh, 2017) that the PPL manages a database that records sales of individual tracks and could therefore perform this function.

The algorithms that are in development will inform the fiscal value attributed to each sample used, which may lead to the Mashup customer generating income even where tracks contain multiple samples. The algorithms have been informed via a consensus of stakeholders (as defined by McLeod and Dicola, 2011) who were required during interview to prioritize what they perceive to be the most compelling facets for a music sample. The facets identified by McLeod and Dicola were honed into a form that would remove subjectivity, generating numerical data that will allow development of a quantitative and fully automated process. Stakeholders were required to consider:

- The extent to which the artist concerned is prominent, as opposed to less well known
- The importance attached to whether the sample has been taken from a verse or a chorus
- The importance attached to whether the sample is of a vocal or another instrument part

Aspects such as the duration and number of repetitions of samples are also easily measured via this approach. In an interview with *Wired* magazine in 2004, Beastie Boys' Adam Yauch summarizes such components neatly when asked how he felt about his own work being sampled, proffering that "it's totally context. And it depends on how much of our song they're using and how much of a part it plays in their song" (Wired, 2004). The process by which the producer who wishes to commercially exploit a stem within their own mix has to report the structural components of their track at the point of upload captures these details, which are reflected in the subsequent division of royalties. The algorithms work in conjunction with user-generated data, such as the declared duration of the sample incorporated and how often it is repeated throughout the track.

This task serves to inform the revised payment function. The aim is to develop a solution that will allow payments to be redistributed quickly, securely and accurately to rights' holders and, where applicable, to creators of sample-based music. Data captured throughout the upload function will inform these payments.

THE AUDIENCE

In relation to the audience, I will consider:

- Changes in how audiences interact with existing digital media products to create new ones
- Changes in how media institutions react to how audiences interact with content
- The role of nostalgia into relation to music samples, looking at how samples communicate with audiences in different ways
- The contemporary phenomenon of user-generated content reaching audiences on an unprecedented scale and with increasing regularity

Although development of Mashupaudio.com had progressed towards beta testing by the end of 2016, it remained to be seen whether audiences might engage. Interviewees expressed their opinions concerning the viability of the platform, but as its developer I remain cognisant of the fact that it would be audiences that would ultimately decide its commercial fate. Label owner Andy Heath expressed skepticism, suggesting that "I think people are quite simple beasts. They don't want to mash things up" (Heath, 2016). To the contrary, Interviewee C noted "Apple Music this week has licensed some mash-ups of user-generated content to be licensed for its platform. Kids have it in the palm of their hand. . . . So I believe the future is going to open this up even wider" (Interviewee C, 2016). So as an idea in principle, expert opinion was divided. More useful to me as a developer, however, was the more detailed insight that might help shape and improve the platform's performance.

Music lawyer Florian Koempel focused his initial thoughts on the imperative of ease of use, asserting

> You need clear terms and conditions on the website. It needs to be like a blurb, five bullet points. . . . I think if you explain this to people, under English law . . . I think that technically it's feasible and from a copyright point of view. If you get the permission everything is possible.
>
> (Koempel, 2016)

Interviewee C also offered a supportive endorsement of the approach, enthusing "that license approach at stage one would be a good idea, and once people become comfortable with that format then I believe it could widen out even further" (Interviewee C, 2016).

But while there appeared to be support of the model, skepticism remained concerned the size of the market and hence the viability of the platform as a business. Publisher John Truelove contested: "what I see generally with this model is getting critical mass, getting enough content there to be interesting enough for all those people to come in and start playing around" (Truelove, 2016). Heath extends this concern further, suggesting that there is a high-cost professional marketplace that already exists for cleared samples alongside a much lower-cost amateur marketplace: "The people who are professional music makers on the whole can get samples cleared. The people who are amateur music makers don't bother, they just go for it and they don't pay. And it doesn't matter, because you don't really want to stifle amateur music makers" (Heath, 2016). Challis suggested that sample-based music had become a permanent subset of the music industry, noting that "you can't transform nothing. And not every-one wants to create from scratch. So there's always going to be that need" (Challis, 2016). But similar to my own assertions relating to whether there might be a market for this, none of the varied stakeholder opinions would help enlighten this facet of development in as informed a manner as a test that involved real fans of real artists.

By the end of 2016, Mashupaudio.com was sufficiently developed as a beta platform to be ready to undergo tests. These were focused on the robustness and reliability of the platform, as well as to evaluate audience engagement. In October 2016, I made contact with Amelia Scivier at Good Soldier Records, which led to a test of the site based on a December 2016 release of a track by one of their artists, Huntar. The label was optimistic that the track, entitled '808 Heartbeat' (Huntar, 2016), might achieve chart success and seemed eager to explore whether Mashup might help heighten the buzz around its release. The label pushed the initiative via their social media, and although there was some encouraging initial engagement, this was limited to 190 page views and 52 actual users over the first month, after which audience/fan activity levels were negligible.

While the site appeared to work from a technical perspective, and the label did not seem disappointed or surprised by this level of traffic, the activity was not sufficient to spike interest from them in pursuing further initiatives of this ilk (Figure 26.1). While questions remained about the size of the market, other concerns persisted regarding the suitability of the platform and the manner in which its customer base might be engaged. Lawyer Mulika Sannie acknowledges "there is so much more content out

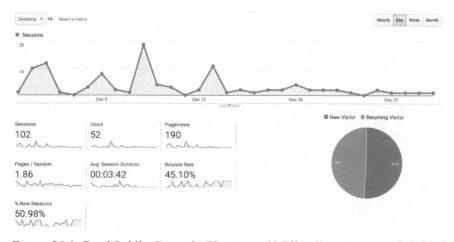

Figure 26.1 Good Soldier Records (Huntar track) Pilot, Engagement at 31/12/16

there now, human brains can only process so much at one time" (Sannie, 2016). The volume of content that Sannie notes suggests that audiences are engaging in great numbers, but they were not engaging in large numbers with the platform that I had developed in this format. During the time that Mashup had been in development, both the internet and the manner in which audiences were engaging had changed rapidly.

I recall in the Summer of 2015, the first time I saw my daughter jumping around our front room with her tablet, lip-syncing her voice to a track that she had downloaded on what transpired to be a popular new music app called Musical.ly. Musical.ly cofounder and co-CEO Alex Zhu declares: "more than 10 million people use the app daily and produce around the same number of videos every single day. All in, 70 million people have registered as Musical.ly users" (Carson, 2016). Widespread audience engagement with music-based apps since then led me to reconsider whether a website-based repository constitutes the best deployment of the concept behind Mashup, and more specifically whether such growth in music-based apps should inspire a different approach altogether.

The decision to review the functionality of the platform, in response to the trial with Good Soldier Music, led to a broader consideration of what the most engaging user experience would look like. The user experience for mining samples from a web-based repository might be engaging to some, but popular apps such as Musical.ly seemed to function more like games. There is also a significant social media component, motivated by the sharing of content and experiences with friends. Consumers appear prepared to engage with (and often pay for) *experiences*, especially those within which there is an interaction with their peers. Consumers pay for experiences at live music events and for interactive experiences such as via online gaming. Through ticket costs or online subscriptions embellished by regular one-off payments, the live music and online gaming industries have flourished during the digital era. So a viable solution in this context would have to have accessibility and ease of use at the center of its engaging user experience. Design of the user experience would need to incorporate elements that contemporary app users find engaging in order for this initiative to appeal to its intended audience in an increasingly cluttered app marketplace, as Sannie notes: "what happens when everybody on the planet gets online? How how are we going to be able to police or track that?" (Sannie, 2016). What can happen with well-designed and engaging music-based apps that are effectively marketed is that they quickly become adopted by audiences on a huge scale, as has been the case with the likes of Spotify, Deezer and Musical.ly in recent years. On this premise, the next iteration of this recorded music-based initiative should exploit value that might be added via components focused on interactivity and user experience.

REFERENCES

Anon. (2017). In focus: Stream-ripping. *PRS for Music, M: Members Music Magazine*, (65) (Sept.), p. 10.

Anon. IFPI. (2017). The evolving market. *Global Music Report 2017—Full Report: Data and Analysis*, pp. 16–23.

Anon. IFPI. (2017). Rewarding creativity: Fixing the value gap. *Global Music Report 2017—Full Report: Data and Analysis*, pp. 24–27.

Anon. Interviewee, A. (2016).

Anon. Interviewee, B. (2016).

Anon. Interviewee, C. (2016).

Beastie Boys. (1989). *Paul's boutique* [CD]. New York: Capitol.

Capital Xtra. (2014). The hip-hop songs you didn't know were samples but really should. *Capital Xtra* [online]. Available at: www.capitalxtra.com/xplore/lists/famous-hip hop-samples-loops/#Py7tqHUjHV2pwUry.97 [Accessed 16 Apr. 2015].

Carson, L. (2016). How a failed education startup turned into Musical.ly, the most popular app you've probably never heard of. *Business Insider*. [online]. UK. Available at: http://uk.businessinsider.com/what-is-musically-2016-5 [Accessed 16 Oct. 2017].

Caulfield, K. (2014). *CD album sales fall behind album downloads, is 2014 the year digital takes over?* [online]. Billboard. Available at: www.billboard.com/biz/articles/news/digital-and-mobile/5901188/cd-album-sales-fall-behind-album-downloads-is-2014-the [Accessed 10 Sept. 2014].

Challis, B. (2016). Interviewed by Stephen Partridge, 26 Sept.

Davies, N. (2017). Interviewed by Stephen Partridge, 10 Aug.

Demers, J. (2006). *Steal this music: How intellectual property law affects musical creativity*. Georgia: University of Georgia Press.

DNA featuring Suzanne Vega. (1990). *Tom's diner* [CD]. New York: A&M.

Edwards, A. (2016). Who will build the music industry's global rights database? *Music Business Worldwide* [online]. Available at: www.musicbusiness-worldwide.com/who-will-build-the-music-industrys-global-rights-database/ [Accessed 19 Oct. 2017].

Ellis-Petersen, H. (2017). How streaming saved the music: Global industry revenues hit £12bn. *The Guardian*. [online]. Available at: www.theguardian.com/business/2017/apr/25/2016-marks-tipping-point-for-music-industry-with-revenues-of-15bn [Accessed 10 Oct. 2017].

Frith, S. (1998). Copyright and the music business. *Popular Music*, 7(1) (Jan.), pp. 57–75. Available at: http://journals.cambridge.org/action/displayAbstract?fromPage=online&aid=2630340&fileId=S0261143000002531 [Accessed 9 Jan. 2015].

Google. (2014). *How Google fights piracy* [online]. Google. Available at: https://drive.google.com/file/d/0BwxyRPFduTN2NmdYdGdJQnFTeTA/view [Accessed 24 Nov. 2014].

Google. (2017). *How content ID works* [online]. Google. Available at: https://support.google.com/youtube/answer/2797370?hl=en-GB [Accessed 13 Oct. 2017].

Harrison, A. (2011). *Music: The business—the essential guide to the law and the deals*. Chatham: Virgin.

Heath, A. (2016). Interviewed by Stephen Partridge, 29 Sept.

Huntar. (2016). *808 heartbeat* [download]. London, UK: Good Soldier Music.

Iansiti, M. and Lakhami, K. (2017). The truth about blockchain. *Harvard Business Review*. [online]. Available at: https://hbr.org/2017/01/the-truth-about-blockchain [Accessed 15 Oct. 2017].

Jay-Z. (2003). *The black album* [CD]. New York: Roc.

Katz, M. (2004). *Capturing sound: How technology has changed music.* Los Angeles, CA: University of California Press.

Koempel, F. (2016). Interviewed by Stephen Partridge, 8 Sept.

Kusek, D. (2005). *The future of music: Manifesto for the digital music revolution.* Boston: Berklee.

Kyles, Y. (2014). *No hip hop albums among top ten sellers of 2014* [online]. All Hip Hop. Available at: http://allhiphop.com/2015/01/02/no-hip-hop-albums-among-top-ten-sellers-of-2014/ [Accessed 16 Apr. 2015].

Lee, D. (2014). Google changes "to fight piracy" by highlighting legal sites. *BBC News.* [online]. Available at: www.bbc.co.uk/news/technology-29689949 [Accessed 24 Nov. 2014].

McLeod, K. and Dicola, P. (2011). *Creative license: The law and culture of digital sampling.* London: Duke University Press.

Milosic, K. (2015). *The failure of the global repertoire database* [online]. Hyperbot. Available from: www.hypebot.com/hypebot/2015/08/the-failure-of-the-global-repertoire-database-effort-draft.html [Accessed 19 Oct. 2017].

Need, P. (2017). *Pop-a-nomics. PRS for Music, M: Members Music Magazine,* (65) (Sept.), pp. 24–27.

Public Enemy. (1988). *It takes a nation of millions to hold us back* [CD]. New York and Columbia: Def Jam.

Sannie, M. (2016). Interviewed by Stephen Partridge, 29 Sept.

Scholz, B. (2015). *Clyde stubblefield: Samples of funk* [online]. All About Jazz. Available at: www.allaboutjazz.com/clyde-stubblefield-samples-of-funk-by-ben-scholz.php [Accessed 26 Nov. 2015].

Silver, J. (2013). *Digital medieval: The first twenty years of music on the web . . . and the next twenty.* La Vergne, TN: Xstorical Publications Media.

Soul, De La. (1989). *Three feet high and rising* [CD]. New York: Tommy Boy, Warner Bros.

Stein, D., aka DJ Fresh. (2017). Interviewed by Stephen Partridge, 3 July.

The Beatles. (1968). *The beatles* [vinyl album]. London: EMI.

Thompson, K., aka DJ Krust. (2017). Interviewed by Stephen Partridge, 7 July.

Titcomb, J. (2016). Spotify's revenue hits €2bn, but when will it make money? *The Telegraph.* [online]. Available at: www.telegraph.co.uk/technology/2016/05/24/spotifys-revenue-hits-2bn-but-when-will-it-make-money/ [Accessed 10 Oct. 2017].

Truelove, J. (2016). Interviewed by Stephen Partridge, 10 Oct.

Tsort. (2015). *Album chart* [online]. Tsort. Available at: http://tsort.info/music/albums.htm and http://tsort.info/music/faq_album_sales.htm [Accessed 16 Apr. 2015].

Where's Eric. (2015). *Gold/platinum/diamond record awards: Certification process.* [online]. Where's Eric. Available at: www.whereseric.com/the-vault/awards/gold-platinum-diamond-record-awards-certification-process [Accessed 16 Apr. 2015].

Wired. (2004). The remix masters: Hip hop pranksters. Pop culture giants. Digital music pioneers. A conversation with the Beastie boys. *Wired* [online]. Available at: http://archive.wired.com/wired/archive/12.11/beastie.html [Accessed 12 Mar. 2015].

Yarbrough, C. (1975). *Take Yo'praise* [from the vinyl album *The Iron Pot Cooker*]. New York: Vanguard.

Appendix 1

The CD Baby Royalty Split Sheet

A PUBLISHING SPLIT SHEET FOR CO-SONGWRITERS IN THREE EASY STEPS

Step 1—Fill out all the info about this specific song.

Song Title: Date:
Recording Artist or Band: Label:
Studio Name: Studio Address: Studio Phone Number:
Samples in the Song?: **Y/N** (circle one)
Album & Artist Sampled:

> (*Samples require permission/clearance from the original Sound Recording Owner and Songwriter/Publishers.*
> *Original creators/owners will become songwriters for the new composition that uses the sample.*)

Step 2—Complete the following info about each and every songwriter that contributed to this specific song. Cross out any blank spaces.

Writer #1:
Address:
Phone: Email:
Publishing Company:
% of Song Represented by Publishing Company:
Additional Publishing Company or Publishing Administrator:
% of Song Represented by Additional Publishing Company:
PRO Affiliation: **ASCAP BMI SESAC SOCAN PRS** Other:

Writer Ownership %:
Writer Signature:

Writer #2:
Address:
Phone:
Email:
Publishing Company:
% of Song Represented by Publishing Company:
Additional Publishing Company or Publishing Administrator:
% of Song Represented by Additional Publishing Company:
PRO Affiliation: **ASCAP BMI SESAC SOCAN PRS** Other:
Writer Ownership %:
Writer Signature:

Writer #3:
Address:
Phone:
Email:
Publishing Company:
% of Song Represented by Publishing Company:
Additional Publishing Company or Publishing Administrator:
% of Song Represented by Additional Publishing Company:
PRO Affiliation: **ASCAP BMI SESAC SOCAN PRS** Other:
Writer Ownership %:
Writer Signature:

☐ Check this box if there are additional writers. If so, attach a sheet listing
    Additional Writers that follows this exact same format and is signed by
    all other writers in the attachment.

***Step 3—Make a copy and give it to each writer that is signing
this split sheet.***

*Collect all your publishing royalties with CD Baby Pro Publishing! Learn
    more at **CDBabyPublishing.com***

27

(Re)Engineering the Cultural Object

Sonic Pasts in Hip-Hop's Future

Michail Exarchos (a.k.a. Stereo Mike)

INTRODUCTION

Hip-hop music has largely depended upon the phonographic past for its function and sonic aesthetic. Borne out of the prolongation and manipulation of funk breaks, and originally performed on turntables, the practice gradually evolved through the deployment and refinement of sample-based practices that took advantage of increasingly affordable sampling technologies (Chang, 2007; Katz, 2012). A genealogy of sample-based producers solidified hip-hop's *Golden Era* aesthetic, characterized as "a sublime 10-year period from 1988 to 1998 . . . in which sampling hit a dizzying new depth of layered complexity and innovation" (Kulkarni, 2015, p. 78). The period was "defined by a solid 'boombap' sound that was shaped by the interactions between emerging sampling technologies and traditional turntable practice" (D'Errico, 2015, p. 281). This "dizzying" creative explosion was powered by an initially unregulated legal landscape with respect to the use of phonographic samples, but the moment this changed, practitioners had to revise their approach and come face to face with a number of aesthetic conundrums. The subsequent creative reactions have given way to a rich spectrum of stylistic variations (subgenres) within the wider hip-hop umbrella, bringing to the fore a set of important questions. These concern the—present and—future of the sample-based aesthetic (in rap music and beyond), the reign of highly synthesized hybrids (such as Trap) making up the lion share's of rap releases in recent years, and the tireless pursuit by contemporary hip-hop practitioners for alternative sources, innovative methods to facilitate a sample-based process and effective methods to render new recordings aesthetically worthy of inclusion into a sample-based modus operandi.

The pursuit of sonic impact and an 'authentic' hip-hop aesthetic bring into focus the unquantifiable 'magic' of phonographic samples, the unique "sonic signatures" (Zagorski-Thomas, 2014) resulting from past production practices, their effect on contemporary sample-based composition, but—also—the sonic limitations that characterize new recordings when these are utilized within a sampling context. From hip-hop's birth and performative tradition prior to the availability of sampling technology, to

437

Dr. Dre's 'interpolation' practices in the 1990s, and to The Roots' live-performed hip-hop, practitioners have continued to create, navigating their way through the minefield of ethics, pragmatics and the legal context surrounding sampling. Some have even resorted to using license-free content sourced from mass-produced sample libraries, a practice that although frowned upon by purists (Schloss, 2014) can increasingly be heard in contemporary productions—such as Kendrick Lamar's recent Pulitzer-winning album *DAMN.* (2017).[1] This article, however, focuses on the alternative practices deployed in pursuit of a *phonographic* sample-based aesthetic. Although the use of commercial sample libraries is acknowledged as one of the contemporary alternatives to using copyrighted samples, it is the function of the (phonographic) past that will be investigated in current practices (an investigation of the stylistic 'legitimacy' of sample libraries in hip-hop production is indeed worthwhile, but beyond the scope of this article). The limitations imposed on the practice by copyright implications have, thus, inspired a variety of responses, simultaneously dictating a power-dependent dynamic. The current landscape—ever since the infamous case involving Biz Markie's *Alone Again* in 1991[2]—can be summarized as one where only major producers have been able to afford the high sample-clearance premiums enabling sampling practice, underground producers have continued to sample operating under the mainstream radar, while all other practitioners have been starved from access to (and legitimate use of) phonographic material in their work (Marshall, 2006).

Furthermore, there is growing concern among hip-hop producers regarding the shrinking pool of worthwhile phonographic sampling content and the increasing reuse of previously featured sampling material. Cited in Schloss (2014, p. 164), producer Domino claims: "I just think that, now, you're getting to the point where . . . you're running out of things to find. And so a lot of the best loops have been used already". Producer No I.D. (cited in Leight, 2017) notes his own heavy-handed use of Nina Simone samples in his recent collaboration with Jay-Z for album *4:44* (2017): "We can't use two Nina Simones! We can't use Stevie Wonder!" Conversely, legendary groups such as De La Soul find their back catalog stuck in "Digital Limbo", due to retrospective licensing complications affecting streaming (Cohen, 2016), resorting to making *And the Anonymous Nobody* (2016) their "first album in 11 years. . . [out] of 300 hours of live material" (De La Soul, 2017).

But if artists are resorting to these means, it is worth examining the variables that enable an effective interaction between the creation of new source content and the sample-based hip-hop process. This forms one of two major quests in this chapter. The second one relates to the significance of the 'past' in facilitating a hip-hop aesthetic. In other words, it is worth questioning how the past manifests itself in the sample-based process, how past sonic signatures interact with sampling techniques, and how these findings inform the construction of feasible (i.e. aesthetically and referentially useable) new material. The sonic materiality and cultural referentiality of samples may lead to a philosophical dead-end if one attempts to separate them from each other, but this very spectrum of possibilities is frequently wrestled with in practice, and records containing these tensions *do* get released as a result (for example, the aforementioned De La Soul release, and productions by J.U.S.T.I.C.E. League and Frank Dukes, which will be discussed in more detail below). This investigation, therefore, first reviews the literary and practice-based context surrounding these questions, before exploring the function of nostalgia in hip-hop (and pop production at large), and the level of historicity and stylization required to facilitate it. As such, the research aims to develop a preliminary theoretical framework aiding practitioners in their future work, while examining the essential tension between historicity and what is regarded as 'phonographic' at the heart of sample-based hip-hop.

LITERATURE | PRACTICE

Much has been written in the literature about sampling as composition (Demers, 2003; Harkins, 2009, 2010; Moorefield, 2005; Morey and McIntyre, 2014; Rodgers, 2003; Swiboda, 2014), the legality and ethics of sampling (Collins, 2008; Goodwin, 1988; McLeod, 2004), and sampling as a driver of stylistic authenticity in hip-hop (Marshall, 2006; Rose, 1994; Schloss, 2014; Williams, 2010). A number of scholars have also dealt with the historicity of samples from a number of perspectives. In her book *Black Noise: Rap Music and Black Culture in Contemporary America*, Tricia Rose (1994, p. 79) effectively demonstrates how hip-hop producers consciously quote from a musical past they resonate with as a form of cultural association: "For the most part, sampling, not unlike versioning practices in Caribbean musics, is about paying homage, an invocation of another's voice to help you say what you want to say".

In his vast ethnographic study, *Making Beats: The Art of Sampled-Based Hip-Hop*, Joseph G. Schloss (2014) reveals a complex ethical code shared by 1990s Boom-bap (Golden era) hip-hop practitioners, with strict rules about the époques, records and particular content that may or may not be sampled. It is worth stating that sampling ethics in hip-hop have much more to do with adherence to (sub)cultural codes of practice than copyright law.

On the other hand, Simon Reynolds states in *Retromania: Pop Culture's Addiction to its Own Past* that:

> It's curious that almost all the intellectual effort expanded on the subject of sampling has been in its defence. . . . A Marxist analysis of sampling might conceivably see it as the purest form of exploiting the labour of others.
>
> (Reynolds, 2012, pp. 314–315)

Despite taking a critical stance towards the politics, ethics and economics of sampling, Reynolds here highlights a number of important problems. From the perspective of a critic who does not enjoy sample-based artifacts—and therefore self-admittedly fails to understand the popularity of musics such as hip-hop—Reynolds, however, focuses our attention on the multi-dimensionality inherent in the complex phenomenon of 'recordings within recordings':

> Recording is pretty freaky, then, if you think about it. But sampling doubles its inherent supernaturalism. Woven out of looped moments that are like portals to far-flung times and places, the sample collage creates a musical event that never happened; . . . Sampling involves using recordings to make new recordings; it's the musical art of ghost co-ordination and ghost arrangement.
>
> (Reynolds, 2012, pp. 313–314)

With this observation, the author provides an eloquent description of the sample-based phonographic *condition*. As such, the associated problems become key considerations for any practice attempting a process of reverse-engineering: how could an awareness of this exponential or 'supernatural' multi-dimensionality inform alternative practices that pursue a sample-based aesthetic? Morey and McIntyre (2014) criticize Reynolds for ignoring the contribution of the sampling composer in this position, thus adding to the complexity of this creative equation. The tension between their position and Reynolds' is perhaps a symptom of a larger philosophical problem: in attempting to serve a sample-based aesthetic through re-construction, a practitioner may become aware of the irony between materiality and cultural referencing. Does a short sound contain history, 'style', a unique sonic signature? When is this historicity motivic, i.e. relating to melody, rhythm and performance? Conversely— when not—what are the inherent sonic manifestations that infuse phonographic 'resonance' to a minute sonic segment? Can these be re-created? Albin J. Zak III concludes in his book, *The Poetics of Rock: Cutting Tracks, Making Records*:

> The overall resonant frame amplifies, as it were, the smallest nuances with which records are filled. . . . [Record collections] represent historical documents and instruments of instruction that provide both ground and atmosphere. . . . Collectively, records present an image of a cultural practice whose conceptual coherence is assured . . . by the shared perception that its works possess the power of resonance.
>
> (Zak, 2001, pp. 195–197)

The author here supports the idea that cultural resonance can be embedded within the sonic grain of a record and consequently hints at a matrix of inter-relationships situated between phonographic artifacts of different eras. A related aspect informing this investigation is that of inter-stylization. It can be argued that hip-hop is inherently inter-stylistic, its process resulting in new musical forms out of the manipulation of past ones, while morphing into numerous subgenres, due to the speed and power of the dissemination and interaction afforded by digital technology. Sandywell and Beer theorize convincingly on this phenomenon

in their article "Stylistic Morphing: Notes on the Digitisation of Contemporary Music Culture":

> It seems that there is no such thing as genre. . . . Under further scrutiny
> canons prove to be complex configured collections of stylistic signi-
> fiers traversing cultural fields and interwoven with cultural objects.
> Against this paradoxical conclusion we suggest that genre is more
> than a technical or theoretical term. It is also a practitioner's term
> invoked in the recognition, consumption, and production of musical
> performances.
>
> (Sandywell and Beer, 2005, p. 119)

This is a useful description of the creative flux facilitated by digital tools
from the perspective of practitioners, and it has the potential to inform
the theoretical framework behind a reverse-engineering process. Although
the majority of rap practitioners may be *reacting* creatively to the cultural
and legal context surrounding them—rather than first theorizing about
it—a number of telling positions towards sampling, characteristic of different rap eras, shed light onto the spectrum of creative possibility. The
performative tradition of isolating, repeating, elongating and juxtaposing
sections from phonographic records on turntables by DJ pioneers such as
Kool Herc, Grandmaster Flash and Afrika Bambaataa signifies the DNA
of the art form, long before it could actually be committed phonographically; conversely, the first hip-hop releases utilized live disco, funk and
soul session musicians in order to provide the instrumental backing under
proto-rap vocal performances. Kulkarni (2015, p. 37) informs us that: "In
late 1982 and early 1983, hip-hop records didn't sound like hip-hop. They
were essentially R'n'B records with rapping on them, created by bands,
session players and producers".

Grandmaster Flash was the first of the DJ pioneers to provide a phonographic 'exception' in the form of *The Adventures of Grandmaster Flash on
the Wheels of Steel* (1981), when he committed the performative tradition of
'turntablism' to record. The importance of the release is that it carries an early
manifestation of what Reynolds (2012, pp. 313–314) describes as the process of "using recordings to make new recordings". At this point in hip-hop
history though, it had only been through turntable performance that the 'citation' and manipulation of previously released records could be committed
phonographically, which explains why the majority of non-live Old School
rap releases utilized synthesizers and drum machines to provide the electro-
rap instrumentals that functioned as an alternative to live performance.

Fast-forwarding to the mid-to-late 1980s—and the availability of affordable sampling technology—a number of seminal releases leveraged sample-
based composition and arrangement, taking advantage of the record
industry's initial inertia in (legally) reacting to the creative manifestations
afforded by sampling. Records such as *It Takes a Nation of Millions to Hold
Us Back* (1988) and *Fear of a Black Planet* (1990) by Public Enemy, and
Paul's Boutique by Beastie Boys (1989), are rumored to contain hundreds

of samples of previously released phonographic content, signifying maximal masterpieces of the sample-based art form that are often compared to a kind of rap musique concrète (LeRoy, 2006; Sewell, 2013; Weingarten, 2010). And yet, by 1991, the shift in the legal landscape kick-started a case of legal necessity becoming the driver of sonic innovation.

A notable reaction can be observed in Dr. Dre's process of 'interpolation'. Dre's initial success with N.W.A. afforded him access to an era of musicians he revered—musicians that he could invite into the studio to (re)play elements of their own records, facilitating his sampling endeavors. Using the original players, instruments and technology enabled the acquisition of authentic sonics from a different era, but without the need to pay high sampling premiums to record companies (who were holding the mechanical copyright). His heavy dependence on P-funk sonics was so impactful that it birthed a geographical divergence in hip-hop known as West-Coast Rap (or G-funk)—one that was diametrically opposed to New York's East Coast aesthetic, remaining synthesizer-heavy (and often sample-averse in the mechanical sense). Further reactions to the legal landscape, and the decreasing creative opportunities for phonographic sampling, can be summarized in three overarching approaches:

1. Live-performed hip-hop
2. The construction of content replacing samples referenced/used in hip-hop production
3. The creation of original but era-referential content that can act as new sampling material

A number of practitioner case studies are discussed in the following section, exemplifying these practices.

CASE STUDIES

Live Hip-Hop

The Roots are perhaps the most famous case in point for a predominantly live-performing (and recording) hip-hop band, but the distinction of their outputs to proto-rap, live-based instrumentals are a result of exhaustive research on sample-based utterances and sonics, manifested in their performance practices, choice of instruments and studio approaches. In "Giving up Hip-Hop's Firstborn: A Quest for the Real after the Death of Sampling", Wayne Marshall delineates their approach in the following way:

> [T]he degree to which the Roots' music indexes hip-hop's sample-based aesthetic serves as a crucial determinant of the group's "realness" to many listeners. At the same time, the Roots' instrumental facility affords them a certain flexibility and freedom and allows them to advance a unique, if markedly experimental, voice within the creative constraints of "traditional" hip-hop's somewhat conservative conventions.
>
> (Marshall, 2006, p. 880)

And yet what sparks stylistic criticisms directed at The Roots by the hip-hop community at large is the fact that their live sonics and musicality do not interact sufficiently with sampling processes and their resulting artifacts.

Creating Sample-Replacement Content

J.U.S.T.I.C.E. League, on the other hand, are a production duo responsible for a plethora of contemporary rap hits (for artists such as Rick Ross, Gucci Mane, Drake and Lil' Wayne) who deploy methods that lie somewhere between interpolation and a convincing re-interpretation (and then manipulation) of referenced samples. In an interview with Hotnewhiphop. com they shed light on the specifics of their process:

> Ok, we have a guitar—what kind of guitar was it? What was the pre-amp? What was the amp? What was the board that it was being recorded to? What kind of tape was it being recorded to? What kind of room was it in?
>
> (Law, 2016)

Interviewer Carver Law (2016) asserts that once they have "all the information available about the original sample, they begin . . . recreating every aspect . . . down to the kind of room it was recorded in." J.U.S.T.I.C.E. League's process reveals the importance of the sonic variables that lend a sample its particular 'aura'. Their meticulous re-engineering attempts to infuse convincing (vintage) sonics onto their referential, yet newly recorded, source content.

Creating New Content for Sampling

There are also increasing contemporary cases where practitioners create content infused with referential—stylistic and historical—attributes, but without direct semblances to previously released compositions. Producer Frank Dukes meticulously records sonically referential, but musically original, vintage-sounding material, to facilitate his sample-based production process. When this level of reverse-engineering is applied to completely original creations, the potential exists for musical innovation that, nevertheless, adheres to the sonic requirements of the sample-based aesthetic. Interviewed in *Fader* magazine, Adam Feeney (a.k.a. Frank Dukes) explains in his own words: "I'm still using that traditional approach, but trying to create music that's completely forward-thinking and pushing some sort of boundary" (cited in Whalen, 2016). Extrapolating on this approach in relation to Feeney's production *Real Friends* (Kanye West, 2016), interviewer Eamon Whalen explains:

> the song's "sample," a delicate piano loop that sounds like it's lifted from a dusty jazz record, but that Dukes found without having to dig for anything, because he made it himself. . . . Manipulating his own

compositions like they were somebody else's is a technique that has brought Feeney—an avowed crate-digger turned self-taught multi-instrumentalist—from relative obscurity to a go-to producer for the industry's elite.

(Whalen, 2016)

THEORIZING

The Function of Nostalgia (*Aesthetic Problem #1*)

In all the practitioner approaches described here lies a conscious approach to navigate the legal landscape safely, while establishing links with the past, either through motivic referencing (Dre's interpolation) or via sonic referencing (The Roots in their instrumental/studio choices—Frank Dukes and J.U.S.T.I.C.E. League in their meticulous recreation of vintage sonic signatures). Hip-hop celebrates 44 years at the time of this writing (Google. com, 2017), so could its obsession with the past be regarded as a metaphor for approaching a stylistic middle-life crisis? Or is this form of sonic nostalgia a wider symptom in popular music, as Reynolds claims, which becomes exponential in a form of music that owes its very inception, architecture and DNA to previous music forms? The website Metamodernism. com has published the following criticism on Reynolds' position:

> Simon Reynolds states that popular (music) culture is suffering from retromania, an incurable addiction to its own past . . . his analysis is based on a nineteenth century—and therefore very modern—notion of 'authenticity'. It makes himself a symptom of that which he criticizes: retromania.

(Van Poecke, 2014)

Perhaps a metamodern predisposition is not essential for the criticism to stand: the problem with a mono-dimensional diagnosis of an aesthetic 'fault' (in this case, solely attributed to nostalgia) is that it is using the symptom as both diagnosis *and* condition. From antique hunters to fashion designers to phonographic 'crate-diggers', it appears that a certain distance from the past allows the human mind the benefit of retrospective appreciation. But to avoid oversimplification, hip-hop is a complex phenomenon that deserves more thorough analysis. Socio-economic and technological factors are entangled in its history, development and sonic genetics, so nostalgia alone appears an easy escape notion, distracting from a meaningful investigation of the conditions shaping this more complex sub-cultural phenomenon. In *Can't Stop Won't Stop: A History of the Hip-Hop Generation*, Jeff Chang (2007, p. 13) aptly summarizes that: "if blues culture had developed under the conditions of oppressive, forced labor, hip-hop culture would arise from the conditions of no work".

Chang goes on to explain that social engineering, Kool Herc's Jamaican-derived sound-system mentality, the withdrawal of funding for instrumental musicianship in New York schools, and a technically trained but unemployed young generation became the conditions for hip-hop's 'big bang'. As a result, sample-based hip-hop was borne out of improbable factors colliding and, as a result, old funk breaks became the instrumental bed for a generation that needed to dance, rap, come together, party or rebel. From this point onwards, the DJ-as-performer had begun 'jamming' with musicians from the past, reacting to their utterances, interacting with their recorded performances, collaborating (in non-real-time) and manipulating their recordings live (just like King Tubby had previously done with Dub multitracks in the recording studio environment)—a trait that has been reproduced by sampling producers ever since via their interaction with (affordable) sampling technology.

It is not a stretch to consider that performing with turntables became a solitary alternative to improvising with a band—only, one recorded in the past—for a generation that was largely deprived of instrumental tuition and opportunity. Fast-forwarding to the condition of the current bedroom producer, one can observe a parallel in the solitary state of collaboration with the past: a plethora of historic audio segments residing in the hard disks, memory banks and sampler pads of a contemporary hip-hop studio setup—providing the 'live-musician' resonances for a solitary performer/writer/producer to interact with. As a result, the sample-based hip-hop process could be described as a *jam across time* with pre-recorded musicians from the past, afforded by digital sampling technology (and initially turntables): in other words, a 'hip-hop time-machine'. This can be represented schematically with the following 'equation' (see Figure 27.1).

The end result may sound nostalgic because of its obsession with the past, but it is really a manifestation of an inherent genetic trait that defines its very function and aesthetic. The sample-based condition has occupied such a large ratio of hip-hop outputs in its 44-year-long lifetime that it has elevated and celebrated the morphing, synthesis and interaction of old and new music to the forefront of its modus operandi. Of course, hip-hop has had an undeniable effect on other popular musics too, so perhaps 'nostalgia' is an afterthought or post-scriptum on a rhizome with a very real history, birth and raison d'être. Furthermore, the techniques hip-hop adopted—for a while unilaterally—have by now been inherited by mainstream pop producers, so Reynold's nostalgic generalization may be suffering from a misunderstanding of the phenomenon in its wider cross-genre implications.

Does a consideration of nostalgia, then, have any significance for the current rap practitioner? It could be argued that facing it critically brings to the forefront the *past-present* binary inherent in the sample-based

$$HIP\text{-}HOP = \frac{\text{improvised performance} \times \text{digital sampling technology}}{\text{pre-recorded musicians} \times \text{time}}$$

Figure 27.1 A Schematic Representation of the "Hip-Hop Time-Machine Equation"

aesthetic. Additionally, understanding the problematic may be helpful in drawing a line with nostalgia (i.e. the past), isolating it as a variable, so that the process of creating new sample content to serve future hip-hop development can focus on further factors.

How Much Historicity is Needed? (*Aesthetic Problem #2*)

This brings about the question of how much historicity needs to be 'embedded' within a sample for the hip-hop aesthetic to function. It is a question that can drive a retrospective investigation of sample-based content, but also one that can inform future (re)construction. As such, it becomes theoretically important, and practically essential, should future sample-based hip-hop continue to utilize newly constructed source content. Consequently also, defining the necessity and degree of source-content historicity will help inform the practice in a more scholastic fashion.

But what are the facets of sonic historicity that can be observed in a sample-based context? The case studies discussed here highlight a number of sonic/musical examples that help define the manifestations of this historicity in a systematic manner. There have been detailed previous attempts to provide sampling typologies (Ratcliffe, 2014; Sewell, 2013), but the focus here is somewhat different. On the one hand, the purpose of the investigation is to inform future practice, so the focus is on observing traits that are reproducible; on the other hand, this is not an attempt to account for every type of sample use, but to do so from the perspective of what qualities infuse 'historicity' in a sample.

In the first instance, sample duration becomes an important parameter in this exploration. The longer a phonographic sample is, the more motivic information it contains. It could be argued that, conversely, a short sample—often described as a single 'shot', 'hit' or 'stab' in hip-hop practice—focuses our attention to the sonic, granular or layered phonographic instance. Philosophically, this binary allows for a theoretical delineation between the sample as sonic instance and the sample as obvious musical or phonographic 'citation'.

Samplists may be able to 'chop' (i.e. truncate) individual instrumental sounds from records (should they appear in isolation in the mono or stereo master) or opt for layered instances (such as a momentary combination of kick drum, bass note, harmonic chord and horn stab). Access to the original multi-track data of previously released recordings has become more commonplace recently (with artists openly inviting remixers to interact with their content), and a number of hip-hop producers source their samples in this fashion. In all of these cases, the sample contains sonic information that—to the trained ear, crate-digger or avid hip-hop fan—may point to specific sources (single or layered instrumentation), real or artificial ambience captured during recording or applied in postproduction, and unique sonic artifacts resulting from the recording signal flow, media used, and mixing, mastering and manufacturing processes. Multiple sub-variables can be associated with these top-level sonic characteristics. Figure 27.2 provides a schematic representation of essential qualities (with variables in parentheses applicable to longer segments).

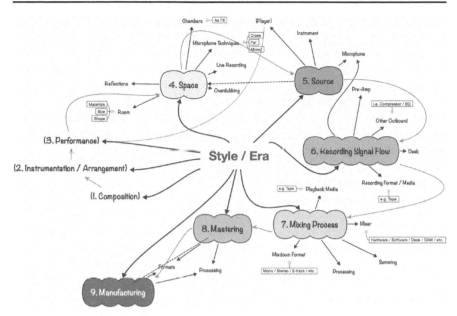

Figure 27.2 A Schematic Representation of Essential Phonographic Sonic Characteristics and Related Variables

Therefore, the period that the phonographic sample was captured in becomes 'communicated' even for short excerpts, because of the type of sources, spaces, equipment and media used, but also the engineering and production processes applied that were typical of particular studios, production teams, record labels and eras. Longer samples, on the other hand, may reveal all of the above, but also contain musical, rhythmical, performed and composed utterances, which also become audible in the new context of the subsequent hip-hop production process. These add further layers of historicity to a sample, such as stylization expressed by the compositional and arrangement choices, but also the performing idioms and musicianship carrying additional era signifiers.

How are these observations useful to the practitioner creating new sample content that is meant to serve a sample-based aesthetic? Before even tackling the practical implications, the very process of purposely infusing historicity into—new—samples has to be analyzed. Undoubtedly, there is an inherent irony in this proposition but, at the same time, from Frank Dukes to De La Soul, the notion *is* indeed practiced, which necessitates a theoretical investigation.

On Phonographic "Magic" (*Aesthetic Problem #3*)

The process of digital sampling can, of course, be applied to any recording, old or new, phonographic or directly recorded from an instrumental source into a digital sampler. Tellef Kvifte (2007) provides four definitions for sampling as it has been used in literature: from analog-to-digital conversion; to the emulation of instruments by samplers; to the 'citation' of an earlier recording within a new composition; and, finally, to corrective splicing and pasting of recorded segments on analog tape or digital formats. For a hip-hop producer, the third definition best describes the sample-based method, particularly because a phonographic sample carries more meaning than simply a digitized acoustic vibration (and the rationale behind the process is predominantly creative rather than emulative or corrective). The distinction, though, is useful in delineating differences in workflow, as a hip-hop production built around a phonographic sample follows a very different creative trajectory from that of a production built independently of a source sample (with live instrumentation later overdubbed on top). So, a new problem that arises is that of how live instrumentation interacts with sampling.

Newly constructed sampling content can range from recordings produced to facilitate a particular project to ready-made content provided by sample libraries for a multitude of potential applications. There is a plethora of sample-library companies that provide wide-ranging content, from drum loops suitable for different subgenres to live instrumentation that may fit particular styles, often accurately replicating vintage sonic characteristics mapped to very specific eras, studios and labels. Furthermore, today's DAWs come pre-packaged with an abundance of neatly cataloged single-shot, looped or motivic samples and, as such, software manufacturers at least partly assume a sample-library function. Although libraries are not explicitly disregarded by hip-hop practitioners, Schloss's (2014) work observes that sample-based producers demonstrate a preference for phonographic content, showing little interest for ready-made content solutions.

Practitioners may be partly adhering to the unwritten code of sampling discussed earlier, but there are other pragmatic and aesthetic considerations to take into account. For this part of the investigation, an autoethnographic approach has been undertaken to shed further light onto these considerations. The methodology has consisted of composing, performing and engineering source content to be subsequently used in a sample-based process, but also researching the historical spaces, tools and practices behind the source references (phonographic records) pursued. In these practice-based pursuits, I have found the indefinable 'magic' of phonographic samples difficult to re-create with new recordings. As part of the historical research conducted, I have visited a number of classic studios related to the eras and records that have previously attracted me as a samplist (Chess Records in Chicago, Stax and Sun in Memphis, J&M studio in New Orleans, RCA B and Columbia in Nashville), attempting to ascertain the conditions of this 'magic': noting the spaces, microphones, signal flows, media and equipment used, but also deciphering clues about the techniques, recording approaches and production philosophies practiced by the teams behind the recordings.

At Columbia in Nashville, staff relayed to me how Toontrack—a well-known sample-library company—utilized the facility to re-create authentic country samples for their *Traditional Country EZX* release (Rekkerd. org, 2016). This highlighted the oxymoron with great clarity: if there are specialists ensuring all sonic variables are adhered to in the creation of legally usable sample content, then why do hip-hop producers opt for phonographic sources, despite the inherent copyright complications. Is the (historicity and) phonographic 'magic' more than the sum of perfectly re-created sonic—and musical—parts? Practitioners and analysts struggle to define the missing link, attributing it to a certain *je ne sais quoi*. Citing Bill Stephney of S.O.U.L. Records, Tricia Rose exemplifies this phonographic lure in producers' sampling rationale:

> [Rap producers have] . . . tried recording with live drums. But you really can't replicate those sounds. Maybe it's the way engineers mike, maybe it's the lack of baffles in the room. Who knows? But that's why these kids have to go back to the old records.
>
> (Rose, 1994, p. 40)

In Zak's *Poetics of Rock*, there is another telling account of how the phonographic process can result in such unique sonic 'ephemera', responsible for drawing the samplist in:

> The guitar's sound was bleeding into other instruments' microphones, but it had no focused presence of its own. Spector, however, insisted: this was to be the sound of "Zip-A-Dee-Doo-Dah". For it was at this moment that the complex of relationships among all the layers and aspects of the sonic texture came together to bring the desired image into focus.
>
> (Zak, 2001, pp. 82–83)

This demonstrates contextual 'happy accidents', which are difficult to imagine outside of an actual record-making engagement. If there is a philosophical lesson to be acknowledged here that can then inform (re)construction, it is that of phonographic *context*. This is where sample libraries and new recordings fall short in facilitating the sample-based art form, process and aesthetic.

The Irony of Reconstruction: A Metamodern 'Structure-of-Feeling' *(Aesthetic Problem #4)*

Phonographic context therefore appears as an essential condition in rendering samples useful to the (sample-based) hip-hop aesthetic. It could be argued that hip-hop is borne out of the interaction between sampling processes and past phonographic content (see Figure 27.3).

But is the past—manifested as nostalgia and historicity—essential in this 'equation'? Or could convincing phonographic context—i.e. a newly

HIP-HOP = sampling processes ✖ phonographic (past) signatures

Figure 27.3 A Schematic Representation of Hip-Hop as the Interaction Between Sampling Processes and (Questioning the Notion of the Past in Sonic Signatures Present in) Phonographic Content

constructed record—suffice as usable content? Some contemporary hip-hop has indeed started quoting from more recent phonography, but the lion's share of sample-based releases focus on a more distant past. This is no surprise, as the 44-year lifespan of the style has had such a dispro-portionately long dependence on the sonic past that it continues to project this (past-present) temporal juxtaposition on the majority of its outputs, almost as stylistic dogma. As hip-hop evolves, the past may become less essential as an aesthetic qualifier, and phonographic context may become prioritized as the driver behind (the creation of) suitable sampling content. But for now, it appears that most practitioners resort to stylization and son-ics referential to past eras, in order to infuse their raw sonic materials with substantial potential for forthcoming sample-based processes.

The obvious irony observed here is that this reconstructive proposition sees the practitioner pursuing new musical content—which is (mechani-cally) copyright-free—while artificially infusing it with vintage sonic characteristics. This is both forward-thinking and pragmatic, but also nos-talgic and 'retro-manic'. Hip-hop producers who practice this conscious duality are sonically oscillating between analog nostalgia and digital futurism. As such, they demonstrate an awareness of hip-hop's addiction to the phonographic past—they adhere to its nostalgic romanticism and honor this naivete—while both constructing and re-constructing: con-structing new music, but *re*-constructing vintage sonic signatures. This is both naive and cynical; it puts faith in future development while paying homage to the dogma of historicity; and it simultaneously represents mul-tiple dualities in a mixed, juxtaposed and synthesized fashion, consisting of all of these polarities at once. Vermeulen and Akker (2010, p. 56) have described such a "discourse, oscillating between a modern enthusiasm and a postmodern irony, (as) metamodernism". And while the retrospective necessity in the aesthetic condition of sample-based hip-hop is to a certain degree explained by its own historic development, technical processes and phonographic dependences, Vermeulen and Akker's (2010) observation of a new 'structure of feeling' across architecture, art and film, (even politics) puts this (re)constructive proposition within a wider, contemporary multi-arts context.

Consequently, the notion does not simply provide a legitimization for the conscious practice of 'irony' in this context; it rather appreciates the process as an artistic invention—or creative solution—borne out of pragmatic necessity, as *part* of an interdisciplinary movement, universal condition or 'structure of feeling'. It could therefore be argued that, if sam-ple-based hip-hop was postmodern, *reconstructive* sample-based hip-hop is metamodern. The methodological paradigms discussed in the case stud-ies embrace further manifestations of metamodernism, such as: exercising multiple practitioner 'personalities' as part of the process (composer and engineer, performer of past styles and contemporary remixer); expressing romantic compositional freedom within an Afrological, cyclic sensibility; synthesizing technical precision with—and towards—the poetics of an envisioned sonic; collapsing 'time' through the juxtaposition of multiple sonic époques; removing the historical 'distance' afforded by samples; and creating cross-genre work that offers synchronous opportunities for inter-stylistic morphing.

CONCLUSION

The discussion commenced from a hypothesis that the sample-based hip-hop aesthetic is borne out of the interaction of sampling processes with past phonographic signatures, before questioning the nature and degree of the manifestation of the past as a variable in this creative equation. Acknowl-edging a number of contemporary approaches where hip-hop practitioners create new content in order to facilitate sample-based processes, the inves-tigation has consequently examined the variables that enable an effective interaction between newly created content and the sample-based process. The aim has been to theorize on this dynamic, arming future practitioners with a better understanding of the aesthetic implications of dealing with both phonographic and newly created sampling content, so that referential sonic 'objects' (Moylan, 2014) can be (re)constructed, aiding the future evolution of the (sub)genre.

Looking at representative practitioners deploying a number of alter-native contemporary approaches as a way to innovate and negotiate the pool of available sampling material, the examination has theorized on four areas of aesthetic concern: (a) the function of sonic/musical nostal-gia, (b) the infusion of 'historicity' onto source content, (c) the notion of phonographic 'magic', and (d) the identification of this reconstructive proposition as metamodern practice. In comparing phonographic samples to newly recorded source material or sample-library content, a number of differences have become apparent, which point to the techno-artistic processes deployed in the construction of vintage sonic material. These, nevertheless, can arguably be re-constructed to close proximity as exem-plified by the meticulous reverse-engineering of both sample-library com-panies and practitioners alike.

Therefore, the missing link in explaining sample-based producers' pref-erence for phonographic content, and the unquantifiable 'draw' towards it, may be located in the ephemeral manifestations of cultural resonance that result from the complex interactions and chaotic dynamic of the record-making process: sounds and utterances resulting from phonographic *con-text*. Perhaps the most promising potential for the future of a sample-based approach (in hip-hop and beyond) lies in the exponential promise of 'mak-ing records within records', consciously putting phonographic context to the square, and synthesizing the paradigm with that of a metamodern 'structure of feeling'. Far from simply adopting a fitting reconstructive perspective, the empowerment for the practitioner in this synthesis stems from the realization that the simultaneous irony of (postmodern) recon-struction merged with the enthusiasm of (modern) creation represents a universal and interdisciplinary cultural paradigm. As a result, a subset of contemporary rap artists and producers may just be engineering new or future cultural objects, while consciously *and* naively entertaining their nostalgic predisposition toward stylizations resulting from the interaction of sampling processes with phonographic ephemera. This way, the oscillation between sonic pasts and hip-hop futures may result in a collapse of time and historical 'distance' via the very synthesis of vintage production techniques and sample-based processes. As Whalen (2016) identifies in the work of Frank Dukes: "By reverse-engineering the art of flipping samples, Feeney is looking at the past, present and future simultaneously".

NOTES

1. For example, track *FEEL.* (2017) makes use of two license-free samples (*COF_125_Am_LaidOut_Underwater* and *COF_134_B_Changed_Dopey*) taken from Loopmasters' *organic future hip hop* (2016) library, as well as a phonographic (copyrighted) sample taken from track *Stormy* (1968) by O.C. Smith (Whosampled, 2018).
2. The lawsuit involving Biz Markie's *Alone Again* resulted in the complete banning of the record and its withdrawal from retail; as a result, the ruling resonated loudly within the hip-hop community, inadvertently affecting producers' future practices and styles in response (Collins, 2008; McLeod, 2004; Sewell, 2013).

BIBLIOGRAPHY

Chang, J. (2007). *Can't stop won't stop: A history of the hip-hop generation.* London: Ebury Press.

Cohen, F. (2016). *De La Soul's legacy is trapped in digital limbo.* Available at: https://nyti.ms/2jINoyG [Accessed 1 Feb. 2017].

Collins, S. (2008). Waveform pirates: Sampling, piracy and musical creativity. *Journal on the Art of Record Production*, 3.

D'Errico, M. (2015). Off the grid: Instrumental hip-hop and experimentation after the golden age. In: J. A. Williams, ed., *The Cambridge companion to hip-hop.* Cambridge: Cambridge University Press, pp. 280–291.

Demers, J. (2003). Sampling the 1970s in hip-hop. *Popular Music*, 22(1), pp. 41–56.

Goodwin, A. (1988). Sample and hold: Pop music in the digital age of reproduction. *Critical Quarterly*, 30(3), pp. 34–49.

Google.com. (2017). *44th anniversary of the birth of hip hop.* Available at: www.google.com/doodles/44th-anniversary-of-the-birth-of-hip-hop [Accessed 11 Aug. 2017].

Harkins, P. (2009). Transmission loss and found: The sampler as compositional tool. *Journal on the Art of Record Production*, 4.

Harkins, P. (2010). Appropriation, additive approaches and accidents: The sampler as compositional tool and recording dislocation. *Journal of the International Association for the Study of Popular Music*, 1(2), pp. 1–19.

Katz, M. (2012). *Groove music: The art and culture of the hip-hop DJ.* New York: Oxford University Press.

Kulkarni, N. (2015). *The periodic table of hip hop.* London: Ebury Press.

Kvifte, T. (2007). Digital sampling and analogue aesthetics. In: A. Melberg, ed., *Aesthetics at work.* Oslo: Unipub, pp. 105–129.

Law, C. (2016). *Behind the beat: J.U.S.T.I.C.E. league.* Available at: www.hot-newhiphop.com/behind-the-beat-justice-league-news.23006.html [Accessed 6 Sept. 2017].

Leight, E. (2017). *"4:44" producer No I.D. talks pushing Jay-Z, creating "500 Ideas".* Available at: www.rollingstone.com/music/features/444-producer-no-id-talks-pushing-jay-z-creating-500-ideas-w490602 [Accessed 6 Oct. 2017].

LeRoy, D. (2006). *Paul's boutique.* New York: Bloomsbury Academic.

Marshall, W. (2006). Giving up hip-hop's firstborn: A quest for the real after the death of sampling. *Callaloo,* 29(3), pp. 868–892.

McLeod, K. (2004). *How copyright law changed hip hop: An interview with public enemy's Chuck D and Hank Shocklee.* Available at: www.alternet.org/story/18830/how_copyright_law_changed_hip_hop [Accessed 1 Apr. 2015].

Moorefield, V. (2005). *The producer as composer: Shaping the sounds of popular music.* London: The MIT Press.

Morey, J. and McIntyre, P. (2014). The creative studio practice of contemporary dance music sampling composers. *Dancecult: Journal of Electronic Music Culture,* 6(1), pp. 41–60.

Moylan, W. (2014). *Understanding and crafting the mix: The art of recording.* 3rd ed. Oxon: Focal Press.

Ratcliffe, R. (2014). A proposed typology of sampled material within electronic dance music. *Dancecult: Journal of Electronic Dance Music Culture,* 6(1), pp. 97–122.

Rekkerd.org. (2016). *Toontrack releases traditional country EZX.* Available at: http://rekkerd.org/toontrack-releases-traditional-country-ezx/ [Accessed 2 Oct. 2017].

Reynolds, S. (2012). *Retromania: Pop culture's addiction to its own past.* London: Faber and Faber.

Rodgers, T. (2003). On the process and aesthetics of sampling in electronic music production. *Organised Sound,* 8(3), pp. 313–320.

Rose, T. (1994). *Black noise: Rap music and black culture in contemporary America.* Middletown: Wesleyan University Press.

Sandywell, B. and Beer, D. (2005). Stylistic morphing: Notes on the digitisation of contemporary music culture. *Convergence,* 11(4), pp. 106–121.

Schloss, J. G. (2014). *Making beats: The art of sampled-based hip-hop.* 2nd ed. Middletown: Wesleyan University Press.

Sewell, A. (2013). *A typology of sampling in hip-hop.* Unpublished PhD Thesis. Indiana University Press.

Soul, De La. (2017). *Facebook,* 24 Feb. Available at: www.facebook.com/wearedelasoul/ [Accessed 25 Feb. 2017].

Swiboda, M. (2014). When beats meet critique: Documenting hip-hop sampling as critical practice. *Critical Studies in Improvisation,* 10(1), pp. 1–11.

Van Poecke, N. (2014). *Beyond postmodern narcolepsy: On metamodernism in popular music culture.* Available at: www.metamodernism.com/2014/06/04/beyond-postmodern-narcolepsy/ [Accessed 5 Sept. 2015].

Vermeulen, T. and Van Den Akker, R. (2010). Notes on metamodernism. *Journal of Aesthetics & Culture,* 2(1), pp. 56–77.

Weingarten, C. R. (2010). *It takes a nation of millions to hold us back.* New York: Bloomsbury Academic.

Whalen, E. (2016). *Frank Dukes is low-key producing everyone right now*. Available at: www.thefader.com/2016/02/04/frank-dukes-producer-interview [Accessed 6 Sept. 2017].

Whosampled.com. (2018). *Kendrick Lamar FEEL*. Available at: www.whosampled.com/Kendrick-Lamar/FEEL./ [Accessed 8 May 2018].

Williams, J. A. (2010). *Musical borrowing in hip-hop music: Theoretical frameworks and case studies*. Unpublished PhD thesis. University of Nottingham.

Zagorski-Thomas, S. (2014). *The musicology of record production*. Cambridge: Cambridge University Press.

Zak, A. J. (2001). *The poetics of rock: Cutting tracks, making records*. Berkeley: University of California Press.

DISCOGRAPHY

Beastie Boys. (1989). *Paul's Boutique* [CD]. Capitol Records, Beastie Boys Records CDP 7 91743 2.

Biz Markie. (1991). *I Need a Haircut* [Vinyl]. Europe: Warner Bros. Records 7599–26648–1.

De La Soul. (2016). *And the Anonymous Nobody* [CD]. AOI Records AOI001CDK.

Dr. Dre. (1992). *The Chronic* [CD]. Europe: Interscope Records 7567–92233–2.

Grandmaster Flash. (1981). *The Adventures of Grandmaster Flash on the Wheels of Steel* [Vinyl]. Sugar Hill Records SH 557.

Jay-Z. (2017). *4:44* [File, FLAC, Album]. Roc Nation.

Kanye West. (2016). *The Life of Pablo* [18xFile, FLAC, Album]. US: G.O.O.D. MUSIC, Def Jam Recordings.

Lamar, K. (2017). *DAMN*. [14xFile, AAC, Album]. Top Dawg Entertainment, Aftermath Entertainment, Interscope Records.

Public Enemy. (1988). *It Takes a Nation of Millions to Hold Us Back* [CD]. UK and Europe: Def Jam Recordings 527 358–2.

Public Enemy. (1990). *Fear of a Black Planet* [CD]. Europe: Def Jam Recordings, CBS 466281 2.

Smith, O. C. (1969). *For Once in My Life* [vinyl]. CBS S 63544.

Stereo Mike. (2007). *Xli3h* [CD]. Greece: EMI 50999 513333 2 4.

The Roots. (1993). *Organix* [CD] Remedy Recordings, Cargo Records CRGD 81100.

28

Anticipating the Cryptopirate

"Don't Bury Treasure" and Other Potential Preventative Measures

Patrick Twaddle

INTRODUCTION

Bitcoin, the first entity based on blockchain technology, emerged in January 2009 and has become more recognizable—perhaps notorious—than the programming which underpins it. It realized the dream of a cryptocurrency, a distributed ledger and exchange system of purely digital assets, as envisioned by cypherpunk Wei Dai (1998). Bitcoin launched with little credibility through an unidentifiable party named "Satoshi Nakamoto" and an online community of cryptography enthusiasts, far removed from financial or government institutions. In fact, part of its inspiration was seemingly declared as a response to the perceived culpability of banks and authorities in precipitating and mismanaging the 2008 financial crisis. It has since inspired copious varieties of cryptocurrencies—or "crypto" as they are commonly called—some of which have also gained considerable value and acceptance. Bitcoin's spectacular—and extremely volatile—performance on exchanges, along with its sense of novelty and self-governance, has served to intensify the "gold rush" towards cryptocurrencies and other blockchain applications. Within nine years, Bitcoin went from having no tradable economic worth to headline-grabbing runs, at one point peaking at over $19,000 USD. Nonetheless, its true and enduring value will likely be found not through financial speculation or exchange rates. Its prime significance comes from how it proved the viability of a leap in technology, as a pioneer from which advanced applications and more impactful systems and enterprises are likely to arise.

While a plethora of blockchain start-ups and initiatives are underway, hope is mounting that this technology holds the answers to many of the problems that plague the media industries, particularly the recording industry. However, as with any disruptive technology, there is inevitable skepticism, insecurity and defensiveness from some vested parties. This is especially the case given the current commercial dynamic which has recently found a savior in a different mechanism for improving its monetization woes: on-demand streaming subscriptions. According to the prevailing narrative, it is tempting for the sound recording industry to, figuratively, breathe a giant sigh of relief. A decade-and-a-half of heavy contraction has come to an end. Recent years have witnessed some impressive gains. The explosion of digital music delivery is finally realizing significant rebounds in overall revenue generation. The pirates who plundered the industry's musical treasures and dispersed the "stolen" goods across the cybersphere are seen to be ceding territory. However, the story is far from complete. Blockchain technology will likely have a profound effect on how sound recordings are managed, even how music businesses and streaming platforms are constructed and remunerated. But it may not always involve content which is properly attributed, authorized and accounted for. Where demand is not met in the licensed and above-board market, alternative forms of trade will help fill the gap.

BLOCKCHAIN BASICS

Somewhere between 2016 and 2017, blockchain technology reached "the peak of inflated expectations", according to global IT firm Gartner (Panetta, 2017). While initiatives initially coalesced around financial services, blockchain solutions are being planned and constructed across virtually every sector, from consumer products and manufacturing, to food distribution and health care, to media and telecommunications. At its most basic, blockchain technology is a method of computer programming, but it represents a radical advancement in coding protocols using cryptographic proofs. It provides for the creation of peer-to-peer (P2P) networks, which run fully distributed ledgers and continuous sequences of recordkeeping blocks. These blocks are fixed and may only be appended; updates are only possible through consensus approval among peers in the network. Not only is the responsibility for recordkeeping dispersed and shared, but it can be publicly accessible at all times and use public-key cryptography for robust encryption. It removes the inconsistencies, vulnerabilities and silo effect of consolidated databases. Blockchains overcome irregularities and malicious behavior by participants—known as *node fault*—by coordinating agreement among honest nodes and continuing to operate despite the potential presence of inaccurate *Byzantine nodes* (creating fault tolerance). Consensus algorithms create uniformity by forcing nodes to decide on the same figures, replicate the exact same data, and repeatedly seal decisions in a sequential chain of blocks. Full, auditable accounting can be performed and confirmed at a remarkable pace through always-running, unbreakable, collective recordkeeping.

Blockchain coding is sometimes referred to as "The Trust Protocol", an idea grounded on "The God Protocols", which was initially propagated by computer scientist and legal scholar Nick Szabo (1997). The concept centers on the need for a trusted third party—especially in the digital sphere—that has all (or almost all) parties' best interests at heart and can be relied upon for processing without revealing unnecessary information. In an ironic twist, a lack of trust is precisely one of the aspects that the cryptosphere continues to wrestle with the most. As with any tool, results depend on whose hands in which the tool is found and how it is wielded.

For many adherents of the recording industry, blockchain technology is being hailed as a positive disruptor. In a manner that echoes many of the enthusiastic predictions which greeted Internet-based digital music two decades ago, many artists and entrepreneurs are envisioning a reorientation of the industry, one that empowers the under-leveraged and oft-exploited "little guys" who form the foundational and most abundant strata of the commercial music ecosystem. However, there is a dark side to blockchain technology that is garnering far less attention.

Growing acceptance and enthusiasm for cryptocurrencies and blockchain technology has served to obscure many of the risks associated with the misuse and misappropriation of the technology and its appeal from the beginning to various criminals and antipathetic socio-political factions. The blockchain space is a ripe breeding ground for agency that ignores or consciously avoids established structures of neo-liberal economy and trade, including intellectual property regimes (IPR). Bitcoin and other cryptocurrencies have attracted a plethora of rogue agents and opportunists since their infancy, many taking advantage of the systems' alternative means of economic exchange and value storage without the apparent necessity for middlemen or external oversight. Having earned a "wild west" reputation, the cryptosphere is littered with scams, unverifiable identities, dubious ventures and gang operatives, and remains at high risk from fiscal, legal and administrative uncertainties. Moreover, the dominant political adherence within the global crypto-community is libertarianism. In a 2013 survey done by Lui Smyth at University College London, 47% of the Bitcoin users surveyed self-identified as libertarian, with another 7% claiming an anarchist ideology (Bohr and Bashir, 2014, pp. 96–97). These are key constituencies which seek to circumvent or impose radical reform upon IPR.

WHAT IS CRYPTOPIRACY?

Piracy is nothing new to the recording industry, having been in existence in some form or another since at least the 1920s. It is not a clearly definable phenomenon and, instead, tends to encompass a variety of activities that may infringe on perceived or codified moral, ethical, legal and economic rights attributed to creators and rights' holders. The dialectics around piracy tend to focus on these rights and dimensions, becoming highly parsed and contested. Arguments are often leveled to delegitimize and frame piracy as a detriment—even an existential threat—to musical creation in both an artistic and economic sense. Rights holders and other defendants of intellectual property, particularly copyright, have typically opted for legal and technological action, along with appeals to morality and social consciousness, to deter participation in the umbrella of piracy.

Shortly before the appearance of Nakamoto's white paper which laid the foundation for Bitcoin, Tim Werthmann (2008) invoked the term "cryptopiracy" simply as "a conjunction of cryptography and piracy" in his examination of the historical usage and benefits of cryptography in both aiding and combatting media and software piracy. In the blockchain movement, this term and its related forms have acquired new meaning. The prefix "crypto-" is being applied to entities and activities related to digital currencies and other blockchain systems designed around cryptographic protocols. In this context, "cryptopirates" emerges primarily as a reference to parties which employ or receive meaningful support from blockchain technology for the unauthorized reproduction, distribution or exploitation of intellectual property. In theory, any form of piracy that employs cryptographic aids would be cryptopiracy, but current semiotics imply the involvement of a systematized, exchange-enabling application of cryptography as facilitated through blockchains.

WHAT DRIVES CRYPTOPIRACY?

There appear to be three main drivers of cryptopiracy: (1) hacker ethic or informationalism, (2) the current proliferation and advancement of blockchain coding and platforms, and (3) dispersed parties or collectives which develop or coalesce around sets of piracy-enabling protocols. The term "hacker" often has negative connotations, but it is foremost an ardent programmer who believes in the social good of information sharing. The theories surrounding the "hacker ethic", particularly as examined by philosopher Pekka Himanen (2001), are closely related to Manuel Castells' work (1996, 2004) on the "informationalism" which supplanted 20th-century industrialism. The spirit of informationalism involves a passion to explore the potential of cutting-edge technologies. It seeks to break down informational constraints and—while not necessarily politically rebellious or illicit—poses a threat to established forms of information distribution and control. This spirit is part of what feeds the contemporary economy, helping to drive market competition and innovation, but also the continual development of systems which facilitate piracy.

Blockchain technology is still in its infant stages, but it represents a major leap in programming and is spreading and advancing at a rapid rate. In public or semi-public ("consortium") formats, blockchains can function as extreme peer-to-peer (P2P) networks, using open-source programming and visible ledgers, and requiring no third party or central servers. Blockchains have many potential advantages over earlier types and generations of P2P piracy. They feature the intrinsic incentivization of work through the generation of coins or tokens, the potential for increased security and durability, and maximal decentralization and power distribution. Blockchain protocols could inherently record and reward inputs through clear and transparent recordkeeping, tokenization of positive contributions, and management of rights and privileges. Meanwhile, consensus rules and the complete dispersion and immutability of dealings could offer unprecedented resiliency against takedown efforts, even from coordinated, multijurisdictional parties. Even in a public system with all nodes visible, the people or entities behind some nodes may remain unidentifiable, particularly through advanced masking techniques.

The third key ingredient is the need for online collectives or informal groups to coalesce around the exploitation of existing or new blockchains. The arrival of crytocurrencies opened up a novel, covert method of collecting revenues for pirates, though the crypto user base was initially very small. Adoption is increasing—aided by mainstream attention and user-friendly apps and exchanges—improving network effects and the viability of crypto-based exchanges. Platforms could be adjusted or created to run on the backs of cryptocurrencies, probably in the form of micro-payments or subscriptions for reliable pirated resources. This could more effectively and efficiently deal with the funding issues which often require platforms to raise money through donations or annoying adware or malware. However, a group could also agree to a new set of protocols which would establish a new blockchain specifically to traffic pirated media. Blockchains rely on consensus and a dedicated network of supernodes that do the main processing and verification, but they are compensated for their heavy lifting. Uploading of pirated content could also be better organized and rewarded through blockchain protocols which establish minimum viable data requirements, maintain quality standards, and track participation.

THE EMERGENCE OF CRYPTOPIRACY

We are arguably well into the early stages of blockchain-backed piracy, though the support is largely extrinsic. Existing pirate platforms enabled contributions through cryptocurrencies, most notably with The Pirate Bay accepting Bitcoin in April 2013. Periods of extreme appreciation for Bitcoin would have offered potential windfalls for pirate platforms. Given that the cryptocurrency saw its first massive surge at the end of 2013, retained earnings from the first six months of collection would have quickly appreciated by about 600% (Saint Bitts LLC, n.d.). By fall 2017, The Pirate Bay and several torrenting websites were testing out mining scripts, usually coopting users' CPUs to generate the cryptocurrency Monero. This type of surreptitious behavior is largely agreed to be unethical and unacceptable within the global crypto-community, but it is a revenue solution on which some parties are willing to gamble.

The next major paradigm shift in cryptopiracy should occur as blockchains are specifically constructed or maneuvered to traffic media content, and—depending on the effectiveness of the protocols, designs and collective adherence—may lead to a substantial resurgence or re-entrenchment in music piracy. Presently, there appears to be one major obstacle to the development of pirate blockchains: the current energy demands which become colossal when scaled. Entities like Bitcoin run on power-hungry Proof-of-Work protocols, which require massive amounts of computation and the constant replication of data across over 12,000 nodes worldwide. New protocols, such as Proof-of-Stake, are being developed that should incur small fractions of the existing processing power, making blockchain-based pirate platforms much more economical and viable in the near future.

APPROACHES TO PIRACY AND THEIR LIMITATIONS

Efforts to combat digital music piracy have been reactionary and defensive, largely falling into two categories: legal action and forms of digital rights management (DRM). While legal solutions have had some significantly mitigating effects, they are typically outpaced by technological invention and shifts to alternative platforms. Copyright protection and enforcement varies widely by jurisdiction, and laws are habitually slow or ineffective in dealing with the constant flow of technical developments. Some key software services that facilitated piracy have been shut down or restructured, but enforcement and prosecution often appear arbitrary and uneven. Lawsuits have often failed and programmers have learned from the legal flaws in services such as Napster and Limewire. The recording industry is perpetually stuck in a game of whack-a-mole. Moreover, legal fights can reinforce negative narratives and stereotypes about greed, power and mistreatment in the music industries. The main beneficiaries seem to be lawyers while artists may experience conflicts of interest and their rewards are dubious. In the infamous case of *RIAA v. The People*, the American recording industry targeted over 30,000 residents out of millions of Americans who had offended copyrights in their personal use. Among its scare-tactic efforts, it issued thousands of subpoenas and lawsuits on a seemingly indiscriminate basis against people of all different ages, demographics, economic means and levels of offense. Proceedings sometimes stretched over several years and exacted fines worth thousands of dollars for each illegally downloaded song. Virtually all of these defendants also consumed music through official licensed media, such as radio, TV, CDs and live concerts. It is hard to imagine almost any other industry intimidating their customers in such a sweeping and punitive fashion.

Likewise, DRM has often had poor results and a chilling effect on the legal consumption of recorded music. DRM has shown limited efficacy on a technological level and is almost universally disliked, even despised, by music creators and users. This type of restrictive reaction to the potential of piracy only inflames informationalist sensibilities, offering a technical challenge to hackers to take pride in breaking down barriers. DRM was a thorn in the side of many iTunes users for years, and while it has now been abandoned, its lingering effects remain a potential competitive disadvantage. Such is the discomfort surrounding DRM that one of the earliest and most high-profile blockchain projects for music, Dot Blockchain (dotBC), has been forced to counter the perception that its codec-based system is a reintroduction or new form of DRM.

REFRAMING PIRACY

A common concept that piracy is essentially about getting cultural goods for "free" is misleading. Expectations often center more around content being liberated in the informationalist sense of being made available upon demand, even if remuneration is required or facilitated in some form. In comparison with analog antecedents, digitalized music lends itself more easily to transformational use rather than equivalency ("venal copying"); it is about convenience, portability, customization, repurposing, re-appropriation, and so on (Kernfield, 2011 pp. 125–127, 200–203). In many cases, piracy is simply about obtaining what has not been fully released into the open market, or desiring it in a form where it can be re-contextualized or repurposed for the sake of creative expediency and expression. Many torrent users are simply in search of certain content and would be happy to pay for expedient access through legitimate channels. IPRs are typically complicated and not well understood by laypeople. Many users would be willing to properly license audio material if enabled to do so in a comprehensible and affordable fashion.

Disassociating and vilifying music piracy appears to have undervalued the constituents and needs which it serves. The customer base extends far beyond the ideologues such as the anarchists and libertarian socialists who make up a small fraction of piracy participants. Even branches of the Pirate Party, which have popped up in several countries, have platforms which seek not to abolish copyright, but to limit its terms to much shorter periods. However, piracy tends to be treated solely as a threat and enemy—lumping all sorts of recalcitrant behaviors and disparate participants together— rather than looking for opportunity and market insights. Pirate platforms are competitors to properly licensed platforms, coming with strengths that may be emulated and weaknesses which can be exploited. Keeping pirate channels in operation is costly, relying on innovation and entrepreneurial investment to make it viable on a large scale. They rely on the similar pressures of supply and demand. They struggle for resources and revenues and adherents, just as legitimate channels do. Pirate platforms must cover the costs of fixed infrastructure and variables of data, electricity and labor. These are supported through user donations, advertisements, malware and volunteer labor, with the burden of risk and inputs typically shouldered by a small minority of participants. Piracy can be framed primarily as a market occurrence supported by vunerable business models, which can be challenged through effective strategies and superior offerings by rivals. In this light, cryptopirates may be best kept at bay by addressing gaps and vulnerabilities in licensed outlets.

POTENTIAL PREVENTATIVE MEASURES

My first recommendation in pre-empting a rise in cryptopiracy is "don't bury treasure". Treasure refers foremost to deep, rare and premium catalog. Deletion and restricted availability of historical material acts as a motivator for pirates and reinforces a key value proposition or unique "selling" point of piracy-enabling platforms. In the digital era, catalogs can be maintained and made accessible at a very low marginal cost. Secondarily, "treasure" can signify hot new releases that are severely confined to limited release and extensive windowing. Creating such scarcity, whether intentionally or through neglect, inflates a recording's value, but this is essentially artificial and unsustainable. Scarce artifacts become lucrative targets for unlicensed distribution, whether they are hackers simply trying to liberate them from controlling or negligent hands, or commercial pirates attracted to such precious artifacts for the sake of financial profit. Even if consumers do not turn to pirate platforms, they are likely to turn to user-uploaded content (UUC) and ad-based streaming services, which make remittances back to rights' holders at fractions of the typical rates for digital downloads and subscriptions.

In the informational economy, consumers expect that they will not have to jump through hoops or make large commitments to access content. Withholding back or rare catalog is difficult to justify in the eyes of millennial listeners who have grown up with the convenience of digital music, but also in the eyes of an older fan base who may consider it to be a part of their personal and cultural history. In the case of new releases, restricting and staggering releases can be an important tool in marketing strategy, especially when properly targeted. But the effect tends to be short-lived and can easily backfire. Long ago we left the 20th-century phonographic paradigm. Assets that are or can be digitized are essentially impossible to keep under substantial control. Although not entirely linear, there is evidence of a correlation between scarcity techniques and rates of piracy. As Tidal has one of the smallest subscriber bases of major streaming services, exclusive releases by superstar artists on this platform are normally accompanied by noticeably high rates of piracy. The success of on-demand streaming services is based primarily on major technological improvements in the dimensions of selection, portability, customization and discoverability. The potency of this value proposition is precisely how piracy activities appear inferior and increasingly lose out when music consumers are making cost-reward assessments.

My second solution is for the music industries to embrace the leading edge of the innovation curve. Major players need to partner with legitimate blockchain developers and collaborate in earnest on pilot projects, committing to long-term iterative processes and small-scale tests, all with the goal of preparing for the widespread adoption of market-viable blockchain-enabled platforms. Recent history indicates that resistance to technological innovation is ultimately futile. In the 1990s, out of fear of losing control and competitive advantages, record companies could not agree on broad-based digital vendors. So, hackers created it for them. Napster and other illegitimate platforms made quick use of new capabilities and addressed consumer demands years before attractive market outlets did. They drove wedges into the regular market with competitive advantages that were based not solely on cost, but also offered unprecedented selection and personalization to a listener's repertoire. And then Apple, YouTube, Facebook, Google and Amazon did as well, effectively usurping billions of dollars of value from the historical or conventional music industries and passing on varying degrees of diminished returns.

It can be astonishing how much hackers and bedroom programmers are willing to invest in labor, equipment and electricity bills to innovate. We should be encouraging some of these agents to help develop platforms and operate nodes which assist in the maintenance of authorized content. Blockchains can be designed and governed as distributed ledgers forming peer-to-peer networks which enable work to be done by highly dispersed and remote communities. It seems the sooner artists, companies and collectives can adapt and formulate new services and business models, the less value slippage is likely to occur. Where technical possibilities are found, informationalists will work to make them a reality. Onerous or obstructionist behavior from the industry may mean that blockchains which aid unlicensed content and false copyright claims more than properly attributed and licensed (or licensable) material.

The third lesson that can be drawn is the importance of value-added services and increasing quality. The renown of the invite-only pirate forum called Oink's Pink Palace serves as a potent example of the market failing to adequately address issues of quality, accessibility and information sharing. It was hailed by proponents as the best catalog of music that ever existed. On Oink, consistency and accuracy was maintained through strict protocols and moderators. There were rigid standards for sound fidelity and cataloging. Duplication was banned, and rare and special finds were encouraged. A rich community developed that mainly wanted to share, discover and discuss content as music lovers. Artists like Trent Reznor turned to it because the market simply did not offer anything comparable. And by some measures, it still does not.

Despite the great strides provided by current streaming services and other music-based applications, there is ample room to continue growing and adapting. We should expect that music consumers will continue to demand greater convenience, better curation, better community-building mechanisms, and blockchain technology is well-suited to help address these issues. Metadata and catalog information is not readily available or accurate. Visual elements and the artwork attached to releases is minimal, as is visual customization. Biographical and liner note information is poor or non-existent. Duplicate entries are common, and there are many junk tracks and fake artists. Streaming services are currently oriented to the mass market and are under-differentiated. Diversified and value-added services will help meet the needs of underserved segments of consumers, especially ones who have a high willingness to pay or are so dedicated to the rewards of music that they may try to hack the solutions themselves.

Also worth considering is that, if Oink's Pink Palace had been incorporated into a blockchain system, it would likely still exist. While it was only a BitTorrent tracker and forum and not a centralized repository, unplugging and confiscating some equipment in the UK and the Netherlands killed it a decade ago. On a blockchain, it could be exponentially more difficult to take down, both through legal and jurisdictional authority and by physical location. In the future, why not build catalogs and communities like this on authorized blockchains and monetize the benefits that arise from such superior offerings.

CONCLUSION

Blockchain applications for recorded music are in their infancy, but waiting too long to act will likely have disastrous economic implications for the industry. Blockchain technology offers many real business and organizational solutions, with remarkable potential for value-generating mechanisms and the management of intellectual property. Music makers and music companies have the opportunity to recapture some of the value lost in the transition to online digital music.

Despite the explosive growth of subscription streaming, profitability still seems to elude nearly every streaming service. There are numerous opportunities for the improvement and incorporation of blockchain programming in conjunction with current services or as alternative services. There are dozens of blockchain initiatives and start-ups being inserted into the music and media industries, and an even greater amount are likely to follow. These enterprises will often appeal to the roots of the musical ecosystem, not just end listeners and transformational users. Administration and rights remuneration is typically slow, inadequate and hampered by flawed data, especially limiting the cash flow back to the base of the value chain: creators and first makers. Distributed blockchain ledgers and their ability to facilitate easy and equitable exchanges (including smart contracts), along with their wealth of data, may become indispensable alternatives to the current systems.

Blockchain applications—both authorized and illicit—can pose a threat to contemporary paradigms. Experimentation with, and investment in, legitimate blockchain-based entities will allow for a build-up of knowledge, experience and capital in this disruptive technology. Blockchain start-ups and companies in the early stages of development are plentiful, but statistically, most should be expected to fail. Over the past few years, although a rich mixture of ideation, entrepreneurialism and technical headway has emerged around blockchain-based music applications, there remains a high degree of uncertainty. Institutional forces, legal frameworks, and adjustments to policy and regulation are difficult to predict or influence, but technological developments often greatly inform high-level outcomes. The technology is still in its infant stages, and many of the emerging business models may be either deficient or premature. More appropriate models and applications are likely to emerge, in part, through lessons learned by trial and error. The next several years will likely form a key period in exploring the true extent and nature of the benefits and downsides that may be derived from this paradigm shift. Continuing to understand and improve the evolving nature of value management of phonographic goods and services will be key to sustaining creators and music enterprises. If legitimate organizations fail to create viable market solutions, it is almost inevitable that a new wave of unauthorized distribution will arise based on the more robust form of P2P networking offered by distributed ledgers and blockchain technology.

BIBLIOGRAPHY

Arvanitakis, J. and Fredriksson, M. (eds.). (2014). *Piracy: Leakages from modernity*. Sacramento, CA: Litwin Books.

Bohr, J. and Bashir, M. (2014). Who uses Bitcoin? An exploration of the Bitcoin community. In: A. Miri, U. Hengartner, N. Huang, A. Jøsang and J. Garcia-Alfaro, eds., *Twelfth annual international conference on privacy, security and trust*, 23–24 July. Toronto and New York: Institute of Electrical and Electronics Engineers, pp. 94–101.

Castells, M. (ed.). (1996). *The rise of the network society*. Cambridge, MA: Blackwell.

Castells, M. (2004). *The network society: A cross-cultural perspective*. Cheltenham, UK: Edward Elgar.

Dai, W. (1998). *B-money* [online]. Available at: www.weidai.com/bmoney.txt [Accessed 23 May 2018].

David, M. (2010). *Peer to peer and the music industry: The criminalization of sharing*. Los Angeles: Sage.

Electronic Frontier Foundation. (2008). *RIAA v. The people: Five years later* [online]. San Francisco, 30 Sept. Available at: www.eff.org/files/eff-riaa-whitepaper.pdf [Accessed 1 June 2018].

European Intellectual Property Office. (2016). *Intellectual property and youth: Scoreboard 2016* [online]. Alicante, Spain, Apr. Available at: https://euipo.europa.eu/tunnel-web/secure/webdav/guest/document_library/observatory/documents/IP_youth_scoreboard_study/IP_youth_scoreboard_study_en.pdf [Accessed 1 June 2018].

Goodell, J. (2003). Steve Jobs: The rolling stone interview. *Rolling Stone*, 25 Dec.

Heylin, C. (2004). *Bootleg: The rise & fall of the secret recording history*. London: Omnibus.

Himanen, P. (2001). *The hacker ethic, and the spirit of the information age*. New York: Random House.

Ingham, T. (2016). *Beyoncé's Lemonade is a piracy smash—but it's taken TIDAL to No.1* [online]. Music Business Worldwide. Available at: www.musicbusinessworldwide.com/beyonces-lemonade-is-a-piracy-smash-but-its-taken-tidal-to-no-1/ [Accessed 28 Jan. 2018].

Karlstrøm, H. (2014). Do libertarians dream of electric coins? The material embeddedness of Bitcoin. *Distinktion: Scandinavian Journal of Social Theory*, 15, pp. 23–36.

Kernfeld, B. D. (2011). *Pop song piracy: Disobedient music distribution since 1929*. Chicago: University of Chicago Press.

Knopper, S. (2009). *Appetite for self-destruction: The spectacular crash of the record industry in the digital age*. New York: Free Press.

Kusek, D. and Leonhard, G. (2005). *The future of music: Manifesto for the digital music revolution*. Boston: Berklee Press.

Miller, M. (2015). *The ultimate guide to Bitcoin*. Indianapolis: Que.

Music Business Worldwide. (2017). *Jay-Z's 4:44, a TIDAL exclusive, illegally downloaded nearly 1m times in 3 days* [online]. London, 3 July. Available at: www.musicbusinessworldwide.com/jay-zs-444-tidal-exclusive-illegally-downloaded-nearly-1m-times-3-days/ [Accessed 1 June 2018].

MUSO TNT. (2016). *Is stream ripping 2017's biggest music piracy threat?* [online]. London, 16 Dec. Available at: www.muso.com/wp-content/uploads/2016/12/MUSO_Stream_ripping_analysis_2016.pdf [Accessed 1 June 2018].

Nakamoto, S. (2008). *Bitcoin: A peer-to-peer electronic cash system* [online]. Available at: https://bitcoin.org/bitcoin.pdf [Accessed 30 July 2017].

Nordgård, D. (2016). Lessons from the world's most advanced market for music streaming services. In: P. Wikström and R. DeFillippi, eds., *Business innovation and the disruption in the music industry*. Cheltenham, UK and Northampton, MA: Edward Elgar, pp. 170–190.

Oram, A. (2001). *Peer-to-peer: Harnessing the benefits of a disruptive technology*. Sebastopol, CA: O'Reilly.

Panetta, K. (2017). *Top trends in the Gartner hype cycle for emerging technologies, 2017*. [online], 15 Aug. Available at: www.gartner.com/smarterwithgartner/top-trends-in-the-gartner-hype-cycle-for-emerging-technologies-2017/ [Accessed 24 May 2018].

Saint Bitts LLC. (n.d.). *Bitcoin price* [online]. Available at: https://charts.bitcoin.com/chart/price [Accessed 24 May 2018].

Sivante, R. (2016). *Music on the blockchain: An overview of the leading developments transforming the industry* [online], 9 Oct. Available at: https://medium.com/@RokSivante/music-on-the-blockchain-an-overview-of-the-leading-developments-transforming-the-industry-757f44ebf365 [Accessed 10 Dec. 2017].

Smith, M. D. and Telang, R. (2016). *Streaming, sharing, stealing: Big data and the future of entertainment*. Cambridge, MA: MIT Press.

Szabo, N. (1997). *The God protocols* [online]. Available at: https://nakamotoinstitute.org/the-god-protocols/ [Accessed 30 May 2018].

Werthman, T. (2008). *Cryptopiracy: A conjunction of cryptography and piracy* [online]. Bochum, Germany: Horst Görtz Institute for IT Security. Available at: www.nds.rub.de/media/nds/attachments/files/2010/11/Cryptopiracy.A.Conjunction.Of.Cryptography.And.Piracy.pdf [Accessed 20 Jan. 2018].

Witt, S. (2015). *How music got free: The end of an industry, the turn of the century, and the patient zero of piracy*. New York: Viking.

29

Disruption as Contingency

Music, Blockchain, Wtf?

Matthew Lovett

INTRODUCTION

In his 2016 TED talk, "How the Blockchain Is Changing Money and Business", the digital strategist Don Tapscott suggested that "it's going to rain on the blockchain for digital content creators" (Tapscott, 2016). Part of the talk focused on the idea that new, so-called friction-free transactions will mean that more revenue will go to artists as a result of people accessing digital content via a blockchain-enhanced music distribution and delivery system. While in theory it may very well be the case that artists could—and even should—see a greater share of the revenue generated by their creative works, it is far from certain that such a system would only serve to recoup value that had been lost in the distribution chain.

My aim with this chapter, therefore, is to use blockchain technology as a means of articulating one of the most current and dynamic developments in what is a long-running debate about the relationship between evolving technologies and music economies, namely the relation between an evolving music economy and the increasingly ubiquitous concept of 'disruptive technologies'.

Clearly, the internet as an emergent technology has forced huge changes in conventional practices surrounding the production, distribution and consumption of music, and in this regard, we could well understand it to be the epitome of a disruptive technology.[1] However, what is interesting is the current vogue for using the word 'disruption' as a by-word for positive, and possibly necessary, change within a given production or service environment. It could be said that there is a certain duplicitousness at work here, or at least something of an occlusion of what the word disruption actually means. Thus, the secondary aim of the chapter is to draw on recent debates that have arisen in speculative and materialist philosophy about the concept of 'contingency' in order to consider disruption as a contemporary manifestation of contingency at work within the music economy.

Initially, I shall discuss a set of developments relating to blockchain technology in order to generate a framework for considering music economies

Hepworth-Sawyer, R, Hodgson, Paterson, J and Toulson, R (eds). *"Innovation In Music: performance, production, technology and business"*

in relation to technology and consider some of the more obvious consequences for music production, distribution and consumption patterns that
could result from a widespread adoption of that blockchain technology.

Clearly, an appetite for change and transformation is part of the current
enthusiasm for disruption, and as such, the second part of the chapter will
focus in more detail on certain definitions and perspectives of the term
disruptive technologies. As a result, we shall then have an opportunity
to consider how blockchain could be considered a disruptive technology
itself in relation to the music ecosystem.

Finally, while I remain suspicious of the way in which disruption has
come to be increasingly weaponized in the service of what, at least on the
surface, appears to be a form of Silicon Valley-style self-aggrandizement,
my intention is to provide more than simply a left-leaning academic riposte
to this neoliberalist co-option of the term, wherein Adam Smith's invisible
guiding and self-correcting hand of the market is subtly replaced by the
invisible hand of technology. Instead, I shall present a set of philosophical
perspectives that will allow us to reconfigure disruption within a context
of necessary contingency so as to generate a more concrete and substantial
questioning of the sense of 'disruption as positive progress' that is becoming increasingly ubiquitous.

In terms of the 'wtf' in the title, a central idea and question that runs
throughout this chapter is a thinking through what blockchain, as a disruptive technology, is doing, or might do, to music. Will it bring improvements? Or will it bring something more complex, and less resolved? Of
course, it is easy to say that 'disrupting' does not necessarily mean the
same thing as 'improving', but by holding disruption up against a 21st-
century philosophical modeling of contingency, it is my intention to at
least root our current sense of uncertainty about the future of music commerce in something more fundamental than simply our response to a rapidly evolving technosphere.

AN EVOLVING MUSIC ECOSYSTEM AND BLOCKCHAIN

During its short life, the 21st century has so far been witness to a series
of paradigm shifts as regards the production, consumption and distribution of music; notably the pervasive digitization of music formats and
the ever-accelerating move away from music ownership towards music
streaming.[2] For our current purposes, it is possible to identify a set of
developments occurring within the music ecosystem that have resulted
from technological innovation. They include the following: the 'Sharing Economy',[3] which largely consists of amateur makers and producers
making digital content for fun or hobby purposes, and who are happy to
give away the results of their labor for free; 'mass innovation', or what
technology writer and business advisor Charles Leadbeater referred to as
"We-think"[4], a phrase that is intended to convey the way in which people
increasingly began to use the web as a co-production platform in order to
make things together; the increasing challenge posed to music as linear,

or fixed product, by a range of other, more open-ended media forms and experiences (for example, games and YouTube tutorials[5]); and of course the more widespread conversation that focuses on ownership vs. rental, or in digital terms, downloading vs. streaming.

As with the shift from Web 1.0 to Web 2.0, wherein Web 1.0 seemed to merely extend existing consumption or purchasing habits by making them more convenient, while Web 2.0 fundamentally changed human behavior and ushered in new paradigms in terms of mass communication, distribution and collaboration, streaming services have brought seismic change to online music practices and culture. In *Platform Capitalism*, Nick Srnicek suggests that applications such as Spotify, which was launched in 2008, are a certain kind of platform—a 'product platform'—which for Srnicek is a means to understand the way in which "companies attempt to recuperate the tendency to zero marginal costs in some goods", and he describes how "the music industry has been revived in recent years by platforms (Spotify, Pandora) . . . that siphon off fees from music listeners, record labels, and advertisers alike" (Srnicek, 2017, pp. 71–72). This is to say that, from Spotify's perspective, enabling users to listen to music has increasingly become only one aspect of the service and experience that it provides. While streaming music is clearly the core business, Spotify's modus operandi is more nuanced and complex, and as the company has evolved, new kinds of scenes and in-platform (as opposed to simply 'online') communities have developed as a significant by-product of their streaming service. Thus, Spotify-as-platform is an indication of the more exponential effect that emergent technology has had on music commerce, such that the difference in culture is more pervasive than simply the difference between downloading and streaming music, as evidenced in the impact on listener behavior and the growth of participatory digital music cultures.

In 2008, when Bitcoin—arguably the most well-known 'cryptocurrency'—made its first appearance in the "Bitcoin Whitepaper" (Nakamoto, 2008) [6], the technology that enabled Bitcoin's anonymous transacting and distributed consensus functions – blockchain – entered a world where an already dynamic music ecosystem was in a process of volatile evolution. In the context of this wave of technology-induced change and development, we can focus on the idea (or the promise) that Tapscott alludes to, which is that blockchain represents a new, and seemingly tamper-proof, way of protecting rights, as a result of its distributed ledger technology.

At its root, blockchain technology is fundamentally a means by which digital content can be indelibly or 'immutably' watermarked. What facilitates this immutability is the fact that a blockchain operates as a distributed ledger, which is to say that any transaction that is registered on a blockchain—which we could think of as a network of thousands of computers[7]—is simultaneously registered on however many thousands or millions of machines that make up that network. This network is decentralized, meaning that no one computer controls the network, and neither are transactions registered and stored on a central server. This is the reason that, in theory, blockchain-based transactions are, or at least should be, unhackable and immutable: the only way to change or tamper with the provenance of a digital record within a blockchain would be to simultaneously alter 'the entire history of commerce on that block-chain'—in other words, every single computer on that network—which, as Tapscott reports, is "tough to do" (Tapscott, 2016).

As a result of this new-found way to protect rights, a potential step-change that blockchain could facilitate is a future music economy that again creates and redistributes value for digital content, which, after an internet revolution that has so far led to a hemorrhaging of revenue out of traditional music markets, is clearly alluring for early adopters and enthusiastic proponents of blockchain technology.

Decentralization is thus a fundamental component of the Bitcoin-blockchain paradigm, because it tells us that no single computer, or single agency that might own that computer, can control, manipulate or shut down a decentralized network. Such a network will continue to exist even if various computers on that network (or nodes) blink in and out of existence. A block thus comprises a set of transactions that have been validated by peers on a network, and a blockchain, which is a linked chain of these blocks shared across the network, contains the entire and immutable history of all the transactions made on that network. In this regard, blockchain technology is a way of re-thinking how content can be hosted on the internet, where, rather than thinking in terms of music being hosted on a server, we can instead think of digital content being hosted on a blockchain, where rights can be protected in this new, immutable manner.

While this chapter is not intended to be an exhaustive account of blockchain functionality, it is also worth reflecting briefly on smart contracts, which, as a subsidiary function of blockchain technology, also feature heavily in the image of the brave new world of digital commerce. In its most basic form, a smart contract is a means by which ownership of an artifact can be transferred. To an extent, it could be said that a vending machine operates on a smart contract principle: a customer inserts the right amount of money into a coin slot and the machine presents them with a cup of coffee. If the customer inserts too little money, then the machine is able to recognize that there are insufficient funds to execute the contract, which results in no drink being delivered. According to Siraj Raval, "A smart contract is a piece of code that lives in a blockchain" (Raval, 2016, p. 7), which is to say that, within the context of blockchain, the principles of smart contracting remain similar to those of our vending machine, although the contracts themselves have become increasingly complex.

The music "think and do tank" (Mycelia, 2017), Mycelia, founded by the musician Imogen Heap, in collaboration with the blockchain-based music start-up Ujo, developed a proof of concept model in 2015, wherein every contributor to a musical track could be properly credited and reimbursed for their efforts. The track *Tiny Human*, released by Ujo, used different payment scales depending on whether a customer was intending to use the track for private or commercial use, and then made use of smart contracting in order to ensure that payments were equitably and rapidly distributed. In this regard, a smart contract is therefore an automated tool for managing transactions that works by implementing the terms of a contract. In a 2014 article entitled, "What Are Smart Contracts? Cryptocurrency's Killer App", Jay Cassano wrote,

> At core, these automated contracts work like any other computer program's if-then statements. They just happen to be doing it in a way that interacts with real-world assets. When a pre-programmed condition is triggered, the smart contract executes the corresponding contractual clause.
>
> (Cassano, 2014)

Although my aim is not to speculate about the future of music as such, the emergence of smart contracting (native to blockchain, rather than server-based frameworks) has significant implications for IP and rights management for digital content, as demonstrated by Heap's *Tiny Human* track; and this does suggest that the production process and distribution for music, already still in recovery in the post-sampling era, may yet experience further radical, and profound, evolution. Furthermore, within a blockchain environment, since decentralization would also be native to all transactions, it may well be that Leadbeater's appetite for mass innovation, which could simply manifest as mass-attribution and distributed authorship, could yet come to increasingly dominate the production process of digital content.

In 2016, a new software platform, *Blockstack*, was released. Described in the "Blockstack Whitepaper" as "a new decentralized internet secured by blockchains", Blockstack heralds a new kind of browser that functions natively within the blockchain environment and provides an opportunity for decentralized app developers to work with "services for identity, discovery, and storage" (Ali et al., 2017).

Blockstack thus purports to offer a new means to protect intellectual property rights—based on decentralized, peer-to-peer technology—while creating an environment within which friction-free transacting—and thus trading—could occur. Blockstack co-founder Ryan Shea's vision is that Blockstack will enable developers and consumers "to come together in a way that's better for both of them; removing the middle men, removing the monopolies [and that] Blockstack is a way for users to own their identity and own their data" (Shea, 2017).

What is compelling about Blockstack is that it is clear evidence of the variety of current activity that is directed towards achieving a redesign of the internet, building in authorship and identity protection from first principles, rather than having to retrospectively create new legislation and technological work-arounds as new platforms emerge. To a degree, the Blockstack project follows in the wake of what the technologist Jaron Lanier discussed in the book *Who Owns the Future*, where one of the central concerns was that the root cause of the widespread devaluation of digital music and other online content was that, fundamentally, the internet had been wrongly designed. Because of this design flaw, and since money in the digital age is simply another form of information, companies like Google and Facebook

(Lanier refers to such organizations as "Siren Servers"), who control the vast majority of the information flow on the internet, are propagating what he termed "information asymmetry": the root cause of the ongoing destabilization of at the very least, creative economies on the net. Lanier's diagnosis of our contemporary attitude towards value was to suggest that "it has become commonplace to expect online services (not just news, but 21st century treats like search or social networking) to be given for free" (Lanier, 2014, p. 10). In terms of music as digital content, Lanier's point is that while there is now a widespread expectation that music should be free, or at least for a negligible price, this does however come at the cost of us continuously allowing the Siren Servers—the Facebooks, Googles, Apples, Airbnbs and Amazons of the world—to harvest our valuable information. Lanier's grim conclusion is that this state of affairs can only continue if, ultimately, we are prepared to do away with careers in music, journalism and photography, to name but a few (Lanier, 2014, p. 16).

Lanier's solution to this problem was to redesign the internet with "two-way links" (Lanier, 2014, p. 227). This would mean that all of the connections made online (which in simple music terms could be a repost of a YouTube video, streaming or downloading a track) could all be tracked and that information, rather than languishing on a Siren Server's computers, waiting to be sold on to advertisers and other interested parties across the internet, could instead be captured by the content creator—so that a musician themselves could use that information to recoup income for their track and use the information for future marketing. Lanier's point was that, in an increasingly hobbyist music ecosystem, where many creatives have already given up on making money from music—by accepting that it is more rewarding having their music heard for free than have it sitting in silence behind a paywall, and are instead making use of YouTube, Soundcloud, Bandcamp etc. as a means to showcase their creative output, without really expecting this to turn into a long-term sustainable career plan—this is in fact the result of a problem with the design of the internet, rather than a question about whether music should be free.

One of Lanier's key concerns in *Who Owns the Future* was that wealth (re)distribution could be made more equitable via an evolution in technology. For Lanier, instead of a small number of individuals and organizations making huge economic gains within the digital domain (which he refers to as a "winner-takes-all star system"), a two-way links system would allow for a bell-curve distribution of wealth. From what Shea and the Blockstack development team describe in the Blockstack Whitepaper, we may very well be moving closer towards a realization of Lanier's vision, such that a blockchain-based digital infrastructure could provide just such a bell-curve distribution of revenue from music content, where, for example, musicians who currently post their music for free on Soundcloud and Bandcamp, and who are thus already part of a wider music economy, could find that content being monetized within a blockchain-enhanced internet.

As such, Blockstack—whose native function is to run all digital content within a blockchain framework, not unlike Imogen Heap's Mycelia project—is the promise of a new type of music economy. In this new economy, all contributors to digital content are properly paid, and creatives are able to track and manage their work across the internet, accessing payments when payment is due.

DISRUPTION AS A FORCE FOR POSITIVE CHANGE

Blockchain's promise to both protect copyright and remove intermediaries is perhaps the most significant reason for understanding it as a very contemporary form of disruptive technology. From the perspective of the music economy, it is a promise that suggests that widespread wealth will come from ridding ourselves of an entire commercial infrastructure that has evolved around rights protection as well as the distribution and sale of music.

The idea of disruptive technology, as it is presently understood, was introduced in 1995 in an article in the *Harvard Business Review* called "Disrupting Technologies: Catching the Wave". Within that article, authors Joseph Bower and Clayton Christensen made the case for the way in which disruptive technologies can often enter a market unnoticed; undercutting established products in terms of price and providing alternatives that may not have the performance of existing and established products, but are able to offer unforeseen benefits.

Their case study was the development of disk drives during the 1970s and early 1980s. The article described the way in a new generation of disk drives, which admittedly did not perform as well as those made by established companies such as IBM, nevertheless disrupted not only the market for disk drives, but went on to create the conditions for the birth of the personal computer industry. This new generation of drives brought down costs because they were smaller, and although they did not offer IBM's functionality, they required less power to operate, were more portable and were therefore suited to home rather than commercial use. The point that Bower and Christensen were making was that companies who continue to cling to their traditional working practices will find themselves being disrupted by new technologies. Equally, companies who carry on listening to their existing customer base will also find themselves being superseded, because disruptive technologies introduce new approaches that neither producers nor the marketplace necessarily understand.

Technology journalist Jamie Bartlett, in *The Secrets of Silicon Valley*, a set of television documentaries charting the wave of technological, economic, political and cultural disruption emanating from Silicon Valley through 2016 and 2017, has offered some useful contemporary perspectives on disruptive technology:

> Silicon Valley's philosophy is called disruption, breaking down the way we do things and using technology to improve the world [and that] the mantra of Silicon Valley is "Disruption is always good . . . that through smartphones and digital technology we can create more convenient, faster services, and everyone wins from that".
>
> (Bartlett, 2017)

While the positivist inferences of the concept of disruption have no doubt been evolved and amplified since Bower and Christensen first began to re-engineer its meaning in the 1990s, given that the thrust of their work was to create an association between the emergence of a particular piece of technology and the explosion of the home computer industry (which in itself was seen as a positive development), clearly, a precedent was set that signaled disruption as a force for positive change. In this sense, any negative impacts were either offset by the scale of positive development and improvement to a system or ecosystem overall (in terms of the IBM case study, then the creation of a new industry, along with the cultural change that home computing brought clearly mitigated any adverse effects to the commercial disk drive industry of the 1970s), and any collateral damage caused by disruption was on reflection a necessary change that had simply not yet been identified as something that needed to change (again, although the smaller disk drives did not initially meet IBM's performance standards, on reflection, smaller and more efficient disk drives clearly represented the direction that product development needed to take).

So powerful is the contemporary sense of disruption as a force for progress, in *Secrets of Silicon Valley*, Brian Chesky, the co-founder of Airbnb, gamely tells Jamie Bartlett, "To be disruptive means you're changing the world" (Bartlett, 2017). Similarly, in his recent book, *The Disruption Dilemma*, Joshua Gans discusses the way in which this positive re-assignment of disruption has become endemic, quoting Netscape founder Marc Andreessen who says, "To be AGAINST disruption is to be AGAINST consumer choice, AGAINST more people being served, and AGAINST shrinking inequality" (Gans, 2016, p. viii, capitalization in the original). Here we see disruption being equated with freedom and a moral certitude that comes from giving consumers exactly what they want, and any attempt to prevent (or disrupt?) that is not only standing in the way of technological progress, but is again a question of ethics and morality.

Gans also provides his own interpretation of disruption. First, he shows that the *Encyclopaedia Britannica* was not simply disrupted and superseded by a product or by a piece of software, but by the computer itself. So, while the historical timeline might show that it was Microsoft's Encarta—a CD-ROM launched in the 1980s that easily outstripped *Encyclopaedia Britannica*'s information storage capacity—followed by Wikipedia—a potentially limitless library of information, as well as a new a peer-to-peer means of gathering and verifying that information—that led to the physical encyclopedia's downfall, in fact, the real agents of change were the computer itself, followed by the internet. For Gans, Encarta was, and Wikipedia is, simply an expression of these more powerful and pervasive technology platforms. Gans also explores what he refers to as "containerisation" to show that the development of shipping containers in the 1950s completely disrupted and overhauled the cargo industry. His mapping of this change in the industry describes how not only ships, but the ports themselves, had to be redesigned in order to deal with moving containers around, and he concludes by saying that "finally, the entire logistics, information flow, and contracting space had to be reengineered" (Gans,

2016, p. 5). This is at the heart of Gans's analysis of disruptive technology; that disruption is not simply a matter of implementing improvements or efficiencies that have already been identified, but that it introduces a wholesale change to what an industrial or commercial environment is.

In response to these examples, Gans goes on to state that "disruption occurs when successful firms fail, because they continue to make the choices that drove their success" and that "the more a firm is focused on the needs of its traditional customers, the more likely it will fall prey to disruption" (Gans, 2016, pp. 9–10). In both instances, Gans's contention is that historical or behavioral precedent can rapidly become problematic for businesses adapting to the challenges of disruption. This might manifest as an inability to adapt to a new technological environment, or as a previously unrecognised aversion to risk: not only on the part of a company, but also, potentially, on the part of a company's consumer base. In the case of the *Encyclopaedia Britannica*, Gans's point is that not only did the encyclopedia's publisher and then Microsoft fail to adapt to the rapidly evolving digital environment of the 1980s and 1990s quickly enough, but that as containerization suggests, disruption forces adaptation across an industry in ways that are both unprecedented and unforeseeable.

However, Gans suggests that this still does not go far enough to really describe what disruption is. While Gans's re-positioning of disruption is in itself worthy of considerable academic analysis and examination, for our current purposes, we need simply to understand that the step-change that he introduces in terms of addressing what disruption is and how it works is to split the concept into two components, which for his own purposes allow him to reflect more fully on the evolution of encyclopedias and the notion of containerization.

In his reading of "Disrupting Technologies: Catching the Wave", not only does Gans acknowledge the issues that Bowers and Christensen raise in relation to the disruption of IBM market dominance, but he goes further to develop a more nuanced reading of their work. In order to do this, he introduces a "demand-side" theory of disruption along with a "supply-side" theory (Gans, 2016, p. 10). On one hand, demand-side disruption describes the relationship between a company and its customers, where a business is reluctant to stop giving their customers what they want (or at least what they think their customers might want), while on the other, supply-side disruption is more concerned with the framework—what he calls the 'architecture'—of a product (Gans, 2016, p. 10).

With this bifurcated version of disruption in mind, we can now begin to bring our focus back towards music. We have seen that Gans's approach allows us to think how the changes that disrupted businesses and technologies go through do not simply result from a set of decisions that are intended, for example, to enhance a particular product's performance, but instead because everything around them shifts. As we saw earlier, streaming caused huge disruption to both music distribution and consumption patterns, and labels and producers—and even listeners and fans to a degree—might find it difficult to accept that music's linear identity—as a commodity that can be bought and sold—is fast disappearing. While formats and consumption habits can remain all too wedded to outmoded trends and now increasingly anachronistic technologies—even vinyl's recent resurgence may in time appear to be the swan song of and paean to a culture now in practice long gone, although its ghost remains—in practice, the distribution, marketing and sales framework that surrounds music is becoming all but unrecognizable. What is curious in this regard is that a technology futurist such as Don Tapscott is still suggesting that the blockchain is a panacea that will bring value back into music simply by introducing efficiencies that are designed to counteract the wholesale change to value and consumption practices that the internet brought to music. In terms of Gans's supply-side and demand-side theory of disruption, Tapscott may well be right, losses might be recouped and a bell-curve of wealth distribution across the music economy could emerge; but it is far from certain that music as a linear commodity will survive in its current form.

In this regard, it is therefore worth pausing to remind ourselves that blockchain was not designed to increase efficiencies and profit margins within the music economy, since it grew out of the cryptography and cypherpunk communities, who were more concerned with protecting anonymity and solving the double-spend problem that continues to blight digital finance, and thus, any benefits that might accrue to music are not inherent to blockchain, but are simply by-products of another set of design principles and agendas. This suggests that music is already 'behind the curve', as it were, which is to say that music's future within a blockchain environment may be far from secure, and that Tapscott's enthusiasms for the 'rain' that will fall on content creators could well benefit from some additional flood and storm warnings.

DISRUPTION AS DISRUPTION

While Joshua Gans certainly furnishes us with a useful update to Bower and Christensen's ideas, in order to arrive at a more conclusive reading of disruption as actual disruption—wherein disruption speaks of exponential and unforeseeable change, rather than a filtered and weaponized version of the word that serves to affirm the appetites of market entrants as would-be giant-killers, who improve things by tearing them down—we can now turn to a recent set of philosophical perspectives that will open this clearly contentious term to a further set of interpretations.

The philosopher Quentin Meillassoux in his landmark book *After Finitude* discusses the way in which, through a process of logical reasoning, it is possible to show that we can develop a new way of understanding our relationship with the world around us, a world that is presented to us via our human faculties of perception and understanding. Meillassoux's work is concerned with the nature of contingency and its presence in our consciousness of the world. He goes as far as to name it a 'necessary contingency', and it is just such a contingency that sits at the heart of a notion of disruption as actual disruption. Meillassoux's contingency is one that absolutely must form part of the way in which we come to understand our presence in the world, must shape how our knowledge of that world might work, and must be the underlying cause for how anything comes to happen in the world, which in simple terms, is for absolutely no reason at all. Meillassoux's contention is that that there is absolutely no reason that anything happens at all, informing us that "Everything is possible, anything can happen—except something that is necessary, because it is the contingency of the entity that is necessary, not the entity" (Meillassoux, 2009, p. 65). In his essay "Anything Is Possible: Review of Quentin Meillassoux, After Finitude", Peter Hallward condenses Meillassoux's ideas into the phrase "nothing is necessary, apart from the necessity that nothing be necessary" (Hallward, 2011, p. 130), which can serve us as a useful point of reference. Things may get torn down, and things may improve, but fundamentally, disruption is always a process that goes beyond any attempt to capture it in the name of 'improving' the world; the only aspect of disruption that can really be said to have any actual, or necessary, presence is its ultimately contingent nature; everything else is just wishful thinking.

Meillassoux's approach is to argue against a long-standing philosophical perspective known as idealism, which he reframes as "correlationism" (Meillassoux, 2009, p. 5), which according to Meillassoux is a claim that we can have no knowledge or grasp of the world as it is in itself, only in a form that is presented to us via our capacity to comprehend and experience that world: "By 'correlation' we mean the idea according to which we only ever have access to the correlation between thinking and being, and never to either term considered apart from the other" (Meillassoux, 2009, p. 5). Thus, there are no 'real' experiences, simply our bodies' and brains' interpretations of the world around us, if indeed, we can be sure that there even is a world beyond our senses.

Meillassoux's response to this position is to absolutely turn it on its head by turning it against itself. The correlationist's problem is that if it is true that we can only ever think about the world in terms of a correlation between what we experience and what we are able to experience—in other words, if the correlation is 'true'—then such a statement immediately cancels out itself out. This is because any claims about the truth of a correlationist perspective are not objectively true, they are simply statements, as we would say, 'for me'—which is to say that they are made from a subjective and correlationist perspective—and they therefore lack any means of being absolutely, or objectively, verifiable.

Meillassoux supports this logic by using the concept 'facticity', which is his way of referring to our inability to know what underpins knowledge. He claims that it is 'true' that there are certain things that we cannot know, or, in the philosopher Ray Brassier's words, facticity "pertains . . . to the principles of knowledge themselves, concerning which it makes no sense to say either that they are necessary or that they are contingent, since we have no other principles to compare them to" (Brassier, 2007, p. 66), which is to say that we are unable to say anything about how it is that we know what we know, because we have no means of holding this knowledge up against any other point of reference. Meillassoux's conclusion is therefore to say that only one of these positions can be objectively true, facticity or the correlation: either we accept that it must be true that there are certain things that we cannot know, and that lack of knowledge is absolutely 'true', or that it must be 'true' that everything we experience must be in terms of our capacity to experience it. But as he says, this latter position is self-negating, because a subjective truth is not a truth—simply a perspective. For Meillassoux, the logical consequence to all of this is to say that the truth about our lack of knowledge is in itself a form of knowledge about the unknown; in other words, we absolutely know that we cannot know something: that which underpins our world must always go beyond our capacity to understand it and is therefore outside the category of knowledge; it is absolutely contingent. He tells us that "Facticity . . . forces us to grasp the 'possibility' of that which is wholly other to the world, but which resides in the midst of the world as such" (Meillassoux, 2009, p. 40), which is to say that we know that contingency exists right there in front of us, all of the time, and we can do nothing about it.

Meillassoux's thought is rich and complex, and the correlation and facticity are part of his wider philosophical project which is intended to show not only that what lies beyond human thought cannot be conceptualized, but more importantly, that the things that happen in the lived world are grounded on what he calls a "necessary contingency", where "contingency is such that anything might happen, even nothing at all" (Meillassoux, 2009, p. 62). This, then, is the underlying contention of Meillassoux's argument, that instead of nothing existing but our sense of having an experience (the idealist-correlationist position), only absolute contingency can be seen to necessarily exist.

While Meillassoux's philosophical reasoning was not necessarily intended to articulate the contingent nature of disruptive technologies, we can nonetheless refocus his outcome of his thought and use it to consider that, as phenomena that are part of a world that we know to be pinioned on a set of principles that we cannot know—in other words, contingency—disruptive technologies are in themselves fundamentally contingent. So saying, perhaps we would be better served to understand disruptive technologies as 'contingency technologies': technologies that must always go beyond our capacity to understand or anticipate them—not because they are too complicated for us to understand, but because, just as with everything else in the world, there are aspects to them that we know that we cannot know.

While Meillassoux's logic may appear to be slightly deflationary, in that his position suggests that we shall forever be at a disadvantage as regards our knowledge of the world, there is also a very affirming aspect to his thought which allows us to recognize that if nothing else, change and adaptation are constants. Even the seeming hegemony of Lanier's Siren Servers, which Blockstack and Mycelia are looking to topple, are themselves subject to the same inherent instability and necessary contingency that Meillassoux introduces us to. In this sense, all technologies are contingent and therefore disruptive, not simply because of their tendency to disrupt supply and demand–led economic ecosystems, but because, as contingency technologies, they cannot do otherwise.

NOTES

1. Mark Mulligan, the music analyst at Midia Consulting, continues to provide a range of valuable perspectives and insights on his Music Industry Blog and to document the changes and threats that confront embedded and traditional approaches to thinking about music commerce and music as a linear product.
2. Although it was reported in 2016 that "Apple is now preparing to completely terminate music download offerings on the iTunes Store, with an aggressive, two-year termination timetable actively being considered and gaining favor" (Resnikoff, 2016), others, including Mark Mulligan, are more sanguine, and although it is now generally accepted that that streaming will indeed lead eventually to the complete removal of downloads from the mainstream music marketplace, Apple's two-year plan may be too ambitious, even for them, although it is worth noting that Mulligan himself is reported to have put a five-year runout of Apple's iTunes download store (Blake, 2016).
3. In the book, *Making Is Connecting*, David Gauntlett makes the case for what can be termed the Sharing Economy, mapping a range of amateur digital content production practices, largely, although not exclusively, enabled by YouTube, and which have determined the shift from Web 1.0 to Web 2.0.
4. Leadbeater's 2009 book, *We-Think: Mass Innovation, Not Mass Production*, mapped a wave of change that internet practices were bringing to the production and distribution of both ideas and products. One of Leadbeater's key assertions was that digital networks fundamentally undermined what he saw as the principles and frameworks that supported, in his words, "industrial era organisations" by bringing his focus to bear on the architecture of Wikipedia and World of Warcraft.
5. Mulligan has made much of the way in which YouTube content has posed one of the biggest threats to music's potential to generate revenue from online audiences and consumers (Mulligan, 2016).
6. Satoshi Nakamoto is the pseudonym of the – at the time of writing – as-yet-anonymous creator of Bitcoin software, and therefore by default, the underlying blockchain protocol.
7. For example, Imogen Heap's music start-up, Mycelia—whose mission is "empower a fair, sustainable and vibrant music industry ecosystem involving all online music interaction services" (Mycelia, 2017)—makes use of the Ethereum blockchain in order to protect and monetize its artists' intellectual property. Ethereum regularly publishes the number of computers, or 'nodes', that are connected on its network. A recent count showed a global total of 20,567 active nodes (Ethernodes, 2017).

REFERENCES

Ali, M., Shea, R., Nelson, J. and Freedman, M. J. (2017). *Blockstack: A new decentralized internet*. Whitepaper Version 1.0.1. Available at: https://blockstack.org/whitepaper.pdf [Accessed Oct. 2017].

Bartlett, J. (2017). *The Secrets of Silicon Valley*. Available at: www.bbc.co.uk/programmes/b0916ghq [Accessed Oct. 2017].

Blake, E. (2016). *Apple won't end music downloads in 2 years, but it will some-day*. Available at: http://mashable.com/2016/05/12/apple-itunes-kill-down-loads-music/#SZwCM6Vwb8qQ [Accessed Oct. 2017].

Brassier, R. (2007). *Nihil unbound: Enlightenment and extinction*. Basingstoke and New York: Palgrave MacMillan.

Cassano, J. (2014). *What are smart contracts? Cryptocurrency's killer app. Available at:* www.fastcompany.com/3035723/smart-contracts-could-be-crypto-currencys-killer-app [Accessed 25 Sep. 2017].

Ethernodes. (2017). *Network*. Available at: https://ethernodes.org/network/1 [Accessed Oct. 2017].

Gans, J. (2016). *The disruption dilemma*. Cambridge, MA and London: The MIT Press.

Gauntlett, D. (2011). *Making is connecting: The social meaning of creativity, from DIY and knitting to YouTube and Web 2.0*. Cambridge and Malden: Polity Press.

Lanier, J. (2014). *Who owns the future?* New York: Simon & Schuster.

Leadbeater, C. (2009). *We-think: Mass innovation, not mass production*. London: Profile Books.

Meillassoux, Q. (2009). *After finitude: An essay on contingency*. London: Continuum.

Mulligan, M. (2016). *The labels still don't get YouTube and it's costing them*. Available at: https://musicindustryblog.wordpress.com/2016/01/08/the-labels-still-dont-get-youtube-and-its-costing-them/ [Accessed Oct. 2017].

Mycelia. (2017). Available at: http://myceliaformusic.org/ [Accessed Oct. 2017].

Nakamoto, S. (2008). Bitcoin: A Peer-to-Peer Electronic Cash System. Available at: https://bitcoin.org/bitcoin.pdf [Accessed Oct. 2017].

Raval, S. (2016). *Decentralized applications: Harnessing Bitcoin's blockchain technology*. Sebastopol: O'Reilly Media Inc.

Resnikoff, P. (2016). *Apple terminating music downloads "Within 2 years"*. Available at: www.digitalmusicnews.com/2016/05/11/apple-terminating-music-downloads-two-years/ [Accessed Oct. 2017].

Shea, R. (2017). *A new blockstack internet*. Available at: https://blockstack.org/videos/blockstack-summit-2017-a-new-blockstack-internet [Accessed Oct. 2017].

Tapscott, D. (2016). *How the blockchain is changing money and business*. Available at: www.youtube.com/watch?v=Pl8OlkkwRpc [Accessed Oct. 2017].

30

Can I Get a Witness?

The Significance of Contracts in an Age of Musical Abundance

Sally Anne Gross

> *Virtually every artist who reaches the charts has partnered with a record company.* **They do so by choice, in a landscape that offers artists more ways to release their music than ever before.** *They choose this route for good reason: to gain the experience, expertise and significant investment that a record deal brings.*
> —*IFPI IIM Report*, 2016

The music industries continue to be characterized as highly competitive but essentially meritocratic, equally accessible to anybody who is hard-working and talented. These optimistic claims are supported by arguments that the digital transformations across the music supply chain have arguably made it easier for aspiring artists and music entrepreneurs. Yet despite earlier predictions that major record companies would become things of the past, it would seem that these companies are still very much in control and the contracts they offer still highly prized. Added to this recent research this chapter will also question the earlier positive claims for participatory culture and digital labor.

Across the music supply chain, there is a widening debate about how these changes are impacting on all music stakeholders. There has been a growing questioning of the material realities of what many have called the cultural turn (McRobbie, 2016) that has included the expansion of higher education provision (Banks, 2017), the gig economy and the pressures on musical workers' health (Gross and Musgrave, 2017).

In this setting of musical abundance, which denotes the ability for anyone to access music, anywhere at any time, if they have the right equipment, this is fantastic for the users and producers of music, but it has also created many challenges.

This chapter is going to explore how the consequences of musical abundance exacerbate the precariousness of musical work by increasing competition in all directions. First, by examining the contribution made by the French economist Jacques Attali, who suggested that by observing a society's relationship to music, one could predict its economic future. Central to his thesis is the role played by music entrepreneurs as a driver of democracy and modernization. Second, by evaluating this line of thinking through the work of media theorist Jodi Dean's concept of communicative capitalism. Third, by exploring the impact of this market fragmentation on current commercial music practices, as Simon Reynolds suggests in his book *Retromania*.

In so doing we will argue that evaluating musical activity solely in an economic sense diminishes and contradicts both the potential of the musical object and also the potential of participatory culture. This economic reductionism acts to reinforce the dominant position of the digital media giants and major music companies while independent music entrepreneurs struggle for recognition of any kind.

Within this setting the act of contracting offers the music producers momentary stability and validation. Such recognition, however fleeting, is still significant across the music supply chain. Yet exclusive contracts also remain sites of tension around inequality and inequity, specifically of access and unequal bargaining power. And despite the positive rhetoric being put forward around digital labor, recent research suggests that the conditions of cultural work reveal continuing inequalities of access and undermines the meritocratic claims of the culture industries. (Banks, 2017; McRobbie, 2016).

> Being a huge artist you are always going to need a record company because of the infrastructure, the investment and the team of people.
> —A&R Major Label, London, University of
> Westminster—March 2018

MUSIC, LIKE WATER, COMES AT A COST

> When we go looking for unity inside a music industry, we should instead assume a polymorphous set of relations among radically different industries and concerns, especially when we analyze economic activity around or through music. There is no "music industry." There are many industries with many relationships to music.
> —(Stern, 2014)

The age of musical abundance has led to a situation of hyper-competition as music stakeholders compete not only with each other but also with multiple other forms of entertainment that is now readily available. It is no longer the case that music or entertainment options are expensive and in short supply; instead, it is the time and attention of users that has become scarce.

The convergence of media has meant that there are multiple opportunities and places where music is a component part but may not be the central feature of these new media environments (Stern, 2014), for example, the gaming industries or social media platforms like YouTube. In these new settings, the concept of work has been significantly challenged (Fuchs, 2014). Being a professional in the music sector often means working for free, and it is equally clear that many 'non-professionals' also work with music. In this ambiguous space, music producers are increasingly reliant on the contracts that define their activities, remuneration, obligations and the end product. And that in a contradictory and confusing fashion, the goal of becoming an 'independent artist' in the music industry has become about signing contracts and increasingly 'partnering' with third parties to increase one's chances of establishing a career, musical or otherwise. This is in stark contrast to the imagined ideal of new digital DIY artists that had been predicted, way back in the time of Myspace just after the turn of the 21st Century.

> There is a lot of confusion about being an independent artist these days, particularly in rap and grime. Everyone talks about "owning" their master rights, but all the big artists have signed deals with major label backing. So they have all licensed their rights to them—one way or another—that is not really independent, it's just reality.
>
> A&R Major Label, London, University of
> Westminster—March 2018

We will be considering how echoes of the pre-existing music and music industries practices can still be clearly heard in this new media setting. Specifically, how music captures time and also uses up time and how the sociality of music can easily be attached to other things, how music communicates but does not fix 'meanings' so that its messages can easily be subverted or recycled, recoded and redistributed, and that this sociality, this communicative aspect of music rather than its economic capacity, makes it so attractive.

The political media theorist Jodi Dean has named this new media setting one of communicative capitalism. She argues that it has defining characteristics, "the change in the form of our utterances from messages to contributions, the decline of symbolic efficiency and the reflexive trap in the circuits of drive" (Dean, 2013).

In doing so we will shed light on the contradictions that are thrown up in the new media landscape when it is contrasted with the optimistic claims for the future of the music industries, the democratization of music-making and participation in culture in general. As Jason Toynbee points out:

> the challenge is to look beyond the substantive field of cultural work to identify the thorough-going contradictions that extend from the larger domain of work in general into the labor of making cultural goods.
>
> Toynbee (2008, p. 87)

MUSIC AS A PREDICTIVE INSTRUMENT AND THE ROLE OF THE ENTREPRENEUR

In this first section we shall explore how music as a cultural phenomenon has been represented as having predictive abilities or as a cultural product that is somehow ahead of or a driver of societal change.

In 1976 the French economist Jacques Attali published a monograph entitled *Noise: The Political Economy of Music* (first translated into English in 1985), in which he suggests that our relationship to music is borne out of the fear of chaos and death. It is a complicated thesis and not without limitations, particularly his Judeo-Christian bias. However, Attali constructs an interesting argument that clearly resonates with our current setting. He argues that by examining how music was used in the establishment of (European) societies, we can explain how these historic economic and social conditions developed. Attali ignores the sociality of music in favor of aligning music to forms of domination and power. Furthermore, he argues that by outlining music's progression throughout 'civilization' in four stages—'Sacrificing', 'Representing', 'Repeating' and 'Composing'— he can show how our relationship to music actually predicts these developments. Starting from music's ritual use value as part of religious practices (Sacrificing), he aligns musical practices with different forms of cultural power, from church to state power (Representing), to the birth of capitalism and emergence of the bourgeoisie, which brings with it the rise of the entrepreneur and individualism (Repeating). In this third stage, he argues that music's value will be set by systems of ranking by popularity because of the possibility recordings offer. In his final stage, Composing, individuals become music-makers (here he was highlighting the then-new development of synthesizers and electronic methods of music-making). He goes on to say that in the future music will become for the most part free from market economics, save for a very few elite musicians who will continue to be able to make money from their labor. He also observes that music has a clear relationship to time and that in the future it will be the scarcity of our time that will shape our social conditions. By 2001 his work was receiving renewed attention, and he was asked to update *Noise*, in light of the transformations in the music industries. By 2014, he was claiming that he had predicted not only the crisis within the music industry but also that this change in the use of music actually foreshadowed the international monetary crisis of capitalism in 2008 (BBC News, 2015). Although Attali's work received much attention in the academic world at the time, it had little to no direct impact in the music industries sphere.

However, in 2005, a very different style of book written by Gerd Leonhard and David Kusek, called *The Future of Music: Manifesto for the Digital Music Revolution*, was widely read and talked about across the music industries; it, too, predicted some of the same outcomes—namely, the end of record labels and a world of free and limitless music. It enthusiastically embraces internet technology and trashes all the 'old myths' it claims the record label dinosaurs are projecting to shore up their sorry, outdated ideas. This book is an exemplar of all the hundreds of digital music books it spawned, replete as it is with the 'Top-10 Truths' you need to know *now*. A year later in 2006 *Wired* Magazine's editor-in-chief Chris Anderson's published his book *The Long Tail*, which was equally enthusiastically received. Both of these books presented a survival strategy for the future for all music stakeholders. Significantly, their ideas appealed to the independent and DIY music-makers and producers, offering as they did a future in which niche and specialist music would not only survive but also flourish.

None of these authors at the time of writing, including Attali, were concerned with any possible negative implications that the changes they were describing might have on people's lived experience beyond of course the possibilities to enjoy more music and the increase in potential market opportunities and new technological 'improvements'.

It is important to note that all of the above authors underplayed the tensions of competing music industry stakeholders in favor of promoting this new configuration as a much more open and accessible playing field. They ignored the reality that major record labels and music publishers were able to represent their interests and defend their copyrights as part of global entertainment corporations. The cultures and practices around the music industries meant that discussions around the problems of new media technologies centered around and condensed into copyright and intellectual property rights, illegal downloading and sharing of music rather than the practices and conditions of music-making and production. The need to reestablish a system of legal exchange of music which included revenue streams became paramount, and the 'suffering' of the big music companies dominated all new strategies and rationalized the introduction of increasingly restrictive contracts for artists and the development of contentious streaming models; for example, the much-talked-about 360% deal, which is now commonly accepted as part of the 'solution' to this crisis. Although a 360% contract may reduce the risk for the investor (usually a music label), for the artist it means being signed into an exclusive relationship in which all their revenue streams will be used to recoup the investment before the artist will see any income, and at the same time limits any possible gains from contracting with different companies; for example, by signing to XL Recordings and Universal Music Publishing, the artist might gain from being part of two competing stakeholders rather than having all one's eggs in the same basket.

For artists, musicians and music producers, copyright is not only a legal system that 'protects' their work and allows them to trade, but it is also a system that defines and differentiates the character of the work they do and assigns its value in a hierarchical manner across a chain of rights (Toynbee, 2008). That starts with the original work or primary right continuing to secondary rights or neighboring rights. So, for example, a songwriter who writes and performs her own material, like Charlie XCX, will have more copyrights to trade than a performer who performs another songwriter's work, like Ellie Goulding or One Direction.

Across the music supply chain, much of the literature and commentary about the changes in the music industries concerned copyright and the loss of the unit value of music alongside the problems of the growing 'freemium' economy, where by-products are offered to users for free, with additional extra packages then available for a price, for example, Spotify (Tschmuck, 2017). Until very recently, there had been a continued tendency to ignore the working conditions of those engaged in the creation and production of music, the notable exceptions being Shank (1994),

McRobbie (1999), Hesmondhalgh (2002), Banks (2017), Toynbee (2008) and Hesmondhalgh and Baker (2011). As these academics point out, this is partly because of the contradictory thinking around music and creativity, but it is also to do with the emerging significance of the creative industries in general, as part of the changing nature of neo-liberal capitalism—and specifically what is commonly referred to as the 'knowledge economy'.

MEDIA THEORY AND COMMUNICATIVE CAPITALISM

Music industries are embedded within much larger complex networks of telecommunications, technology and media industries (Hesmondhalgh, 2013). These new industries share much in common with the old creative industries and create what is known in media theory alongside economic power both 'soft power' and 'cultural capital'. Soft power being a type of influence that can be exerted without force, that is said to influence social behavior by cooperation. An example of this in the music field might be the Grime artist encouraging young people to vote in the general election. Cultural capital, on the other hand, is a kind of power or influence generated by the production of cultural goods that can be accumulated to improve the position of a person or group of people, or a company or state that hold it, so an example of a display of cultural capital was the opening and closing ceremony of the London Olympics in 2012. These types of power are central to the development of media theory and complex network theories around the proliferation of the creative industries and technology.

Music has always been media in the sense that it has always belonged to the expression and articulation of both personal (private, intimate and localized) communication and political (public, regulatory, controlled, global) communication (Gilbert, 2014).

Although Dean's argument is complex and multilayered, her core analysis identifies that communicative capitalism is how profit is generated in the current economic setting. For the sake of the argument here, communicative capitalism describes the way in which our social behavior in the digital economy is communicative and, in so being, each of the messages we create and share becomes units of profit in the complex networked web of global telecommunications. Dean identifies this process as the transformation of our messages into contributions. Whereas before a message was a communication between a sender to another person—the receiver, now this process has changed so that our messages become contributions to a wider flow of messages, a setting in which a receiver is no longer necessary (Dean, 2010). In a musical setting, this can be understood in terms of numbers of shares or plays across multiple platforms and/or number of songs or tracks, regardless of whether the listeners download or repeat the play, what is always important is the number, rather than the response, which could be negative or positive. The amount of activity, rather than the quality, is what becomes significant.

It is not only what we actually produce by making posts or songs or selfies for SoundCloud, Instagram or Snapchat, but it also includes the time and energy we use making them, because we are using and paying for data, equipment, technology and importantly our time and our lived experiences. As we do so, we create volumes of data that can be used in numerous ways by different stakeholders within this network, and although we do this often for our own enjoyment, we are also producing and are subjects of massive data collections.

> Communicative capitalism is that economic-ideological form wherein reflexivity captures creativity and resistance so as to enrich the few as it placates and diverts the many.
>
> (Dean, 2013)

Dean explains that her idea for communicative capitalism, like Attali's, also draws on Theodore Adorno's critique of the production and desire for information in mass culture. She suggests that Adorno's criticism is even more applicable today to contemporary media and entertainment networks, as we are now flooded with messages, music and images, and under this deluge there is no hope to listen, read or see let alone understand and connect with it all. It is this process of fragmenting, deconstructing and reconstructing and reconfigurations that Dean suggests reduces all this information to circulating contributions and that these fragments resist developing or progressing into something more demanding. It is noticeable how well her analysis works to describe what is happening in the contemporary music space, where artists, music entrepreneurs and major music companies vie for the attention and time of consumers.

In this new configuration, music is an encoded unit of information that carries content—the contribution. This content is attached to something else, and whatever the music is attached to is what sets its exchange value, which means what it is worth in economic terms. An obvious example of this would be one's mobile phone or one's internet provider or even headphones; in these examples, what one is paying for is not necessarily the content these devices can deliver but the devices and services they provide. In these technology settings, it is important to remember that for the producers of these items and services, the unit value of music is not as important, albeit the cheaper it is, the better their profit margins. If we look at a more traditional music partner, we can see how the different elements combine together to set the value of the music's use; for example, a new hair product that will be advertised across several different platforms, from terrestrial television like Channel 4 to internet ads on YouTube. L'Oréal is the producer of the product, and if the artist performing in the advert is Dua Lipa (BBC News, 2018)—who has been nominated for five Brit Awards this week, which will increase her appearance fee—then if the song is one of her songs, its use has to be licensed from both her publisher and her record label, who are both major companies. Each element of this production will include a specific fee, and every extension of time, territory and broadcast platform will again have a new fee. The price for using the song will usually be around 10–20% of the overall cost of the production. However, now, these advertising campaigns are often planned to coincide with ongoing promotional activities of the artist. Companies like L'Oréal are entering directly into contractual relationships with record music labels or artists. And within this mix of varying media content, from the product, the advert and the star who acts as a brand 'ambassador', both L'Oréal and Dua Lipa and her record company are hoping to generate more data and more indirect revenue and increase their singular cultural capital.

Brand associations and synchronization deals are not new to the music industries. However, they are now pivotal activities within the development of an artist's career and more significant to their earning potential than their music. As such, the networks that artists and musicians are partnered with play an increasing part in not only securing these deals but also generating them. Now promotional activities in the social media landscape never stop; the new mode of marketing at the top end of the music industries is now driven by this model of continuous flow of information and activity. Both these opportunities and this level of intensive production from posts on Instagram to new videos and music demand high levels of support in terms of human creativity, time, labor power and financial investment.

> Artists used to be "in" the market and then disappear and you would not hear from them but now you have to "present" all the time all the time.
>
> A&RMajor Label, London, University of
> Westminster—March 2018

In communication theory, media and entertainment industries are described as complex networks (Dean, 2010). In complex networks, users always have options: they are free to choose whatever they want to listen to, what concert to go to, and this is known as 'preferential attachment'. Yet this free choice is also affected by who else chooses or listens to the same thing or goes to the same concert—the validation by popularity that Attali described, too. In this situation the more people that like something, the more it potentially grows and so on; this is the basis of all word-of-mouth or viral marketing and social media storms. This 'preferential attachment' follows what is known as a power law rule and is also often referred to in music circles as the 80:20 rule. The consequence of this rule is that it creates a small group of significant winners and a large pool of losers. In the current setting, there is much evidence that the 80:20 rule is becoming even more extreme. If we apply this to the music streaming environment, we can see very clearly how the 'winners' from North America dominate the streaming numbers. This pattern, this logic of numbers in a global environment, is having a significant impact on local music industries (Tschmuck, 2017).

Dean's work usefully illuminates the current setting regardless of our personal musical relationships, experiences or ambitions. Her analysis highlights how claims for participatory culture's potential to liberate music-makers and producers from the global entertainment industries

networks may in fact actually be serving to reinforce the latter's powerful position and, rather than increasing the options for those with musical
ambitions, is in fact reducing them by diverting their productivity into this
endless flow of new technological products and media content. And in this
situation, music-makers of all kinds are using up their time and not being
paid; in this situation, a contract for work or an opportunity to improve
one's musical position becomes increasingly significant, and bargaining
power of those seeking contracts ever decreases.

RETROMANIA AND THE LOSS OF MEANING IN MUSIC

The second part of Dean's analysis involves a merging of sophisticated
arguments about how symbols contain, create and transfer meaning. So
what do we mean when we say that musical objects are symbols so that
songs, albums, music videos and music artists themselves all contain,
create and convey meaning? For example, if we saw an image of four
young white men wearing jeans and leather biker jackets, and one is sitting behind a drum kit, we would immediately think rock band, and if we
showed the same image to somebody in Japan, they would most likely
come to the same conclusion. However, if we changed just one member of
the band by either race, ethnicity or gender or age, we might have to think
a bit harder or ask questions of it, but if their outfits remained the same
(i.e. all leather biker jackets), we could confidently stick to our original
belief. The biker jackets, the jeans and the drums are enough: they efficiently contain the meaning and can easily be translated to a Japanese
setting. Very simply, that is what symbolic efficiency refers to. A decline
in symbolic efficiency thus describes the opposite effect, so if a group of
young girls in Japan start wearing jeans and biker jackets, although they
might be using the 'rock band' symbol to give meaning to their clothes,
they will also be 'diluting' the efficiency of the symbol. This idea is core to
the music industries business model, creating symbols that can travel and
communicate to the biggest audience, but in order for that to happen on a
global scale, symbols have to be efficient to work.

Despite the complexity behind this idea, the way in which it operates
and impacts on the music industry is very easily observed. In 2012 the
music critic Simon Reynolds published his book *Retromania*. Although
his starting point is not media theory but refers to a common observation from the 1970s 'pop will eat itself effect', whereby popular music
regurgitates again and again the same formula, and this coincides exactly
with what Attali was also saying. Reynolds identifies music practices that
recycle earlier styles of music in an attempt to mitigate the loss of meaning
and as a shortcut to musical recognition, as a trigger or what is commonly
called a hook. Yet the cycle never stops as the spiral of musical ideas turns
inward, and these practices simultaneously amplify the loss of symbolic
efficiency because the further they travel away from the thing they sym-
bolize, the weaker they become.

The format informs the music—people are making Spotify songs now—the first few bars are either loops or samples because they help draw in the listener.

A&R Major Label, London, University of
Westminster—March 2018

However, Reynolds insists, we must remember that digital music techniques originally gave birth to the most exciting and new form of music that used recycling and reconfiguration as a central method of its originality—hip-hop! When hip-hop emerged, it was heavily censored and attacked for its new language and its sampling techniques. The legal system was most definitely not on the side of these new young urban symbol creators. The use of and recycling of 'older' styles and the reimagining of them in a new current context can be very exciting. The most perfect and immediate use of this form of message 'trigger' can be found in sampling, where the chosen sample or loop seeks to create a short-circuit to a message, a symbol that roots the new offering in a 'real' past of musical value. Similarly, the practice of piggy-backing stars in collaborative tracks that join together fan bases serves to reinforce symbols and impact. The power law of musical memory, this is cultural capital in action.

More pessimistically, Reynolds observed that rather than creating new music, the music industry and musicians and artists are mitigating risks by reproducing or imitating sounds from the popular music canon. So Adele and Amy Winehouse sound like singers from the 1960s and 1970s, and new dance music sounds like classic dance hits from the 1990s. If you listen to BBC Radio 6 music, the apparent home of new music, it is always difficult to tell if they are playing an old record or a new record that is produced to sound old.

This type of targeted production and target marketing demands efficient demarcation of genres and markets. Thus, a proliferation of choice gives us the exact opposite, and is where we find the weakness of the Long Tail Solution, in a winner-takes-all environment.

As a decline in symbolic efficiency further exacerbates this problem of a struggling middle ground, of how to get and keep attention, Dean suggests that symbols join together to support the meaning, so she suggests that photographs speak louder than words (Dean, 2010) In the music environment, music becomes attached to a moving image. Let us look at this in hyper-action, in Taylor Swift's video for her single 'Bad Blood'—it is no longer a music video but rather a mini-blockbuster action movie replete with an all-star cast and full cinematic credits! Within a year, Beyoncé would release her album 'Lemonade' as a film, inspired by feminism, Black Panthers and her husband's adultery. That is a perfect manifestation of every aspect of communicative capitalism.

Where 'crossover' was once the bridge to mainstream success, it is now a drawbridge across a great void; the price of crossing the bridge is the investment and power provided by the major entertainment companies and third-party alignments working to consolidate one's position.

Another problematic observation is that in the new media setting, rather than increase access and open up diversity, as is commonly argued by the supporters of the Internet democracy, it has served to narrow and silence voices of difference and niche and new musical styles.

Here we can see how, although the digital music sphere continues the DIY model, which emphasizes the need and desire to be in 'control' of one's musical destiny, the reality is, as Dean points out, very much in favor of the power holders. The result is that an ever increasingly small proportion of musical power players and technology giants rule the music ecosphere, and coupling and contracting with the powerful becomes the sought-after thing to do.

Significantly for Dean's argument, her third characteristic of communicative capitalism is "the reflexive trap of the circuits of drive", in which she identifies what she calls "a strange accommodation" of this kind of thinking from a leftist and anarchist position. This is a very complex analysis based on Marxist theory of reflexivity. However, on a simplistic level, it serves to help us understand how so many thinkers and creative people might have seen this new media environment as a place of possibility for resisting the marketization of so many parts of human experience.

This 'alternative' or rather "point of accommodation" (Dean, 2013) centers around Marxist ideas about autonomy, creativity and productive labor (Hardt, Negri, 2000). It is an important observation that illuminates why so many see digital environments as opportunities, and is particularly relevant to prosumer and maker communities and is underwritten with DIY principles of not only individual entrepreneurs but also equally with communities wanting to establish themselves and increase their power and influence:

> Communicative capitalism designates the strange convergence of democracy and capitalism in networked communication and entertainment media.
>
> (Dean, 2013)

Even the way these positions polarize is a significant feature of communicative capitalism, whereby the middle ground has become a great divide. The hierarchical ways in which structures and customs across the music supply chain have been established, starting with copyright (Towse, 2004), cements difference and emphasizes competing interests. We can see this by looking at all the different interest groups that have been established from the Featured Artist Coalition to the Music Managers Forum, the growth of PRS for Music, and the much-weakened Musicians Union (Hesmondhalgh, 2013). Instead of recognizing commonalities between other creative industries and creative workforces, they have fought to retain their specialist's status. Rather than recognizing what they have in common, they are invested in emphasizing their differences. This notion of establishing one's unique selling point is central to much thinking around music and cultural production, linked as it is to notions of originality, novelty and the holy grail of qualities in the digital world: authenticity. Indeed, Attali identified this characteristic for accumulation of value through systems of ranking, in his stage three 'Repetition'. This ranking process is described by immaterial labor theorists as validation (Lazaratto, 1997). Validation and systems of ranking become ever more significant once we enter the world of data accumulation, where numbers come to rule above content. In the hierarchy of validation, the 'contract' symbolizes that point of commitment, the seal of approval, joined as it is by consideration and union of free and equal parties—laissez-faire contracting. Becoming the commodity has become the goal and is literally the reverse implication of fleeing from the machine; instead, you want to be hooked into it. The point of commodification has a chain of symbolic stages that each as they are arrived at are gathered together, add on and validate the 'artist' or music—by removing doubt and cementing validation.

What we are looking at here is not only the economic value (although that is always implicated) but rather how the processes of copyright, contract and commodification intersect to impact on the validation cycles so as to deal with the 'problem' of oversupply.

CONCLUSION

There is a new wave of optimism at the top end of the music industries, specifically for the large catalog holders. However, there is also growing concern about the working conditions in the music sector that echo many of the problems of the 'gig' economy. Recent research has drawn attention to problems of inclusivity and representation in the music and wider culture industries. These findings contradict the original claims that the internet would eliminate such issues by giving everybody a voice. As Attali predicted, music is freely available, but now it is time and attention that are scarce. In this setting, we need to think beyond the interests of the dominant music stakeholders if we are to address the challenges thrown up by these conditions, so that we might achieve the kind of creative justice (Banks, 2017) that is accessible to all.

Dean's naming of the digital media landscape as one of communicative capitalism lays out the setting in which musical workers produce and work as they compete for attention. In this setting, the conditions of exploitation are amplified, leading to an ever-widening gap between the powerful and the dispossessed. This is important for democracy and participatory culture.

It is not only developments in new technologies that have impacted on the exchange value of music and with it the subsequent increasingly harsh economic conditions in the music landscape. Rather, as suggested here, these extreme conditions are the consequence of music sector practices, right across the vertical music and creative industries supply chain.

However, the problems that genre fragmentation and market segmentation compound in the current configuration is that, rather than aid the development of healthy markets, the abundance of choice has led to the loss of a sustainable middle ground, thus heightening the disparity between the heard and the unheard, the paid and the great unpaid.

BIBLIOGRAPHY

Attali, J. (1976). Tra.1985. *Noise: The political economy of music*. Minneapolis and London: University of Minnesota Press.

Banks, M. (2017). *Creative justice-cultural industries, work and inequalities*. London: Rowman & Littlefield International Ltd.

Barney, D. (2007). *We nation under Google*. The Hart House Lectures.

BBC News. (2015). *The pop star and the prophet*. Available at: www.bbc.co.uk/news/magazine-34268474.

BBC News. (2018) *Dua Lipa makes Brit awards history*. Available at: www.bbc.co.uk/news/entertainment-arts-42677238.

Billboard. Available at: www.billboard.com/articles/business/7495408/album-release-strategy-beyonce-frank-ocean.

Billboard. Available at: www.musicbusinessworldwide.com/3-ominous-problems-the-uk-music-business-must-overcome-in-2017/.

Dean, J. (2010). *Blog theory*. Cambridge: Polity Press.

Dean, J. (2013). *The Limits of the Web in an Age of Communicative Capitalism*. [Online Video]. 30 June 2013. Available from: https://www.youtube.com/watch?v=Ly_uN3zbQSU. [Accessed: 10 December 2017].

Fuchs, C. (2014). *Digital labour and Karl Marx*. London: Routledge.

Gilbert, J. (2014). *Common ground*. London: Pluto Press.

Gross, S. and Musgrave, G. (2017). *Can music make you sick*. London: Music-Tank, University of Westminster.

Hardt, M. and Negri, A. (2000) *Empire*. Cambridge: Harvard University Press.

Hesmondhalgh, D. afnd Baker, S. (2011). *Creative Labour: Media work in three cultural industries*. London: Routledge.

Hesmondhalgh, D. (2002). *The culture industries*. 3rd ed. London: Sage Pub.

Hesmondhalgh, D. (2013). *Why music matters*. Oxford: Wiley-Blackwell Pub.

Kusek, L. and Leonhard, G. (2005). *The future of music- manifesto for the digital revolution*. Boston: Berklee Press.

Lazaratto, M. (1997). *Immaterial labour*. Available at: www.generation-online.org/c/fcimmateriallabour3.htm.

McRobbie, A. (1999). *In the culture society—art, fashion and popular music*. London: Routledge.

McRobbie, A. (2016). *Be creative: Making a living in the new culture industries*. Cambridge: Polity Press.

Reynolds, S. (2011). *Retromania-pop culture's addiction to its own past*. London: Faber and Faber Ltd.

Rogers, J. (2014). Canary down the mine: Music and copyright at the digital coalface. *Socialism and Democracy*, 28(1), pp. 34–50. doi:10.1080/08854300.2013.869875.

Shank, B. (1994). *Dissident Identities: The Rock 'n' Roll Scene in Austin, Texas*. Hanover, NH: Wesleyan University Press.

Stern, J. (2014). There is no music industry. *Media Industries Journal*, 1(1).

Towse, R. (2004). *Music and copyright*. 2nd ed. (Edited by S. Frith and L. Marshall). Edinburgh.

Toynbee, J. (2008). *How special-cultural work, copyright and politic*.

Tschmuck, P. (2017). *The economics of music*. Newcastle upon Tyne: Agenda Publishing.

UK Music Measuring Music. (2016). Available at: www.ukmusic.org/research/measuring-music-2016/.

Wikstrom, P. (2009). *The music industry*. Cambridge: Polity Press.

31

The End of a Golden Era of British Music?

Exploration of Educational Gaps in the Current UK Creative Industry Strategy

Carola Boehm

INTRODUCTION

The creative industries, and particularly our UK Music industry, are perceived as healthy, resilient and strong. However, with the ongoing policy changes in secondary and higher education, as well as the continued cuts to council budgets and the ongoing lack of commitment to wealth distribution and even investment in the whole nation, this golden era of the creative industries in the UK may not last. In my latest articles, I explore critical themes relevant for the UK Music industry and the UK creative sector as a whole. Current national policy expressions often omit to address these themes, which are necessary to safeguard our future creative resilience. In writing this article, much relevance will be drawn from making connections to recent public debates on what universities are for and what their role is within the creative economy. Attention is given to considering current governmental industry strategies critically and their relevance for the music industry, together with their sector responses.

AREN'T WE DOING WELL IN THE MUSIC INDUSTRY?

Over the last two decades, the music industry, and more generally the creative and cultural industry as a whole, has had a successful period of growth and expansion (UK Music, 2017b; Creative Industries Council, 2017). It has in recent times been heralded as one of the few sectors in growth while the productivity of the rest of the economy seems to have stalled; thus, the music and creative industries are outgrowing the UK economy. UK Music has pointed out that there are still areas that need to be addressed to ensure continued growth in the future, specifically relating to the increase of digital cultures and its related issues of fair remuneration to artists and rights' holders as music consumption changes from physical ownership to streaming (UK Music, 2017b). Relevant to these themes, Brexit and its effect on the creative sector has also given rise to new anxieties about the ability to continue our golden area of creativity within the UK political trajectories (CIF, 2017).

With the sector's success as the fastest growing part of the UK economy, consecutive governments have been increasingly afforded to consider it explicitly in their economic strategies. The political agendas have moved from realizing the potential of the creative industries back in 2001 (DCMS, 2001), to policy imperatives in 2008 (DCMS, 2008), to industrial strategy (BIS, 2017), which is currently influenced and informed by both an independent review of the creative industries (Bazalgette, 2017) and the Arts Council's commissioned report on the value of the creative economy (CEBR, 2017). Back in 2001 the Government's vision was of "a Britain in ten years' time, where the local economies in our biggest cities are driven by creativity", and this was—I suggested in a series of articles—to be a major shift in the perception of the value of arts and creativity (Boehm, 2009, 2014, 2015b, 2016b; Patterson and Boehm, 2001).

However, although the creative industries are perceived as healthy, resilient and strong (ACE, 2017a), recent policy expressions and their interventions often neglect vital aspects (or barriers) that are likely to hamper future productivity growth within the creative industries, including the music industry. These themes include:

- connectivity of place, people, cultures (including open-innovation partnership models)
- continued policy emphasis for our learning organizations on knowledge content (and learning outcomes) rather than on educational environments for developing creative life-long learners within a knowledge economy context
- significance of Culture 3.0 concepts with its possible future of a vanishing creative industry as a demarcated sector, as user/producer divides disappear and co-production models take over

This article will focus more on the first two themes, the first section covering creative sector strategies in relation to national infrastructures, such as how to cope with the north-south economic divide, its effects on our creative economy. The second theme focuses on recent policy interventions both at secondary and tertiary educational levels and their direct impact on development of creative talent. The third theme, Culture 3.0, although relevant for the subject matter of this chapter, has been developed in depth in some recent articles (Boehm, 2016a; Sacco, 2014), but in short is based on a new type of cultural engagement where the need for large intermediaries, such as labels, diminishes and consumer/producer divides vanish with technologically immersive and immediate environments.

FRAGMENTATION AND DISCONNECTEDNESS: CONNECTIVITY OF PLACE, PEOPLE AND CULTURES

The creative and cultural economy is highly fragmented (characterized by a high proportion of what the media associates with the gig economy) and is populated by a large number of 'one-man bands', e.g. micro-cultural producers, including sole traders and self-employed. The "gig economy", or more positively framed as "new forms of more flexible working", is becoming a characteristic work practice in an increasing number of fields, including music and the creative industries. This was the topic of a short conference in London, on June 13, 2017, where it was suggested that

> the number of self-employed workers in Britain has grown by 1 million between 2008 and 2015 (Office for National Statistics, 2016), while so-called zero-hours contracts have also reached a record high. This "uber-ification" of the workplace signals a transformation in labor relations and structures of employment that raises pressing questions about the future of work.
>
> (CAMRI, 2017)

This mirrors the music sector, where "the vast majority of music businesses are small or micro companies, and the music industry has a higher proportion of sole traders and freelance workers than the average sector" (UK Music, 2012). The UK Music report goes on to highlight that their income is often not from one single source, and unless VAT registered and employing workers using PAYE, they may not appear on business registers. By 2017, the UK Music report does not mention sole traders or self-employed, but just denotes measures related to individual types of professions as part of its rigorous definition of what the music industries are (UK Music, 2017a). Thus, in its 2017 report, the largest contribution (2bn GBP of a total of 4.4bn GBP) for the total GVA in 2016 was that of musicians, composers and songwriters.

This fragmentation provides a high amount of sector resilience, even in times of economic upheaval, as the aftermath of the 2008 financial crisis and subsequent recession has demonstrated. However, this high fragmentation also carries with it a disproportionate risk and vulnerability to the smallest denominators. Micro-cultural producers and the small end of the Small and Medium Enterprises (SMEs) are highly vulnerable on an individual basis, and a lack of sufficient wealth distribution on a national basis affects the productivity of this sector. So, where statistics are available, they demonstrate that creative producers in today's world are less likely to survive solely from their earnings. Thus, a 2017 report for literature suggested that in 2005 approximately 40% of authors survived solely from the earnings they made through their writing, and this was reduced to 11.6% eight years later in 2013 (ACE, 2017c).

Additionally, where the middle classes are squeezed, productivity based on SMEs—where the critical mass of start-ups and small businesses are initiated—is affected, and this creates also a geographical disconnect, or as we experience in the UK, a substantial north-south economic divide.

The north-south productivity/wealth divide has given rise to devolution arguments, and were recognized as being significantly influential in the 2015 general election outcomes. Paul Mason's Channel 4 blog presented an interesting map for the time around the general election of 2015.

The digitally aged paper map, still available at the open democracy website on Hanson's Blog (www.opendemocracy.net/uk/steve-hanson/

northern-powerhouse-as-real-mirage), overlays the UK cleverly onto the Finnish geographies, with Scotland being "Southern Scandinavia" taking up the rural, most northern part of Finland on the map; Greater London is depicted as part of the "asset-rich southlands", geographically overlaying areas that would theoretically include Helsinki and Turku; and the Finnish Archipelago being labeled as the "Post-Industrial Archipelago", which, as Hanson suggests, represents

> the Detroitified, abandoned middle, drawn as spiky red islands. Like the "new" north-south split, there is nothing new about this map either. It has just gone public for the first time, as it suddenly matches voting swings more closely than it did before.
>
> (Hanson, 2015)

A more fact-based graph can be seen in Booner's report for HEFCE on University-Industry Collaboration (Bonner et al., 2015), where most of the graphs indicate that economically London is doing very well, and the rest of the country is not. This picture is reiterated in various metrics— productivity, income, employment and when considering the music industry, a similar picture emerges with most of the SME indies being located in the London area.

Some might consider this to be the "natural order of things", but the central focus of assets and resources, as well as governance, is far from being inevitable, and the UK is the most centralized government of the G7. The success in other European countries is often based on much wider, devolved and distributed governance (compare Germany), investment and wealth distribution models.

In the UK, this resulted in a long-standing lack of commitment to investments in infrastructures that provide connectivity throughout the nation, and not just to the London metropole. The theme of the under-connected northern powerhouse has been subject to increasing expression for a stronger commitment to building more devolved high-speed railways, e.g. HS3 before HS2 or amending HS2 to include HS3 elements. It would thus connect the northern part of the UK before building yet another improvement to London's connectivity (Wand, 2017; Shaw, 2017; RTM Rail Technology Magazine, 2017; IPPR North, 2012, 2017). Improving the connectivity of the musical powerhouses of Liverpool, Manchester and Leeds would help not only to close the creative productivity gap between the north and the south but also to unleash a further productivity potential held by this creative nation, which is held back by the bottlenecks (and the high cost of living) that London represents in a centric nation. Specifically, for the music and creative industry, regional infrastructure could be considered as vital. However, the lack of commitment to addressing the geographical disconnects are continually evident, such as in November 2017, when Transport Minister Grayling announced that plans were being accelerated to reopen the railway line from Oxford to Cambridge, ahead of any other transport infrastructure plan in the North (Topham, 2017).

There is also the perceived disconnect between the cultural creative for-profit and not-for-profit sector, which is having a big effect on the creative industries in a geographically biased manner. Where there is insufficient continued investment in the cultural sector, specifically considering that the creative industries are much closer to the not-for-profit third sectors and cultural not commercially driven actors, there tends to be a lack of SME resilience. Thus, both the former and current CEOs of the Arts Council (Bazalgette and Henley) have spoken out on this matter of regional cultural deficits outside of London. Bazalgette put the blame of regional cultural deficits firmly on consecutive council cuts (Thorpe, 2014), and Henley reiterated the London versus the regions issue as one of the most pressing arts policy debates in England, and with his taking up the CEO mantle promised "more of a two-way street" (Hanley in Brown, 2015). When Darren Henley joined the Arts Council in 2015, he put in place initiatives that will ensure that 75% of ACE funding goes to regions outside London by 2018, and a map of the newly funded portfolio organizations, published in 2017, demonstrated the implemented commitment to attain these targets (ACE, 2017b).

In a similar vein, and considering university-industry collaborations, the 2015 HEFCE report (Bonner et al., 2015) indicated that whereas the average SME Creative Industries sector lies at around 8.7%, London has a share of 18.8% of Creative Industries SMEs. Some of the areas with the lowest employment figures demonstrate a below-average level of enterprise and creative industries, e.g. Stoke-on-Trent is highlighted (p. 15) as the area with the lowest share of SME Creative Industries. Stoke-on-Trent is an interesting example here, as with its pitch in 2017 for becoming City of Culture 2021 it has recontextualized itself as the ceramic city (besides being Soul-on-Trent, the "home of northern soul", with a long tradition of supporting a flourishing music club scene) that has both a long-standing industrial heritage based on a creative industry sector and a flourishing future creative cluster. But as the crafts sector, and with it, the whole micro-producing ceramics sector, was left out of NESTA's 2016 Geography of Creativity (Mateos-Garcia and Bakshi, 2016), because craft business "does not lend itself easily to the approach used in the report" as "evidence shows that 88% of craft businesses are sole traders". So NESTA and the Crafts Council can clearly learn something from the UK Music and the Music sector, which have managed to devise a methodology to continually try to ensure that micro-cultural producers and their value to the economy are counted. This is increasingly becoming more important, as governments base their policy decisions on these commissioned reports, with (perhaps) an over-reliance on their contained metrics.

Thus, for instance, when the initial call for the £45million Creative Industries Clusters Programme was launched by the AHRC as part of its Industrial Strategy Challenge fund, it originally emphasized the use of NESTA data to ensure that only pre-existing clusters mentioned in NESTA would be eligible. Thus, as an example of the challenge of high fragmentation and its measurability, Stoke-on-Trent—one of the oldest and deepest creative clusters, albeit predominantly made up of micros and SMEs—did not feature in NESTA's report on the creative industries (Mateos-Garcia and Bakshi, 2016). On the other hand, Crewe and Cheshire East were mentioned, but they appeared due to the high spill-out of the publishing industry from Manchester into the Macclesfield region. These are typical pitfalls of sectors that have a very high fragmentation, which make it difficult to get a sense of what is going on. I laud again the transparent and rigorous methodology published by UK Music (UK Music, 2017a), which can potentially set a standard for other highly fragmented sectors when attempting to measure their economic value, and provide visibility and effective advocacy.

It is also worthwhile mentioning that with the demise of free access to higher education, another pressure on what Germans call the "Mittelstand" (closest equivalent is the concept of SMEs) has been mentioned—the need for the middle class to increasingly absorb student debt, which hampers enterprise start-up, productivity and/or growth (Newfield, 2016). Student debt, of course, leads to the next point—educational environments.

THE RISE OF STANDARDIZATION

Perhaps resulting from the dissolution of the UK government's department of Business, Innovation and Skills during the overhaul of Whitehall in July 2016 by new prime minister Theresa May, with Higher Education (HE), Further Education (FE) and apprenticeship joining the government Department of Education, a new conceptual distance between innovation and education allowed a second major theme to be not sufficiently addressed by the governmental industrial strategy: that of arts education in secondary and tertiary education. Higher Education was brought together within a department encultured in standardization and regulation, and this had an immediate impact on the HE sector. As Ken Robinson put it even more bluntly: "If you run an education system based on standardization and conformity that suppresses individuality, imagination and creativity, don't be surprised if that's what it does" (Robinson and Aronica, 2015).

As the UK Music advocacy group argued: "future talent will never get the chance to shine if we continue to see cuts in music in schools and closures in venues where artists need to learn their craft in the first place" (UK Music, 2017b). But the governmental strategy from November 2017 still seems to rely on the need for technical education above anything else. It suggests, "We will also update school and college performance measures to ensure that students can make an informed choice between technical or academic education in time for the introduction of the first T levels, recognizing them as equally valued routes" (BIS, 2017, p. 102).

In relation to arts provision, Bazalgette pointed out in his independent review that "industry should develop . . . curriculum materials to broaden and deepen the talent pipeline that starts at school" (Bazalgette, 2017, p. 10), as if schools have not been involved in creative education for the last decade, despite increasing numbers of interventions by government to cut arts provision for the sake of STEM, and this despite all evidence of its effectiveness to the contrary. So rather than introducing industry

intervention into the school sector (after decades-long governmental interventions and affordability of university-school interventions), it might
be useful to consider other national school models that seem to support
learner excellence, as measured by PISA metrics (Programme for International Student Assessment). It is hard not to mention the Finnish model
of school education (Sahlberg, 2017) in this context, which has not only
allowed school performance to be steadily improved over the last 20 years
but has produced this with a high amount of equity and well-being, and
less standardization or competition.

On the opposite scale, in the UK, we are just average performers in PISA
metrics, and the mental health and well-being of both school and university based learners is critically low; the productivity levels in our economy
are the worst they have been for a long time. On top of this, student debt
is at an all-time high and social mobility at an all-time low. Simultaneously, head teachers up and down the country are expressing their dismay
at having to close or to stop more creative subject provision, such as music,
drama and art, within their schools (Ratcliffe, 2017; Savage, 2018).

The result of the recent decades of standardization, competition and corporatization (and the following quote seems to fairly accurately describe
both secondary and tertiary education systems) includes

> alarming rates of non-graduation from school and colleges, the lev-
> els of stress and depression—even suicide—among students and their
> teachers, the falling value of a university degree, the rocketing costs of
> getting one, and the rising levels of unemployment among graduates
> and non-graduates alike.
>
> (Robinson and Aronica, 2015)

We need to do something different in our educational systems. Rather
than another governmental intervention into curriculum, or new teacher/
staff/institutional performance metrics, what is needed is a complete
rethink at local/regional levels with associated freedoms to implement
what learners need in our knowledge economy. The focus here, as many
have argued, should be on learning environments, rather than a focus on
knowledge content. The basic question that needs to be asked here, both at
secondary and tertiary levels, is if we live in a knowledge economy with
knowledge being all around us, what is the role of schools and universities
and how should they support the development of learners to confidently
navigate, critically reflect, creatively produce and significantly contribute
to our society's future?

In a 2010 keynote speech at the Royal Society of Arts (Robinson, 2010),
Robinson gave a short summary of our still current school challenges:

> Every country on the earth is at the moment reforming education.
> There are two reasons for this. The first of them is economic. They
> are trying to find out how do we educate children to take their place
> in the economies of the 21st century—how do we do that—given
> that we can't anticipate what the economy will look like at the end of

next week—as the recent turmoil has demonstrated. How do we do that? . . . The second is cultural: Every country on the earth is trying to figure out how we educate our children so they have a sense of cultural identity, how do we pass on the genes of our culture, while being part of globalization. How do we square that circle.

(Robinson, 2010)

As mentioned in 2016 (in Boehm), Robinson suggests that our learning organizations still prioritize a very particular way of academic thinking that excludes many children and young people. The existing system, he argues (Robinson and Aronica, 2015), is still based on an outmoded industrial revolution model, based on standardization and essentially the principles of factory production. It is still an "industrial character of public education" and thus is deeply flawed.

This industrial heritage created an ingrained believe in standardization, based on the need to provide a mass public schooling system. This "modernistic model of prioritizing highly specialized knowledge causes problems for the divide between the practice-based (or vocational, but the different terms have different connotations) and the academic" (Boehm, 2014). This is evident in secondary education, and the presence of the divide continues right into tertiary education. As Robinson suggests, the concept of dividing the vocational from the academic is based on a series of assumptions about social structure and capacity and a very specific intellectual model of the mind.

[This] was essentially the enlightenment view of intelligence. That real intelligence consists of this capacity of a certain type of deductive reasoning and a knowledge of the classics, originally. What we came to think of as academic ability. And this is deep in the gene pool of public education, that there are two types of people, academic and non-academic. Smart people and non-smart people. And the consequence of that is that many brilliant people think they are not, because they are being judged against this particular view of the mind.

(Robinson, 2010)

Robinson goes on to suggest that this has caused some of the perceived misery in our school education systems:

The problem is that they're trying to meet the future by doing what they did in the past. And on the way they are alienating millions of kids, who don't see any purpose of going to school. When we went to school, we were kept there with a story, which if you worked hard and did well, and got a college degree you would have a job. Our kids don't believe that. And they are right not to—by the way. You're better having a degree than not, but it is not a guarantee anymore. And particularly not if the route to it marginalizes everything you think is important about yourself.

(Robinson, 2010)

Thus, our (secondary) educational institutions have been increasingly afforded to standardize the curriculum, standardize teaching and standardize assessment. And the biggest threats from the recent political climates within a neo-liberal encultured political norm are the strategies taken by successful government to attain standards: by standardization, by introducing competition, and by allowing corporatization.

> The typical reform story goes like this: A high-performing education system is critical to national economic prosperity and to staying ahead of our competitors. Standards of academic achievement must be as high as possible, and schools must give priority to subjects and methods of teaching that promote these standards. Given the growth of the knowledge economy, it's essential that as many people as possible go on to higher education, especially four-year colleges and universities.
>
> Because these matters are too important to be left to the discretion of schools, government needs to take control of education by setting standards, specifying the content of the curriculum, testing students systematically to check that standards are being met, and making education more efficient through increased accountability and competition.
>
> (Robinson and Aronica, 2015)

And when standards are not improving, the unfailing belief that the system is inherently the right one leads to greater efforts in raising standards through introducing even more competition and even more accountability. This standards movement is allegedly making systems more efficient and accountable, but the result is that we have one of the lowest productivity levels in decades, low innovative entrepreneurial output in international terms, and a mental health crisis linked to the stresses of constant performance measurement.

Most countries now have a national curriculum of some form, often specifying year-level knowledge content. This is true for England, France, Germany, China and many others. Somewhat looser frameworks exist in Scotland, and notably Finland (Sahlberg, 2017) and, so far, the US and Singapore. Pasi Sahlberg has written extensively on the success of the Finnish education model, which he suggests is based on being a "cultural outlier" (Sahlberg, 2017, p. xxi). Finland had school policies that were almost the opposite of those introduced in the Anglo-American sectors and much of the rest of the world. Thus, it could be seen as a story of the impact of alternative education solutions, those based on a core belief in cultivating trust, enhancing autonomy and supporting diversity. Key drivers for this success are suggested to be found in the greater amount of local school and teacher autonomy, and the lack of census-based standardized testing, test preparation and private tutoring. "In Finland, teachers teach less and students spend less time studying both in and out of school, than their peers in other countries" (Sahlberg, 2017, p. 14).

In Sahlberg's view, standardizations stand in opposition to creativity. In educational systems where standardization is perceived to be the key, there is often the suggestion that what is needed to attain standards are introductions of competition, test-based accountability, but with a resulting perception of de-professionalization, and (as he suggests) an addiction to reform, together with a belief in the notion of "excellence". Alternatively, systems that tend to do better are ones that allow creative flexibility of content and personalized learning, including aspects of collaboration, and a belief that this is based on trust-based responsibility, along with experiential professional leadership and a firm belief in equity, all resulting in a sustained improvement.

This belief in equity is, of course, something that does not come easy to us in an England still entrenched in class cultural divides. Even less accepted is the consideration that the prioritization of equity is essential for economic and social well-being, and that there should be the same provision of access to education for everyone. In his post-Brexit-vote article, Dutchman Luyendijk suggested that the Brexit vote was a "logical outcome of a set of English pathologies" (Luyendijk, 2017), and that "a nation that gave the world the term 'fair play' sees the fact that rich children receive a better education than poor ones as a perfectly natural thing". Luyendijk's article is a very painful read for those who do not believe that money should give you the right to access better education, and whereas this belief has simply been put aside in the Finnish system with a comprehensive buy-in into equity for all, Britain's ruling majorities—and more specifically England's—sees this as a fair, neo-liberal, market-driven and competitively more productive system in the long run—despite all evidence to the contrary.

But with this belief, creativity and our UK passion for music-making is in danger of being squeezed out, despite all indicators pointing toward the need to include more creative practice in everyday school environments. Standardization has also made this more difficult:

> It is clear to me that one of the main obstacles in focusing more on real learning, giving more room to music and arts in American schools, building learning in schools around curiosity, creativity and exploration of interesting issues, is standardized testing.
>
> (Sahlberg, 2017)

Sahlberg goes on to hammer the US systems using an increasing amount of standardization, to suggest that "perhaps most importantly, I don't know any other OECD country where cheating and corruption are so common in all levels of the school system than it is in the U.S., only because of the dominance of standardized tests".

Until quite recently, universities were out of the spotlight from successive governments, but have now become the newest scapegoat of choice, as David Sweeney suggested at a 2017 SRHE conference keynote. Over the last 18 months, various government officials seemed to have washed their hands of the responsibility for the mess in which our nation finds itself, as George Monbiot has stated in his recent journalistic explorations of class, inequality, environment, growth obsessions and financial crises (Monbiot, 2016). Universities have been asked by politicians to take responsibility for (a) growing economic productivity, (b) increasing social mobility, (c) solving the challenge of our failing school systems, (d) meeting the increasing expectations of student consumers, (e) reducing immigration and (f) doing all that with minimal public funding and simultaneously being increasingly forced to allow market forces to regulate their work.

The view currently often portrayed by ministers and therefore the media is that universities are a separate entity, still an ivory-tower-like structure. But what the government does not seem to understand is that universities are as much a part of society as all our communities are. The same state structures that allow the growing gap between the richest and the poorest in society is driving a similar gap between the smallest salary earners and the richest within academia. The government's newest university scapegoat is Vice-Chancellors' salaries, and although the debate between the richest and the poorest is a valuable debate to be had in terms of debating social inequalities, to focus on universities here is a confusing message: on the one hand, the government introduced market competition in the HE sector, deregulation and freedom to innovate, but then on the other hand, they simultaneously ask for the highest amount of scrutiny, public accountability and comparison with the public sector—and this to a degree that few private sectors experience (Rushforth, 2017). English universities are thus being torn asunder—on the one hand asked to act as businesses while on the other having to undergo intense public-accountability processes. This constantly feels like being knotted tightly into a public-accountability straightjacket, with hands and feet tied behind your back, while being thrown into a competitive free-market shark tank. The only movement left is squirming, and that is certainly what the HE sector is doing at the moment.

Subsequent policy interventions resulted in our HE sector being less diverse than ever before, as every institution is afforded to hunt after the same performance indicators that the government continues to throw down at its feet, from REF, to NSS, to DLHE, to TEF to now the upcoming KEF (or written out: Research Excellence Framework, National Student Survey, Destination of Leavers in Higher Education Survey, Teaching Excellence Framework and Knowledge Exchange Framework). This is also standardization, but here in HE it is driven predominantly by standardization of key performance indicators representing what the government perceives to be teaching quality, graduate outcomes, research excellence and knowledge exchange metrics. This has created a risk-averse, neo-liberal, overly managerial-reliant system that is inefficient in its excessive need to justify every part of its process.

There is a Wiley Miller political cartoon from 2012 that still holds true today. A row boat contains eight managers at a table with graphs and metrics informing the single rower of his performance metrics and telling him how to row faster. Productivity just does not work when the focus is more on getting the performance correctly measured in order to instruct the dwindling group of front-line employees. But the government interventions have increasingly made it necessary for the number of professionals involved in accountability, scrutiny, quality assurance, data analysis and justification of resources to be steadily increased. With the coming into force of the Higher Education and Research Act 2017, and the introduction of a sector regulator, the increasing focus on value for money is about to become even more complicated (CDBU, 2017).

For students, English universities represent not only the most expensive higher education system in the world, but also a system with one of the highest administrative costs. This is undoubtedly primarily a result of governmental interventions (see also Boehm, 2016a). Thus, universities are increasingly forced to compete with each other in a climate where the need to make an institution more nationally competitive within its own HE sector, also through league table positioning, takes priority over the socioeconomic benefit to a region. England is only one of a few countries where the stance of commercially conceptualized HE has been implemented to such an extreme: a "university market" selling education as a consumer good.

Now private providers do not necessarily represent a bad thing, and the music sector is one of the only creative industry sectors with flourishing private for-profit and private not-for-profit new HE entrants. Students have flourished and successfully made their careers based on degrees from BIMM, FutureWorks, ICMP, SAE and others, with each provider able to contribute specific strengths to the market. However, there is a different role that these providers have within society compared to universities that are still—according to the industrial strategy—anchors in their region and supporting economic growth and society as whole, by having a multi-purpose holistic remit with a civic mission. The fact that things will undoubtedly go wrong more often with private for-profit providers with their inherent larger potential of conflicts of interest has been outlined in detail in *The Great University Gamble: Money, Markets and the Future of Higher Education* by Andrew McGettigan (McGettigan, 2013).

For McGettigan the issue of marketization of Higher Education is also about democratic deficits, and these can be found not only in the way that a country is governed (with our First-Past-The-Post non-proportional electoral system) but also in the way university departments choose to appoint their Heads or Deans (elected vs. permanent managers). This movement from flat-structured networked knowledge organizations to the belief that universities need strong, central decision-makers also focused employees on those leaders to whom they are accountable to, rather than to look to the surrounding society (Wright and Shore, 2017, p. 78). This intentional increase of these democratic deficits has led to a path toward what Newfield calls the "The Great Mistake: how we wrecked public universities and how we can fix them" (Newfield, 2016) and is a follow-up to his "Unmaking of the Public University" (Newfield, 2008). In the US, a system often seen to be further advanced in the marketization narrative, this has led to locked-in economic inequality and systemic lack of student attainment while society must cope with student debt.

For me personally, there are glimmers of hope that we, as a society, and we, as a sector, are starting to be more assertive in our arguments that the

English HE solution may not necessarily be the way to design the future. Yes, we live in a different world than just ten years ago, our knowledge society and our knowledge economy has arrived, and this does mean that knowledge institutions have to consider how this affects them. New models of educational frameworks are needed. Over the past two years, I have had the privilege to be involved with entrepreneurial creative communities that are planning either not-for-profit but private music higher education provision or cooperatively owned university provision. An increasing number of professionals and academics are considering alternative futures for higher education and with it the role of universities in the future. Old and new universities are increasingly beginning to (re-) emphasize their civic mission, and there have been increasing calls for revisiting the concept of what universities are for, what a public university should be and the reiteration of the need for societally engaged universities with an institutional and individual conscience that break the ivory tower concepts once and for all (Levin and Greenwood, 2016; Collini, 2012; Watson, 2014). The implications of universities as anchors with a focus on the knowledge economy is explored in Perry and May's *Cities and the Knowledge Economy: Promise, Politics and Possibilities* (May and Perry, 2017), and the threats of not having a public university system is explored in a recent volume, *Death of the Public University?* (Wright and Shore, 2017). Pedagogical underpinnings are revisited and newly proposed, from students as producers (Neary and Winn, 2009; Neary, 2010), to a focus on learning environments moving away from outcome-based learning (Thomas and Brown, 2011; Davidson, 2017), to research-embedded learning as part of a cohesive discovery-based learning framework (Fung, 2017).

There are new initiatives to explore the viability of the first UK cooperative universities (Bothwell, 2016; Cook, 2013; Winn, 2015). "New old" models of HE are being explored, focusing back on private vs. common vs. public good, including alternative models such as trust universities (Boden et al., 2012; Wright et al., 2011; Wright and Shore, 2017), and also, more relevant for the creative sector, my own expressions of the role of universities in the creative economy and society (Boehm, 2015b, 2014, 2015a, 2016a, 2016b; Boehm et al., 2014).

A healthy debate has emerged, and just in the two months of November and December 2017, there were four conferences mixing policymakers, educators and researching academics, all concerned with focusing on the role of universities in contemporary society. On the November 6–7, WONKHE's Wonkfest17 took place, with the fabulous strapline "Revenge of the Experts". On November 9, there was the Coop College's inaugural "Making the Co-operative University: New Places, Spaces and Models for Learning". The same week saw the Centre for Higher Education Futures (CHEF)'s inaugural international conference "The Purpose of Future Universities" in Aarhus. And in December, SRHE's annual conference in Newport had the strapline: "Higher Education Rising to the Challenge: Balancing Expectations of Students, Society and Stakeholders".

CONCLUSION

In conclusion, I do see a glimmer of hope in the increasing articulations of alternatives to the neo-liberal conceptualized models of (unsustainable) sustainability. These articulations go far beyond the music industry, in which this publication's readership is contextualized, but they have particular relevance to it. And as music is one of the oldest creative practice–based disciplines in the academy and in schools, its communities are at the forefront of considering what it means to be involved in education, research and development for a creative society.

We know that if we do not attend to our secondary and tertiary education infrastructures and fail to include music and creative provision in our curricula, we will not have a creative future in which music-making in all its forms can flourish. We might find ourselves looking back and realizing that in the last 25 years in Britain, we had lived in a golden age of music, culture and the arts, having fostered a whole generation of artists who had led the way within a global cultural community, driving our very own diversity-rich, international, but also very uniquely British creativity, in music and the arts. If we don't ensure that music and arts feature as elements within our learning institutions, this will be an era whose end has just started with the 2017 general elections, Brexit and the subsequent educational reforms, with their ongoing focus on standardization squeezing out any notion of creative freedom for young talent.

Universities play a large role here, but the climate in which British HE finds itself has its own challenges. So, before we even start connecting the dots of the challenges described here, we need to understand the context that universities find themselves in and the challenges that this context provides in devising effective learning provision. This will justify the needed move from formalized and structured learning objects to formalized structured learning environments, and this journey has only just begun. There are examples where this has always happened in practice, specifically in music and the arts. These examples provide lessons to be learned for those universities who truly want to be connected. Music and creativity have a large part to play here, especially because "culture is not simply a large and important sector of the economy, it is a 'social software' that is badly needed to manage the complexity of contemporary societies and economies in all of its manifold implications" (Sacco, 2014).

REFERENCES

ACE (2017a). *Creative sector is taking a leading role in boosting the UK economy | press release.* London: Arts Council England.

ACE (2017b). *Interactive map: Our funded organisations, 2018–2022 | report* (Arts Council England).

ACE (2017c). *Literature in the 21st century: Understanding models of support for literary fiction | report* (Arts Council England. Commissioned report by Canelo). Arts Council.

Bazalgette, P. (2017). *Independent review of the creative industries—GOV.UK.*

BIS (2017). Industrial strategy: Building a Britain fit for the future—GOV.UK. In: *Department for business energy & industrial strategy*. London: BIS.

Boden, R., Ciancanelli, P. and Wright, S. (2012). Trust universities? Governance for post-capitalist futures. *Journal of Co-operative Studies*, 45, pp. 16–24(9).

Boehm, C. (2009). 2084-brave creative world: Creativity in the computer music curriculum. In: *Proceedings of the international computer music conference*. Montreal, ICMA.

Boehm, C. (2014). A brittle discipline: Music technology and third culture thinking. In: E. Himoinoides and A. King, eds., *Researching music, education, technology: Critical insights. Proceedings of the Sempre MET2014*. London: University of London.

Boehm, C. (2015a). Engaged universities, mode 3 knowledge production and the impact agendas of the REF. In: S. Radford, ed., *The next steps for the research excellence framework*. London: Westminster Forum Projects.

Boehm, C. (2015b). Triple helix partnerships for the music sector: Music industry, academia and the public. In: R. Hepworth-Sawyer, ed., *International conference on innovation in music 2015*. Cambridge: KES Transactions.

Boehm, C. (2016a). Academia in culture 3.0: A crime story of death and birth (but also of curation, innovation and sector mash-ups). *REPERTÓRIO: Teatro & Dança*, 19, pp. 37–48.

Boehm, C. (2016b). Music industry, academia and the public. In: *International computer music conference*. Ghent, International Computer Music Association.

Boehm, C., Linden, J. and Gibson, J. (2014). Sustainability, impact, identity and the university arts centre. A panel discussion. In: ELIA, ed., *Location aesthetics. 13th ELIA biennal conference*. Glasgow, European League of Institutes of Arts.

Bonner, Karen & Hewitt-Dundas, Nola & Roper, Stephen. (2015). *Collaboration between SMEs and universities – local population, growth and innovation metrics*. 10.13140/RG.2.1.4145.9365.

Bothwell, E. (2016). Plan to "recreate public higher education" in cooperative university. *Times Higher Education*. London, THE.

Brown, M. (2015). *Arts council England to increase ratio of funding outside London*. @guardian.

Camri (2017). *Creative Industries and beyond: is the "gig economy" the way forward? Policy Observatory of the Communication and Media Research Institute (CAMRI)*. Conference Announcement for 13 June. London: University of Westminster.

CDBU (2017). *Value for money are students right to complain?* London: Council for the Defense of British Universities.

CEBR (2017). *Contribution of the arts and culture industry to the UK economy | (Arts Council commissioned report)*.

CIF (2017). *Global talent report* (Creative Industries Federation).

Collini, S. (2012). *What are universities for?* London and New York: Penguin.

Cook, D. (2013). *Realising the co-operative university*. A consultancy report for The Co-operative College.

Creative Industries Council (2017). *News: Creative industries add £87.4bn to UK economy*.

Davidson, C. N. (2017). *The new education: How to revolutionize the university to prepare students for a world in flux*. New York: Basic Books.

DCMS (2001). *Creative industries mapping document*. London: Department for Culture Media and Sports.

DCMS (2008). *Creative Britain: New talents for the new economy*. London: Department for Culture Media and Sports.

Fung, D. (2017). *A connected curriculum for higher education*. London: UCL Press.

Hanson, S. (2015). *The Northern powerhouse as "Real Mirage": The new maps of Britain, and how we should understand them. | openDemocracy. @ opendemocracy*.

IPPR North (2012). *Northern rail priorities statement: five priorities for immediate action and investment*. @IPPR.

Ippr North (2017). *New transport figures reveal London gets £1,500 per head more than the North—but North West powerhouse "catching-up"*. @IPPR.

Levin, M. and Greenwood, D. J. (2016). *Creating a new public university and reviving democracy: Action research in higher education*. New York: Berghahn Books.

Luyendijk, J. (2017). *How I learnt to loathe England*. @prospect_uk.

Mateos-Garcia, J. and Bakshi, H. (2016). *The geography of creativity in the UK*. London: Nesta.

May, T. A. and Perry, B. A. (2017). *Cities and the knowledge economy: Promises, politics and possibilities*. London: Routledge.

McGettigan, A. (2013). *The great university gamble: Money, markets and the future of higher education*. London: Pluto Press.

Monbiot, G. (2016). *How did we get into this mess? Politics, equality, nature*. London and Brooklyn, NY: Verso.

Neary, M. (2010). *Student as producer: Research engaged teaching and learning at the University of Lincoln user's guide 2010–11*. Lincoln: University of Lincoln.

Neary, M. and Winn, J. (2009). The student as producer: Reinventing the student experience in higher education. In: L. Bell, H. Stevenson and M. Neary, eds., *The future of higher education: Policy, pedagogy and the student experience*. London: Continuum.

Newfield, C. (2008). *Unmaking the public university: The forty-year assault on the middle class*. Cambridge, MA: Harvard University Press.

Newfield, C. (2016). *The great mistake: How we wrecked public universities and how we can fix them*. Baltimore: Johns Hopkins University Press.

Patterson, J. and Boehm, C. (2001). Circus for beginners. In: G. Boehm and Schuter, eds., *CIRCUS 2001—new synergies in digital creativity. Proceedings of the conference for content integrated research in creative user systems*. Glasgow: University of Glasgow.

Ratcliffe, R. (2017). *Cuts, cuts, cuts. Headteachers tell fo school system "that could implode"*. @guardian.

Robinson, K. (2010). Changing education paradigms. In: *Annual conference for the royal society for arts*. London: Royal Society for Arts.

Robinson, K. and Aronica, L. (2015). *Creative schools: The grassroots revolution that's transforming education*. New York: Viking.

Rtm Rail Technology Magazine (2017). *Majority rules: We must build HS3 first*. @rtmnews.

Rushforth, J. (2017). *VCs' salaries—go compare and justify—association of heads of university administration*. John Rushforth, Executive Secretary of the Committee of University Chairs (CUC), gives his personal views on the debate around Vice-Chancellors' pay and benefits and discusses the justification of salaries of HE senior leaders.

Sacco, P. L. (2014). *Culture 3.0*. ELIA Keynote Talk, European League of Institutes of the Arts, Glasgow.

Sahlberg, P. (2017). *Finnish lessons 2.0: What can the world learn from educational change in Finland?* New York: Teacher's College Press.

Savage, M. (2018). *Top musicians unite in call for all pupils to have the right to learn an instrument*. @guardian.

Shaw, J. A. (2017). *Building HS3 before HS2—should happen*. Retrieved from: https://www.constructionnews.co.uk/markets/sectors/infrastructure/build-hs3-before-hs2-major-think-tank-says/10009614.article.

Thomas, D. and Brown, J. S. (2011). *A new culture of learning: Cultivating the imagination for a world of constant change*. CreateSpace Independent Publishing Platform.

Thorpe, V. (2014). *Peter Bazalgette on regional arts funding: "blame lies with council cuts"*. @guardian.

Topham, G. (2017). *Rail services lost under 1960s beeching cuts may reopen*. @guardian.

UK Music (2012). *The economic contrubution of the core UK music industry*. UK Music.

UK Music (2017a). *Measuring music 2017—methodology*.

UK Music (2017b). *Measuring music 2017—UK music*.

Wand, M. (2017). *The case for building an HS3 before HS2*. Retrieved from: https://shouldhappen.com/building-hs3-before-hs2/

Watson, D. (2014). *The question of conscience: Higher education and personal responsibility*. London: Institute of Education Press.

Winn, J. (2015). The co-operative university: Labour, property and pedagogy. *Power and Education*, 7, pp. 39–55.

Wright, S., Greenwood, D. and Boden, R. (2011). Report on a field visit to Mondragón university: A cooperative experience/experiment. *Learning and Teaching: The International Journal of Higher Education in the Social Sciences (LATISS)*, 4, pp. 38–56.

Wright, S. E. and Shore, C. E. (2017). *Death of the public university? Uncertain futures for higher education in the knowledge economy*. Oxford: Berghahn Press.

FURTHER READING

Bazalgette, P. (2017). *Independent review of the creative industries—GOV.UK*.

BIS (2017). *Industrial strategy: Building a Britain fit for the future—GOV. UK*. London: Department for Business Energy & Industrial Strategy.

(One of the current key reports covering recommendations for the creative industries)

Boehm, C. (2014). A brittle discipline: Music technology and third culture thinking. In: E. Himoinoides and A. King, eds., *Researching music, education, technology: Critical insights. Proceedings of the Sempre MET2014*. London: University of London.

(An article covering the challenges of interdiscplinarity in Higher Education in the area of music technology)

Boehm, C. (2015). Triple helix partnerships for the music sector: Music industry, academia and the public. In: R. Hepworth Sawyer, ed., *International conference on innovation in music 2015*. Cambridge: KES Transactions.

(An article covering a new partnership model paradigm, its evolution and challenges for the future)

Boehm, C. (2016). Academia in culture 3.0: A crime story of death and birth (but also of Curation, Innovation and Sector Mash-ups). *Repertório: Teatro & Dança*, 19, pp. 37–48. CIF (2017) Global Talent Report (Creative Industries Federation).

(An article covering in details the concept of Culture 3.0 and its relevance for the creative and cultural sector)

Robinson, K. and Aronica, L. (2015). *Creative schools: The grassroots revolution that's transforming education*. New York: Viking. SDG (2017) Exploring the role of arts and culture in the creative industries—Exploring the role of arts and culture in the creative industries.

(An extended treatment on the benefits of keeping creative subjects in school and the evolution of educational policy that made this increasingly difficult).

NOTES ON CONTRIBUTOR

Carola Boehm is Associate Dean (Students) and Professor of Arts and Higher Education in the School of Creative Arts and Engineering at Staffordshire University. Her research areas include music technology education, interdisciplinarity in Higher Education and the role played by universities in the creative economy.

Index

Note: **Boldface** page references indicate tables. *Italic* references indicate figures.